Essentials of Fluid Dynamics

Essentials of Fluid Dynamics

Edited by
Fay McGuire

Willford Press
www.willfordpress.com

Published by Willford Press,
118-35 Queens Blvd., Suite 400,
Forest Hills, NY 11375, USA

ISBN: 978-1-68285-389-4

Cataloging-in-Publication Data

Essentials of fluid dynamics / edited by Fay McGuire.
p. cm.
Includes bibliographical references and index.
ISBN 978-1-68285-389-4
1. Fluid dynamics. 2. Computational fluid dynamics. 3. Turbulence. 4. Fluid mechanics.
I. McGuire, Fay.
QC151 .E77 2017
532.05--dc23

For information on all Willford Press publications
visit our website at www.willfordpress.com

WILLFORD PRESS

Printed in China.

Contents

Preface

The main aim of this book is to educate learners and enhance their research focus by presenting diverse topics covering this vast field. This is an advanced book which compiles significant studies by distinguished experts. This book addresses successive solutions to the challenges arising in the area of application, along with it; the book provides scope for future developments.

The study of the movement of liquids and gases is known as fluid dynamics. This book on fluid dynamics deals with the applied and computational methods of fluid dynamics. This field follows the basic laws of motion and force to measure and predict the various flows that occur to a body, either at motion or rest. Fluid dynamics is applied to a wide variety of fields such as hydrology, limnology, aeronautics, astronomy, etc. This book is a compilation of chapters that discuss the most vital concepts and emerging trends of this field. It is meant for students who are looking for an elaborate reference text on fluid dynamics. This text is a complete source of knowledge on the present status of this important field.

It was a great honour to edit this book, though there were challenges, as it involved a lot of communication and networking between me and the editorial team. However, the end result was this all-inclusive book covering diverse themes in the field.

Finally, it is important to acknowledge the efforts of the contributors for their excellent chapters, through which a wide variety of issues have been addressed. I would also like to thank my colleagues for their valuable feedback during the making of this book.

Editor

1

A cost-effective FE method for 2D Navier–Stokes equations

A. C. Kengni Jotsa[a] and V. A. Pennati[b]*

[a]*African Institute For Mathematical Sciences (AIMS-Cameroon), PO Box 608 Limbe Crystal Gardens, South West Region;* [b]*Università Degli Studi dell'Insubria, Dipartimento Di Scienza e Alta Tecnologia, Via Valleggio 11, 22100 Como, Italy*

A cost-effective approach to the solution of 2D Navier–Stokes equations for incompressible fluid flow problems is presented. The aim is to reach a good compromise between numerical properties and computational efficiency. In order to achieve the set goal, the nonlinear convective terms are approximated by means of characteristics and spatial approximations of equal order are performed by polynomials of degree two. In this way, the computational kernels are reduced to elliptic ones for which solution very efficient techniques are available. The time-advancing is afforded by a fractional step method combined with a stabilization technique suitably simplified, so that the inf-sup condition is easily overcome. The algebraic systems generated by the new technique are solved by an iterative solver (Bi-CGSTAB), preconditioned by means of a suitable Schwarz additive scalable preconditioner. The properties of the new method have been confirmed from the comparison among the results obtained by it, and those obtained from other methods in the solution of some well known test problems. The obtained results, both in terms of accuracy and computational efficiency, make realistic the possibility to extend the method to 3D problems and to develop a multidomain approach.

Keywords: 2D Navier–Stokes; fractional step; finite element; stabilization; Schwarz preconditioner

1. Introduction

Nowadays in fluid dynamics, the transient incompressible Navier–Stokes equations are widely accepted as a mathematical model for the description of incompressible viscous flows. Their application to the solution of real problems, both 2D and 3D, is more and more widespread and requires a great computing effort, see for example (Cavazzutti, Corticelli, Masina, & Saponelli, 2013; Kumar, Kumar, & Kumar Das, 2009; Stickland, Fabre, Scanlon, Oldroyd, Mickelson, & Astrup, 2013). The system of equations is made up of the momentum equations and the incompressibility relation, and it is well known that the main difficulties in its solution reside in the approximation of the convective term and in the calculation of the pressure. A large amount of numerical methods have been developed particularly in the framework of finite elements (FE) (Baker, Dougalis, & Karakashian, 1982; Choi, Choi, & Yoo, 1997; Codina & Soto, 2004; Plana Fattori, Chantoiseau, Doursat, & Flick, 2013; Sun, Zhang, & Ren, 2012), of cell-centred finite volumes (Deponti, Pennati, & De Biase, 2006; Deponti, & Pennati, et al., 2006; Feraudi & Pennati, 1997; Kim & Moin, 1985; Pai, Prakash, & Patnaik, 2013) or cell-vertex finite volumes (Hookey & Baliga, 1998; Malan, Lewis, & Nithiarasu, 2002; Trittthart & Gutknecht, 2007; Vrahliotis, Pappou, & Tsangaris, 2012) and finite differences (Ali, Fieldhouse, & Talbot, 2011; Kumar, Dass, & Dewan, 2010; Shih & Tan, 1989). In the finite difference (FD) and finite volume (FV) schemes some methods for time-advancing are based on the so-called SIMPLE method, and its developments (for example, SIMPLER, SIMPLEC, SIMPLEX, PISO, etc.). The SIMPLE method (Patankar, 1980) can be regarded as quite a rough (but simple) preconditioner obtained by an approximation of the upper triangular matrix **U** (usually the principal diagonal) of the **LU** factorization of the global system matrix. Two important drawbacks are inherent the SIMPLE preconditioner, namely: the low rate of convergence and the global approach to the solution of the system of equations (particularly computationally heavy in 3D problems) (Acharya & Moukallad, 1989; Van Doormal & Raithby, 1984). In order to improve the computational efficiency of the solvers used in FD and FV approximations, a huge literature has been developed regarding the multigrid methods. These can be very efficiently formulated if based on structured grids (the so called geometric multigrids), because the spectral analysis of the stiffness matrices allow us to choose optimal smoothers (Donatelli, Molteni, Pennati, & Serra Capizzano, In press). A potential drawback of the geometric multigrid solvers is their difficulty to solve problems defined in general domains, in fact, in this case a transformation from the physical space to the computational one has to be performed (Kumar et al., 2010). Another important point involved in the time-advancing is the approximation of the convective term. It is nonlinear, and for its approximation several techniques have been developed, for example, the implicit one (based

*Corresponding author. Email: VPennati@teletu.it

on a Newton cycle requiring at every iteration the solution of a linear system), the semi implicit one (in which the nonlinear term is approximated explicitly) and the Picard approximation (in which the transferring term is "frozen" at time n and the gradient of velocity at time $n+1$). The first approximation is of second order but very heavy with respect to the computing time, the second one requires a severe restriction on Δt in order to guarantee the stability, the third one is only first order in time. In transient problems, besides the schemes commonly used for advancing in time, there is another approach called the fractional steps; it is based on the splitting of the differential operator in the sum of two suitable differential operators. This last technique combined with a Lagrangian approximation of the convective term is the base of our time-marching scheme.

In this paper, a numerical approach for the solution of 2D incompressible Navier–Stokes equations is presented with the aim of obtaining an accurate solution while maintaining a low computational effort. This choice is very important because we are interested in generalizing the presented method to problems with moving boundaries (in particular to 3D problems). The new method can be considered optimal in the sense that its formulation attempts to obtain a good compromise between numerical properties (accuracy and stability) and computational efficiency. The main features of the method are: using a triangular (curved edges) FE spatial approximation with polynomials of degree two for velocity and pressure (and temperature); time-advancing by a fractional step approach in which the nonlinear convective term is approximated by characteristics and the pressure correction is calculated by solving a Poisson equation (Codina & Badia, 2006; Gervasio & Saleri, 2006; Han, Zhou, & Tu, 2013; Shirokoff & Rosales, 2011); addition of a suitable stabilization technique in order to overcome the instabilities inherent in the equal-order choice. A well-known technique of stabilization is the one based on the definition of subscale velocities formulated by Codina and Soto (2004). Unfortunately, this technique depends on a weight parameter τ whose estimation generally is not easy. In section 4, a very simple rule for calculating the τ value is suggested and its efficiency verified by solving several numerical tests. In the new technique the most expensive computational kernels are reduced to elliptic ones so that, in order to reduce as much as possible the computational effort, an iterative method (Bi-CGSTAB) preconditioned by an additive Schwarz preconditioner is used. Such preconditioners are endowed with very interesting properties, in particular the scalability and the computational flexibility. The latter is dependent on the number of subdomains in which the domain is divided, and on the level of *fill-in* chosen for the incomplete **LU** factorization of the local stiffness matrix. The multidomain approach, besides being a necessary tool for creating an efficient Schwarz preconditioner, could be also considered a convenient starting point for developing distributed memory parallel computing algorithms (Corti & Pennati, 1997).

The content of the paper is organized as follows. Section 2 introduces the governing equations completed by the energy equation. Section 3 describes the time-advancing to be used. In Section 4, the issue of the approximation of the 2D Navier–Stokes equations with stabilization is considered. The main result of this section consists of the fractional step scheme (58)–(63) that has been implemented in our in-house code. Finally, in Section 5 the flexibility and the efficiency of the method are illustrated by four examples: the two dimensional unsteady flow of decaying vortices, two lid-driven problems: the first one with analytical solution and the second one with constant boundary condition and, finally, some thermal convection problems in a square cavity. The last section is devoted to some concluding remarks.

2. Governing equations

Let Ω be a bounded domain of $\mathbb{R}^2$ occupied by the fluid, with Lipschitz boundary $\partial\Omega$ and $[0, T]$ the time interval of analysis. The 2D Navier–Stokes problem for incompressible flows consists in finding a velocity $\boldsymbol{u}$, a pressure p, and a temperature T such that:

$$\frac{\partial \boldsymbol{u}}{\partial t} + \boldsymbol{u} \cdot \nabla \boldsymbol{u} - \nu \Delta \boldsymbol{u} + \nabla p = \boldsymbol{f} \text{ in } \Omega, \quad t \in (0, T) \tag{1}$$

$$\nabla \cdot \boldsymbol{u} = 0 \text{ in } \Omega, \quad t \in (0, T) \tag{2}$$

$$\boldsymbol{u} = 0 \text{ on } \partial\Omega, \quad t \in (0, T) \tag{3}$$

$$\boldsymbol{u}|_{t=0} = \boldsymbol{u}_0 \text{ in } \Omega \tag{4}$$

$$\frac{\partial \mathrm{T}}{\partial t} + \boldsymbol{u} \cdot \nabla \mathrm{T} - \Gamma \Delta \mathrm{T} = \boldsymbol{S} \text{ in } \Omega, t \in (0, T) \tag{5}$$

$$\mathrm{T} = 0 \text{ on } \partial\Omega, \quad t \in (0, T) \tag{6}$$

$$\mathrm{T}|_{t=0} = \mathrm{T}_0 \text{ in } \Omega \tag{7}$$

where $\boldsymbol{f}$ denotes a volumetric body force such as gravitation, ν the kinematic viscosity, $\boldsymbol{u}_0$ and T_0 the initial velocity and temperature respectively. We have considered homogeneous Dirichlet boundary conditions (3) and (6) for simplicity. We need to introduce some useful notations, e.g., to denote by $\mathrm{H}^1(\Omega)$ the Sobolev space of functions whose first derivatives belong to $\mathrm{L}^2(\Omega)$, and by $\mathrm{H}_0^1(\Omega)$ the subspace of $\mathrm{H}^1(\Omega)$ of functions with zero trace on $\partial\Omega$.

3. Time-advancing scheme

As the nonlinear convective term of the momentum equation (1) needs to be treated with care, and as we would like to reduce the computational effort significantly, we use the second-order characteristics scheme (Pironneau, 1982) that computes the values of the variable of interest at the foot of the trajectory. In fact, assume for instance we have the equation:

$$\frac{DQ}{Dt} = \Psi \tag{8}$$

where $\frac{DQ}{Dt}$ represents the total derivative, that is $\frac{\partial Q}{\partial t} + \boldsymbol{u} \cdot \nabla Q = \Psi$ with $\boldsymbol{u}$ known velocity.

If the total derivative is approximated with a backward Euler scheme:

$$\frac{DQ}{Dt} \cong \frac{Q^{n+1}(\mathbf{x}) - Q^n_{ft}}{\Delta t} \tag{9}$$

then an implicit semi-Lagrangian scheme for (8) is:

$$\frac{Q^{n+1}(\mathbf{x}) - Q^n_{ft}}{\Delta t} = \Psi^{n+1}(\mathbf{x}) \tag{10}$$

Knowing the velocity field $\boldsymbol{u}$, the foot of trajectory can be calculated by:

$$\tilde{\boldsymbol{x}}_{ft} = \mathbf{x} - \Delta t \boldsymbol{u}^{n+1}(\mathbf{x}) \tag{11}$$

The index *ft* shows that the relevant variable is calculated (at instant t^n) at the foot of a trajectory associated to the velocity field $\boldsymbol{u}$. Being given $\mathbf{x} \in \Omega$ and $\lambda \in [0, T]$, the trajectories are the functions $\tilde{\boldsymbol{X}}(t; \lambda, \mathbf{x})$ such that:

$$\frac{d\tilde{\boldsymbol{X}}(t; \lambda, \mathbf{x})}{dt} = \boldsymbol{u}(t; \tilde{\boldsymbol{X}}(t; \lambda, \mathbf{x})) \tag{12}$$

$$\tilde{\boldsymbol{X}}(\lambda; \lambda, \mathbf{x}) = \mathbf{x} \tag{13}$$

The application of a similar approach to a convective-diffusive equation gives:

$$\frac{\mathrm{T}^{n+1}(\mathbf{x}) - \mathrm{T}^n(\tilde{\boldsymbol{x}}_{ft})}{\Delta t} - \Gamma \Delta \mathrm{T}^{n+1}(\mathbf{x}) = S^{n+1}(\mathbf{x}) \tag{14}$$

where Γ is the diffusion parameter and S a source term. Of course, also a second-order approximation in time could be used for equation (14), but in order to reduce the computational effort we decided to use a first-order approximation. The interpolation of T^n is easily calculated after having located the element to which the foot $\tilde{\boldsymbol{x}}_{ft}$ belongs.The generalization of the above technique to the momentum equation is direct and reads (vectorially):

$$\boldsymbol{u}^{n+1} - \Delta t \nu \Delta \boldsymbol{u}^{n+1} = \boldsymbol{u}^n(\tilde{\boldsymbol{x}}_{ft}) + \Delta t\,(-\nabla p^n + \boldsymbol{f}^{n+1}) \tag{15}$$

In (15) it can be seen that the gradient of pressure is evaluated at the instant n, therefore considering instead of $\boldsymbol{u}^{n+1}$ a provisional velocity $\tilde{\boldsymbol{u}}^{n+\frac{1}{2}}$, applying to it the divergence operator and considering the incompressibility relation, we can write a Poisson equation for the pressure correction $\tilde{p}^{n+1}$:

$$-\Delta \tilde{p}^{n+1} = -\frac{1}{\Delta t} \nabla \cdot \tilde{\boldsymbol{u}}^{n+\frac{1}{2}} \tag{16}$$

Finally the adjourned values of velocity and pressure are given by:

$$\boldsymbol{u}^{n+1} = \tilde{\boldsymbol{u}}^{n+\frac{1}{2}} - \Delta t \Delta \tilde{p}^{n+1} \tag{17}$$

$$p^{n+1} = p^n + \tilde{p}^{n+1} \tag{18}$$

The following important remarks can be made:

a) the difficulties inherent in the nonlinearity of the convective term are overcome, infact, the stability is granted under mild conditions (characteristics are unconditionally stable for linear problems);
b) the incompressibility is granted by construction;
c) the most expensive computational kernels stem from elliptic problems.

4. Approximation of the 2D Navier–Stokes equations with stabilization

In order to obtain a weak formulation of problem (1) – (4), we formally multiply (1) by a suitable function ψ and by integration in Ω we obtain:

$$\int_\Omega \frac{\partial \boldsymbol{u}}{\partial t} \cdot \psi \, d\Omega + \int_\Omega [(\boldsymbol{u} \cdot \nabla)\boldsymbol{u}] \cdot \psi \, d\Omega - \nu \int_\Omega \Delta \boldsymbol{u} \cdot \psi \, d\Omega + \int_\Omega \nabla p \cdot \psi \, d\Omega = \int_\Omega \boldsymbol{f} \cdot \psi \, d\Omega \tag{19}$$

Using the Green formula we have:

$$-\nu \int_\Omega \Delta \boldsymbol{u} \cdot \psi \, \mathrm{d}\Omega = \nu \int_\Omega \nabla \boldsymbol{u} \cdot \nabla \psi \mathrm{d}\Omega - \nu \int_{\partial\Omega} \frac{\partial \boldsymbol{u}}{\partial \boldsymbol{n}} . \psi d\gamma \tag{20}$$

$$\int_\Omega \nabla p \cdot \psi \, d\Omega = -\int_\Omega p \nabla \cdot \psi \, d\Omega + \int_{\partial\Omega} p \psi \cdot \boldsymbol{n} \, d\gamma \tag{21}$$

Keeping into account the homogeneous boundary conditions, and by substitution of the last two relations in the momentum equation, we obtain:

$$\int_\Omega \frac{\partial \boldsymbol{u}}{\partial t} \cdot \psi \, d\Omega + \int_\Omega [(\boldsymbol{u} \cdot \nabla)\boldsymbol{u}] \cdot \psi \, d\Omega + \nu \int_\Omega \nabla \boldsymbol{u} \cdot \nabla \psi \, d\Omega - \int_\Omega p \, \nabla \cdot \psi \, d\Omega = \int_\Omega \boldsymbol{f} \cdot \psi \, d\Omega + \int_{\partial\Omega} \left(\nu \frac{\partial \boldsymbol{u}}{\partial \boldsymbol{n}} - p\boldsymbol{n} \right) \cdot \psi \, d\gamma \qquad \forall \psi \in \mathbf{V} \tag{22}$$

By multiplying (2) by a suitable test function $\varphi \in Q$, we obtain:

$$\int_\Omega \varphi \, \nabla \cdot \boldsymbol{u} \, d\Omega = 0 \qquad \forall \varphi \in Q \tag{23}$$

The spaces $\mathbf{V}$ and Q can be chosen $\mathbf{V} := [\mathrm{H}^1_0(\Omega)]^2$ and $Q := \mathrm{L}^2(\Omega)$, respectively.

Equations (22) and (23) are the weak vectorial forms of the equations that we are interested in solving. Rewriting the fractional step (15)–(18) with respect to each component of the velocity field $\boldsymbol{u} = (u, v)$ and applying the weak formulation to the velocity correction, the problem becomes:

to find $\boldsymbol{u}^{n+1} = (u^{n+1}, v^{n+1}) \in L^2(\mathbb{R}^+; [H^1(\Omega)]^2) \cap C^0(\mathbb{R}^+; [L^2(\Omega)]^2)$ and $p^{n+1} \in L^2(\mathbb{R}^+; H^1(\Omega))$ such that:

$$\frac{1}{\Delta t}\int_\Omega \tilde{u}^{n+\frac{1}{2}}\psi_1 d\Omega + \nu\int_\Omega \nabla\tilde{u}^{n+\frac{1}{2}}\cdot\nabla\psi_1 d\Omega = -\int_\Omega \frac{\partial p^n}{\partial x}\psi_1 d\Omega + \int_\Omega f_u^{n+1}\psi_1\, d\Omega + \frac{1}{\Delta t}\int_\Omega u^n(\tilde{\boldsymbol{x}}_{ft})\psi_1\, d\Omega \quad \forall\,\psi_1 \in V_1 \tag{24}$$

$$\frac{1}{\Delta t}\int_\Omega \tilde{v}^{n+\frac{1}{2}}\psi_2 d\Omega + \nu\int_\Omega \nabla\,\tilde{v}^{n+\frac{1}{2}}\cdot\nabla\psi_2 d\Omega = -\int_\Omega \frac{\partial p^n}{\partial y}\psi_2\, d\Omega + \int_\Omega f_v^{n+1}\psi_2\, d\Omega + \frac{1}{\Delta t}\int_\Omega v^n(\tilde{\boldsymbol{x}}_{ft})\psi_2 d\Omega \quad \forall\psi_2 \in V_2 \tag{25}$$

$$\int_\Omega \nabla\tilde{p}^{n+1}\cdot\nabla\varphi d\Omega = -\frac{1}{\Delta t}\int_\Omega\left(\frac{\partial\tilde{u}^{n+\frac{1}{2}}}{\partial x} + \frac{\partial\tilde{v}^{n+\frac{1}{2}}}{\partial y}\right)\varphi\, d\Omega \quad \forall\,\varphi \in Q \tag{26}$$

$$p^{n+1} = p^n + \tilde{p}^{n+1} \tag{27}$$

$$\int_\Omega u^{n+1}\varphi\, d\Omega = \int_\Omega \tilde{u}^{n+\frac{1}{2}}\varphi\, d\Omega - \Delta t\int_\Omega \frac{\partial\tilde{p}^{n+1}}{\partial x}\varphi\, d\Omega \quad \forall\,\varphi \in Q \tag{28}$$

$$\int_\Omega v^{n+1}\varphi\, d\Omega = \int_\Omega \tilde{v}^{n+\frac{1}{2}}\varphi\, d\Omega - \Delta t\int_\Omega \frac{\partial\tilde{p}^{n+1}}{\partial y}\varphi\, d\Omega \quad \forall\,\varphi \in Q \tag{29}$$

Let $T_h = \{K\}$ be a suitable triangulation of the domain Ω. By using equal-order velocity-pressure approximation (i.e., $\psi_1 = \psi_2 = \varphi \in X_h^2$) the system of equations (24)–(29) can be rewritten as follow:

$$\sum_K\left[\frac{1}{\Delta t}\sum_i \tilde{u}_i^{n+\frac{1}{2}}\int_K \varphi_i\varphi_j dK + \nu\sum_i \tilde{u}_i^{n+\frac{1}{2}}\int_K \nabla\varphi_i\cdot\nabla\varphi_j dK\right] = \sum_K\left[-\sum_i p_i^n\int_K \frac{\partial\varphi_i}{\partial x}\varphi_j dK + \sum_i (f_i^{n+1})_u\int_K \varphi_i\varphi_j dK + \frac{1}{\Delta t}\sum_i u_i^n(\tilde{\boldsymbol{x}}_{ft})\int_K \varphi_i\varphi_j dK\right] \tag{30}$$

$$\sum_K\left[\frac{1}{\Delta t}\sum_i \tilde{v}_i^{n+\frac{1}{2}}\int_K \varphi_i\varphi_j dK + \nu\sum_i \tilde{v}_i^{n+\frac{1}{2}}\int_K \nabla\varphi_i\cdot\nabla\varphi_j dK\right] = \sum_K\left[-\sum_i p_i^n\int_K \frac{\partial\varphi_i}{\partial y}\varphi_j dK + \sum_i (f_i^{n+1})_v\int_K \varphi_i\varphi_j dK + \frac{1}{\Delta t}\sum_i v_i^n(\tilde{\boldsymbol{x}}_{ft})\int_K \varphi_i\varphi_j dK\right] \tag{31}$$

$$\sum_K\left[\sum_i \tilde{p}_i^{n+1}\int_K \nabla\varphi_i\,\nabla\varphi_j\, dK\right] = -\frac{1}{\Delta t}\sum_K\left[\sum_i \tilde{u}_i^{n+\frac{1}{2}}\int_K \frac{\partial\varphi_i}{\partial x}\varphi_j dK + \sum_i \tilde{v}_i^{n+\frac{1}{2}}\int_K \frac{\partial\varphi_i}{\partial y}\varphi_j dK\right] \tag{32}$$

$$p_i^{n+1} = p_i^n + \tilde{p}_i^{n+1} \tag{33}$$

$$\sum_K\left[\sum_i u_i^{n+1}\int_K \varphi_i\varphi_j\, dK\right] = \sum_K\left[\sum_i \tilde{u}_i^{n+\frac{1}{2}}\int_K \varphi_i\varphi_j dK - \Delta t\sum_i \tilde{p}_i^{n+1}\int_K \frac{\partial\varphi_i}{\partial x}\varphi_j dK\right] \tag{34}$$

$$\sum_K\left[\sum_i v_i^{n+1}\int_K \varphi_i\varphi_j\, dK\right] = \sum_K\left[\sum_i \tilde{v}_i^{n+\frac{1}{2}}\int_K \varphi_i\varphi_j dK - \Delta t\sum_i \tilde{p}_i^{n+1}\int_K \frac{\partial\varphi_i}{\partial y}\varphi_j dK\right] \tag{35}$$

The algebraic form of equations (30)–(35) is:

$$\frac{1}{\Delta t}\mathbf{M}\tilde{\boldsymbol{u}}^{n+\frac{1}{2}} + \nu\mathbf{A}\tilde{\boldsymbol{u}}^{n+\frac{1}{2}} = -\mathbf{B}^T\boldsymbol{p}^n + \mathbf{M}\boldsymbol{f}_u^{n+1} + \frac{1}{\Delta t}\mathbf{M}\boldsymbol{u}^n(\tilde{\boldsymbol{x}}_{ft}) \tag{36}$$

$$\frac{1}{\Delta t}\mathbf{M}\tilde{\boldsymbol{v}}^{n+\frac{1}{2}} + \nu\mathbf{A}\tilde{\boldsymbol{v}}^{n+\frac{1}{2}} = -\mathbf{C}^T\boldsymbol{p}^n + \mathbf{M}\boldsymbol{f}_v^{n+1} + \frac{1}{\Delta t}\mathbf{M}\boldsymbol{v}^n(\tilde{\boldsymbol{x}}_{ft}) \tag{37}$$

$$\mathbf{A}\tilde{\boldsymbol{p}}^{n+1} = -\frac{1}{\Delta t}\left[\mathbf{B}^T\tilde{\boldsymbol{u}}^{n+\frac{1}{2}} + \mathbf{C}^T\,\tilde{\boldsymbol{v}}^{n+\frac{1}{2}}\right] \tag{38}$$

$$\boldsymbol{p}^{n+1} = \boldsymbol{p}^n + \tilde{\boldsymbol{p}}^{n+1} \tag{39}$$

$$\mathbf{M}\boldsymbol{u}^{n+1} = \mathbf{M}\tilde{\boldsymbol{u}}^{n+\frac{1}{2}} - \Delta t\,\mathbf{B}^T\tilde{\boldsymbol{p}}^{n+1} \tag{40}$$

$$\mathbf{M}\boldsymbol{v}^{n+1} = \mathbf{M}\tilde{\boldsymbol{v}}^{n+\frac{1}{2}} - \Delta t\,\mathbf{C}^T\tilde{\boldsymbol{p}}^{n+1} \tag{41}$$

where $\mathbf{A} = \sum_K\int_K \nabla\varphi_i\cdot\nabla\varphi_j dK$ is the stiffness matrix, $\mathbf{M} = \sum_K\int_K \varphi_i\varphi_j\, dK$ is the mass matrix, $\mathbf{B} = \sum_K\int_K \frac{\partial\varphi_i}{\partial x}\varphi_j\, dK$ is the matrix of the x-component of the gradient operator and $\mathbf{C} = \sum_K\int_K \frac{\partial\varphi_i}{\partial y}\varphi_j\, dK$ is the matrix of the y-component of the gradient operator.

Remark 1. It is easy to see that all the algebraic systems of (36)–(41) are reduced to elliptic-like ones so that they are optimal from a computational point of view, in particular the solutions can be calculated by an iterative method preconditioned by a Schwarz scalable additive preconditioner.

Remark 2. If the domain Ω does not change during the transient, the matrices to be built (just once) are **M, A, B, C.**

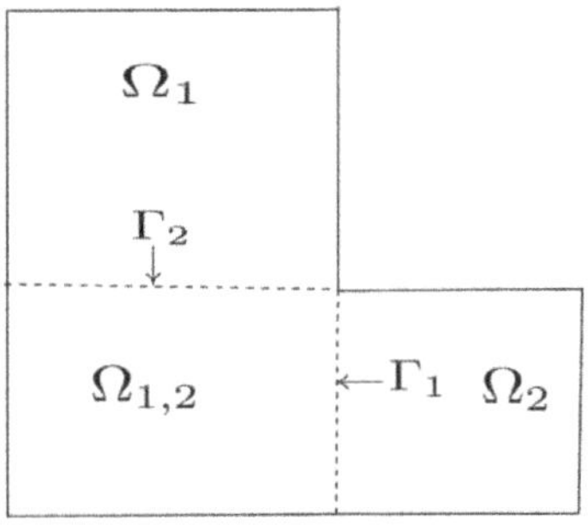

Figure 1. Example of overlapping subdivision with two subdomains.

Remark 3. In order to apply the Projection Method, the variables $\tilde{u}^{n+\frac{1}{2}}$, $\tilde{v}^{n+\frac{1}{2}}$ and $\tilde{p}^{\,n+1}$ have to belong to the spaces:

$$\tilde{u}^{n+\frac{1}{2}},\ \tilde{v}^{n+\frac{1}{2}}\ \epsilon\ \mathrm{H}_{0,\,div}$$
$$= \{\boldsymbol{u}^{n+1} \epsilon [\mathrm{L}^2(\Omega)]^2 : \nabla\cdot\boldsymbol{u}^{n+1} = 0,\ \boldsymbol{u}^{n+1}\cdot\boldsymbol{n} = 0 \text{ on } \partial\Omega\} \tag{42}$$

and $\tilde{p}^{n+1}\ \epsilon\ \mathrm{H}^1(\Omega)$

4.1. *Schwarz scalable additive preconditioner*

We give a brief description of the Schwarz preconditioner for elliptic problems introduced in Quarteroni and Valli (1999) and adopted in this paper. We consider, for example, the domain Ω as decomposed in two overlapping subdomains Ω_1 and Ω_2, see Figure 1, and we wish to solve the Poisson problem with homogeneous boundary conditions:

$$-\Delta u = f \ \text{ in } \Omega \tag{43}$$
$$u = 0 \ \text{ on } \partial\Omega \tag{44}$$

Let $\partial\Omega$, $\partial\Omega_1$, $\partial\Omega_2$ be the boundaries of Ω, Ω_1 , Ω_2 respectively, and Γ_1 , Γ_2 the internal boundaries of Ω_1 and Ω_2 respectively. Let u_i^k denote the approximate solution, after k iterations, on $\bar{\Omega}_i = \Omega_i \cup \partial\Omega_i$ with i = 1 and 2, and $u_1^k|\Gamma_2$ (respectively $u_2^k|\Gamma_1$) the restriction of u_1^k (respectively u_2^k) on Γ_2 (respectively Γ_1). Let u^0 an initial function defined in Ω and vanishing on $\partial\Omega$ and set $u_2^0 = u^0|\Omega_2$.Therefore, the Schwarz approach consists of defining two sequences u_1^{k+1} and u_2^{k+1} with $k \geq 0$ and solving iteratively the two boundary-value problems for u_1^{k+1} and u_2^{k+1} :

$$-\Delta u_1^{k+1} = f \ \text{ in } \Omega_1 \tag{45}$$
$$u_1^{k+1} = u_2^k \text{ on } \Gamma_1 \tag{46}$$
$$u_1^{k+1} = 0 \text{ on } \partial\Omega_1 \cup \partial\Omega \tag{47}$$

and

$$-\Delta u_2^{k+1} = f \ \text{ in } \Omega_2 \tag{48}$$
$$u_2^{k+1} = u_1^k \text{ on } \Gamma_2 \tag{49}$$
$$u_2^{k+1} = 0 \text{ on } \partial\Omega_2 \cup \partial\Omega \tag{50}$$

In order to define the variational formulation of the Schwarz problem at the discrete level, we set $\mathrm{V} := \mathrm{H}_0^1(\Omega)$,$\mathrm{V}_i^0 := \mathrm{H}_0^1(\Omega_i)$. We define the finite element space:

$\mathrm{V_h} := \{\mathrm{v} \in \mathrm{V} : \mathrm{v}|_K \in Pr(K), \forall K \in T_h\} \subset \mathrm{V} = \mathrm{H}_0^1(\Omega)$ where Pr are polynomials of degree $\leq r$. We know that a FE approximation to the original Poisson problem (43–44) consists in finding $u_h \in \mathrm{V}_h \subset \mathrm{V}$ such that:

$$a_h(u_h\ ,\ v_h) = (f\ ,\ v_h), \quad \forall\, v_h \in V_h \tag{51}$$

where $a_h(\cdot,)$ is the bilinear form given by:

$a_h(u_h\ ,\ \mathrm{v}_h) = \int_\Omega \nabla u_h \,.\, \nabla v_h\, d\Omega$ and $(f\ , v_h)$ the linear form defined as $(f,\ v_h) = \int_\Omega f\ v_h d\Omega$. Then, the Schwarz method at the discrete level consists of assigning $u^0 \in \mathrm{V}_h$ and solving for each $k \geq 0$:

$$w_{1.h}^k \epsilon \mathrm{V}_{1,h}^0 : \ a_1(w_{1.h}^k, v_{1.h}) = (f, v_{1.h})_{\Omega_1} - a_1\,(u_h^k,\ v_{1,h})$$
$$\forall v_{1,h} \in \mathrm{V}_{1,h}^0 \tag{52}$$
$$w_{2.h}^k \epsilon \mathrm{V}_{2,h}^0 : \ a_2(w_{2.h}^k, v_{2.h}) = (f, v_{2.h})_{\Omega_2} - a_2\,(u_h^k,\ v_{2,h})$$
$$\forall v_{2,h} \in \mathrm{V}_{2,h}^0 \tag{53}$$
$$u_h^{k+1} = u_h^{k+1} + \widetilde{w_{1,h}^k} + \widetilde{w_{2,h}^k} \tag{54}$$

Where $\widetilde{w_{i,h}^k}$ is the function that extends $w_{i.h}^k$ by zeros in $\Omega\backslash\Omega_i$, $a_i\ (\cdot,\cdot)$ denotes the restriction of $a_h(\cdot,\cdot)$ to Ω_i and $\mathrm{V}_{i,h}^0 \subset \mathrm{V}_i^0$.

For the algebraic formulation of the Schwarz method we can think that the domain Ω is subdivided by M subdomains Ω_i with $i = 1, ..., M$. Let $\mathrm{V}_i^* = \{v \in \mathrm{V} : v = 0$ in $\Omega\backslash\Omega_i\}$ and $\mathrm{V}_{i,h}^*$ be a finite dimensional subset of V_i^*. We define the prolongation operator $\mathbf{R}_i^T : \mathrm{V}_{i,h}^* \rightarrow \mathrm{V}_h$, which extends by zeros the degree of freedom outside Ω_i. The restriction operator $\mathbf{R}_i$ is defined as the transposition of $\mathbf{R}_i^T$. By denoting $\mathbf{A}_h$ and $\mathbf{A}_i$ the stiffness matrices with respect to the global and local bilinear forms $a_h\ (\cdot,\cdot)$ and $a_i\ (\cdot,\cdot)$ and defining the operator $\mathbf{Q}_i$ like the matrix $\mathbf{Q}_i = \mathbf{R}_i^T\mathbf{A}_i^{-1}\mathbf{R}_i$, then after some algebraic manipulations the one level additive Schwarz iterative method reads for $M \geq 2$ subdomains:

$$\mathbf{u}_h^{k+1} = \mathbf{u}_h^k + \left(\sum_{i=1}^{M} \mathbf{Q}_i\right)(\mathbf{b}_h - \mathbf{A}_h\mathbf{u}_h^k), \quad k \geq 0 \tag{55}$$

where $k \geq 0$ is the index of iteration. Regarding (55) as a fixed-point relation, we can define the Schwarz method simply as a Richardson iterative method with preconditioner $\mathbf{P}_{as} = \left[\sum_{i=1}^{M} \mathbf{Q}_i\right]^{-1}$. Unfortunately, the convergence of the one level Schwarz method deteriorates when the number M of subdomains becomes larger because of the lack of information between subdomains Ω_i. In order to overcome this drawback, we have introduced a global coarse mesh as a particular subdomain Ω_0, so that communications among all the subdomains are guaranteed. Therefore, the final form of the preconditioner is $\mathbf{P}_{asc}^{-1} = \left(\sum_{i=0}^{M} \mathbf{R}_i^T \mathbf{A}_i^{-1} \mathbf{R}_i\right)$. Since the final algebraic system (51) is $\mathbf{A}_h \mathbf{u}_h = \mathbf{b}_h$, the preconditioned two level additive Schwarz system is written:

$$\mathbf{P}_{asc}^{-1} \mathbf{A}_h \, \mathbf{u}_h = \left(\sum_{i=0}^{M} \mathbf{R}_i^T \mathbf{A}_i^{-1} \mathbf{R}_i\right) \mathbf{A}_h \mathbf{u}_h = \left(\sum_{i=0}^{M} \mathbf{R}_i^T \mathbf{A}_i^{-1} \mathbf{R}_i\right) \mathbf{b}_h \tag{56}$$

A well-known theorem (Dryja & Widlund, 1989), under suitable conditions, provides a limit for the condition number $K(\mathbf{P}_{asc}^{-1}\mathbf{A}_h)$ of the Schwarz preconditioned matrix. Called h and H the diameters of the elements of the fine and coarse triangulations respectively, then we have:

$$K\,(\mathbf{P}_{asc}^{-1}\mathbf{A}_h) \leq c\left[1 + \frac{H}{h}\right]^2 \tag{57}$$

where c is a positive constant independent of H and h. Thus the presence of the coarse partition makes the convergence rate independent of both h and H. This property (scalability) is very desirable especially when the method is implemented on massively parallel architectures, and when the number of subdomains is big. We recall that the computational kernels of (36–41) are of elliptic kind so that a scalable Schwarz additive overlapping multidomains preconditioner is worthwhile to be adopted. The algorithm of the implementation of the Schwarz additive preconditioner can be seen in Kengni Jotsa (2012).

Remark 4. The number M of the subdomains in which Ω is divided, decides the efficiency of the Schwarz preconditioner. In fact, the number of nodes belonging to each subdomain depends on M and, of course, also on the subdomain form; both the choices should be as shrewd as possible in order to reach a good balance among the dimension of the local stiffness matrices $\mathbf{A}_i$ and the optimal computational time. Another choice influences the efficiency of the **P***as* and **P***asc* preconditioners, namely the level of *fill-in* allowed in the process of **LU** factorization to calculate the $\mathbf{A}_i^{-1}$ matrices, the so-called **I**ncomplete **LU** factorization (**ILU**) process.

4.2. The inf-sup condition and a stabilization

In a study (Guermond & Quartapelle, 1998) it is affirmed that the fractional step method based on the Poisson equation for pressure overcomes the inf-sup condition under a suitable limitation for Δt and for P_1 /P_1 linear approximation; without respecting the temporal limitation or for equal-order P_2 /P_2 approximation, the numerical solution is unstable. Since we are using equal-order P_2 /P_2 approximation, stabilization techniques needed in order to complete the formulation (36–41). Among the wide spread stabilized methods, we can quote the Galerkin least square technique (Franca & Hughes, 1993; Franca & Stenberg, 1991; Hughes, Franca, & Balestra, 1986), the least square method for first-order system such as (Bochev, Cai, Manteuffel, & McCormick, 1998) and the least square method for second-order scheme (Fortin & Boivin, 1990; Silvester & Kechkar, 1990) and other methods (Brezzi & Douglas, 1988; Douglas & Wang, 1989). Codina (2001; Codina & Soto (2004)) built a stabilization technique based on the definition of the "subscale velocities" that are responsible for the instability due to violation of the inf-sup condition. The algebraic terms able to govern the effects of the subscale velocities (SSV) are obtained by:

(i) imposing that the SSV satisfy a momentum-like equation to which is added a suitable function;

(ii) orthogonalizing the space for the approximation of the SSV to the space V_h in which the physical velocity $\boldsymbol{u}_h$ has to be approximated.

This technique allows us in particular to deal with convection dominated flows and to use equal-order velocity pressure interpolations. We apply this technique to complete the formulation $(36-41)$.

Final algebraic stabilized system including the energy equation

Thus, in order to stabilize $(36-41)$, we have added the suitable algebraic terms to equations of the momentums and the Poisson equation of the pressure, to get:

$$\begin{aligned}&(\mathbf{M} + \nu\Delta t\mathbf{A})\tilde{\boldsymbol{u}}^{n+\frac{1}{2}}\\ &\quad= \Delta t\{-\mathbf{B}^T\boldsymbol{p}^n + \mathbf{M}\boldsymbol{f}_u^{n+1} - \tau(\mathbf{A} - \mathbf{B}^T\mathbf{M}_L^{-1}\mathbf{B}^T)\boldsymbol{p}^n\}\\ &\qquad+ \mathbf{M}\boldsymbol{u}^n(\tilde{\boldsymbol{x}}_{ft})\end{aligned} \tag{58}$$

$$\begin{aligned}&(\mathbf{M} + \nu\Delta t\mathbf{A})\tilde{\boldsymbol{v}}^{n+\frac{1}{2}}\\ &\quad= \Delta t\{-\mathbf{C}^T\boldsymbol{p}^n + \mathbf{M}\boldsymbol{f}_v^{n+1} - \tau(\mathbf{A} - \mathbf{C}^T\mathbf{M}_L^{-1}\mathbf{C}^T)\boldsymbol{p}^n\}\\ &\qquad+ \mathbf{M}\boldsymbol{v}^n(\tilde{\boldsymbol{x}}_{ft})\end{aligned} \tag{59}$$

$$\begin{aligned}&[(\Delta t + \tau)\mathbf{A} - \tau(\mathbf{B}^T\mathbf{M}_L^{-1}\mathbf{B}^T + \mathbf{C}^T\mathbf{M}_L^{-1}\mathbf{C}^T)]\boldsymbol{p}^{n+1}\\ &\quad= \Delta t\mathbf{A}\boldsymbol{p}^n - \left[\mathbf{B}^T\tilde{\boldsymbol{u}}^{n+\frac{1}{2}} + \mathbf{C}^T\tilde{\boldsymbol{v}}^{n+\frac{1}{2}}\right]\end{aligned} \tag{60}$$

where τ is a weight parameter and $\mathbf{M}_L$ is the lumped mass matrix. We also have the equations for the correction of the velocity fields, given by:

$$\mathbf{M}\boldsymbol{u}^{n+1} = \mathbf{M}\tilde{\boldsymbol{u}}^{n+\frac{1}{2}} - \Delta t\mathbf{B}^T\tilde{\boldsymbol{p}}^{n+1} \tag{61}$$

$$\mathbf{M}\boldsymbol{v}^{n+1} = \mathbf{M}\tilde{\boldsymbol{v}}^{n+\frac{1}{2}} - \Delta t\mathbf{C}^T\tilde{\boldsymbol{p}}^{n+1} \tag{62}$$

With the energy equation written in terms of the temperature and its corresponding boundary conditions (5)–(7) to the final algebraic stabilized formulation (58)–(62) of the problem (1)–(7), we can add the equation:

$$(\mathbf{M} + \Gamma\Delta t\mathbf{A})\mathbf{T}^{n+1} = \Delta t\mathbf{M}\mathbf{S}^{n+1} + \mathbf{M}\mathbf{T}^n(\tilde{\boldsymbol{x}}_{ft}) \tag{63}$$

Remark 5. We recall that in our stabilization process we do not have to include the convective terms.

4.3. *Choice of the weight parameter* τ

According to Codina and Soto (2004), a scheme is suggested that expresses τ in terms of the kinematic viscosity ν, the modulus of the velocity $\boldsymbol{u}^*$ (which is the sum of the computed velocity $\boldsymbol{u}_h$ and of the subscale velocity) and a characteristic length h. The formula of the weight τ is:

$$\tau = \left(c_1 \frac{\nu}{h^2} + c_2 \frac{|\boldsymbol{u}^*|}{h} \right)^{-1} \quad (64)$$

where c_1 and c_2 are suitable constants; it can be simplified as:

$$\tau = c_3 \frac{h^2}{\nu + |\boldsymbol{u}^*| h} \quad (65)$$

The characteristic length h can be estimated as the ratio between the modulus of two vectors $\boldsymbol{h}$ and $\boldsymbol{k}$, and represents the modulus of the Jacobian of the transformation of the generic element of the partition into the reference element, multiplied for a suitable constant c_4 (for triangular elements $c_4 = 0.7$). In order to simplify the procedure we choose for vector $\boldsymbol{k}$ the hypotenuse of the reference triangle, and for vector $\boldsymbol{h}$ the corresponding edge of the generic element. In this case $c_4 = 0.5$. Another simplification of the estimation of $\boldsymbol{h}$ could be by choosing the smallest edge of the elements of the partition and multiplying its length by 0.5. In the section relevant to the numerical tests a sensitivity analysis of τ is carried out.

4.4. *Numerical procedure*

In summary, the fractional step works following these steps:

i) the velocity, pressure and temperature fields are given at time $t = t^n$;
ii) compute the intermediate velocity field from Equations (58) and (59);
iii) compute the pressure field at time $t = t^{n+1}$ from Equation (60);
iv) correct the intermediate velocity field from Equations (61) and (62);
v) compute the temperature field at time $t = t^{n+1}$ from Equation (63);
vi) go to step (i) and repeat the above procedure until the desired solution is obtained.

5. Test cases

5.1. *Two dimensional time-dependent problem*

We have used an analytical solution of the Navier–Stokes equations similar to that of Zang, Street, and Koseff (1994) as benchmark to our approach (and also to check the performance of our in-house code). The following is the solution to the two dimensional unsteady flow of decaying vortices:

$$u = -\cos(\pi x)\sin(\pi y)exp(-2t) \quad (66)$$

$$v = \sin(\pi x)\cos(\pi y)\exp(-2t) \quad (67)$$

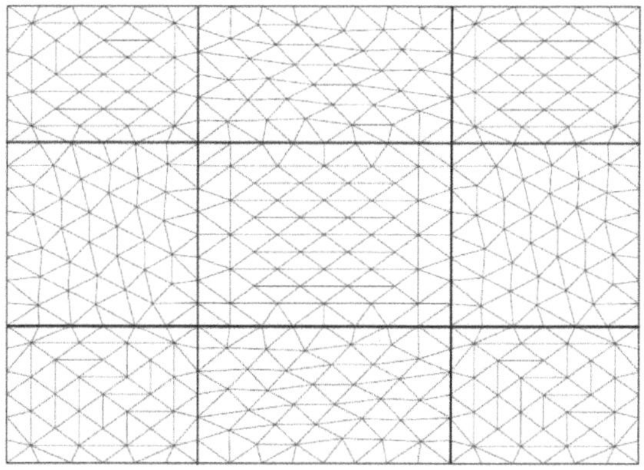

Figure 2. Problem 5.1: mesh with 1569 nodes and nine subdomains for the Schwarz preconditioner.

Table 1. Problem 5.1, L^∞-error norm for $\tau = 1.000\text{E}-5$, $\Delta t = 5.000\text{E}-3$, $N_{TS} = 200$, $u_{max} = 1.353\text{E}-1$, $v_{max} = 1.353\text{E}-1$, $p_{max} = 9.157\text{E}-3$ and various *Re*.

Re	1.000 E+2	4.000 E+2	1.000 E+3	5.000 E+3
e_u	9.407 E-4	2.374 E-3	3.811 E-3	7.217 E-3
e_v	7.098 E-4	1.538 E-3	3.845 E-3	6.354 E-3
e_p	7.964 E-3	6.452 E-3	5.403 E-3	4.921 E-3

$$p = -\frac{1}{4}[\cos(2\pi x) + \cos(2\pi y)]\exp(-4t) \quad (68)$$

$$f_u = 2(1 - \nu\pi^2)\cos(\pi x)\sin(\pi y)\exp(-2t) \quad (69)$$

$$f_v = 2(-1 + \nu\pi^2)\sin(\pi x)\cos(\pi y)\exp(-2t) \quad (70)$$

where f_u and f_v are the source terms. The initial and boundary conditions are obtained from the analytical solution. Instead of $\Omega = [0, \pi]^2$, we have chosen the domain $\Omega = [0, 1]^2$ and the computational mesh used has 1569 nodes, as represented in the Figure 2. We have taken the viscosity $\nu = \frac{1}{Re}$ where *Re* is the Reynolds number and various Reynolds numbers have been used ($Re = 100$, 400, 1000 and 5000). The length of the smallest edge is $l = 4.600\text{E}-2$ so that the estimation of characteristic length h, based on the second rule, is $h = 2.300\text{E}-2$; the time increment used is $\Delta t = 5.000\text{E}-3$and we used 200 time steps to reach the final time $t = 1$. This problem has also been used by previous researchers such as Feraudi and Pennati (1997), Kim and Moin (1985), Patankar (1980), to test the accuracy of their numerical methods. By denoting e_u, e_v and e_p the L^∞–error norms of the two components of the velocity and of the pressure respectively, u_{max}, v_{max} and p_{max} the maximum of the analytical velocity and pressure, N_{TS} the total number of time increments, we present in Table 1 the accuracy obtained at the final time $t = 1$. Figure 3 shows the streamlines and Figure 4 depicts the directions of the flow inside the domain, while Figure 5 illustrates the pressure contours (both for analytical and numerical solutions at time $t = 1$ for $Re = 5000$). Table 2 shows the performance of the Schwarz preconditioner for

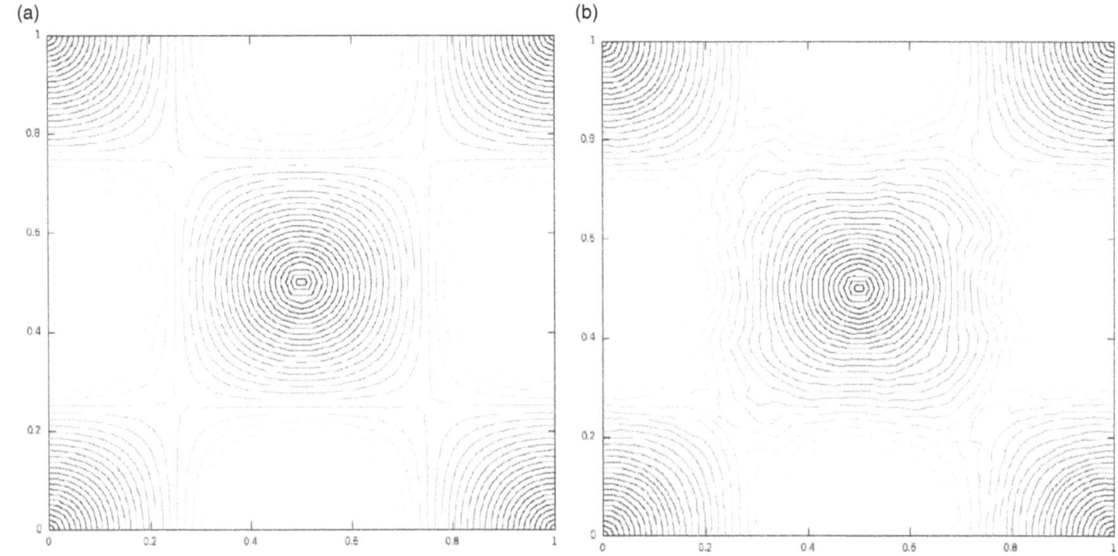

Figure 3. Problem 5.1 $Re = 5000$: streamlines of the analytical (a) and numerical (b) solutions.

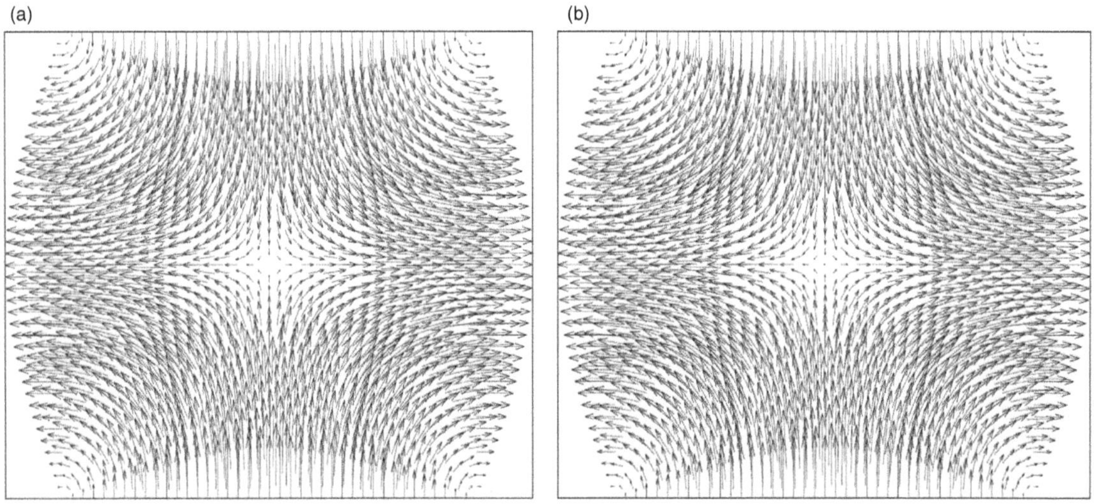

Figure 4. Problem 5.1 $Re = 5000$: velocity vectors of the analytical (a) and numerical (b) solutions.

the case resolved. In all the four tests, in order to simplify the calculations, we chose $\tau = 1.000\text{E} - 5$. This value was not calculated by the formula in (65), but by choosing in an heuristic way a suitable value for c_3. Nevertheless we should keep in mind that in the formula (64) there are the two constants c_1 and c_2 and that the good results obtained for all the numerical tests with the same value of τ, show the great flexibility in the choice of τ. In the next problem a broader analysis about the influence of τon the results will be given.

Remark 6. In all the tests solved (both in this problem as in the others problems) the Schwarz preconditioner was used with, substantially, the same reduction of the iteration number.

5.2. Lid-driven cavity problem with analytical solution

This problem has been used by Shih and Tan (1989) and it represents the isothermal recirculation flow in a square cavity generated by the uniform translation of the upper surface (lid) of a cavity whose exact solution is given by:

$$u = 8f(x)g'(y) = 8(x^4 - 2x^3 + x^2)(4y^3 - 2y) \qquad (71)$$

$$v = -8f'(x)g(y) - 8(4x^3 - 6x^2 + 2x)(y^4 - y^2) \qquad (72)$$

$$p = \frac{8}{Re}[F(x)g''(y) + f'(x)g'(y)] + 64F_1(x)\{g(y)g''(y) - [g'(y)]^2\} \qquad (73)$$

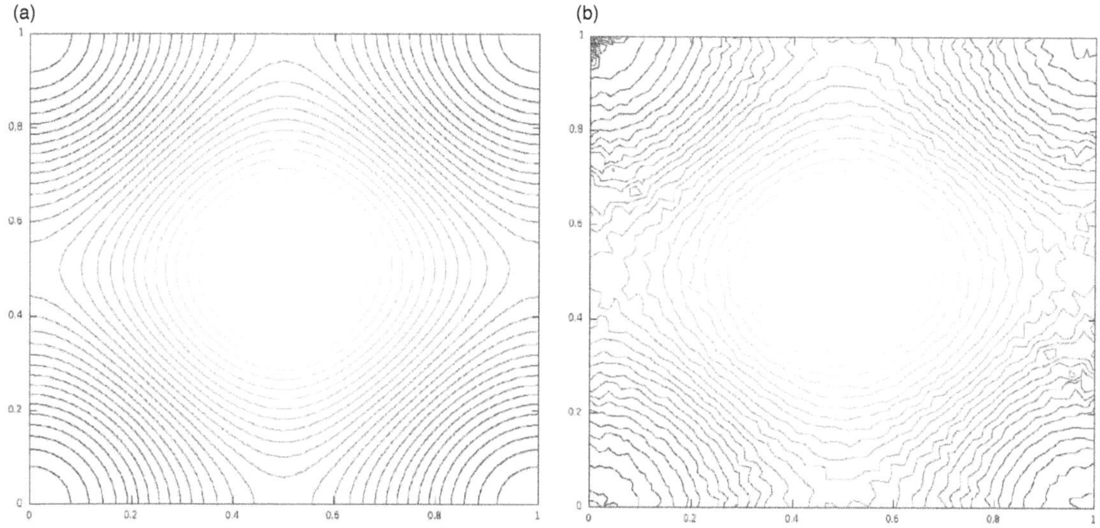

Figure 5. Problem 5.1 $Re = 5000$: pressure contours of the analytical (a) and numerical (b) solutions.

Table 2. Problem 5.1, Condition numbers K ($\mathbf{A}_h$) and K ($\mathbf{P}_{asc}^{-1}\mathbf{A}_h$) of the not preconditioned and of the preconditioned matrices, and iteration numbers ≠ Iters and ≠ Iters_p for the solution of the non-preconditioned and preconditioned systems for various values of *Re*.

Re	$\mathbf{A}_h$of variables	K ($\mathbf{A}_h$)	K ($\mathbf{P}_{asc}^{-1}\mathbf{A}_h$)	≠ Iters	≠ Iters _p
100	$\tilde{u}^{n+\frac{1}{2}}, \tilde{v}^{n+\frac{1}{2}}$	5.945E+0	3.576E+0	13	4
	p^{n+1}	5.687E+8	4.011 E+4	256	22
	u^{n+1}, v^{n+1}	6.238E+0	3.623E+0	16	5
400	$\tilde{u}^{n+\frac{1}{2}}, \tilde{v}^{n+\frac{1}{2}}$	5.667E+0	3.599E+0	16	5
	p^{n+1}	5.687E+8	4.011E+4	259	22
	u^{n+1}, v^{n+1}	6.238E+0	3.623E+0	16	5
1000	$\tilde{u}^{n+\frac{1}{2}}, \tilde{v}^{n+\frac{1}{2}}$	5.923E+0	3.603E+0	13	5
	p^{n+1}	5.687E+8	4.011E+4	258	22
	u^{n+1}, v^{n+1}	6.238E+0	3.623E+0	16	5
5000	$\tilde{u}^{n+\frac{1}{2}}, \tilde{v}^{n+\frac{1}{2}}$	6.150E+0	3.606E+0	13	5
	p^{n+1}	5.687E+8	4.011E+4	259	22
	u^{n+1}, v^{n+1}	6.238E+0	3.623E+0	16	5

where the source term with respect to u is:

$$f_u = 64 f(x) f'(x)\,[(g'(y))^2 - g(y)g''(y)] - \frac{8}{Re}(f''(x)g'(y) + f(x)g^{(3)}(y)) - \frac{8}{Re}[F'(x)g^{(3)}(y) + f''(x)g'(y)] - 64F_1'(x)\{g(y)g''(y) - [g'(y)]^2\} \tag{74}$$

the source term with respect to v is:

$$f_v = 64g(y)g'(y)[-f(x)f''(x) + (f'(x))^2] + \frac{8}{Re}(f^{(3)}(x)g(y) + f'(x)g''(y)) - \frac{8}{Re}[F(x)g^{(4)}(y) + f'(x)g''(y)] - 64F_1(x)\{g'(y)g''(y) + g(y)g^{(3)}(y) - 2g'(y)g''(y)\} \tag{75}$$

Where: $f(x) = x^4 - 2x^3 + x^2, \quad g(y) = y^4 - y^2,$

$$F(x) = 0.2x^5 - 0.5x^4 + \frac{x^3}{3},$$

$$F_1(x) = 0.5[f(x)]^2$$

and the primes of *f(x)* and *g(y)* denote the differentiation with respect to x and y, respectively. In Figure 6 we present the domain and the boundary conditions for the velocity, at the pressure we assign homogeneous Neumann conditions. We have chosen the domain $\Omega = [0,1]^2$ and the viscosity $v = 0.01$.We solved the stationary problem by an iterative process starting with the rest condition, advancing along a fictitious transient and ending with a suitable stopping criteria (in other words, was built an iterative cycle with relaxing parameter $\Delta t = 0.05$, and at every iteration a fractional step scheme was applied so that the resulting algebraic systems were solved by the preconditioned

Table 3. Problem 5.2 $Re = 100$: L^{∞}-error norm for various τ. $u_{max} = 1.000E+0$, $v_{max} = 3.838E-1$ and $p_{max} = 6.399E-2$.

τ	2.000 E-3	2.000 E-4	2.000 E-5	2.000 E-6
e_u	2.823 E-2	2.491 E-2	2.605 E-2	2.614 E-2
e_v	2.330 E-2	1.753 E-2	1.834 E-2	1.839 E-2
e_p	2.615 E-1	2.119 E-1	2.849 E-2	2.808 E-2

iterative method Bi-CGSTAB). We have chosen to stop the iterative process when the differences of the maximum values of the velocities and pressure at time instant n and $n+1$ are less than a fixed value ϵ (e.g., $\epsilon = 10E-6$). We have computed the L^{∞} – error norms of the numerical solutions generated by the new method and have reported them in the Table 3. We would point out that as the analytical solution is known, a strict control of the numerical results is possible. Figure 2 shows the mesh of 1569 nodes used, and Figures 7 and 8 present the streamlines and the vectors of the analytical and numerical velocities respectively. In Figure 9 the contours of pressure are shown. The value calculated by the simple rule (65) for $c_3 = 1.$ is $\tau = 1.500E-2$. It is interesting to note that the analytical velocity field is not directly influenced by the Reynolds parameter, but the Reynolds value changes the expressions of the pressure and the source term. Keeping this in mind, the graphics of Figures 7, 8 and 9 seem regular and coherent. From Table 3 we can deduce that the variation of τ affects (like expected) the accuracy but not the stability (at least for a quite large range of its values) of the results and that the errors of the velocity components are quite low.

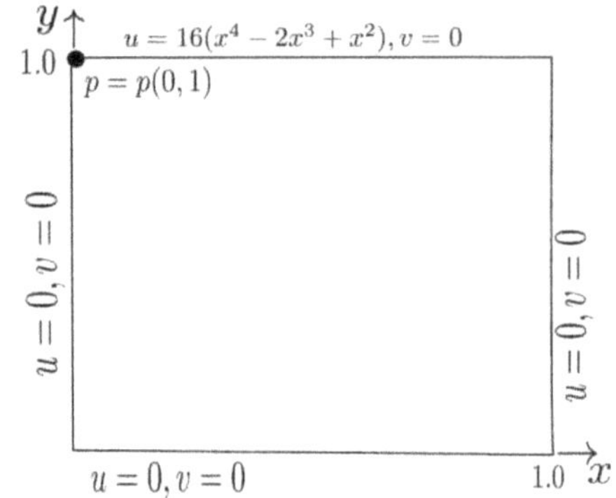

Figure 6. Problem 5.2: boundary conditions.

5.3. *Lid-driven problem with constant boundary conditions*

This problem has been used by Bruneau and Saad (2006) and it is similar to problem 5.2 except for the boundary condition of $\boldsymbol{u}$ on the top boundary where it takes the constant value 1. The analytical solution of this problem is unknown, however, it is a classical test for checking the efficiency of new algorithms. In fact, the lid-driven cavity problem is considered the prototypical recirculation flow because the streamlines pattern exhibits a central primary vortex and for high Reynolds number ($Re \geq 1000$) the possible formation of counter-rotating secondary vortices at the bottom corners of the cavity. In order to check the accuracy and the computational efficiency of the proposed method, we compare our results of problem 5.3 (with $Re = 1000$) with those found in four recent papers (Han et al., 2013; Kumar et al., 2010; Kumar et al., 2009; Pai et al.,

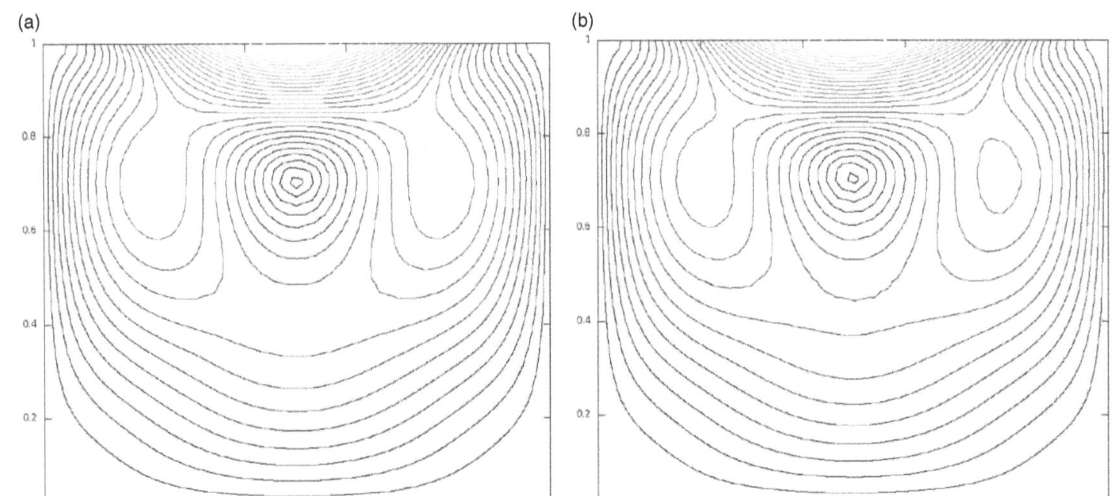

Figure 7. Problem 5.2 τ = 2E-5: streamlines of the analytical (a) and numerical (b)solutions.

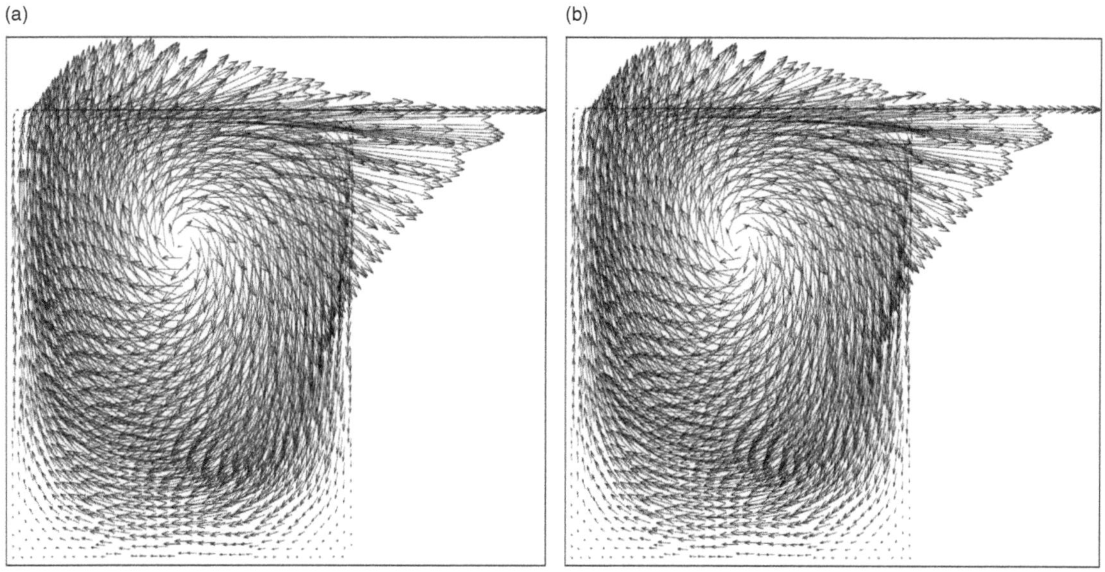

Figure 8. Problem 5.2 $\tau = $ 2E-5: velocity vectors of the analytical (a) and numerical (b)solutions.

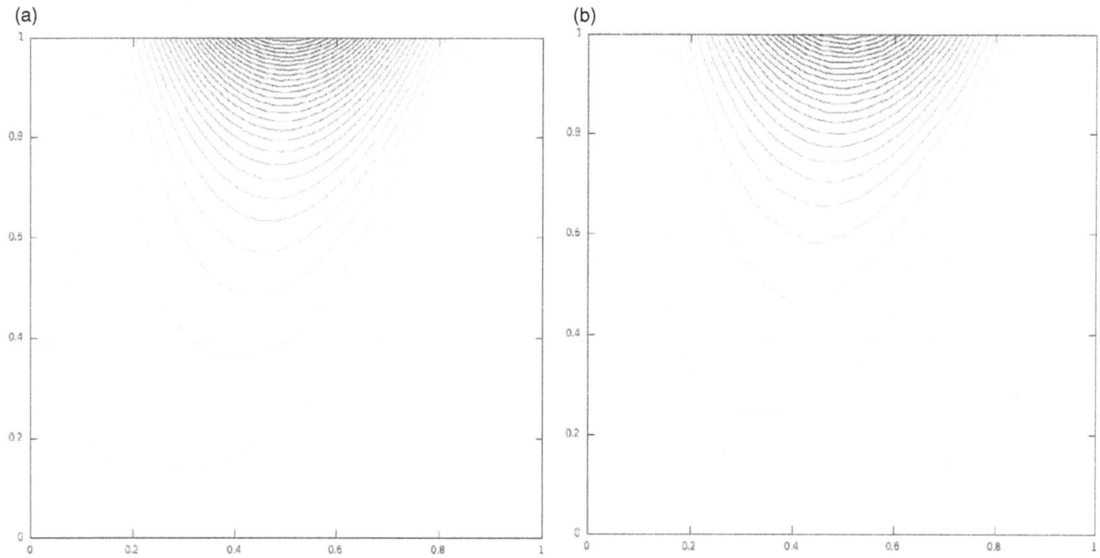

Figure 9. Problem 5.2 $\tau = $ 2E-5: pressure contours of the analytical (a) and numerical (b)solutions.

2013). In these papers the problem 5.3 (and other problems) was solved by using different numerical techniques, with different meshes or grids. A short presentation of the tests developed in the papers is as follows. In the first paper (Kumar et al., 2009) the spatial discretization was realized with colocated grids and by cell-centred FV where the convective flux was approximated by the QUICK scheme and the diffusion flux by the central second-order FD formula. To solve the coupled system, a SIMPLEC algorithm was applied and in order to reduce the computing time and obtain a grid independent convergence rates, a multigrid method with full approximation scheme and four grid levels was adopted. In order to see the secondary vortices a uniform grid of 513 $\times$ 513 volumes (263,000 grid points) was necessary, because the solution obtained by a grid of 129 $\times$ 129 volumes (16,641 grid points) did not show the secondary vortices. The computational effort of the multigrid *V-cycle* was of 32 work units, and the number of iterations (for the grid of 129 $\times$ 129 volumes) required to reach the grid convergence with a fixed tolerance $\epsilon = 10E-8$, was 1500 (without multigrid 14,500 iterations were necessary to reach the convergence). In the second paper (Kumar et al., 2010) the target to save computing time while maintaining a high accuracy, was sought by using nonuniform staggered grids in the physical space and clustering the grid points in the regions of large gradients. Since the spatial approximation was carried out by FD formulas, a transformation into a computational space with uniform grids was made. Unfortunately the transformed equations show a greater number of terms than the primitive equations. The time-advancing was obtained via a second-order non incremental fractional-step method. To improve the computational efficiency, the pressure Poisson equation (that is the most expensive computational kernel) was solved at every time step by a multi grid technique. The *V-cycle* of four level grids, with a computational effort of 36 work units, produced a good time-wise gain for the lid-driven problem at $Re = 3200$, discretized with a 129×129 grid. In the study of the lid-driven problem at $Re = 1000$, the 129×129 grid and $\Delta t = 0.001$ were used. In some figures the time evolutions of the streamline patterns, until the steady state, are shown. Through them we can deduce that at time $t = 5$ two recirculation zones start to be present, at $t = 10$ two secondary vortices arise at the bottom corners, at $t = 40$ the two secondary vortices are almost completely shaped. In the third paper (Han et al., 2013), the authors solved the 2D Navier–Stokes equations in primitive variables by means of a FE approach in which the elements are nine-node quadrilaterals both for velocities and pressure. The convective term was approximated by a semi-implicit form of three steps Taylor-characteristic-based-split scheme, while the diffusive term by the usual FE technique (namely by the stiffness matrix). The time-advancing is declared to be an incremental fractional-step method (even if from the equations (26) and (27) it can be deduced that is of non-incremental kind). The domain of the lid-driven test was partitioned with 20×20 elements in a structured adaptive way; for the marching temporal discretization $\Delta t = 0.005$ was chosen and the computations were stopped when the L^2-norm error was lower than $10E-5$. For the case $Re = 1000$, the streamlines velocity figure is not given, while the vertical and horizontal profiles of the velocities at the cavity center at the final state of the transient are shown. No information about the solution of the algebraic systems is reported. In the fourth paper (Pai et al., 2013), the lid-driven cavity problem was used like a precursor model for studying mixer designs. The 2D Navier–Stokes equations were solved by the commercial code FLUENT which employs FV for the discretization in space and the SIMPLE algorithm for time-advancing. The nonuniform grid used was finest near the no-slip wall boundary and it had 100 grid-points along walls for a total number of 17,730 volumes. The value of Δt, considered good enough to capture the physics of the problem, was $\Delta t = 0.001$. We solved, by the same approach as followed in the previous problem, the case with $Re = 1000$. The mesh used is represented in Figure 10, it was created by the TRIANGLE package (Joe, 1991) respecting an adaptive criteria, that is, using elements with smaller diameter in the regions where higher gradients are expected. It has 3009 nodes, with $hmin = 2.1E-2$, $hmax = 7.9E-2$. The other parameters used were $\tau = 2.0E-5$, $\Delta t = 0.05$ and $\epsilon = 10E-8$. In Figure 11 are represented the velocity vectors at $t = 80$ and in it is possible to single out the central primary vortex and, at the bottom corners, the well-known secondary vortices. The zooms of the secondary vortices are represented in Figure 12. In Figure 13 are plotted the profiles of the u and v velocity components along the vertical and horizontal centerlines. The comparison of the graphics shown in Figures 12 and 13 and those in some of the above papers (Han et al., 2013; Kumar et al., 2010; Pai et al., 2013) point out some differences; in particular the velocity profiles appear to be affected by viscous effects. All that should not be surprising because the graphics are built by the interpolation of the nodal values calculated by means of a quite coarse mesh (it has only 3009 nodes). In the numerical simulations the maximum values of Courant number was $C = 2.5$ and of Peclet number was $P_e = 79$. The computational efficiency is confirmed by the particularly low number of iterations of the Schwarz preconditioned Bi-CGSTAB iterative solver used to solve the momentum equations, the pressure equation and the velocity and pressure correction equations (5 and 5, 21, 5 and 5 iterations, respectively). The CPU time corresponding to each of the above iterations was 0.024s for the momentum equation, 1.005s for the pressure equation and 0.020s for the velocity and pressure correction equations, on a Sony VAIO Series EA46FMW personal computer.

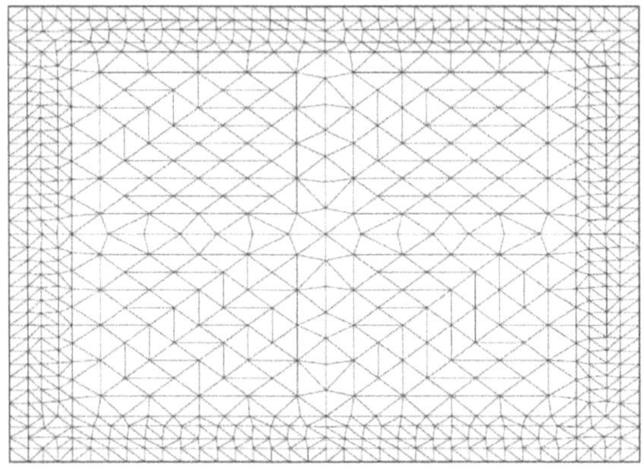

Figure 10. Problem 5.3: mesh with 3009 nodes.

Some comments regarding our results are appropriate. The use of an adaptive mesh (not requiring any transformation) and the numerical schemes used in our approach make possible to capture the physics of the problem (in particular the presence of secondary vortices at $Re = 1000$) also with a small number of nodes. The approximation of the convective terms by characteristics allows to obtain good results even in presence of very high Peclet and Courant numbers; moreover, the solution of the algebraic systems calculated by means of preconditioned iterative

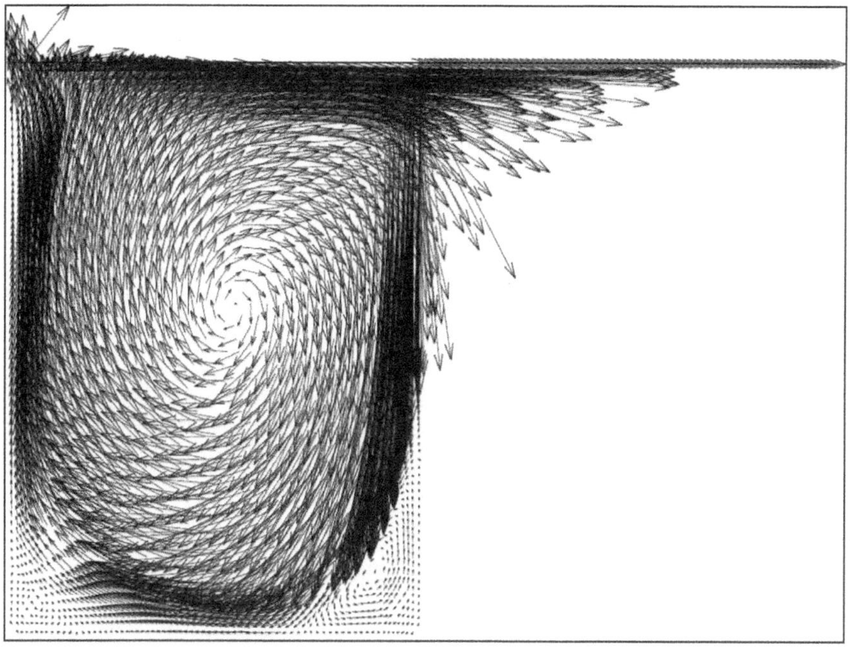

Figure 11. Problem 5.3: velocity vectors at $t = 80$.

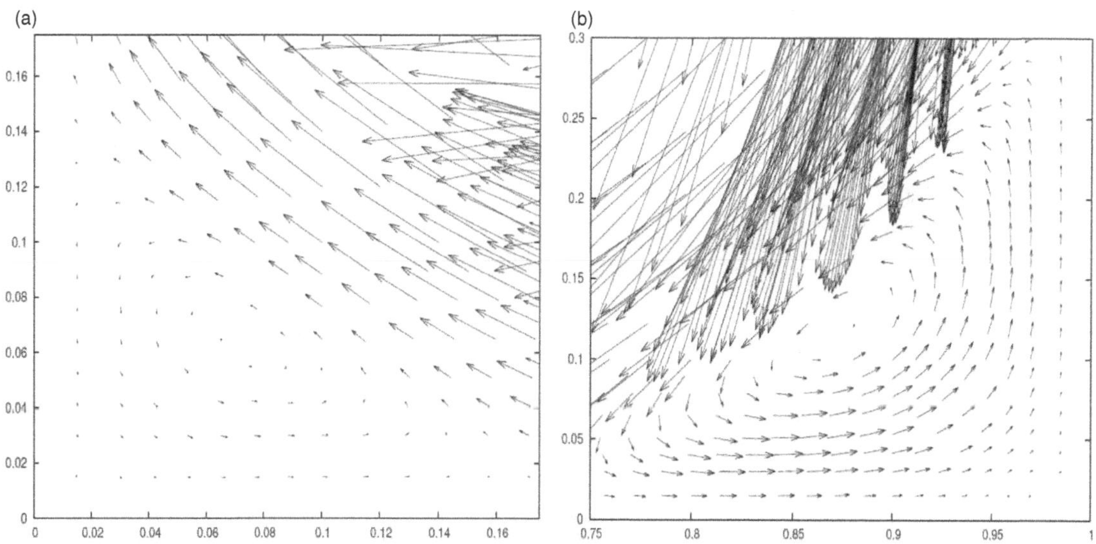

Figure 12. Problem 5.3: zoom of the secondary vortices, left (a) and right (b) corner.

solvers allows to compare the computational efficiency of our approach with that of the multigrid solvers.

5.4. Thermal convection in a square cavity

In this section, we consider a benchmark flow problem where no analytic solution is known, but considered very important by the researchers. The aim is to compare our results to some of well-established schemes present in the literature for a natural convection in a square cavity problem. In particular in De Vahl Davis (1983) is reported the definition and the results of a large number of tests along with some significant reference values. In the problem, is examined the flow of a fluid inside a square cavity for which the top and the bottom walls are kept to be adiabatic and the vertical walls are kept to be isothermal at temperatures $T_c = -0.5$ and $T_h = 0.5$ respectively (see Figure 15). Initially the fluid is assumed

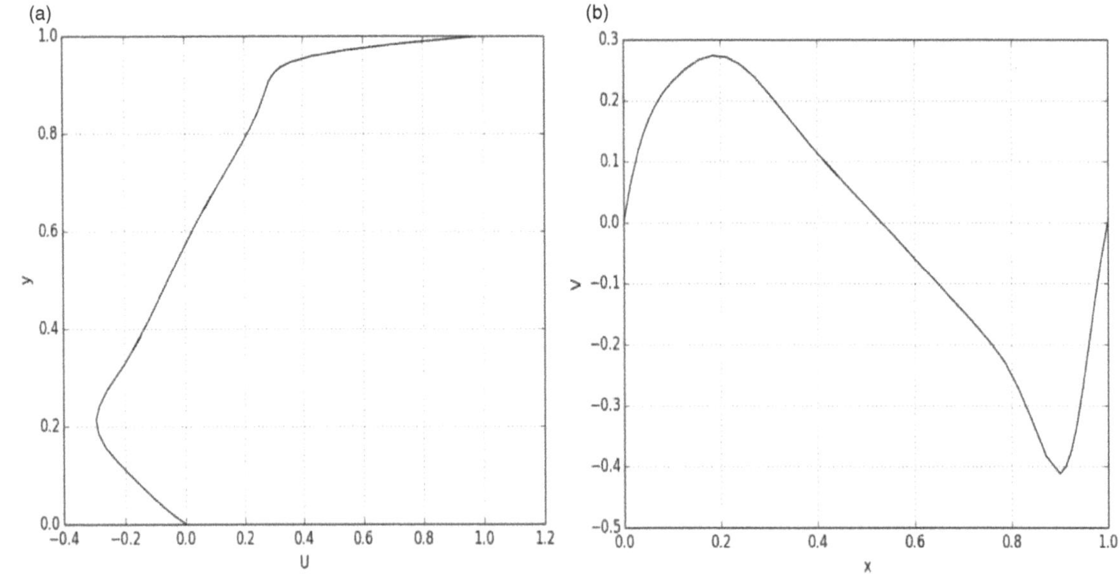

Figure 13. Problem 5.3: profiles of *u* and *v* velocity components along the vertical (a)and horizontal (b) centerlines.

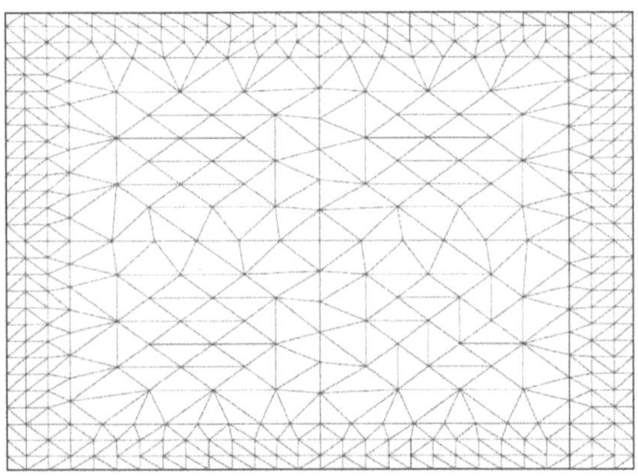

Figure 14. Problem 5.4: mesh with 1761 nodes.

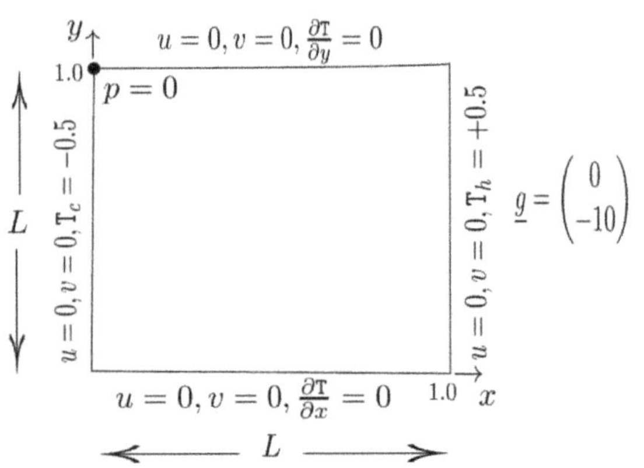

Figure 15. Problem 5.4: boundary conditions.

to be at rest at temperature $T=0$; then the temperature at the vertical walls begins to change and the fluid is subjected to a phenomenon of convection due to the thermal gradient. We assumed that the fluid is incompressible such that the Boussinesq approximation holds, that is $\boldsymbol{f} = \rho \boldsymbol{g} \alpha\,(T - T_r)$. By this benchmark problem, we want to check the behavior of the solution taking into account the estimation of τ in different situations both with respect to the physical and the computational situations. Therefore, for the mesh of Figure 14 with 1761 nodes, the length of the smallest edge is $l = 3.600\text{E}-2$ so that $h = 1.800\text{E}-2$, for the mesh of Figure 10 with 3009 nodes, the length of the smallest edge is $l = 2.1\text{E}-2$ so $h = 1.05\text{E}-2$. The results obtained after 200 steps with $\Delta t = 5.000\text{E}-3$ for various values of Rayleigh (*Ra*), Reynolds (*Re*) and Prandtl (*Pr*) numbers on the meshes of Figures 10 and 14 are reported in Tables 4 and 5. In Figures 16 and 17 are plotted the velocity vectors and the temperature contours for the case with $Ra = 3.5\text{E}+5$. Through them it can be seen that both the distributions are very close to those expected. The results presented in Table 4 show that the solutions of problem 5.4 are strongly dependent on the physical parameters (besides, obviously, the mesh rates), for example one can see that for $Ra = 10\text{E}+4$ and $Pr = 7.1\text{E}-1$ the velocity components double their values with respect to the case with $Pr = 7.1\text{E}-2$. Moreover, it can be seen that the dependence from the physical parameters becomes more and more important for increasing *Ra* values, a clear example of this fact is the variation of the velocity values for the cases $Ra = 10\text{E}+5$, $Pr = 7.1\text{E}-1$ that give $Re = 1.260\text{E}+2$, and $Ra = 3.510\text{E}+5$, $Pr = 4.0\text{E}-2$ that give $Re = 1.755\text{E}+3$.

Table 4. Problem 5.4: u_{max} and v_{max} for various *Ra, Re, Pr*, τ_c is the computed value of τ from formula (65) with $c_3 = 1.$ and τ_u the used value of τ.

	Mesh 10			Mesh 14		
τ_u	1.100 E-4	1.100 E-4	2.500 E-4	1.000 E-5	3.500 E-5	3.300 E-5
τ_c	2.400 E-4	9.900 E-5	2.500 E-4	2.700 E-4	3.300 E-4	3.200 E-4
Pr	5.000 E-2	7.100 E-1	4.000 E-2	7.100 E-1	7.100 E-2	5.000 E-2
v	5.000 E-2	7.100 E-1	4.000 E-2	7.100 E-1	7.100 E-2	5.000 E-2
u_{max}	3.234 E+1	4.897 E+1	3.738 E+2	1.624 E+1	3.004 E+1	3.112 E+1
v_{max}	4.344 E+1	7.591 E+1	5.936 E+2	1.923 E+1	4.067 E+1	4.294 E+1
Re	1.083 E+3	1.270 E+2	1.755 E+3	3.500 E+1	7.120 E+2	1.060 E+3
Ra	10 E+4	10 E+5	3.5 E+5	10 E+4	10 E+4	10 E+4

Table 5. Problem 5.4: comparison with available benchmark solutions.

Solutions	Results	*Ra* = 10E +4	*Ra* = 10E +5	*Ra* = 3.5E +5
Proposed	u_{max}	1.624E+1	4.897E+1	3.738E+2
	v_{max}	1.923E+1	7.591E+1	5.936E+2
	Re	3.500E+1	1.270E+2	1.755E+3
	Pr	7.100 E-1	7.100 E-1	4.000 E-2
(De Valh Davis, 1983)	u_{max}	1.617E+1	3.473E+1	
	v_{max}	1.961E+1	6.859E+1	
	Re	3.500E+1	1.082E+3	
(Hookey & Baliga, 1998)	u_{max}	1.620E+1	3.493E+1	
	v_{max}	1.947E+1	6.912E+1	
	Re	3.500E+1	1.090E+3	
(Malan et al., 2002)	u_{max}	1.617E+1	3.495E+1	
	v_{max}	1.980E+1	6.866E+1	
	Re	3.600E+1	1.085E+3	

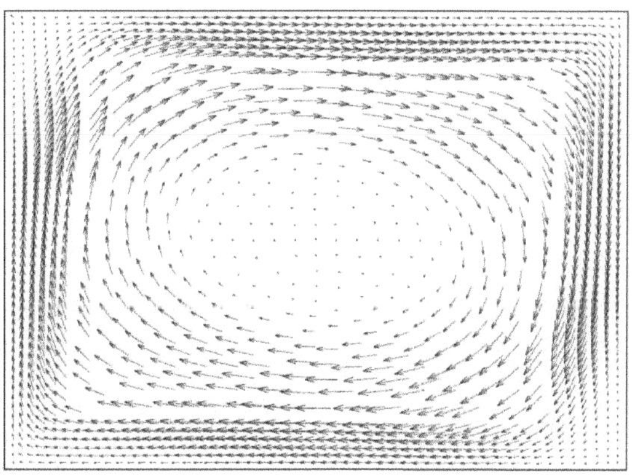

Figure 16. Problem 5.4 $Ra = 3.5E+5$ and $Re = 1755$: velocity vectors at $t = 1$.

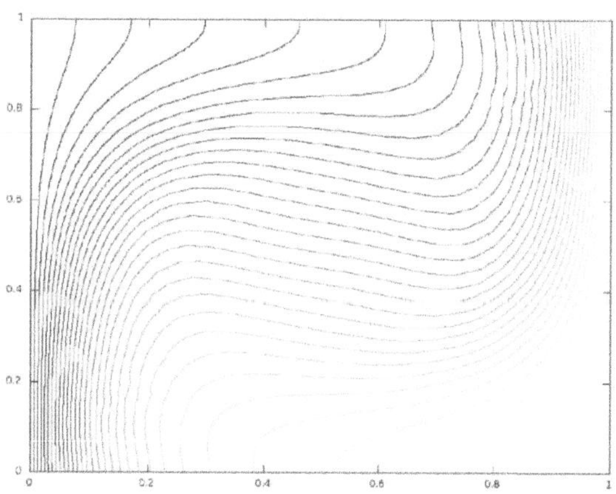

Figure 17. Problem 5.4 $Ra = 3.5E+5$ and $Re = 1755$: temperature contours.

6. Concluding remarks

In this paper, a cost-effective approach for the solution of 2D incompressible Navier–Stokes equations has been addressed. The new method obtains an optimal compromise between numerical properties (stability and accuracy) and computational efficiency (simplicity and flexibility). The most important features of the new model are: using polynomials of degree two (both for velocities and pressure) in an FE spatial approximation; time-advancing by a fractional step approach in which the nonlinear convective term is approximated by characteristics; adding a suitable stabilization technique in order to overcome the instabilities inherent in the equal-order choice. In this way the computational kernels are reduced to elliptic-like ones for whose solution very efficient solution techniques are available. In fact the algebraic systems are solved by iterative solvers preconditioned with Schwarz additive scalable preconditioners. The new model has been tested, by solving several well-known problems, some with analytical solution. This allowed us to check both the accuracy of the results and the effectiveness of the simple rule used for the estimation of the weight τ involved in the stabilization technique. The several numerical experiments carried out prompt one to conclude that the rule is robust, in fact in our test the value of τ was constant during the transient and in every point of the domain, even when anisotropic meshes were used. The formal comparison (i.e., number of iterations used in the solution of the algebraic systems of the test problem 5.3) among the numerical techniques adopted in the presented model and the ones adopted in other similar models, allow us to conclude that the objective of the work has been reached.

Some improvements and generalizations could be developed; for example the spatial approximation based on polynomials of degree two could be substituted with polynomials of degree three, obtaining a better accuracy both for the convective and diffusive terms, or the temporal approximation could be improved by a Crank-Nicolson scheme. Both these choices should not excessively increase the computational time. Moreover, being the computational kernels of elliptic kind, the feasibility to apply well-established *h-adaptive* techniques could be advantageously adopted (Naga & Zhang, 2004).

Furthermore, the presented method could be suitably generalized to the solution of moving boundary CFD problems. In particular, the time approximation based on the marching scheme and the fractional step method, the spatial approximation based on the characteristics and (probably) on FE with serendipity third-degree polynomial basis, and the Schwarz preconditioned iterative solvers of the algebraic systems, make promising the extension of the presented method to 3D problems (e.g., 3D Navier–Stokes equations). The multidomain approach, formally used for the construction of the Schwarz preconditioner, could be considered like a starting point for developing distributed memory parallel computing algorithms, similarly to what made for the solution of 2D shallow water problems (Corti & Pennati, 1997).

A drawback of the presented method is the lack of local conservativity (drawback common to almost every FE approximation); this handicap could become particularly severe for 3D problems. In order to guarantee local conservation in the FE context, an approximation by cell-vertex FV is generally adopted, this choice requires the definition of a dual mesh leading to a high increase of the computational effort (particularly significant for 3D high-order FV). An option for the cell-vertex FV approximation could consist in the construction of conservative finite element methods; the development of such methods has already started (Kengni Jotsa, 2010; Kengni Jotsa, 2012; Kengni Jotsa & Pennati, 2013) and work is in progress along these directions.

Acknowledgements

Authors thank Prof. K.W. Chau and four anonymous reviewers for their valuable suggestions and comments improving the quality of the paper. Authors thank also Dr. J. Nashipu and Dr. B. Kometa for useful grammatical corrections of the manuscript.

References

Acharya, S., & Moukalled, F. H. (1989). Improvements to incompressible flow calculations on a non staggered curvilinear grid. *Numerical Heat Transfer*, *15*, 131–152.

Ali, J., Fieldhouse, J., & Talbot, C. (2011). Numerical modelling of three dimensional thermal surface discharges. *Engineering Applications of Computational Fluid Mechanics*, *5*(2), 201–209.

Baker, G. A., Dougalis, V. A., & Karakashian, O. A. (1982). On a higher order accurate fully discrete Galerkin approximation to Navier-Stokes equations. *Mathematics of Computation*, *39*, 339–375.

Bochev, P., Cai, Z., Manteuffel, T. A., & McCormick, S. F. (1998). Analysis of velocity-flux first order system least-squares principles for the Navier-Stokes equations. Part I. *SIAM Journal of Numerical Analysis*, *35*, 990–1009.

Brezzi, F., & Douglas, J. (1988). Stabilized mixed methods for the Stokes problem. *Numerische Mathematik*, *53*, 225–235.

Bruneau, C. H., & Saad, M. (2006). The 2D lid-driven cavity problem revisited. *Computers and Fluids*, *35*, 326–348.

Cavazzutti, M., Corticelli, M. A., Masina, G., & Saponelli, R. (2013). CFD analyses of syngas-fired industrial tiles kiln module. *Engineering Applications of Computational Fluid Mechanics*, *7*(4), 533–543.

Choi, H. G., Choi, H., & Yoo, J. Y. (1997). A fractional four-step finite element formulation of the unsteady incompressible Navier-Stokes equations using SUPG and linear equal-order finite element methods. *Computer Methods Applied Mechanics Engineering*, *143*, 333–348.

Codina, R. (2001). Pressure stability in fractional step finite element methods for incompressible flows. *Journal of Computational Physics*, *170*, 112–140.

Codina, R., & Badia, S. (2006). On some pressure segregation methods of fractional step type for the finite element approximation of incompressible flow problems. *Computer Methods Applied Mechanics Engineering*, *195*(23–24), 2900–2918.

Codina, R., & Soto, O. (2004). Approximation of the incompressible Navier-Stokes equations using orthogonal subscale stabilization and the pressure segregation on anisotropic finite element meshes. *Computer Methods Applied Mechanics Engineering*, *193*, 1403–1419.

Corti, S., & Pennati, V. (1997). Design environment simulations and implementation of the different parallelization strategies. Third deliverable of the PCECOWATER Project-PCI second initiative, Esprit Project 21039.

De Vahl Davis, G. (1983). A natural convection of air in a square cavity: A benchmark numerical solution. *International Journal for Numerical Methods in Fluids*, *3*, 249–264.

Deponti, A., Pennati, V., & De Biase, L. (2006). A fully 3D finite volume method for incompressible Navier-Stokes equations. *International Journal for Numerical Methods in Fluids*, *52*, 617–638.

Deponti, A., & Pennati, V., et al. (2006). A new fully 3D numerical model for ice dynamics. *Journal of Glaciology*, *52*(178), 365–376.

Donatelli, M., Molteni, M., Pennati, V., & Serra Capizzano S. (In press). Multigrid methods for cubic spline solution two points (and 2D) boundary value problems. *Applied Numerical Mathematics*. doi:10.1016/j.apnum.2014.04.004.

Douglas, J., & Wang, J. (1989). An absolutely stabilized finite element method for the Stokes problem. *Mathematics of Computation*, *52*, 495–508.

Dryja, M., & Widlund, O. B. (1989). Some domain decomposition algorithms for elliptic problems. In L. Hayes & D. Kincaid (Eds.), *Iterative methods for large linear systems* (pp. 273–291). San Diego, CA: Academic Press.

Franca, L. P., & Hughes, T. J. R. (1993). Convergence analysis of Galerkin least-squares methods for advective-diffusive forms of the Stokes and incompressible Navier-Stokes equations. *Computer Methods Applied Mechanics Engineering*, *105*, 285–298.

Franca, L. P., & Stenberg, R. (1991). Error analysis of some Galerkin least-squares methods for elasticity equations. *SIAM Journal of Numerical Analysis*, *28*, 1680–1697.

Feraudi, F., & Pennati, V. (1997). Transporto del calore in fluidi incomprimibili: Un nuovo approccio numerico per problemi bidimensionali non stazionari. *L'Energia Elettrica*, *74*(4), 238–251.

Fortin, M., & Boivin, S. (1990). Iterative stabilization of bilinear velocity-constant pressure element. *International Journal for Numerical Methods in Fluids*, *10*, 1680–1697.

Gervasio, P., & Saleri, F. (2006). Algebraic fractional step schemes for time-dependent incompressible Navier-Stokes equations. *Journal of Scientific Computing*, *27*(1–3), 257–269.

Guermond, J. L., & Quartapelle, L. (1998). On stability and convergence of projection methods based on pressure Poisson equation. *International Journal for Numerical Methods in Fluids*, *26*, 1039–1053.

Han, Z., Zhou, D., & Tu, J. (2013). Laminar flow patterns around three side-by-side arranged circular cylinders using semi implicit three step Taylor characteristic based split (3-Tcbs) algorithm. *Engineering Application Computational Fluid Mechanics*, *7*(1), 1–12.

Hookey, N. A., & Baliga, B. R. (1998). Evaluation and enhancements of some control volume finite-element methods-part 2. Incompressible fluid flow problems. *Numerical Heat Transfer*, *14*, 273–293.

Hughes, T. J. R., Franca, L. P., & Balestra, M. (1986). A new finite element formulation for computational fluids dynamics: V. Circumventing the Babuska-Brezzi condition: A stable Petrov- Galerkin formulation of the Stokes problem accomodating equal-order interpolations. *Computer Methods in Applied Mechanics Engineering*, *59*, 85–89.

Joe, B. (1991). GEOMPACK: A software package for the generation of meshes using geometric algorithms. *Advances in Engineering Software and Workstations*, *13*(5–6), 325–331.

Kengni Jotsa, A. C. (2010). *A new method for elliptic problems in CFD resembling the discontinuous FE approach*. Proceedings MASCOT 2010 Special Session, IMACS/ISGG, IMACS Series in Computational and Applied Mathematics, ULPGC; Las Palmas de Gran Canaria, Spain (pp. 219–228).

Kengni Jotsa, A. C. (2012). *Solution of the 2D Navier-Stokes Equations by a new FE fractional step method* (PhD thesis, Università Degli Studi Dell'Insubria Sede di Como, Italy). Retrieved from http://insubriaspace.cilea.it/handle/10277/477.

Kengni Jotsa, A. C., & Pennati, V. A. (2013). *Solution of 2D convection-diffusion transient problems by a fractional step FE method*. MASCOT11 Proceedings, IMACS Series in Computational and Applied Mathematics 17, IMACS, Rome (pp. 141–150).

Kim, J., & Moin, P. (1985). Application of fractional-step method to incompressible Navier-Stokes equations. *Journal of Computational Physics*, *59*, 308–323.

Kumar, D. S., Dass, A. K., & Dewan, A. (2010). A multigrid accelerate code on graded Cartesian meshes for 2D time dependent incompressible viscous flows. *Engineering Applications of Computational Fluid Mechanics*, *4*(1), 71–90.

Kumar, D. S., Kumar, K. S., & Kumar Das, M. (2009). A fine grid solution for a lid-driven cavity flow using a multigrid method. *Engineering Applications of Computational Fluid Mechanics*, *3*(3), 336–354.

Malan, A. G., Lewis, R.W., & Nithiarasu, P. (2002). An improved unsteady, unstructured, artificial compressibility, finite volume scheme for viscous incompressible flows: Part II. Application. *International Journal for Numerical Methods in Engineering*, *54*, 715–729.

Naga, A., & Zhang, Z. (2004). A posteriori error estimates based on the polynomial preserving recovery. *SIAM Journal of Numerical Analysis*, *42*, 1780–1800.

Pai, S. A., Prakash, P., & Patnaik, B. S. V. (2013). Numerical simulation of chaotic mixing in lid drive cavity: effect of passive plug. *Engineering Applications of Computational Fluid Mechanics*, *7*(3), 406–418.

Patankar, S. V. (1980). *Numerical heat transfer and fluid flow*. New York: Hemisphere.

Pironneau, O. (1982). On the transport-diffusion algorithm and its applications to the Navier-Stokes equations. *Numerische Mathematik*, *38*, 309–332.

Plana Fattori, A., Chantoiseau, E., Doursat, C., & Flick, D. (2013). Two-way coupling of fluid-flow, heat transfer and product transformation during heat treatment of starch suspension inside tubular exchanger. *Engineering Applications of Computational Fluid Mechanics*, *7*(3), 334–345.

Quarteroni, A., & Valli, A. (1999). *Domain decomposition methods for partial differential equations*. Oxford: Oxford Science Publications.

Shih, T. M., & Tan, C. H. (1989). Effects of grid staggering on numerical schemes. *International Journal for Numerical Methods of Fluids*, *9*, 193–212.

Shirokoff, D., & Rosales, R. R. (2011). An efficient method for the incompressible Navier-Stokes equations on irregular domains with no-slip boundary conditions, high order up to the boundary. *Journal of Computational Physics*, *230*, 8619–8646.

Silvester, P. J., & Kechkar, N. (1990). Stabilized bilinear-constant velocity pressure finite elements for the conjugate gradient solution of the Stokes problem. *Computer Methods in Applied Mechanics Engineering*, *79*, 71–86.

Stickland, M., Fabre, S., Scanlon, T., Oldroyd, A., Mickelson, T., & Astrup, P. (2013). CFD techniques for estimating flow distorsion effects on lidar measurements when made in complex flow fields. *Engineering Applications of Computational Fluid Mechanics*, *7*(3), 324–333.

Sun, X., Zhang, J. Z., & Ren, X. L. (2012). Characteristic based split (CBS) finite element method for incompressible viscous flow with moving boundaries. *Engineering Applications of Computational Fluid Mechanics*, *6*(3), 461–474.

Tritthart, M., & Gutknecht, D. (2007). Three dimensional simulation of free-surface flows using polyhedral finite volumes. *Engineering Applications of Computational Fluid Mechanics*, *1*(1), 1–14.

Van Doormal, J. P., & Raithby, G. D. (1984). Enhancements of the SIMPLE metodh for predicting incompressible fluid flows. *Numerical Heat Transfer*, *7*, 147–163.

Vrahliotis, S., Pappou, T., & Tsangaris, S. (2012). Artificial compressibility 3D Navier-Stokes solver for unsteady incompressible flows with hybrid grids. *Engineering Applications of Computational Fluid Mechanics*, *6*(2), 248–270.

Zang, Y., Street, R. L., & Koseff, J. R. (1994). A non-staggered grid, fractional step method for time-dependent incompressible Navier-Stokes equations in curvilinear coordinates. *Journal of Computational Physics*, *114*, 18–33.

2

CFD simulation of a turbulent fiber suspension flow – a modified near-wall treatment

Carla Cotas[a], Dariusz Asendrych[b], Fernando Garcia[a], Pedro Faia[c] and Maria Graça Rasteiro[a]*

[a] *Chemical Engineering and Forest Products Research Centre (CIEPQPF), Chemical Engineering Department, Faculty of Sciences and Technology, University of Coimbra, Rua Sílvio Lima, Pólo II, 3030-790 Coimbra, Portugal;* [b] *Częstochowa University of Technology, Institute of Thermal Machinery, 42-200 Częstochowa, Poland;* [c] *Electrical and Computers Engineering Department, Faculty of Sciences and Technology, University of Coimbra, Portugal*

Turbulent *Eucalyptus* fiber suspension flow in pipes was studied numerically using commercial CFD software. A pseudo-homogeneous approach was proposed to predict the flow behavior of pulp fiber suspensions for medium consistencies and for Reynolds numbers ranging from $4.7 \div 65.3 \cdot 10^3$. Viscosity was introduced into the model as a function of shear rate to represent the non-Newtonian behavior of the pulp suspension. Additionally, the existence of a water annulus was considered at the pipe wall, surrounding the flow core, where viscosity is equal to the water viscosity. The near-wall treatment was modified considering an expression for the logarithmic velocity profile in the boundary layer, similar to the one suggested by Jäsberg (2007). The final model could reproduce the drag-reduction effect resulting from the presence of fibers in the flow. Moreover, the numerical results show that a better fit for pressure drop is obtained when the modified near-wall treatment is used and the Jäsberg adjustable parameters are adapted to take into account the flow conditions.

Keywords: turbulent pulp fiber flow; k-ε turbulence model; near-wall treatment modification; computational fluid dynamics; drag reduction; water annulus

1. Introduction

The flows of pulp fiber suspensions are important in the papermaking industry. The properties of the final product can be different depending on the characteristics of the pulp in the pipes. Furthermore, in different stages of the process an incorrect design of the flow system can lead to an inefficient operation of a pump and to excessive energy consumption. In pulp and paper mills, the design of most process equipment still remains based on empirical correlations that can fail when the process conditions change. There is currently great interest in the experimental and numerical studies of pulp suspension flows as they still remain poor and incomplete. The design of pulp and paper process equipment can be improved by using computational tools, such as, CFD (computational fluid dynamics) methods. Their use requires, however, the input knowledge necessary to develop the numerical models.

The final product in the papermaking industry can be directed for particular applications and be quite different, but the manufacturing processes are basically the same. The chemistry, physics and fluid dynamics are the main important areas of knowledge, in this process (Lundell et al., 2011). However, fiber suspensions are different and more complex than the other solid-liquid systems due to complex interactions between the different pulp and paper components (Ventura et al., 2008).

Three main groups can be distinguished in the fluid mechanics related to the paper making process: (*i*) the modelling of fiber suspensions, (*ii*) experimental methods to obtain the data necessary for tuning and validation of the models, and (*iii*) the knowledge about the coupling between rheology and suspension characteristics (e.g., fiber length, morphology, and concentration) (Lundell et al., 2011).

At the rheological level, the pulp fiber suspension behavior is peculiar and, there are different situations where the flow regimes and pressure loss curves are different from those of typical slurry or Newtonian liquid flow systems. When the shear stress exceeds the yield stress, τ_y, the flow of pulp suspension starts (Figure 1) (Ventura et al., 2007). When the shear stress exerted on the mass of fibers surpasses the fluidization point, τ_d, the network structure becomes totally disrupted and the suspension starts to exhibit a fluid-like behavior (Ventura et al., 2007). At this point, the suspension begins to move in a fully developed turbulent pipe flow with its hydrodynamic properties similar to those of water.

The existing studies report three different regimes in the flow of pulp fiber suspensions: plug, transition and turbulent flow regimes (Figure 2). The flow regimes are different from those for Newtonian fluids or even for other solid suspensions. In each regime there are also sub-regimes

*Corresponding author. Email: mgr@eq.uc.pt

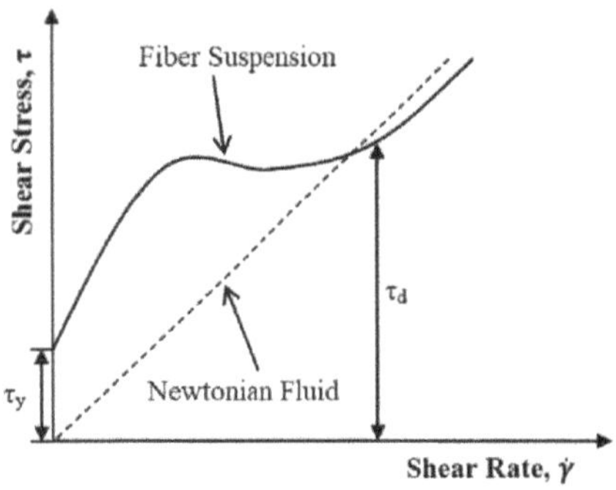

Figure 1. Stress versus shear rate curve for a fibre suspension (adapted from Gullichsen & Härkönen (1981)).

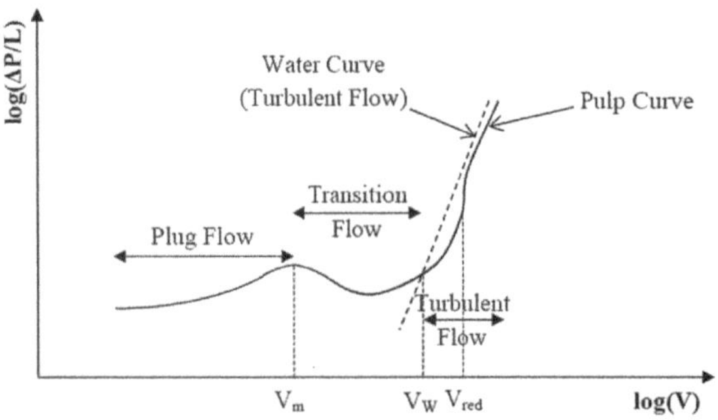

Figure 2. Typical pressure drop curve for pulp suspensions (adapted from Lundell et al. (2011) and Li et al. (2001)).

reported in literature with well-defined shear mechanisms (Fock et al., 2009; Gullichsen & Härkönen, 1981; Ventura et al., 2008). Initially, a plug of fibers is in contact with the pipe wall without any movement, a shear force has to strain the plug and when the yield stress is exceeded the motion begins. At low velocities, the suspension flows as a plug of fibers and water which induces larger values of pressure drop than those for water flow in a pipe. V_m is the velocity corresponding to the maximum of the head loss curve at the plug flow. At high velocities, the pressure loss is lower than the expected for water and all the components are in complex turbulent motion (Ventura et al., 2008). In the transition regime corresponding to intermediate velocities, a turbulent fiber-water annulus surrounds an intact plug. The transition flow regime starts at the velocity V_m (Ventura et al., 2008). The fiber plug size decreases and the turbulence intensity grow with the increase of velocity and more fibers leave the plug. The whole suspension shows a turbulent-like behavior when the velocity exceeds the point where the fiber suspension and water flow curves intersect and become parallel. V_W represents the velocity corresponding to the onset of drag reduction (Ventura et al., 2008) and V_{red} stands for the speed corresponding to the maximum of drag-reduction effect.

The simulation of pulp fiber suspension flows is a recent topic and different models with different complexity levels have been proposed to describe such flows – the pseudo-homogeneous models and the multiphase models, each having a specific domain of application. The modeling strategies which can be employed to simulate papermaking flows are extensively discussed in Hämäläinen et al. (2011). The multiphase models are only appropriate for dilute systems while the pseudo-homogeneous models are suited for concentrated systems. The implementation of the mathematical models to simulate the flow of suspensions with high consistency and high Reynolds numbers can be difficult due to the high computer requirements. In fact, simulation of the fiber suspension flows encounters various problems. Among the most important difficulties it must be mentioned the measurement of suspension apparent viscosity and yield stress (Blanco et al., 2007). Moreover, the heterogeneous mass distribution and formation of depletion layers at flow boundaries (Derakhshandeh et al., 2011) should be taken into account for the successful simulation. Some authors also emphasize the viscoelastic behavior of pulp suspensions which may have impact on the flow modeling (Derakhshandeh et al., 2011; Huhtanen & Karvinen, 2005). The successful example of CFD application to model the pulp suspension flows in various geometries (pipe, converging channel, mixing tank) with the use of single-phase approach is presented in the work of Huhtanen & Karvinen (2005). In order to simulate the drag-reduction effect in the pipe flow due to reduced fiber concentration in the near-wall region they applied the slip boundary condition at the wall. They also tested various non-Newtonian pulp models and found power-law fluid model as the most adequate for turbulent regime. Their findings were confirmed by experimental trials (Huhtanen & Karvinen, 2006). Hammarström (2004) proposed a single-phase continuum model based on his experimental observation by combining the conventional fluid dynamics law, non-Newtonian behavior and turbulence. He also considered the existence of the lubrication layer between the flow core and the pipe wall behaving in Newtonian manner. The works of Huhtanen & Karvinen (2006) and Hammarström (2004) show, however, that successful modeling requires deep knowledge about the fiber properties including their dimensions, surface quality and process history and each simulation requires special care and use of individually determined constants.

The continuum model was also shown to be an adequate tool to reconstruct the unsteady pulp flow in an LC refiner (Kondora & Asendrych, 2013), characterized by the system of spiral secondary flows. Additionally, CFD methods were applied successfully to study the flow of other fluids. The steady state swirling flow in draft tubes was studied by applying k-ε turbulence models (Galván et al., 2011) considering the wall function and a near-wall treatment to model the near-wall region. Solid suspension flow in a cylindrical stirred tank equipped with four side-entering impellers was modeled using CFD models (Fang et al., 2013). A CFD study concerning the turbulent flow of nanofluids over periodic rib-grooved channels is presented in Vatani & Mohammed (2013).

The present study is focused on a mathematical model capable of describing properly the turbulent pipe flow of *Eucalyptus* pulp fibers. The numerical study was made using commercial CFD code with implemented modifications to the built-in model enabling to take into account the presence of fibers in the flow. It applies a pseudo-homogeneous one-phase approach that is appropriate for concentrated fiber suspensions. The classical near-wall model assuming the logarithmic velocity profile was extended by an additional term being dependent on the pulp suspension properties according to the idea of Jäsberg (2007). Moreover, the water annulus free of fibres was assumed to bound the flow core. In this way the inhomogeneity of the suspension in the wall vicinity could be captured, where fiber concentration falls down reaching zero at the wall (Dong et al., 2003). The viscosity of the water annulus is that of water, thus leading to significant reduction of flow resistance. The model validation was made by comparing the numerically obtained pressure drops with experimental data acquired at the reference pilot rig for the same flow conditions (Ventura et al., 2008) and pulp properties (Ventura et al., 2007). The paper is organized in 5 Sections. The current Section introduces the subject of the work, the general concepts and its main objectives. Section "Experimental data" reports the experimental information used in this study. Section "Numerical methodology" describes the mathematical model and the numerical scheme used to simulate the turbulent pipe flow of *Eucalyptus* pulp fibers. Section "Results and discussion" presents the results obtained with the proposed model and provides their interpretation. Section "Conclusions" summarizes the most relevant and important outcomes of this work.

2. Experimental data

The turbulent pipe flow of an *Eucalyptus* pulp (fiber averaged length of 0.706 mm (Ventura et al., 2007, 2008)) was simulated for different experimental flow rates and fiber consistencies. The correlations and data to be introduced in the CFD code were obtained from experimental data (viscosity of the pulp suspensions and head loss in the pipe). This experimental information was used also to validate the mathematical model.

2.1. *Rheological data*

There is no standard procedure or equipment for the rheological characterization of pulp suspensions. Various off-line and in-line techniques like rotational rheometers, conventional rheometers, pipe rheometers, NMRI and UVP-PD (Fock et al., 2009) techniques can be used to acquire rheological data.

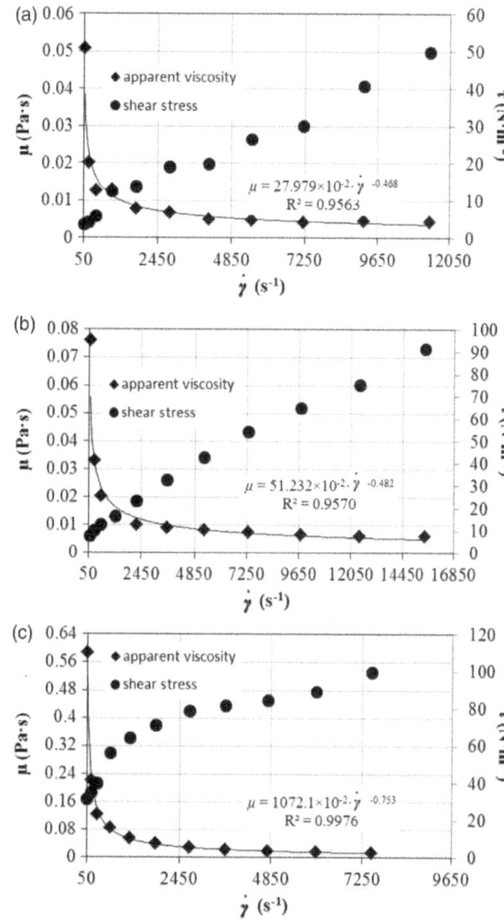

Figure 3. Rheograms and apparent viscosities for a pulp suspension of *Eucalyptus* fibres for (a) c = 1.50%(w/w), (b) c = 1.80%(w/w), and, (c) c = 2.50%(w/w).

The rheological data (Figure 3) from literature was obtained using an off-line equipment, an adapted Searle-type rotational rheometer – plate viscometer, as described in (Blanco et al., 2007; Ventura et al., 2007). Three different *Eucalyptus* pulps with consistency equal to 1.50, 1.80, and 2.50 % (w/w) were studied.

The apparent pulp viscosity μ_{app} was regarded as a viscosity of a Newtonian fluid producing the same resistance to the flow (Ventura et al., 2011):

$$\mu_{app} = \frac{\tau}{\dot{\gamma}} \tag{1}$$

The apparent viscosity of the suspension can be expressed as a function of a shear rate by the

Non-Newtonian power-law formula:

$$\mu_{app} = K(\dot{\gamma})^{n-1} \quad (2)$$

with $n < 1$.

2.2. *Flow data*

The experimental data for pulp flow was adopted from Ventura et al. (2008) and Ventura et al. (2011) for velocities larger than V_{red} to insure that the Reynolds number is high enough and all the suspension are in fully developed turbulent flow. For this case the use of the standard k-ε turbulence model is appropriate (Reynolds number varied between $4.7 \div 65.3 \cdot 10^3$). These data were acquired in a pilot rig properly adapted to study pulp suspension flows. The main information from literature was the pressure drop used to validate the numerical simulations.

The test section in the pilot rig is composed of a horizontal pipe 11.5-m long and 0.0762 m in diameter (Ventura et al., 2008). The flow section system is equipped with an electromagnetic flowmeter, a differential pressure transducer and a thermometer. The pressure taps are 4 m apart (Ventura et al., 2008). An appropriate (4 m) entrance length ensures that the flow is fully developed inside the test section. Downstream that section a sufficiently long pipe segment guarantees that the exit effects can be neglected.

3. Numerical methodology

3.1. *Governing equations*

The basic configuration of the flow domain is schematically represented in Figure 4. The *Eucalyptus* pulp is supplied into the pipe through the inlet section at the left and leaves it through the outlet section at the right. The pipe walls were assumed to be smooth. Pipe dimensions, i.e. a diameter D and its length L, were equal to 0.0762 m and 1 m, respectively (Figure 4). The length of the pipe could be reduced with respect to the test section used in the experiment, as the flow was assumed to be fully developed turbulent pipe flow.

The flow was assumed to be: (*i*) steady state, (*ii*) isothermal, (*iii*) incompressible, (*iv*) non-Newtonian fluid, and, (*v*) 2D with axial symmetry. The extent of the numerical domain was 1 m × 0.0381 m (*length* × pipe *radius*).

The dynamic viscosity of the *Eucalyptus* pulp was expressed by Equation (2) considering it as a function of local shear rate expressed by a power-law formula. The pulp density was considered equal to water density since the density of water and *Eucalyptus* pulp are very similar. The numerical methodology used was based on the solution of the partial differential equations governing the transport phenomena in a single fluid motion – the mass and the momentum conservation equations. In this study, the two-layer standard high Reynolds k-ε turbulence model was used. Therefore, two additional transport equations for the turbulent kinetic energy (k) and its dissipation rate (ε) were considered to create a complete system of governing equations to be solved (ANSYS FLUENT, 2011; Chang et al., 1995; Hsieh & Chang, 1996; Mathur & He, 2013):

Mass Conservation Equation

$$\frac{\partial u}{\partial x} + \frac{\partial v}{\partial r} + \frac{v}{r} = 0 \quad (3)$$

Axial Momentum Conservation Equation

$$\frac{1}{r}\left[\frac{\partial}{\partial x}(r\rho uu) + \frac{\partial}{\partial r}(r\rho vu)\right] = -\frac{\partial p}{\partial x} + \frac{1}{r}\left[\frac{\partial}{\partial x}\left(r(\mu+\mu_t)\frac{\partial u}{\partial x}\right) + \frac{\partial}{\partial r}\left(r(\mu+\mu_t)\frac{\partial u}{\partial r}\right)\right] + \frac{\partial}{\partial x}\left[(\mu+\mu_t)\frac{\partial u}{\partial x}\right] + \frac{1}{r}\frac{\partial}{\partial r}\left[r(\mu+\mu_t)\frac{\partial v}{\partial x}\right] \quad (4)$$

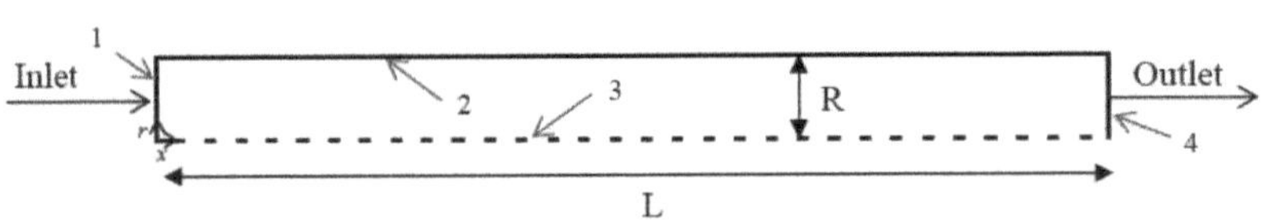

L (m)	R (m)
1	0.0381

Location	Boundary Conditions
Left side –1	Periodic boundary
Right side -4	Periodic boundary
Top –2	Wall
Bottom -3	Axis

Figure 4. Numerical domain and boundary conditions applied.

Radial Momentum Conservation Equation

$$\frac{1}{r}\left[\frac{\partial}{\partial x}(r\rho uv)+\frac{\partial}{\partial r}(r\rho vv)\right]$$
$$=-\frac{\partial p}{\partial r}+\frac{1}{r}\left[\frac{\partial}{\partial x}\left(r(\mu+\mu_t)\frac{\partial v}{\partial x}\right)+\frac{\partial}{\partial r}\left(r(\mu+\mu_t)\frac{\partial v}{\partial r}\right)\right]$$
$$+\frac{\partial}{\partial x}\left[(\mu+\mu_t)\frac{\partial u}{\partial r}\right]+\frac{1}{r}\frac{\partial}{\partial r}\left[r(\mu+\mu_t)\frac{\partial v}{\partial r}\right]$$
$$-2(\mu+\mu_t)\frac{v}{r^2} \quad (5)$$

Turbulent Kinetic Energy Conservation Equation

$$\frac{1}{r}\left[\frac{\partial}{\partial x}(r\rho uk)+\frac{\partial}{\partial r}(r\rho vk)\right]$$
$$=\frac{1}{r}\left[\frac{\partial}{\partial x}\left(r\left(\mu+\frac{\mu_t}{\sigma_k}\right)\frac{\partial k}{\partial x}\right)+\frac{\partial}{\partial r}\left(r\left(\mu+\frac{\mu_t}{\sigma_k}\right)\frac{\partial k}{\partial r}\right)\right]$$
$$+\mu_t\left\{2\left[\left(\frac{\partial u}{\partial x}\right)^2+\left(\frac{\partial v}{\partial r}\right)^2+\left(\frac{v}{r}\right)^2\right]+\left(\frac{\partial v}{\partial x}+\frac{\partial u}{\partial r}\right)^2\right\}$$
$$-\rho\varepsilon \quad (6)$$

Dissipation Rate of Turbulent Kinetic Energy Conservation Equation

$$\frac{1}{r}\left[\frac{\partial}{\partial x}(r\rho u\varepsilon)+\frac{\partial}{\partial r}(r\rho v\varepsilon)\right]$$
$$=\frac{1}{r}\left[\frac{\partial}{\partial x}\left(r\left(\mu+\frac{\mu_t}{\sigma_\varepsilon}\right)\frac{\partial \varepsilon}{\partial x}\right)+\frac{\partial}{\partial r}\left(r\left(\mu+\frac{\mu_t}{\sigma_\varepsilon}\right)\frac{\partial \varepsilon}{\partial r}\right)\right]$$
$$+C_{\varepsilon 1}\mu_t\frac{\varepsilon}{k}\left\{2\left[\left(\frac{\partial u}{\partial x}\right)^2+\left(\frac{\partial v}{\partial r}\right)^2+\left(\frac{v}{r}\right)^2\right]+\left(\frac{\partial v}{\partial x}+\frac{\partial u}{\partial r}\right)^2\right\}$$
$$-C_{\varepsilon 2}\frac{\varepsilon^2}{k} \quad (7)$$

The turbulence model constants σ_k, σ_ε, $C_{\varepsilon 1}$ and $C_{\varepsilon 2}$ were assumed to be equal to their standard values (Ferziger & Peric, 2002). The turbulent viscosity being a feature of the flow was locally computed by combining k and ε as follows (Ferziger & Peric, 2002):

$$\mu_t=\rho C_\mu\frac{k^2}{\varepsilon} \quad (8)$$

where C_μ is equal to 0.09.

3.2. Numerical solution procedure

The system of partial differential Equations (3) to (7) consists of non-linear equations and includes derivative terms of first and second order in space. The equations are discretized by a finite volume method (Ferziger & Peric, 2002). The differential equations are integrated for each individual control volume in the computational domain to deduce the algebraic equations for the unknown discrete variables that can be solved numerically. The linearization of discretized equations results in a linear system of equations and its solution provides updated values for the dependent variables. A second-order upwind scheme was selected to interpolate the face values required for the convection terms in the discretized equations of dependent variables from cell centre values, i.e., it is assumed that the quantities of variables at cell faces are similar to the values in the cell centre that represents a cell-average value. The cell face variable value is equal to the variable value in the cell centre in the upstream cell (Ferziger & Peric, 2002). A pressure equation is solved by manipulating the continuity and momentum equations to extract the pressure field, with the use of the pressure-based approach. The SIMPLE algorithm was chosen to obtain the pressure field (Ferziger & Peric, 2002). An additional condition for the pressure was derived by reformulating the continuity equation and the mass conservation was implemented using a relationship between velocity and pressure corrections.

The mesh used for the computational domain is a uniform mesh with 20 × 54 nodes, according to x and r coordinates, respectively. The total number of cells is equal to 1007 and the number of faces is 2033. The number of iterations considered for the periodic boundary condition is equal to 2 and the relaxation factor equal to 0.5. Numerical tests showed that the mesh selected for this study gives grid-independent solutions for all investigated cases. The first mesh node is placed at $y^+ > 30$ (fully turbulent region) as recommended for the cases where wall functions are used (Xu & Aidun, 2005). The convergence criterion of the iterative process is that the residuals of all equations were less than 10^{-5}.

3.3. Boundary conditions

Boundary conditions are required for the numerical solution of the complete system of equations. The boundary conditions in the present study are related to the inlet flow (1), outlet flow (4), pipe wall (2), and pipe axis (3) (see Figure 4).

The boundary conditions for inlet and outlet sections were specified as periodic boundary conditions. The turbulent pulp flow can be regarded as a periodic flow due to the periodicity of both, the physical geometry and expected pattern of the flow. The periodic boundary condition selected for the cases studied was the pressure-drop periodic flow in which a pressure drop is present across the translational periodic boundary and the flow is regarded as "fully developed" flow. The data to be reproduced numerically were obtained in a pipe section located 4-m downstream an entrance region ensuring fully developed flow. The periodic boundary conditions selected allow the simulation of pipe segment 1-m long, specifying the mass flow rate to each case and to obtain the pressure gradient to compare with the experimental pressure value.

Along the flow direction the flow quantities are periodic (Ferziger & Peric, 2002), which can be expressed as:

$$\frac{\partial(\bullet)}{\partial x} = 0 \tag{9}$$

except the pressure which drops linearly

$$\frac{\partial p}{\partial x} = const \tag{10}$$

The periodic boundary conditions mean that there is no need to specify the turbulence parameters at the inlet boundary which will be computed in all domain to satisfy the flow governing equations. As referred previously the only physical setting specified is the mass flow rate.

The main objective of the present work is to analyze the influence a modified near-wall treatment has on the prediction of the turbulent pipe flow of pulp fibres suspensions. Ventura et al. (2011) adjusted the turbulence parameters according to the pulp consistency, here the authors focused on the modification of the near wall treatment. A future study will be performed to analyze the influence of both the modified near-wall treatment and turbulence quantities supplied in the inlet section which should be considered as *velocity inlet* or *mass flow inlet* instead of the present *periodic* boundary condition.

The following boundary conditions were applied in the model. At the pipe symmetry axis, the axis boundary condition type was set. The wall boundary condition was selected for the boundary representing the pipe wall, where the no-slip condition was chosen for velocity ($u_{Wall} = v_{Wall} = 0$). The shear stress on the pulp at the wall is predicted, in ANSYS FLUENT 13.0, using the properties of the flow adjacent to the wall boundary (ANSYS FLUENT, 2011) considering the momentum wall functions. The near-wall region was modeled with the use of the wall function approach. In this approach, the viscous sublayer is not resolved and wall functions are used to bridge the viscosity affected region. The wall law for mean velocity was modified according to the expressions presented in Jäsberg (2007). The boundary condition for k imposed at the wall is (ANSYS FLUENT, 2011):

$$\left.\frac{\partial k}{\partial n}\right|_{Wall} = 0 \tag{11}$$

The k dissipation rate equation is not solved at the wall-adjacent cells, it is assumed to be equal in the wall-adjacent control volume, computed at the first near-wall point P from (ANSYS FLUENT, 2011):

$$\varepsilon_P = \frac{C_\mu^{3/4} \cdot k_P^{3/2}}{\kappa \cdot y_P} \tag{12}$$

where k_P and ε_P are turbulent parameters at the first near-wall node P.

At the start of the numerical calculation, the initial conditions for the field variables are specified be equal to zero except for the turbulent kinetic energy and its dissipation rate that were initialized equal to 1 $m^2 \cdot s^{-2}$ and 1 $m^2 \cdot s^{-3}$, respectively.

3.4. *Modified wall function*

A peculiar S-shaped velocity profile near the wall was experimentally observed for the pipe flow (Jäsberg, 2007) and channel flow (Xu & Aidun, 2005) of natural flexible cellulose wood fibers. In these studies, mathematical expressions were proposed for the dimensionless velocity profiles as functions of the dimensionless distance. The strategy applied in this work was to change the momentum wall law according to the expressions presented in Jäsberg (2007).

A combination of logarithmic law for the turbulent region and a wake function is presented in Xu & Aidun (2005). An expression based on the standard logarithmic velocity profile for turbulent Newtonian fluid considering three distinct regions in the fully turbulent region is proposed in Jäsberg (2007). The experimental velocity profiles were reported in Jäsberg (2007) for pine pulp suspensions (c = 1% (w/w)) and birch suspensions (c = 2% (w/w)) in a straight pipe with a sudden expansion (turbulence generator system). Figure 5 represents the dimensionless velocity profile approximation presented in Jäsberg (2007). Three distinct regions: *near wall region*, *yield region* and *core region* can be identified in the profile. In the near wall region, there is no difference between the new profile and the profile of a Newtonian flow. In the yield region, the velocity gradient is higher than that of a Newtonian flow. Last, in the core region the velocity gradient form is similar to that of a Newtonian fluid but with a lower slope.

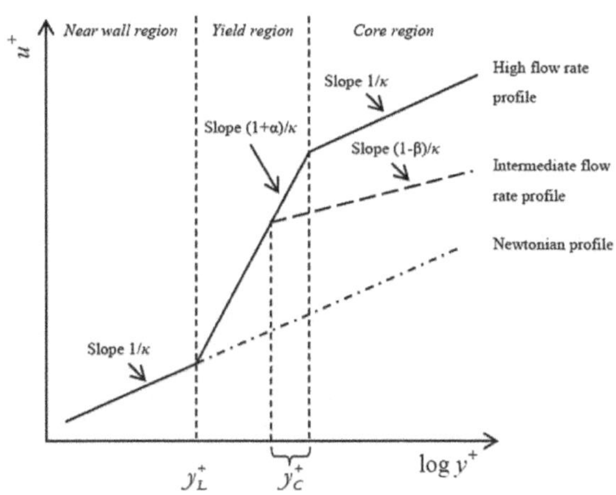

Figure 5. The structure of dimensionless velocity profile in a turbulent boundary-layer modified according to Jäsberg (2007).

Mathematically, the velocity behavior presented in Figure 5 can be expressed by Equation (13). The general expression for the dimensionless velocity profile corresponds to the *standard* logarithmic wall function modified with an additional term Δu^+, (see Equation (13)) depending on the dimensionless distance to the pipe wall and on the parameters related to the slope in the yield and core regions.

$$u^+ = \frac{1}{\kappa} ln(y^+) + B + \Delta u^+ \tag{13}$$

$$\Delta u^+ = \begin{cases} 0, & 0 < y^+ < y_L^+ \\ \frac{\alpha}{\kappa} ln\left(\frac{y^+}{y_L^+}\right), & y_L^+ < y^+ < y_C^+ \\ \frac{\alpha}{\kappa} ln\left(\frac{y_C^+}{y_L^+}\right) - \frac{\beta}{\kappa} ln\left(\frac{y^+}{y_C^+}\right), & y_C^+ < y^+ < R^+ \end{cases} \tag{14}$$

The standard nondimensional wall-layer variables are defined as follows (Jäsberg, 2007):

$$u^+ = \frac{u}{u_\tau} \tag{15}$$

$$y^+ = \frac{\rho}{\mu_W} u_\tau y_P \tag{16}$$

$$R^+ = \frac{\rho}{\mu_W} u_\tau R \tag{17}$$

$$u_\tau = \left(\frac{\tau_W}{\rho}\right)^{1/2} \tag{18}$$

The parameters α and β are adjustable parameters that characterize the slope, in relation to the Newtonian profile value, of the envelope curve in the yield and core regions, respectively. The parameter y_L^+ represents the dimensionless distance between the near wall region and the yield region, y_C^+ indicates the variable position of the plug surface (region between the high flow rate envelope curve and the upper limit of the yield region). The standard values proposed by Jäsberg in his study for the birch pulp flow (c = 2% (w/w)) are presented in Table 1.

Table 1. *Standard* Jäsberg adjustable parameters for birch pulp flow (c = 2% (w/w)) (Jäsberg, 2007).

Parameter	α	β	y_L^+	y_C^+
Value	2.4	0.0	50	320

4. Results and discussion

The experimental conditions and flow data for which the model was applied are listed in Table 2. The non-Newtonian behavior of the *Eucalyptus* pulp was introduced into the CFD model by implementing the viscosity Equation (2) with the rheological parameters given in Table 2. The presence of a water annulus at the pipe wall surrounding the core flow was taken into account with the viscosity dramatically reduced to the water level. As shown in Asendrych & Kondora (2009) the water layer with no fibers present or with very reduced consistency leads to significant changes of the flow in the near-wall region and may have dramatic consequences to the flow resistance. The water annulus thickness was assumed equal to the mean *Eucalyptus* fiber length (0.706 mm). The *standard* law-of-the-wall for mean velocity was modified according to Equation (13). The numerical model has been validated by comparing the numerical pressure drop values with those obtained experimentally and the simulated dimensionless velocity profiles with those calculated directly using Jäsberg equation (Equation (13)).

4.1. *Standard Jäsberg adjustable parameters*

The modified law-of-the-wall for the velocity, according to Jäsberg (2007), has four parameters (Equations (13) and (14)), designated in this work as Jäsberg adjustable parameters: α, β, y_L^+ and y_C^+. The standard values proposed by Jäsberg in his study for the birch pulp flow (c = 2% (w/w)) which were presented in Table 1, were used as initial approximation in present study.

Table 2. Experimental information.

Case	c (% (w/w))	$K^{(a)}$ ($Pa \cdot s^n$)	$n^{(b)}$	$U_{in}{}^{(c)}$ ($m \cdot s^{-1}$)	$Re_W{}^{(d)}$	$\Delta P/L_{exp.}{}^{(e)}$ ($Pa \cdot m^{-1}$)
A	1.50	0.2798	0.532	4.49	40170	829.1
B				6.21	65314	1288.7
C	1.80	0.5123	0.518	4.46	23847	842.6
D				6.23	38914	1203.2
E	2.50	10.721	0.247	4.90	4741	1578.9
F				5.55	6111	1753.8

(a) Consistency coefficient, Equation (2); (b) Flow behavior index, Equation (2); (c) Mean inlet *Eucalyptus* pulp flow velocity; (d) Reynolds number calculated based on the viscosity near the wall, at the interface with the water annulus (Rudman & Blackburn, 2006); (e) Experimental pressure drop value.

Table 3. Pressure drop values (cases presented in Table 1).

	c	U_{in} (a)	$\Delta P/L_{water}$ (b)	$\Delta P/L_{exp.}$ (c)	$\Delta P/L_{st}$ (d)	$\Delta P/L_{num.}$ (e)	δ (f)
Case	(% (w/w))	($m \cdot s^{-1}$)	($Pa \cdot m^{-1}$)	($Pa \cdot m^{-1}$)	($Pa \cdot m^{-1}$)	($Pa \cdot m^{-1}$)	(%)
A	1.50	4.49	1885.2	829.1	1740.5	1324.4	59.8
B		6.21	3419.7	1288.7	3425.4	2146.9	66.6
C	1.80	4.46	1862.2	842.6	1997.1	1359.1	61.3
D		6.23	3440.0	1203.2	3545.3	2219.6	84.5
E	2.50	4.90	2212.6	1578.9	3312.8	2299.3	45.6
F		5.55	2781.1	1753.8	3741.8	2814.3	60.5

(a) Mean inlet *Eucalyptus* pulp flow velocity; (b) Only flow of water in the same pipe; (c) Experimental pressure drop value; (d) Numerical pressure drop value (standard high Reynolds k-ε turbulence model); (e) Numerical pressure drop value (high Reynolds k-ε turbulence model with modified near-wall treatment according to Jäsberg (2007) – standard values Table 1); (f) Relative error between the experimental and numerical pressure drop ($|\Delta P/L_{exp.} - \Delta P/L_{num.}|/(\Delta P/L_{exp.}) \cdot 100\%$).

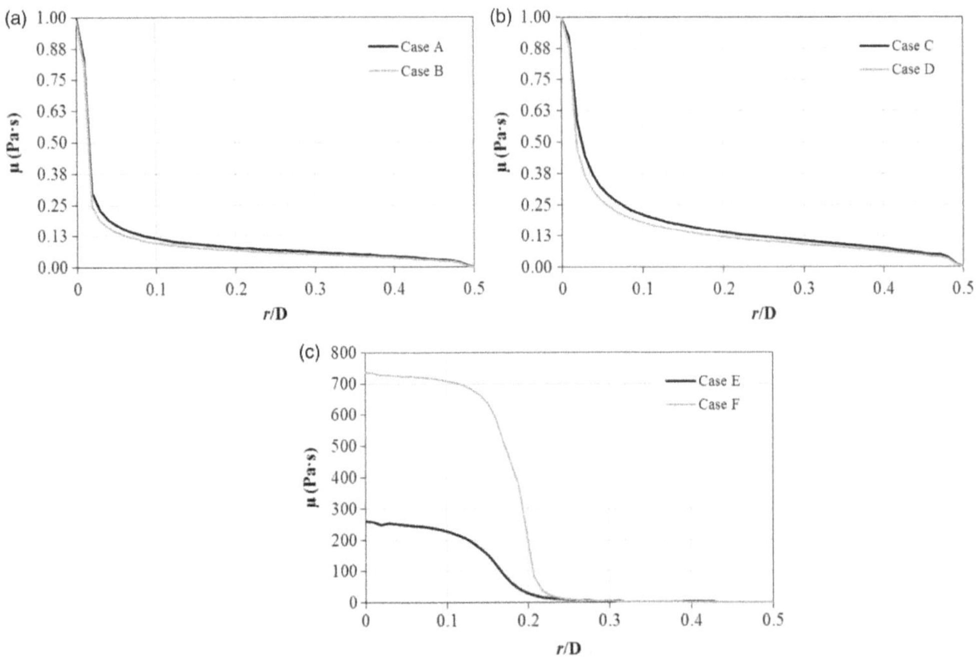

Figure 6. Numerical dynamic viscosity profiles obtained for (a) cases A and B (c = 1.50%(w/w)), (b) cases C and D (c = 1.80%(w/w)), and, (c) cases E and F (c = 2.50%(w/w)).

The numerical pressure drop results obtained with the modified CFD code are presented in Table 3 along with the experimental pressure drops for the turbulent *Eucalyptus* pulp flow and for water flow with the same bulk velocity. As expected, when the mean inlet velocity increases the pressure drop grows. The same tendency was reproduced numerically for all the cases considered. Additionally, a drag-reduction effect could always be observed in the experimental data. The numerical values are always higher than the experimental ones, showing, nevertheless the existence of drag reduction except for the highest consistency. It should be remarked, however, that the simulations were performed using the original values of Jäsberg adjustable parameters presented in Table 1. As in Jäsberg (2007) birch fiber suspension was investigated, this fibers having different properties compared with eucalyptus fibers even if length is similar, that can be the reason for the behavior discrepancies between experimental and simulation results. However, this approach is better to simulate the turbulent pulp flow under study than the standard high Reynolds k-ε turbulence model (Equation (13) considering $\Delta u^+ = 0$) which results on pressure drop values closer to those obtained only when flow of water is considered in the same pipe geometry (see Table 3), thus indicating the presence of a drag reduction effect.

In the tests with *Eucalyptus* pulp of consistency equal to 2.50% (cases E and F), it was expected that the viscosity values should be higher than those for the other two lower consistencies and, also, a higher viscosity gradient would occur across the pipe. Probably, in these cases, a wider core region and higher velocity gradients in the yield region should be expected, corresponding to a higher value for the Jäsberg adjustable parameter α and a lower value for y_C^+ than the standard one.

In this work, the complete set of differential equations was modified by introducing the non-Newtonian behavior of pulp – modification of the diffusion term of differential equations. The dynamic viscosity is a parameter to take into account in the discussion of the results. Dynamic viscosity profiles for the cases reported in Table 3 are presented in Figure 6. In all the cases, the viscosity reaches its maxima in the centre of the pipe and decreases toward the pipe wall, where no fibers are present, resulting in very low flow resistance. A more uniform radial distribution of fibers is present when the fiber consistency is lower and the total number of fibers in the flow is also lower. For the highest consistency, the viscosity profile observed in Figure 6(c) expresses the higher resistance to the flow in the central region of the pipe reflecting the high concentration of fibers forming the core region.

The large difference in the viscosity ranges between the cases in which the pulp consistency is equal to 2.50% (w/w) and cases in which it is equal to 1.50 and 1.80 % (w/w), observed in Figure 6, is in agreement with the difference in viscosity ranges observed in Figure 3 obtained experimentally. The difference in the scales of viscosity between the corresponding cases from Figure 3 and Figure 6 can be due to the fact that the experimental rheological data are not available for shear rates lower than 50 s^{-1} where it is observed the tendency to have a very high variation in viscosity as a function of shear rate. It is evident from the numerical calculations that in the centre of the pipe the shear rate attains lower values than the minimum values reached experimentally.

In Figure 7, the calculated dimensionless velocity profiles using the CFD code with the modified equations and those calculated using directly the Jäsberg Equation (13) (Jäsberg, 2007) are presented. The numerical results obtained with the modified CFD code (Equations (15) and (16)) are referred to as ModelModified, JasbergCalculation corresponds to the profile calculated with the use of Equation (13) for the dimensionless distance obtained by Equation (16). The StandardLogLaw profile designates the profiles obtained with the *standard* wall law (Equation (13) with $\Delta u^+ = 0$).

In all the cases, except when using the standard log law, the yield region and the core region can be easily identified. For the lower consistency cases (cases A to D) the dimensionless velocity profiles are closer to those calculated by applying directly Equation (13). In cases E and F, an abrupt modification in the u^+ profile near the wall can be observed and it can be seen that the slope in the core region is very low. For the higher consistency cases, the diffusion term has more significance in the problem solution. Thus, the core region, where the viscosity gradient is low, presents a more uniform velocity field.

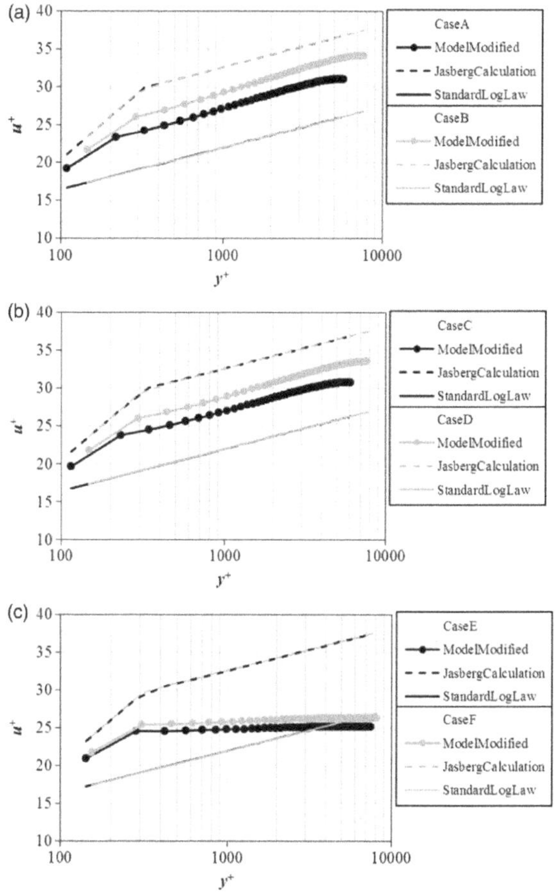

Figure 7. Dimensionless velocity profiles as a function of dimensionless distance to the pipe wall for (a) cases A and B (c = 1.50%(w/w)), (b) cases C and D (c = 1.80%(w/w)), and, (c) cases E and F (c = 2.50%(w/w)).

As referenced in Ventura et al. (2011), the presence of fibers in the flow creates a turbulence damping effect. Figure 8 shows the turbulent kinetic energy profiles along the pipe radius. From there, it can be concluded that when consistency increases, the turbulent kinetic energy increases, contradicting what was to be expected, but in line with the calculated pressure drop values for the highest consistency which did not reveal the existence of drag reduction. In fact, for the higher consistency it is possible that the turbulence damping region was not reached, due to pump limitations, and only tests for the transition region (see Figure 2) could be performed. In the present

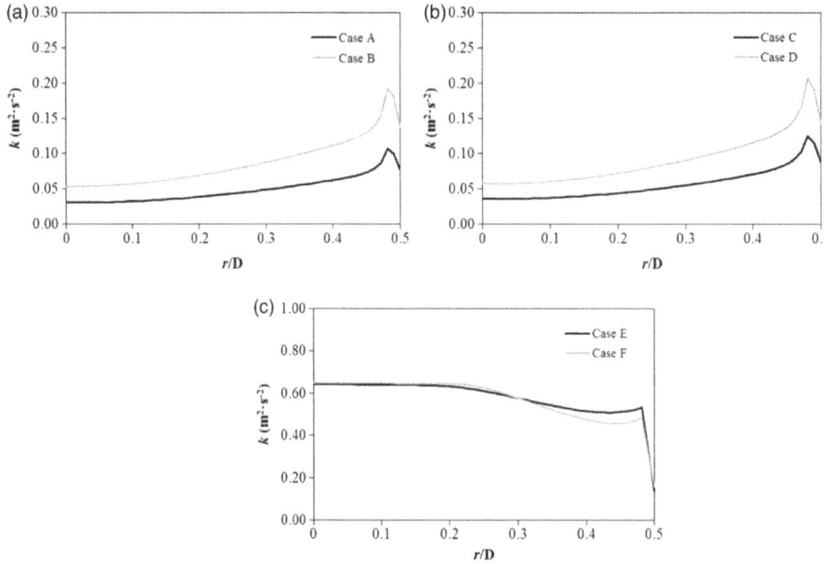

Figure 8. Numerical turbulent kinetic energy profiles for (a) cases A and B (c = 1.50%(w/w)), (b) cases C and D (c = 1.80%(w/w)), and, (c) cases E and F (c = 2.50%(w/w)).

simulation, for the lower consistency, the turbulent kinetic energy increases from the pipe axis to the pipe wall, as expected, reflecting the decrease of viscosity and increase of transverse velocity gradient towards the pipe wall. However, for the highest consistency, Figure 8.c), it can be seen that the turbulent kinetic energy for this case is higher in the pipe axis than in the pipe wall contrarily to what was expected. This behavior reflects the lower velocity gradient and dimensionless velocity gradient observed in Figure 7 for the flow of the higher consistency pulp.

These larger deviations in the turbulent kinetic energy, velocity profiles and viscosity profiles detected for the highest pulp consistency are consistent with the larger discrepancies in the pressure drop values obtained for these two cases (cases E and F). The relative errors are significant in these two cases, which is most likely due to inappropriate values of Jäsberg adjustable parameters used. So, the next step was to analyze the influence the Jäsberg adjustable parameters may have on both the dimensionless velocity profile and the corresponding pressure drop and turbulence kinetic energy profile.

4.2. *Effect of Jäsberg adjustable parameters*

The influence of the Jäsberg adjustable parameters on the dimensionless velocity profile and pressure drop was analyzed for the reference cases B, E and F (Table 4). The α parameter was increased, what corresponds physically to a larger velocity gradient in the yield region. On the other hand, the velocity gradient in the core region was considered lower by increasing β. When the velocity gradient in the yield region increases, the thickness of the core region has to increase, thus y_C^+ is reduced. For all the tests, the thickness of the near wall region was kept constant ($y_L^+ = 50$).

The pressure drop values obtained for the different cases are presented in Table 4. As discussed previously, the drag-reduction effect could not be reproduced in cases E and F. The results presented in Table 4 show that the numerical pressure drop values (*standard* Jäsberg adjustable parameters) were improved, in all cases, by modification of Jäsberg adjustable parameters. The highest improvement was achieved for the pulp consistency of 2.50% (w/w), the drag-reduction effect being now observed.

If the pulp consistency is higher, the thickness of the core region is also higher, corresponding to a lower thickness of the yield region, with a higher velocity gradient (larger slope) in this region. This effect can be seen in the velocity profiles in Figures 9 to 11. In all the cases, the presence of the yield and core regions is observed, therefore confirming the Jäsberg's predictions. The Jäsberg velocity profiles and the calculated velocity profiles become much closer when the slope of the yield region is increased (cases B1, B2, E2, E4, F2, and F4). This is more pronounced for the higher consistency where a better approximation between the calculated and experimental pressure drop was also achieved. For this consistency, the additional increase of β (flatter core region) leads to improvement in both the velocity profile and the calculated pressure drop (cases E4, F2, and F4). When the consistency increases, the increase of the plug region is expected as observed in the numerical results.

Again, as before, a higher viscosity gradient is observed when the consistency is higher (Figures 12b) and c)). In the cases of pulp consistency equal to 2.50% (w/w), the

Table 4. Different Jäsberg adjustable parameters values tested and pressure drop results.

	c	U_{in}	Parameter $y_L^+ = 50$			$\Delta P/L_{water}$	$\Delta P/L_{exp.}$	$\Delta P/L_{num.}$	δ
Case	(% (w/w))	(m·s^{-1})	α	β	y_C^+	(Pa·m^{-1})	(Pa·m^{-1})	(Pa·m^{-1})	(%)
B			2.4	0	320			2146.9	66.6
B1			3.6	0	160			1867.0	44.9
B2	1.50	6.21	3.6	0.5	160	3419.7	1288.7	1867.0	44.9
B3			4.4	0.0	100			1730.8	34.3
B4			4.4	0.9	100			1730.9	34.3
E			2.4	0	320			2299.3	45.6
E1			3.6	0	160			1888.4	19.6
E2	2.50	4.90	3.6	0.5	160	2212.6	1578.9	1840.8	16.6
E3			4.4	0.0	100			1702.6	7.8
E4			4.4	0.9	100			1602.5	1.5
F			2.4	0	320			2814.3	60.5
F1			3.6	0	160			2234.2	27.4
F2	2.50	5.55	3.6	0.5	160	2781.1	1753.8	2233.6	27.4
F3			4.4	0.0	100			2047.0	16.7
F4			4.4	0.9	100			1898.6	8.3

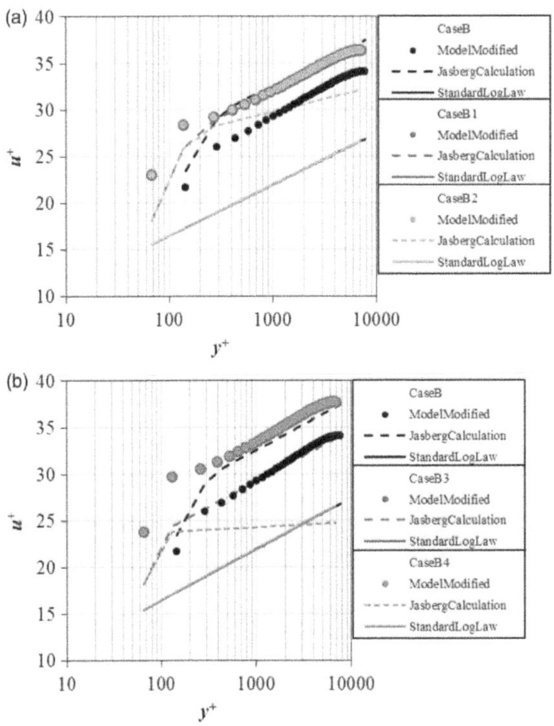

Figure 9. Dimensionless velocity profiles as a function of dimensionless distance to the pipe wall for (a) cases B, B1 and B2, and, (b) cases B, B3 and B4 (c = 1.50%(w/w), U_{in} = 6.61 m·s^{-1}).

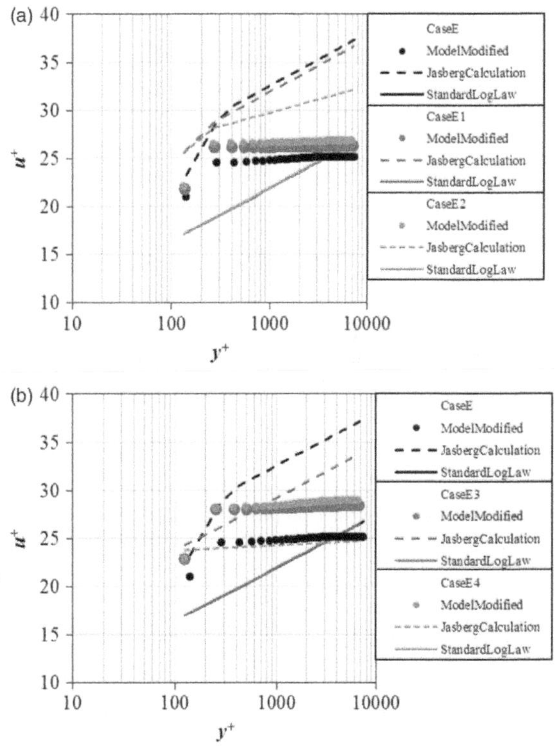

Figure 10. Dimensionless velocity profiles as a function of dimensionless distance to the pipe wall for (a) cases E, E1 and E2, and, (b) cases E, E3 and E4 (c = 2.50%(w/w), U_{in} = 4.90 m·s^{-1}).

non-Newtonian behavior of the pulp suspension is stronger than for the lower pulp consistency studied leading to a smooth velocity gradient in the core region. Comparing the viscosity profiles obtained with the *standard* Jäsberg parameters and with the modified ones, (Figure 12) it can be observed that for the lower consistency the profiles are very similar (Figure 12a)), while for the higher consistency

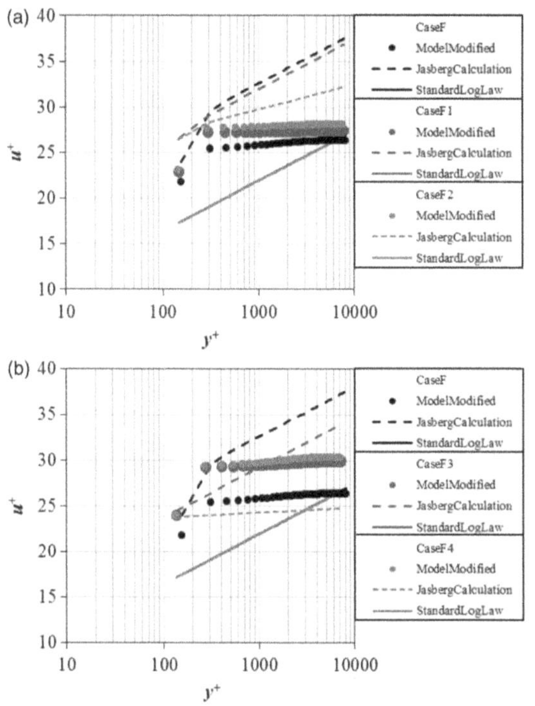

Figure 11. Dimensionless velocity profiles as a function of dimensionless distance to the pipe wall for (a) cases F, F1 and F2, and, (b) cases F, F3 and F4 (c = 2.50%(w/w), U_{in} = 5.55 $m{\cdot}s^{-1}$).

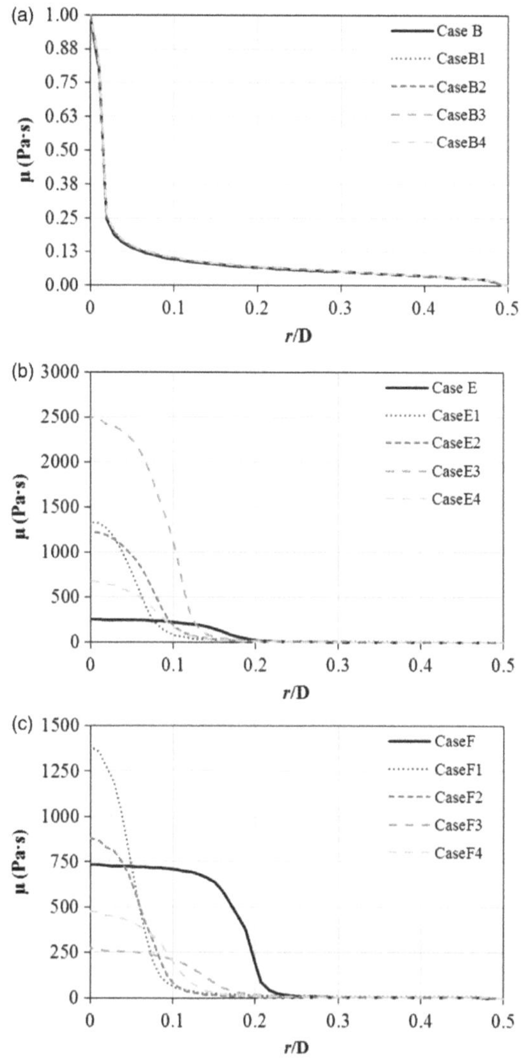

Figure 12. Numerical dynamic viscosity profiles obtained for (a) cases B, B1 to B4 (c = 1.50%(w/w), U_{in} = 6.61 $m{\cdot}s^{-1}$), (b) cases E, E1 to E4 (c = 2.50%(w/w), U_{in} = 4.90 $m{\cdot}s^{-1}$), and, (c) cases F, F1 to F4 (c = 2.50%(w/w), U_{in} = 5.55 $m{\cdot}s^{-1}$).

differences are quite evident (Figures 12b) and c)). Moreover, for cases E4 and F4 viscosity is not only lower than for the *standard* Jäsberg parameters but, additionally, the higher viscosities do not prevail to so close to the wall, this justifying the lower pressure drop values obtained. For cases E4, F2 and F4 it is also observed a decrease on turbulence (Figure 13) and it can be concluded that the parameters α and β have a strong impact on the numerical results. The numerical turbulent kinetic energy has the maximum decrease, compared to the standard cases (cases B, E and F), when the yield region has a lower thickness and the core region has a higher thickness (the slope of the core region is lower (increase of β) and the slope of the yield region is higher (increase of α)). This agrees with what was to be expected since a larger plug of fibers should lead to turbulence damping. Comparing cases B with cases F, pulp consistency is higher and mean inlet velocity is lower in cases F, thus, the more concentrated suspension (cases F) has a higher resistance to the flow and therefore a lower-velocity gradient is observed in the core region.

5. Summary and perspectives

In the present paper a numerical study of *Eucalyptus* pulp flow in a pipe was developed using the commercial CFD software, ANSYS FLUENT 13.0. The CFD code was modified by considering the non-Newtonian behavior of the pulp introduced into the CFD model by viscosity dependence on the local shear rate. The developed model was also modified by considering the logarithmic wall function expression following the proposal of Jäsberg (2007).

The drag-reduction effect was reproduced for various fiber consistencies and flow bulk velocities with the different sets of Jäsberg adjustable parameters. For the highest pulp consistency, the dynamic viscosity has a highest gradient near the pipe wall. For these cases, the existence of the yield and the core regions in the dimensionless velocity profile is more evident. For the cases where the pulp

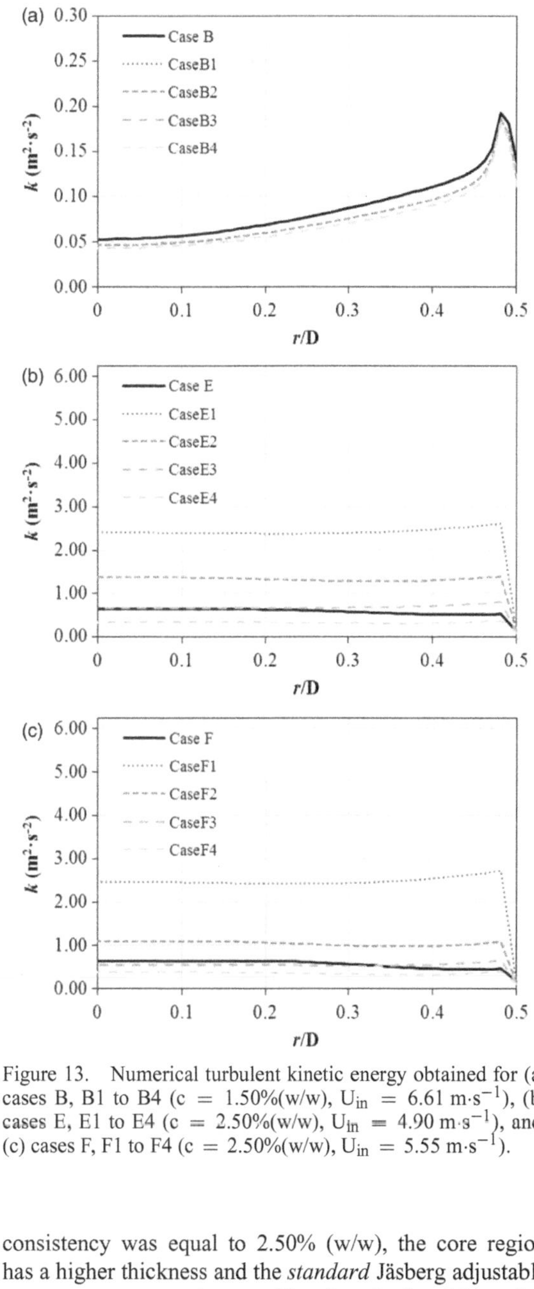

Figure 13. Numerical turbulent kinetic energy obtained for (a) cases B, B1 to B4 (c = 1.50%(w/w), U_{in} = 6.61 $m \cdot s^{-1}$), (b) cases E, E1 to E4 (c = 2.50%(w/w), U_{in} = 4.90 $m \cdot s^{-1}$), and, (c) cases F, F1 to F4 (c = 2.50%(w/w), U_{in} = 5.55 $m \cdot s^{-1}$).

consistency was equal to 2.50% (w/w), the core region has a higher thickness and the *standard* Jäsberg adjustable parameters are not adequate. The slope in the yield region increases and the thickness and the corresponding slope in the core region are reduced. Thus, the Jäsberg parameter α corresponding to the yield region has to increase and the same must happen to the parameter β associated with the core region, this meaning that its slope decreases. A better fit to the experimental pressure drop is obtained in this case. A better fit between calculated and experimental pressure drop is always obtained when the simulated and the calculated Jäsberg dimensionless velocity profiles are similar, what means that the quantitative agreement can only be achieved providing the qualitative correspondence between model and reality exists.

In summary, the strategy followed in this work is able to predict the turbulent flow of concentrated pulp suspensions, but in a future work, it is foreseen to extend the numerical model by considering the presence of fibers in the near-wall region, as shown experimentally by Olson (1996) and predicted by LES simulations of Dong et al. (2003), since that will affect the local viscosity and in turn the transfer of momentum and turbulence in the near-wall region and finally the flow resistance expressed as pressure drop.

Acknowledgements

The present work was developed under the project PTDC/QUE-QUE/112388/2009 and Pest/C/EQB/UI0102/2013, financed by FCT/MCTES (PIDDAC) and co-financed by the European Regional Development Fund (ERDF) through the program COMPETE (POFC) and in connection with COST Action FP1005.

Nomenclature

B	constant in Equation (13) [-]
c	pulp consistency [% (w/w)]
$C_{\varepsilon 1}, C_{\varepsilon 2}, C_{\mu}$	turbulence model constants [-]
D	pipe diameter [m]
k	turbulent kinetic energy [$m^2 \cdot s^{-2}$]
K	consistency coefficient [$Pa \cdot s^n$]
L	pipe length [m]
n	local coordinate normal to the wall [m]
n	flow behavior index [-]
p	pressure [Pa]
r	radial coordinate [m]
Re_W	Reynolds number based on the viscosity near the wall, ate the interface with the water annulus
R^+	dimensionless radius of the core region
u	axial velocity component [$m \cdot s^{-1}$]
U_{in}	mean inlet pulp flow velocity [$m \cdot s^{-1}$]
u^+	dimensionless velocity [-]
V	bulk velocity [$m \cdot s^{-1}$]
v	radial velocity component [$m \cdot s^{-1}$]
x	axial coordinate [m]
y_P	distance from point P to the wall [m]
y^+	dimensionless distance to the pipe wall [-]
y_C^+	dimensionless variable position of the plug surface [-]
y_L^+	dimensionless distance between the near wall region and the yield region [-]

Greek letters

$\Delta P/L$	pressure drop [$Pa \cdot m^{-1}$]
Δu^+	additional term on dimensionless velocity equation (Equation (13)) [-]
α	indicator of the slope in relation to Newtonian profile value of the envelope curve in the yield region [-]
β	indicator of the slope in relation to Newtonian profile value of the envelope curve in the core region [-]
$\dot{\gamma}$	shear rate [s^{-1}]
δ	relative error [%]
ε	dissipation rate of turbulent kinetic energy [$m^2 \cdot s^{-3}$]
κ	von Kármán constant [-]

μ	dynamic viscosity of suspension [Pa·s]
μ_{app}	apparent viscosity of suspension [Pa·s]
μ_t	turbulent viscosity [-]
ρ	pulp density [kg·m^{-3}]
$\sigma_k, \sigma_\varepsilon$	turbulence model constants [-]
τ	shear stress [N·m^{-2}]
τ_w	wall shear stress [N·m^{-2}]

References

ANSYS FLUENT Documentation, Ansys Fluent 14.0. 2011.

Asendrych, D., & Kondora, G. (2009, June 1–4). *Fibre suspension mixing in hydropulper chest.* Proceedings of the Papermaking Research Symposium. Kuopio, Finland.

Blanco, A., Negro, C., Fluente, E., & Tijero, J. (2007). Rotor selection for a Searle-type device to study the rheology of paper pulp suspensions. *Chemical Engineering and Processing, 46*(1), 37–44.

Chang, K. C., Hsieh, W. D., & Chen, C. S. (1995). A modified low-Reynolds-number turbulence model applicable to recirculating flow in pipe expansion. *Journal of Fluids Engineering, 117*(3), 417–423.

Derakhshandeh, B., Kerekes, R. J., Hatzikiriakos, S. G., & Bennington, C. P. J. (2011). Rheology of pulp fibre suspensions: a critical review. *Chemical Engineering Science, 66*(15), 3460–3470.

Dong, S., Feng, X., Salcudean, M., & Gartshore, I. (2003). Concentration of pulp fibres in 3D turbulent channel flow. *International Journal of Multiphase Flow, 29*(1), 1–21.

Fang, J., Ling, X., & Sang, Z. F. (2013). Solid suspension in stirred tank equipped with multi-side-entering agitators. *Engineering Applications of Computational Fluid Mechanics, 7*(2), 282–294.

Ferziger, J. H., & Peric, M. (2002). *Computational methods for fluid dynamics.* Berlin Heidelberg: Springer-Verlag.

Fock, H., Wiklund, J., & Rasmuson, A. (2009). Ultrasound velocity profile (UVP) measurements of pulp suspensions flow near the wall. *Journal of Pulp and Paper Science, 35*(1), 26–33.

Galván, S., Reggio, M., & Guibault, F. (2011). Assessment study of *k-ε* turbulence models and near-wall modeling for steady state swirling flow analysis in draft tube using *FLUENT. Engineering Applications of Computational Fluid Mechanics, 5*(4), 459–478.

Gullichsen, J., & Härkönen, E. (1981). Medium consistency technology. *TAPPI Journal, 64*(6), 69–72.

Hämäläinen, J., Lindström, S. B., Hämäläinen, T., & Niskanen, H. (2011). Papermaking fibre-suspension flow simulations at multiple scales. *Journal of Engineering Mathematics, 71*(1), 55–79.

Hammarström, D. (2004). *A model for simulation of fiber suspension flows* (BSc thesis). KTH Royal Institute of Technology.

Hsieh, W. D., & Chang, K. C. (1996). Calculation of wall heat transfer in pipe-expansion turbulent flows. *International Journal of Heat and Mass Transfer, 39*(18), 3813–3822.

Huhtanen, J. P., & Karvinen, R. J. (2005). Interaction of non-newtonian fluid dynamics and turbulence on the behaviour of pulp suspension flows. *Annual Transactions of the Nordic Rheology Society, 13*, 177–186.

Huhtanen, J. P., & Karvinen, R. J. (2006). Characterisation of non-Newtonian fluid models for wood fibre suspensions in laminar and turbulent flows. *Annual Transactions of the Nordic Rheology Society, 14*, 51–60.

Jäsberg, A. (2007). *Flow behaviour of fibre suspensions in straight pipes: new experimental techniques and multiphase modelling* (PhD thesis). University of Jyväskylä.

Kondora, G., & Asendrych, D. (2013). Flow modelling in a low consistency disc refiner. *Nordic Pulp & Paper Research Journal, 28*(1), 119–130.

Li, A., Green, S., & Franzen, M. (2001). *Optimum consistency for pumping pulp.* Technical Information Paper TAPPI TIP 0410-15.

Lundell, F., Söderberg, L. D., & Alfredsson, P. H. (2011). Fluid mechanics of papermaking. *Annual Review of Fluid Mechanics, 43*, 195–217.

Mathur, A., & He, S. (2013). Performance and implementation of the Launder-Sharma low-Reynolds number turbulence model. *Computers & Fluids, 79*(25), 134–139.

Olson, J. S. (1996). *The effect of fibre length on passage through narrow apertures* (PhD thesis). University of British Columbia.

Rudman, M., & Blackburn, H. M. (2006). Direct numerical simulation of turbulent non-Newtonian flow using a spectral element method. *Applied Mathematical Modelling, 30*(11), 1229–1248.

Vatani, A., & Mohammed, H. A. (2013). Turbulent nanofluid flow over periodic rib-grooved channels. *Engineering Applications of Computational Fluid Mechanics, 7*(3), 369–381.

Ventura, C., Blanco, A., Negro, C., Ferreira, P., Garcia, F., & Rasteiro, M. (2007). Modeling pulp fiber suspension rheology. *TAPPI Journal, 6*(7), 17–23.

Ventura, C., Garcia, F., Ferreira, P., & Rasteiro, M. (2008). Flow dynamics of pulp fiber suspensions. *TAPPI Journal, 7*(8): 20–26.

Ventura, C. A. F., Garcia, F. A. P., Ferreira, P. J., & Rasteiro, M. G. (2011). Modeling the turbulent flow of pulp suspensions. *Industrial & Engineering Chemistry Research, 50*(16), 735–742.

Xu, H., & Aidun, C. (2005). Characteristics of fibre suspension flow in a rectangular channel. *International Journal of Multiphase Flow, 31*(3), 318–336.

3

Effects of different parameters on the flow field of peripheral ported rotary engines

Baowei Fan[a], Jianfeng Pan[a*], Wenming Yang[b], Hui An[b], Aikun Tang[a], Xia Shao[a] and Hong Xue[c]

[a]*School of Energy and Power Engineering, Jiangsu University, Zhenjiang 202013, China;* [b]*Department of Mechanical Engineering, National University of Singapore, Singapore 117576, Singapore;* [c]*School of Mechanical Engineering, California State Polytechnic University, Pomona, CA 91768, USA*

The performance of rotary engines is significantly influenced by the flow field. In this study, a detailed mathematical model was integrated into the simulation software FLUENT to investigate the gas flow field in a peripheral ported rotary engine by including a dynamic mesh model and a turbulent flow model. The models were also validated by experimental data. The basic flow mechanism in the combustion chamber was numerically studied. Meanwhile, the effects of the three major parameters on the flow field inside the combustion chamber, namely, rotating speed, intake shape, and intake angle, were also investigated. Results showed that a constantly changing swirl was formed in the combustion chamber during the intake and compression strokes as a result of the combined effects of the pocket of the rotor and the swirls in the combustion chamber. The swirl eventually broke into a unidirectional flow near the top dead center because of the significant decrease in combustion chamber volume. Furthermore, with the change in rotating speed, intake shape, and intake angle, significant differences in flow speed, inertia, and distribution were observed when the fluid entered the combustion chamber, which, in turn, led to obvious differences in the flow field, volume coefficient, and average turbulence kinetic energy in the combustion chamber.

Keywords: peripheral ported rotary engine; flow field; rotating speed; intake shape; intake angle; CFD simulation

1. Introduction

With increasing concerns about the energy crisis and environmental issues, a cleaner, more efficient, and environmentally adaptable energy system has become a pressing need. The highly advanced Wankel rotary engine is a promising energy system, which is an alternative to a reciprocating piston engine. The advantages of the Wankel rotary engine over a conventional reciprocating engine include a high power-to-weight ratio, a large specific power output from a high allowable engine speed, a simple and compact design resulting from less moving parts and multi-fuel capability, as well as low noise and vibration levels attributed to non-reciprocating components (Muroki, 1984; Shapovalov, 1999). As such, the Wankel rotary engine has been extensively used in various stationary and mobile applications (Antonelli & Martorano, 2012; Badr, Naik, O'Callaghan, & Probert, 1991). Nevertheless, the disadvantages of the Wankel rotary engine, which should not be ignored, include its sealing and leakage problems, lower efficiency, and higher unburned hydrocarbon emissions resulting from the flattened combustion chamber shape. Since the 1980s, several studies have attempted to improve the fuel efficiency and exhaust emissions of the Wankel rotary engine (Abraham & Bracco, 1989, 1991). Recently, some new apex seals were designed for the rotary engine to improve sealing (Warren & Yang, 2013). A direct-injection and turbocharged system was also used to improve the efficiency of the engine (Kagawa, Okazaki, Somyo, & Akagi, 1993; Tashima, Taqdokoro, Okimoto, & Niwa, 1991). Hydrogen was added to gasoline in the engine to improve its thermal efficiency and power output (Amrouche, Erickson, Park, & Varnhagen, 2014). Nevertheless, studies on the three-dimensional flow mechanisms of rotary engines, which are a determining factor of the combustion process and efficiency, remain insufficient. To our knowledge, fuel atomization, mixture formation, and combustion process are closely related to the flow field in the combustion chamber; thus, the importance of studies on the flow field is self-evident (Pariotis, Kosmadakis, & Rakopoulos, 2012; Taghavifar, Khalilarya, & Jafarmadar, 2014). For studies on the flow field inside a rotary engine, the main approach has focused on experimental visualization. In addition, only a few studies have utilized numerical simulations. For the experiments, the reflection-type Schlieren method was used by Hasegawa and Yamaguchi (1993) to study the flow field in the center section of the combustion chamber of side-ported rotary engines, work which made a significant contribution to this area of flow field investigation. However, only the shape of the swirl in the rotor housing central plane was reported, and not the velocity and direction of the swirl. DeFilippis, Hamady, Novak, and Schock (1992) used laser Doppler velocimetry

*Corresponding author. Email: mike@ujs.edu.cn

(LDV) to measure the flow field in the center section of the combustion chamber of a peripheral ported rotary engine. The velocity and direction of the swirl were obtained. However, with only a few measuring points, DeFilippis et al. could not determine the overall internal flow characteristics in the combustion chamber. The underlying reason why a swirl was observed in the rotor housing central plane in the experiment was unclear. Meanwhile, given the limitations of the experimental setup, only the flow field under low-speed conditions could be directly measured and the flow field under normal engine operating conditions ($\geq$ 4000 rpm) could not be tested. Nevertheless, these valuable experimental data could still be used for validating the simulation. In terms of numerical simulations, some studies (Abraham & Bracco, 1988; T. I. Shih, Schock, Nguyen, & Stegeman, 1987) have used two-dimensional and three-dimensional numerical simulations to model the flow and combustion processes in rotary engines. However, given the computing power and limitations of the turbulence models developed in the 1980s, accurately predicting the large-scale swirl-changing process in the combustion chamber was impossible. Although various mathematical models have been significantly improved over the past few years, three-dimensional investigation coupled with a reasonable turbulence model on the flow process inside the combustion chamber of a rotary engine is still rare. The reason why a swirl was observed in the rotor housing central plane in the experiment remains unclear.

In the present study, we will integrate a three-dimensional moving grid and a reasonable turbulence model into the computational fluid dynamics (CFD) simulation software FLUENT. The computed results will be compared with the experimental data of DeFilippis et al. (1992) and our own experimental data to verify the turbulence model used in the rotary engines. In this manner, the variation of the flow field, average speed, and turbulent kinetic energy in a high-speed cold drag engine are reported, which are not easily obtained through experiments. The basic flow mechanism in the combustion chamber is numerically studied. The reason why a swirl was observed in the rotor housing central plane in the experiment in DeFilippis et al. is also determined. Meanwhile, the effect of the rotating speed, intake shape, and intake angle on the flow field are simulated and analyzed. The study on the three-dimensional flow mechanisms of peripheral ported rotary engines provides a theoretical foundation to improve and optimize the design of a rotary engine, with regard to such aspects as fuel atomization, mixture formation, and stratified combustion.

2. Dynamic mesh generation

2.1. Computational domain

In this study, the rotary engine used in the experimental study reported in DeFilippis et al. (1992) is selected as an

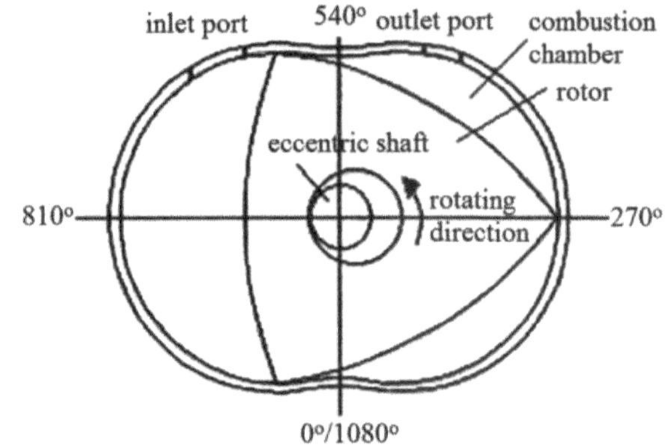

Figure 1. Schematic of the rotary engine.

Table 1. Engine specification.

Engine parameters	Value
Generating radius	105 mm
Eccentricity	15 mm
Chamber width	70 mm
Number of rotors	1
Intake stroke	Advance angle, 447°; Delay angle, 530°
Exhaust stroke	Advance angle, 192°; Delay angle, 336°

object of study. A schematic of the test engine is shown in Figure 1. One cycle (including intake, compression, explosion, and exhaust strokes) corresponds to a crankshaft angle (CA) of 1080° in a rotary engine. As the rotor completes one rotation, during which time the eccentric shaft rotates through three revolutions, each chamber produces one power stroke. The engine specification is provided in Table 1.

2.2. Geometric model generation and meshing

Given that three combustion chambers are evenly distributed in the rotary engine, only one combustion chamber is simulated to save computational time. To our knowledge, a reasonable mesh size is significant in improving computational accuracy and saving central processing unit time. Considering the working characteristics of a rotary engine, the areas of the intake and exhaust channels are considered as a static mesh because they are not moving, whereas that of the combustion chamber is

Table 2. Parameters of four different grid sizes.

Grid size	2 mm	2.5 mm	3 mm	3.5 mm
Grid number at top dead center (TDC)	131,077	74,346	47,361	27,632
Grid number at bottom dead center (BDC)	320,041	193,798	121,188	86,680

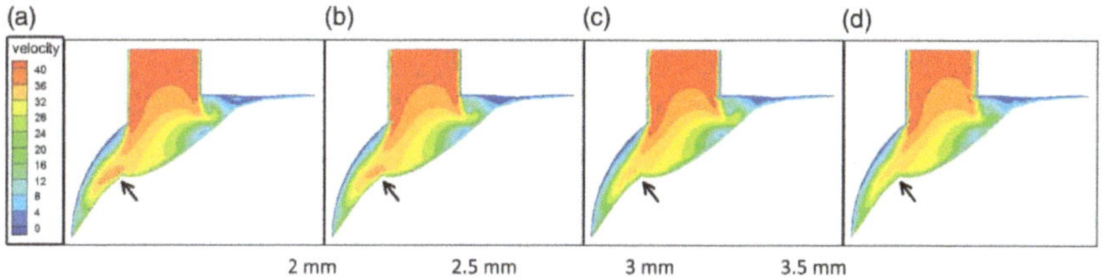

Figure 2. Velocity profile under four different grid sizes (m/s) (450° BTDC).

considered as a dynamic mesh because of the movement of the rotor. For dynamic mesh generation, unstructured triangular grids are adopted. DEFINE_CG_MOTION was used to specify the motion of a particular dynamic zone in ANSYS FLUENT by providing the linear and angular velocities at each time step. For dynamic mesh generation, the more even the mesh size, the better the performance of the mesh movements. Therefore, the near-wall grid is not densified in this study.

To our knowledge, the grid size for engine simulation is typically set to 2 mm to 4 mm. Four different grid sizes are shown in Table 2, namely, 2, 2.5, 3, and 3.5 mm, to check grid sensitivity. The simulation results of the velocity profile under four different grid sizes are shown in Figure 2. Near the bottom of the pocket, which is marked with an arrow in Figure 2, the simulation results for 2 mm and 2.5 mm are better than that for 3 mm and 3.5 mm. That is, the grid sizes of 2 mm and 2.5 mm provide a more detailed description of flow near the rotor wall. Figure 3 shows the average velocity in the combustion chamber for four different grid sizes. As shown in the figure, the average velocity value in the combustion chamber for the grid size of 2.5 mm is close to that for the grid size of 2 mm within the range of 80° CA. However, the average velocity values for the grid sizes of 3 mm and 3.5 mm are significantly smaller than that for the grid size of 2 mm. Meanwhile, the computation time for the grid size of 2.5 mm can reduce calculation time significantly. As such, grid size is set to

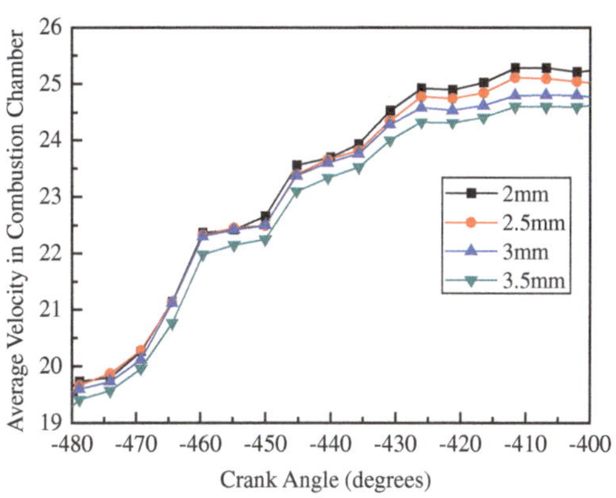

Figure 3. Average velocity in the combustion chamber for four different grid sizes (m/s).

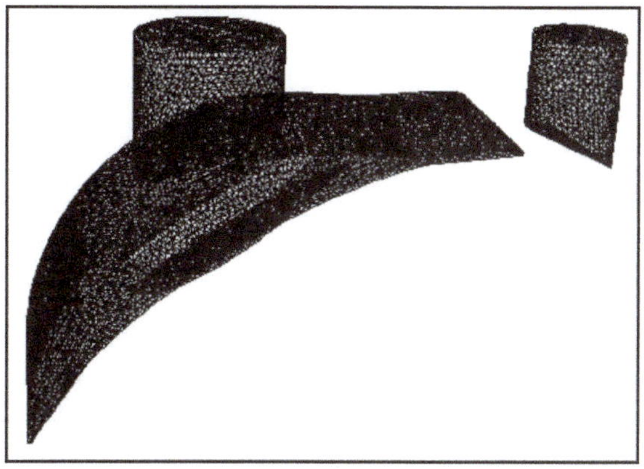

Figure 4. Three-dimensional mesh of the rotary engine at 450° BTDC.

2.5 mm after repeated comparisons. The three-dimensional mesh at 450° before top dead center (BTDC) is shown in Figure 4.

3. Mathematical models and validation

3.1. Mathematical models and boundary conditions

Air is set as the working medium for simulation. The gas flow in a rotary engine combustion chamber is a complex compressible viscous flow; the governing equations of mass, momentum, and energy are solved for unsteady compressible viscous flow. Meanwhile, the governing equations are discretized using the finite volume method and solved using ANSYS FLUENT Release 14.0 (ANSYS, 2011). Fluid density is calculated using the ideal law, and constant thermal conductivity and viscosity of the fluid are assumed. A coupled algorithm is used to deal with pressure–velocity coupling. The first-order upwind scheme is used to discretize the turbulent kinetic energy and turbulent dissipation rate terms. The second-order upwind scheme is used to discretize the energy, density, and momentum terms. Pressure inlet and pressure outlet are defined as the inlet and outlet boundary conditions, respectively. The engine is naturally aspirated; hence, the values of pressure boundary conditions are set to atmospheric pressure. Moreover, the combustion chamber and the rotor are set to be the wall boundary. Given that the engine is motored, the temperature of the wall boundary is set to 300 K. Accurately emulating the effect of turbulence is the key for the flow field simulation because gas flow in a rotary engine combustion chamber is a complex compressible viscous flow and typically has a high turbulence scale. In this study, turbulence is modeled using a two-equation renormalization group (RNG) $k - \varepsilon$ submodel because it considers streamline bending, whirlpool, rotation, and fast-changing tension effects and has a higher potential for complex flow forecasting (Han & Reitz, 1995; T. H. Shih, Liou, Shabbir, Yang, & Zhu, 1995; Wang, Reitz, Pera, Wang, & Wang, 2013). The RNG $k - \varepsilon$ model is derived using a statistical technique called RNG theory (ANSYS, 2011). This model is similar in form to the standard $k - \varepsilon$ model. However, the effect of swirl on turbulence is included in the RNG model, which improves the accuracy for swirling flows and makes it suitable for complex flow field simulation inside a rotary engine.

3.2. *Experimental results and model validation*

The flow field was calculated under the experimental conditions (2000 rpm) described in DeFilippis et al. (1992) to validate the mathematical model. The calculated results were compared with the experimental data from DeFilippis et al., which were obtained by LDV measurement. The measurement locations, as shown in Figure 5, are all in the rotor housing central plane and equidistant from both end housings. The pocket shapes for the experimental test and the simulation were the same. Given that the pocket was located in the middle of the rotor surface, the camera could not photograph it during the experiment. As such, the pocket could not be observed in the experimental results shown in Table 3.

Table 3 shows the comparison of the simulation results with the experimental data in DeFilippis et al. (1992). As shown in the table, the calculated flow profile in the

Table 3. Velocity distribution at different times (m/s).

CA (°)	Experimental results (DeFilippis et al., 1992)	Simulation results
730	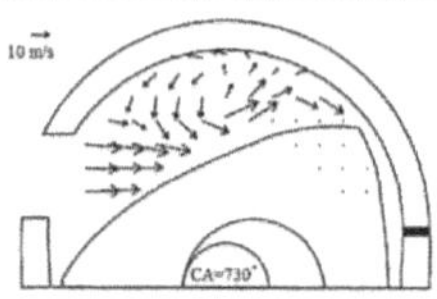	
820	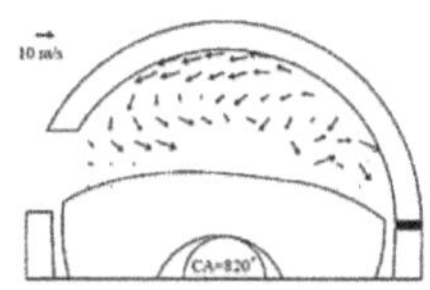	
910	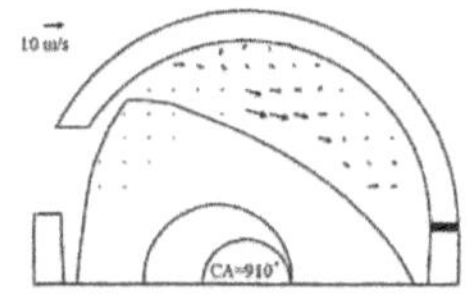	
1000	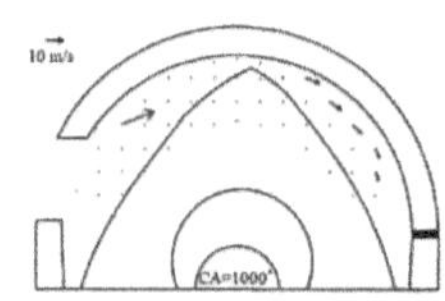	

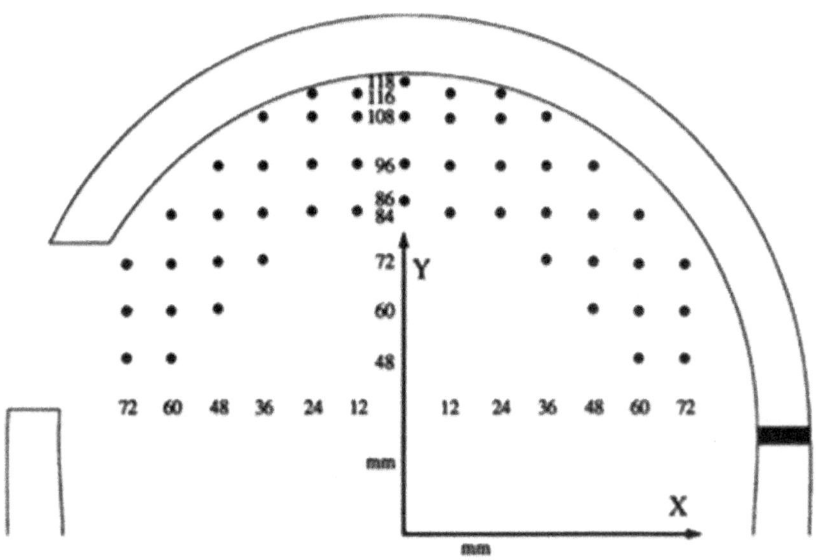

Figure 5. Measurement locations of LDV (DeFilippis et al., 1992).

combustion chamber is similar to the experimental results in terms of the velocity of the flow field and the position of the swirl. For example, when the CA is at 730°, the velocities in the experimental and simulation results near the intake port are approximately equal to 20 m/s and the swirl locations are all located at the front of the combustion chamber. When the CA is at 1000°, the swirl is compressed into a unidirectional flow because of the decrease in combustion chamber volume, and the experimental and calculation results show that the velocity of the unidirectional flow is approximately 10 m/s.

An optical side-ported rotary engine test bed is built and particle image velocimetry (PIV) is used to test the flow field in the rotor housing central plane to further validate the turbulence model. Figures 6 and 7 show the schematic and photograph of the optical rotary engine and the PIV experimental setup in the laboratory, respectively. By controlling frequency and speed modulation, the motor is operated at 600 rpm. The numerical simulations are conducted under the same conditions as the experiment for fair comparison.

Table 4 shows the validity of the mathematical model and the numerical simulation. In most of the central regions, the simulated flow field in the combustion chamber is similar to that of the experimental results in terms of the velocity and position of the vortex. For example, a counterclockwise flow in the rotor housing central plane is observed in the experiment and simulation. Meanwhile, when the CA is at 250° BTDC, the magnitude of velocity for both the experiment and the simulation is approximately equal to 2 m/s and a swirl is located in the middle of the combustion chamber. Meanwhile, Table 5 shows the comparison results of velocity distribution in different turbulence models at 600 rpm (250° BTDC). The simulation result for the RNG $k-\varepsilon$ model can better reflect the experimental results than the standard $k-\varepsilon$ model and the realizable $k-\varepsilon$ model. As such, the RNG $k-\varepsilon$ model is proven to be accurate.

4. Simulation results and discussion

The change in the three-dimensional flow field in the combustion chamber with increasing rotating speed and varying intake shape and angle is difficult to obtain through experiments. In such a case, the three-dimensional dynamic simulation study becomes important. When the intake flow enters the combustion chamber in the intake stroke, it collides with the cylinder and rotor walls. After the collision, the intake flow turns around to form vortices. The vortex whose rotational axis is perpendicular to the cylinder head is called swirl, whereas the vortex whose rotational axis is parallel to the cylinder head is called tumble, in order to clearly express flow motion in the combustion chamber in the rotary engine. Meanwhile, in the streamline figures, high-speed flow through both sides of the intake port is denoted by A. The flow that comes

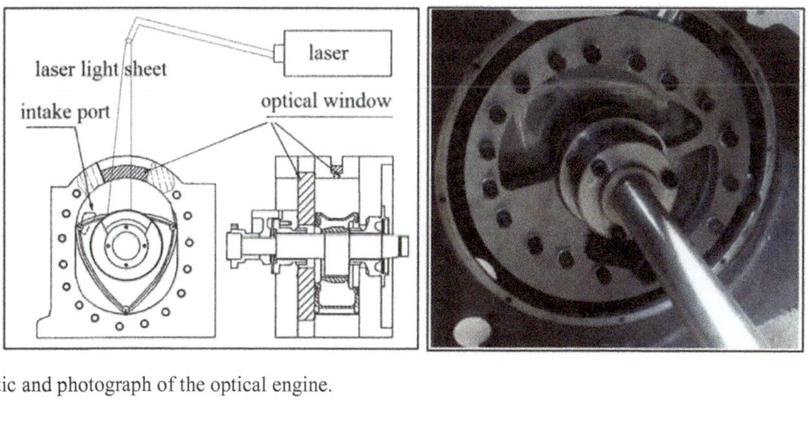

Figure 6. Schematic and photograph of the optical engine.

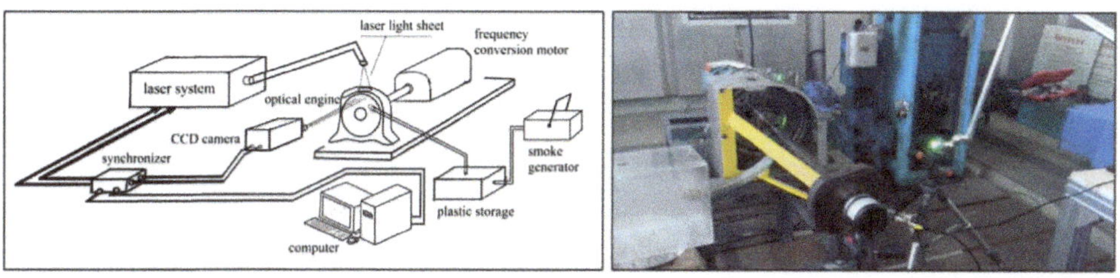

Figure 7. Schematic and photograph of the PIV experimental setup.

Table 4. Comparison of the velocity distribution at 600 rpm.

CAs	CA = 300° BTDC	CA = 250° BTDC	CA = 200° BTDC	CA = 150° BTDC
Testing area	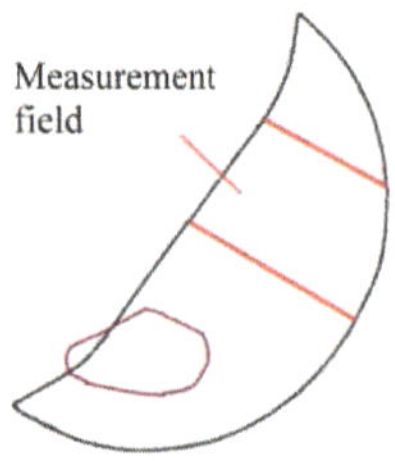	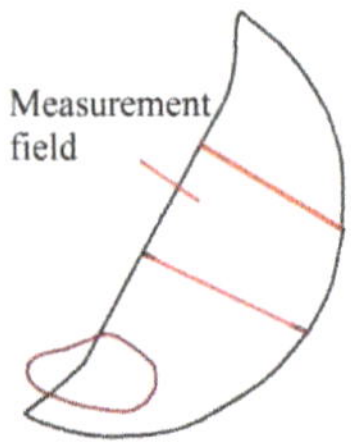	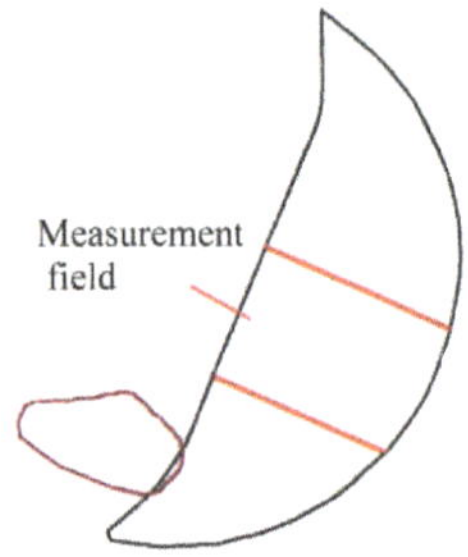	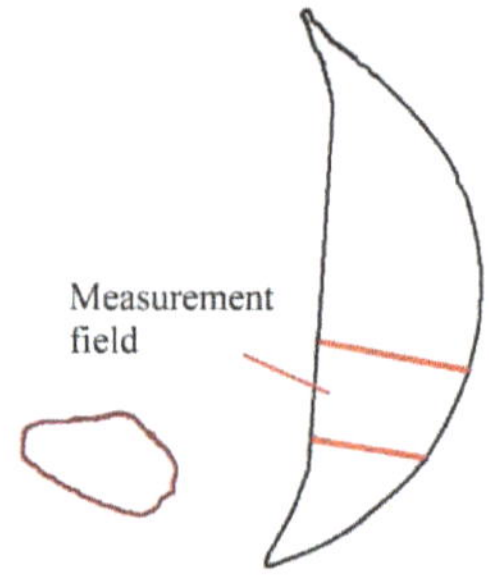
Experimental results (m/s)	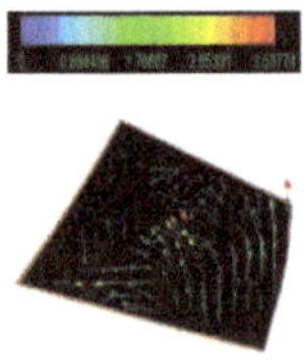	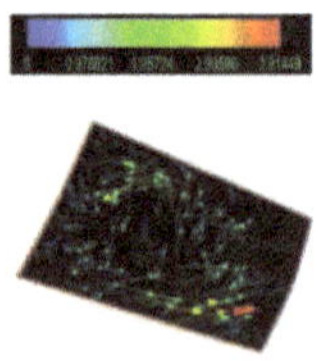		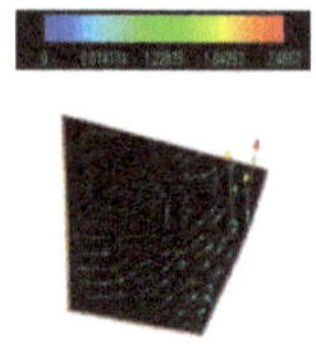
Simulation results (m/s)	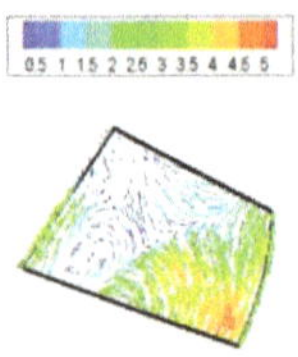	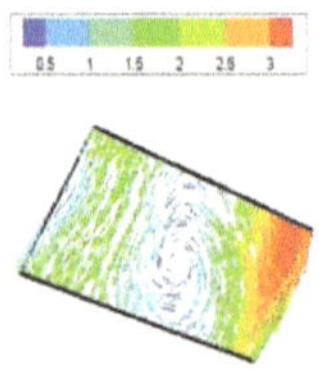	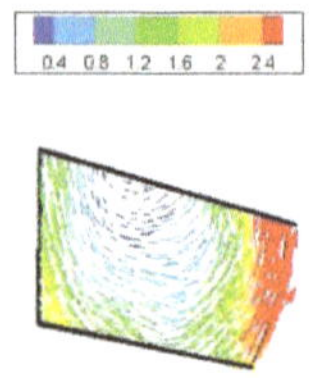	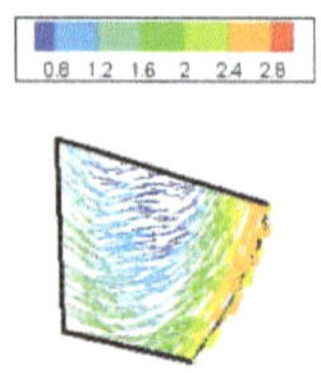

Table 5. Comparison of the velocity distribution in different turbulence models at 250° BTDC under 600 rpm.

Experimental results (m/s)	RNG $k - \varepsilon$ model (m/s)	Standard $k - \varepsilon$ model (m/s)	Realizable $k - \varepsilon$ model (m/s)
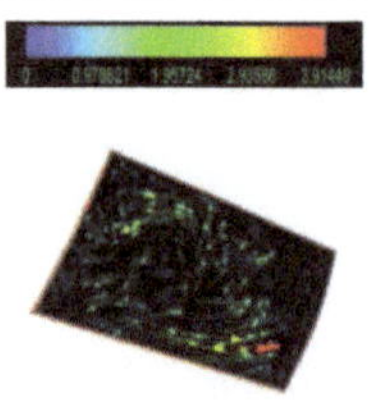	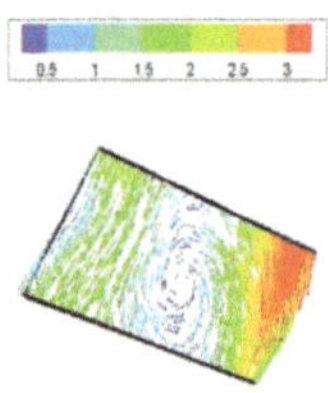	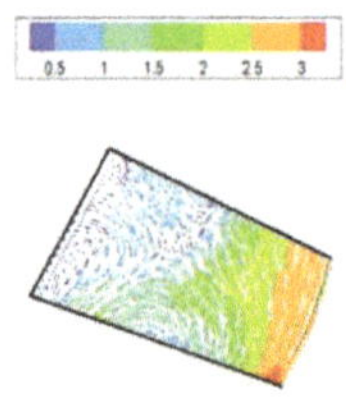	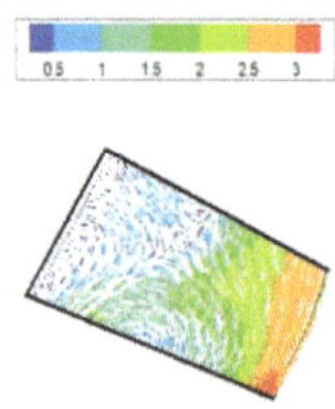

from the rear of the combustion chamber is denoted by B. The reverse flow that comes from the combustion chamber to the intake port is denoted by C.

4.1. Basic flow mechanism in the combustion chamber

The original engine is operated at the normal engine speed of 4000 rpm to illustrate the basic flow mechanism in the combustion chamber. Figure 8 shows the streamlines in the combustion chamber at different CAs.

During the early stage of the intake stroke, two tumbles are located at both sides of the combustion chamber. The tumbles are composed of two parts. One part is the high-speed flow through both sides of the intake port because when high-speed airflow enters the combustion chamber, it will hit the wall surfaces of the rotor and the combustion chamber to finally form the tumble. The other part is the flow that originates from the rear of the combustion chamber and moves toward the front of the chamber because of the thrust of the rotor. However, the high-speed

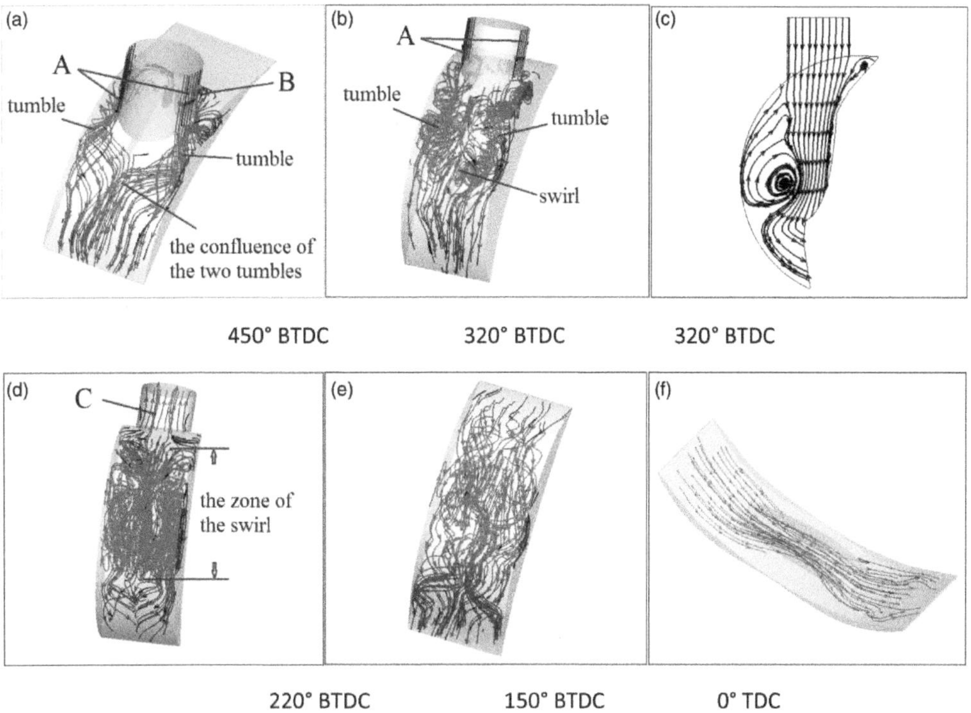

Figure 8. Streamlines in the combustion chamber at different CAs.

airflow from the intake port obstructs the flow from moving forward. When the flow hits the high-speed airflow from the intake port, it will bypass the high-speed airflow and move along both sides of the combustion chamber to form the tumble. In the middle of the intake stroke, a swirl is generated near the bottom of the pocket in front of the combustion chamber, as shown in Figure 8(b), because the two tumbles continue to move forward with the increase in combustion chamber volume. When the two tumbles move close to the bottom of the pocket, they do not readily continue to move forward because of the long and narrow volume in front of the pocket. As such, a swirl is formed when the confluence of the two tumbles moves close to the bottom of the pocket. As shown in Figure 8(c), when the swirl formed in the combustion chamber projects onto the rotor housing central plane, a swirl is also formed in the rotor housing central plane. This is the reason for the formation of the swirl detected in the rotor housing central plane in the experiment conducted by DeFilippis et al. (1992). All of the aforementioned analytical results indicate that swirl generation is the combined effect of the pocket of the rotor and the two tumbles in the combustion chamber. During the last stage of the intake stroke, the reverse flow phenomenon occurs from the combustion chamber to the intake port, as shown in Figure 8(d). As a result of the enlargement of the combustion chamber volume and the reverse flow phenomenon, the swirl begins to diverge and the radius of the swirl begins to increase. During the early stage of the compression stroke, as shown in Figure 8(e), the original swirls have been broken up and the streamlines in cylinder are chaotic. During the last stage of the compression stroke, the flow in the rotor housing central plane eventually breaks into a unidirectional flow near the top dead center (TDC) because the volume of the combustion chamber has a minimum value, as shown in Figure 8(f). Considering the existence of the rotor pocket, the airflow from the rear parts of the combustion chamber flows through the pocket to the front of the combustion chamber.

4.2. *Effects of the rotation speed on the flow field*

The rotary engine has a special structure and operation mode. As its rotor turns one revolution, the output shaft turns three revolutions, which makes the rotary engine more suitable for operating at high rotation speeds than the reciprocating engine. In this study, the change in the flow field is analyzed when the engine speed is less than 30,000 rpm (rotor speed less than 10,000 rpm). Figures 9 and 10 show the change in the flow field at different CAs. In the figures, panels (a) to (d) show the streamlines under the rotating speeds of 4000, 8000, 16,000, and 26,000 rpm, respectively.

During the intake stroke stage, as shown in Figures 9(a) to 9(d), the flow patterns at the combustion chamber are similar under all four different rotating speeds. However,

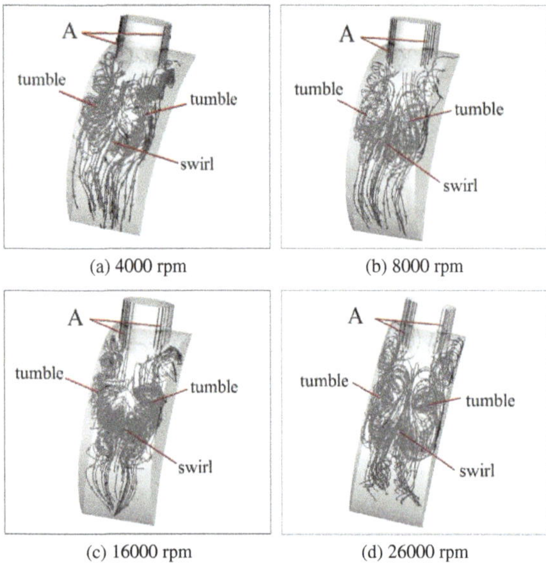

Figure 9. Streamlines in the combustion chamber with different rotation speeds at 320° BTDC.

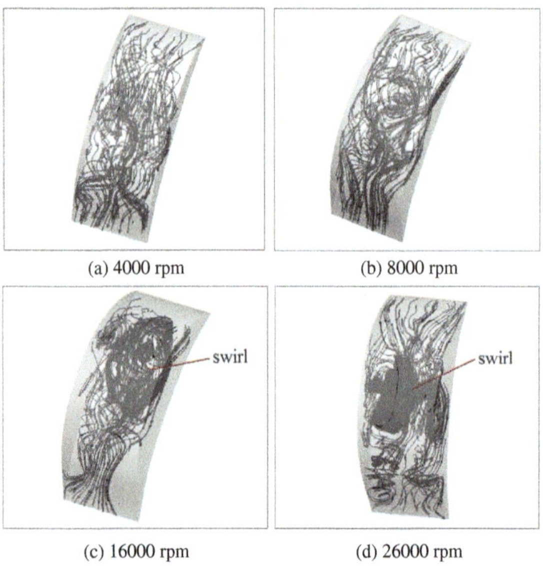

Figure 10. Streamlines in the combustion chamber with different rotation speeds at 150° BTDC.

the vorticity magnitude of the swirl increases with the increase in rotating speed. The largest vorticity magnitudes of the swirl are 7001.8, 15,627.6, 40,196.7, and 47,481 s^{-1} for 4000, 8000, 16,000, and 26,000 rpm, respectively. During the compression stroke stage, as shown in Figures 10(a) to 10(d), the original swirls under the 16,000 rpm and 26,000 rpm conditions still hold up well even when the original swirls under the 4000 rpm and 8000 rpm conditions have broken into several small swirls. This finding indicates that swirl breakup is postponed because the energy and inertia of the swirl increases with the increase in rotation speed. From the overall changes in the flow field during intake and compression strokes, the speed and inertia of the flow in the combustion chamber are increased and swirl breakup is postponed as the rotating speed increases.

Figure 11 shows that the volumetric coefficient initially increases and then decreases with the increase in rotation speed. The highest point is 16,000 rpm because as the rotation speed increases, the speed and inertia of the flow inside the combustion chamber also increase, but intake time constantly decreases under intake stroke. That is, the volumetric coefficient does not always increase with the increase in rotating speed, and 16,000 rpm is the turning point. Compared with the rotating speed of 2000 rpm, a 15.4% increase in the volume coefficient is observed when the rotating speed is 16,000 rpm. Figure 12 shows the average turbulence kinetic energy curves in the combustion chamber under different rotation speeds. To our knowledge, when the intake flow enters the combustion chamber, it collides with the walls of the rotor and the combustion chamber, which disturbs the flow in the combustion chamber. This disturbance increases as the rotation speed

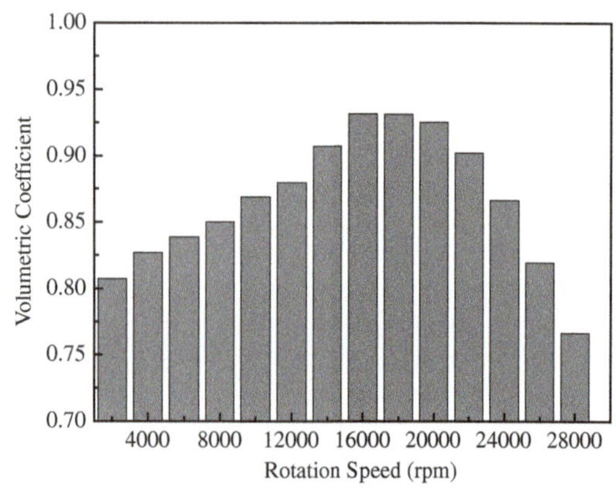

Figure 11. Volumetric coefficient with different rotation speeds.

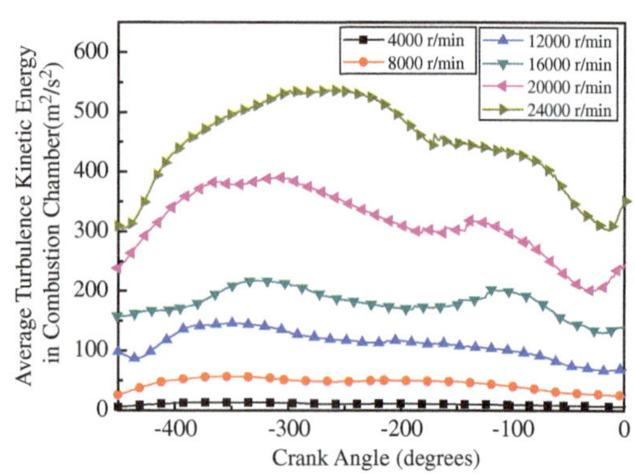

Figure 12. Average turbulence kinetic energy in the combustion chamber with different rotation speeds.

increases; thus, the swirl motion and turbulent kinetic energy increase with the rotation speed.

4.3. *Effects of the intake shape on the flow field*

Four different intake shapes are compared to investigate the influence of intake shape on the flow field in the combustion chamber, as shown in Table 6. The shapes are round intake (the original intake), rectangular intake, regular trapezoid intake, and inverted trapezoid intake. The areas of all the four shapes are set to be the same to eliminate the effects of area.

The changes in the flow field with different intake shapes at 4000 rpm are shown in Figures 13 to 15. In the figures, panels (a) to (d) show the streamlines under the round, rectangular, regular trapezoid, and inverted trapezoid intake shapes, respectively.

During the early stage of the intake stroke, the area near the baseline is larger than that near the topline because of the difference in width of the trapezoid intake shape at different positions, as shown in Figures 13(a) to 13(d). This difference in width causes most of the fluid to flow into the combustion chamber through the baseline side. This area is called the main intake flow area. Correspondingly, the smaller part near the topline is called the secondary intake flow area. As shown in Figures 13(c) and 13(d), the main intake fluid area of the regular trapezoid intake shape is near the center of the combustion chamber. That is, the high-speed fluid from the intake port occupies most of the space in the center of the combustion chamber, which results in a small developing space for the two tumbles. As such, the swirl is squeezed into the rear of the combustion chamber, as shown in Figure 14(c). By contrast, the main intake area of the inverted trapezoid design is far from the

Table 6. Sketches of the different intake shapes.

Round intake (the original intake)	Rectangular intake	Regular trapezoid intake	Inverted trapezoid intake

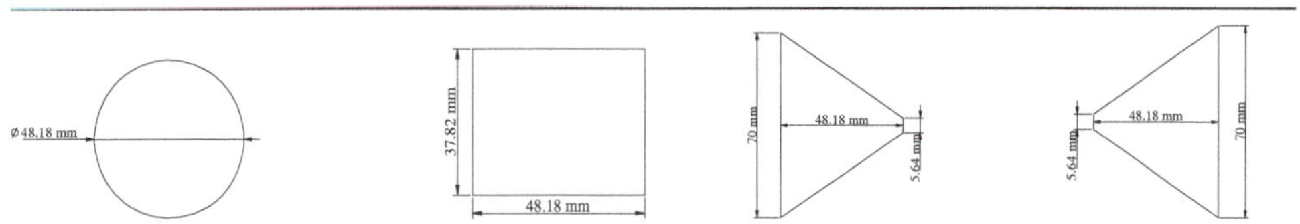

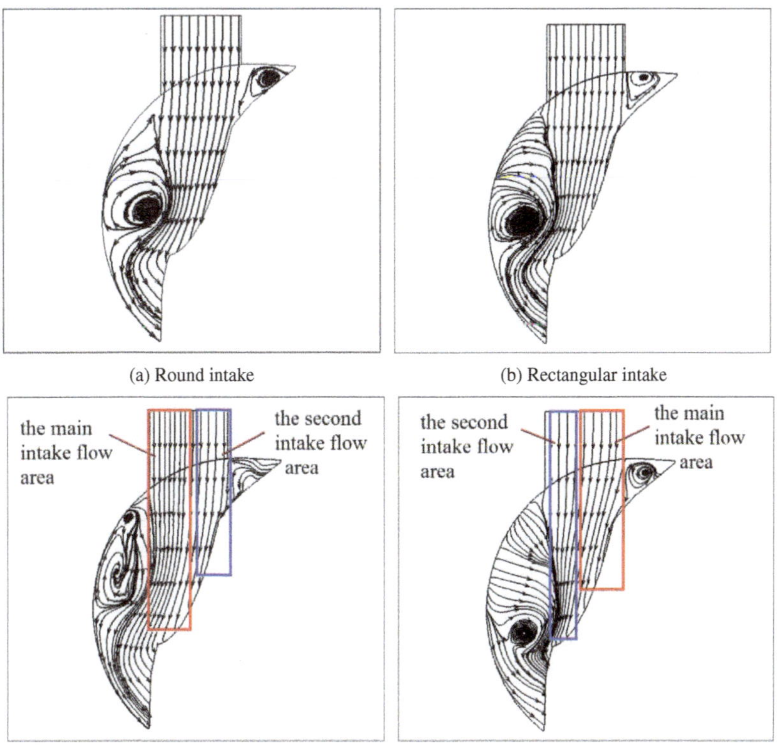

(a) Round intake (b) Rectangular intake

(c) Regular trapezoid intake (d) Inverted trapezoid intake

Figure 13. Streamlines in the rotor housing central plane with different intake shapes at 350° BTDC.

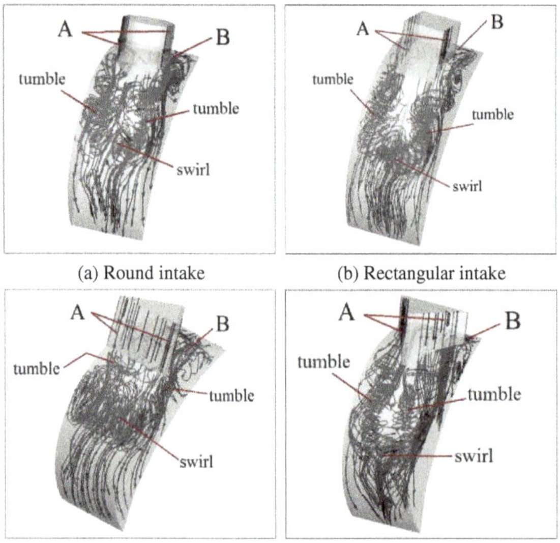

(a) Round intake (b) Rectangular intake

(c) Regular trapezoid intake (d) Inverted trapezoid intake

Figure 14. Streamlines in the combustion chamber with different intake shapes at 350° BTDC.

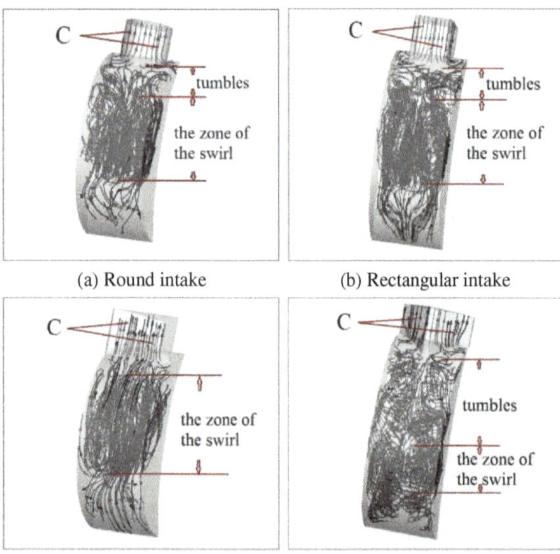

(a) Round intake (b) Rectangular intake

(c) Regular trapezoid intake (d) Inverted trapezoid intake

Figure 15. Streamlines in the combustion chamber with different intake shapes at 250° BTDC.

middle of the combustion chamber, which results in a large developing space for swirls, as shown in Figure 14(d). As such, a swirl is formed at the front of the combustion chamber and the center of the swirl is closer to the rotor wall compared with the round and rectangular intake shapes. During the later stage of the intake stroke, the reverse flow phenomenon occurs in all types of intake shapes. However, different shapes result in different intensities of reverse flow, which decreases as intake shape changes from regular trapezoid to rectangular, round (the original intake), and

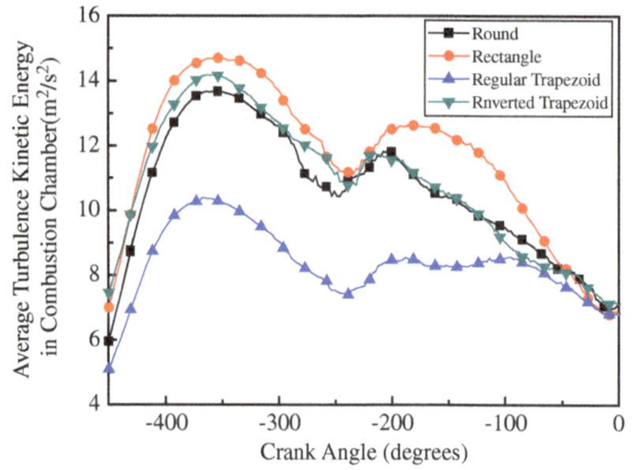

Figure 16. Average turbulence kinetic energy in the combustion chamber with different intake shapes.

inverted trapezoid. Many remnants of the tumbles on top of the swirl for the inverted trapezoid design are observed, whereas no remnant tumbles are observed on top of the swirl for the regular trapezoid intake design.

From the overall changes in the flow field during the intake and compression strokes, we can conclude that the differences in the intake shapes result in different flow distributions when airflow enters the combustion chamber. The effect of reverse flow on the inverted trapezoid intake design is minimal compared with the other intake designs. Moreover, the swirl structure remains the best and swirl breakup is postponed.

The calculated results indicate that the volume coefficients of the four intake shape designs are 0.85293 (inverted trapezoid), 0.82699 (round), 0.81739 (rectangular), and 0.79521 (regular trapezoid). Thus, we can conclude that the volume coefficient of the inverted trapezoid intake design exhibits a 3.1% increase compared with the original intake design (round), primarily because the area and mass of the reverse flow for the inverted trapezoid intake design are minimal compared with those of the other intake designs. Moreover, the volume coefficient of the inverted trapezoid intake has the maximum value. Figure 16 shows the average turbulence kinetic energy in the combustion chamber of the different intake shapes. Evidently, the different intensities of the average turbulence kinetic energy decrease as the intake shape changes from rectangular to inverted trapezoid, round (the original intake), and regular trapezoid. Considering that the effect of reverse flow on the inverted trapezoid intake design is minimal, the intensity of the swirls for the inverted trapezoid intake design is stronger than that of the swirls for the round intake design, which results in a larger average turbulence kinetic energy for the inverted trapezoid intake design than for the round intake design.

4.4. Effects of the intake angle on the flow field

Given that the inverted trapezoid intake design has a better volume coefficient and that the average turbulence kinetic

Table 7. Schematic of different intake angles.

	0°	10°	20°	30°
Schematic of intake angles	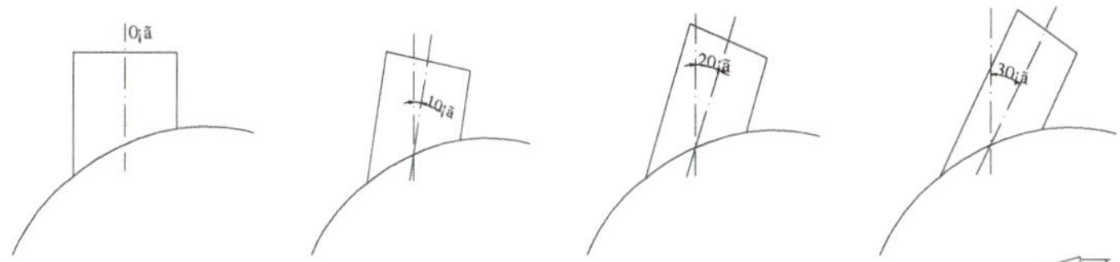			
Schematic of intake	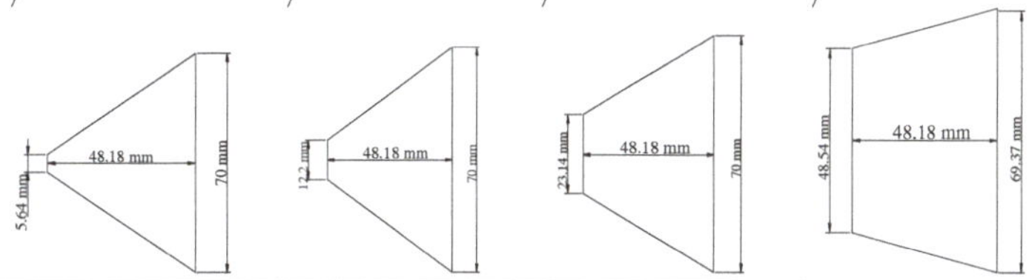			

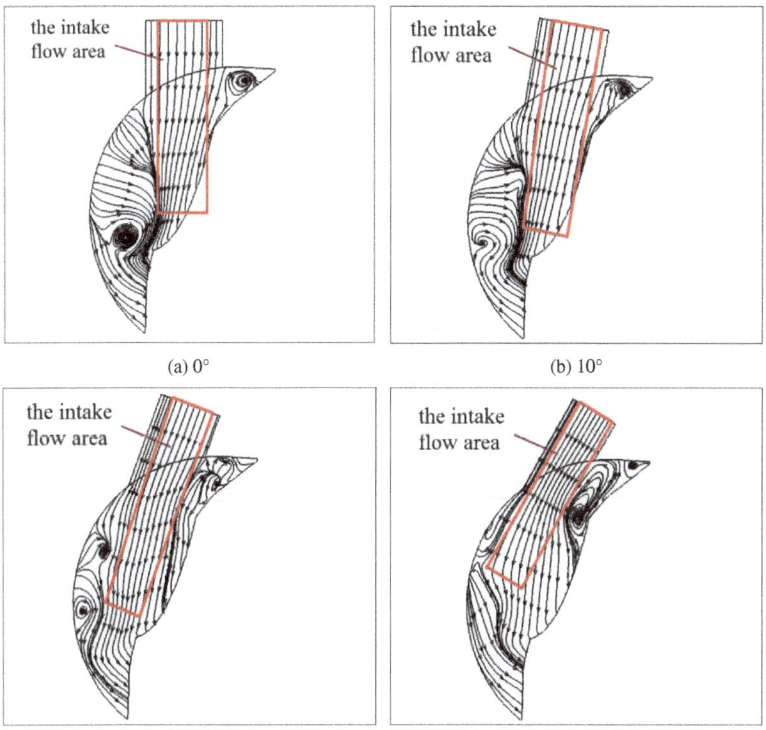

Figure 17. Streamlines in the rotor housing central plane with different intake shapes at 350° BTDC.

energy in the combustion chamber also increases compared with the round intake design (the original intake design), the inverted trapezoid intake design is selected for further optimization. Table 7 shows the shape sketches of different intake angles, namely, 0°, 10°, 20°, and 30°, to analyze the effect of different intake angles on the flow field in the combustion chamber. Meanwhile, the areas of all the intake ports are set to be the same to eliminate the effects of area. As such, the topline lengths of the four types of inverted trapezoid are different.

The changes in the flow field with different intake angles under 4000 rpm conditions are shown in Figures 17 and 18. In the figures, panels (a) to (d) show the streamlines under the intake angles of 0°, 10°, 20°, and 30°, respectively.

With an increase in intake angle, the main flow region of the intake flow shifts away from the rotor wall and toward the combustion chamber wall during the intake stroke stage, as shown in Figures 17(a) to 17(d). Consequently, swirl formation at the front of the

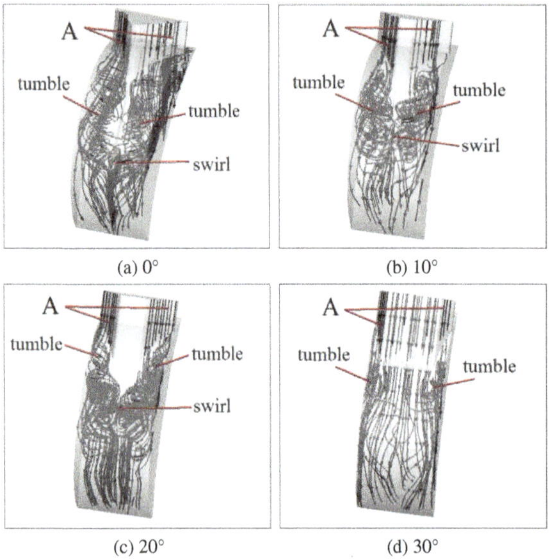

Figure 18. Streamlines in the combustion chamber with different intake angles at 350° BTDC.

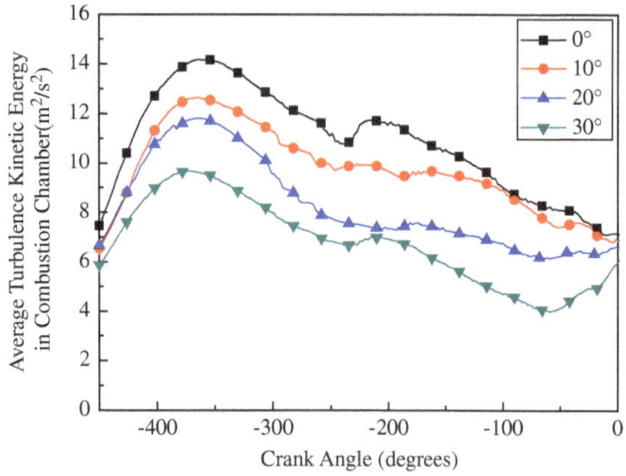

Figure 19. Average turbulence kinetic energy in the combustion chamber with different intake angles.

combustion chamber becomes weaker. The largest vorticity magnitudes of the swirl are 6817, 4196, and 3944.4 s^{-1} for 0°, 10°, and 20° intake, respectively. Moreover, when the intake port angle is 30°, no swirl is formed at all. Figures 18(a) to 18(d) show the main reason for the aforementioned phenomenon. With the increase in intake port angle, more intake air flows directly from the intake port to the front of the combustion chamber. That is, more space in front of the combustion chamber is occupied by intake flow. In addition, the space left for the tumbles to develop decreases. Consequently, with the increase in the intake port angle, the intersection of the two tumbles is constantly pushed toward the intake port. When the intake port angle is 30°, the swirls do not meet one another. From the overall changes in the flow field during the intake stroke, we can conclude that when the flow enters the combustion chamber, various intake port angles result in different flow directions, which leads to obvious differences in the flow field in the combustion chamber. Meanwhile, with the increase in intake port angles, the vorticity magnitudes of the swirl in the combustion chamber are constantly reduced.

The calculated results indicate that the volume coefficients of the four intake angles are 0.8529 (0°), 0.8477 (10°), 0.8422 (20°), and 0.8312 (30°). This finding indicates that intake resistance is enhanced when the intake port has a certain tilt angle. That is, intake resistance increases with the increase in intake port angle, which results in the decrease in volume coefficient. Compared with the intake angle of 0°, a 2.5% decrease is observed in the volume coefficient when the intake angle is 30°. From the average turbulence kinetic energy curves in the combustion chamber of different intake angles shown in Figure 19, the average turbulence kinetic energy in the combustion chamber decreases with the increase in intake port angle. The two reasons for this phenomenon are the decreasing volume coefficient and the increasing restraints of swirl motion.

5. Conclusions

In this study, a detailed mathematical model was integrated into the simulation software FLUENT by including a dynamic mesh and turbulent flow models. After validation by experimental test, further simulations were conducted to investigate the basic flow mechanism and the effect of various parameters on the flow field of a peripheral ported rotary engine. The following conclusions have been obtained:

(1) The basic flow mechanism is that a constantly changing swirl is formed in the combustion chamber during the intake and compression strokes as a result of the combined effects of the pocket of the rotor and the swirls in the combustion chamber. Near the TDC, the swirl breaks into a unidirectional flow because of the significant decrease in combustion chamber volume.

(2) When the swirl formed in the combustion chamber projects onto the rotor housing central plane, a swirl is also formed in the rotor housing central plane. That is the formation reason of the swirl detected in the rotor housing central plane in the experiment conducted by DeFilippis et al. (1992).

(3) With an increase in rotation speed, swirl breakup is postponed as the flow speed and average turbulence kinetic energy in the combustion chamber increase. Moreover, the volumetric coefficient initially increases and then decreases with the increase in rotation speed. The highest point is 16,000 rpm. Compared with the rotating speed of

2000 rpm, a 15.4% increase in the volume coefficient is observed when the rotating speed is 16,000 rpm.

(4) For the inverted trapezoid intake design, the effect of reverse flow is minimal compared with those for the other intake designs. Consequently, the swirl structure remains the best and swirl breakup is postponed. Meanwhile, the average turbulence kinetic energy increases compared with that of the round intake design, and the volume coefficient is at maximum among all intake designs. Compared with the original intake design (round), a 3.1% increase in the volume coefficient is observed.

(5) With an increase in the intake port angle, more space at the front of the combustion chamber is occupied by intake flow. In addition, the space left for the tumbles to develop decreases. Consequently, swirl formation becomes weaker. Moreover, the volume coefficient and the average turbulence kinetic energy in the combustion chamber decrease constantly. Compared with an intake angle of 0°, a 2.5% decrease in the volume coefficient is observed when the intake angle is 30°.

Disclosure statement

No potential conflict of interest was reported by the authors.

Funding

This work was supported by the National Science Foundation of China [51376082]; the Natural Science Grant of Jiangsu Province [BK20131253]; the Jiangsu 'Six Personnel Peak' Talent-Funded Project [2011-ZBZZ-27]; and the Postgraduate Scientific Research and Innovation Project in Jiangsu Province [CXZZ13-0671].

References

Abraham, J., & Bracco, F. V. (1988). Comparisons of computed and measured mean velocity and turbulence intensity in a motored rotary engine (SAE Paper 881602).

Abraham J., & Bracco, F. V. (1989). Fuel-air mixing and distribution in a direct-injection stratified-charge rotary engine (SAE Paper 890329).

Abraham J., & Bracco, F. V. (1991). 3-D computations of premixed-charge natural gas combustion in rotary engines (SAE Paper 910625).

Amrouche, F., Erickson, P., Park, J., & Varnhagen, S. (2014). An experimental investigation of hydrogen-enriched gasoline in a wankel rotary engine. *International Journal of Hydrogen Energy*, *39*, 8525–8534.

ANSYS Inc. (2011). ANSYS Fluent 14.0: theory guide. Canonsburg, PA 1531.

Antonelli, M., & Martorano, L. (2012). A study on the rotary steam engine for distributed generation in small size power plants. *Applied Energy*, *97*, 642–647.

Badr, O., Naik, S., O'Callaghan, P. W., & Probert, S. D. (1991). Wankel engines as steam expanders: Design considerations. *Applied Energy*, *40*(3), 157–170.

DeFilippis, M., Hamady, F., Novak, M., & Schock, H. (1992). Effects of pocket configuration on the flow field in a rotary engine assembly (SAE Paper 920300).

Han, Z., & Reitz, R. D. (1995). Turbulence modeling of internal combustion engines using RNG $k-\varepsilon$ models. *Combustion Science and Technology*, *106*, 267–295.

Hasegawa, Y., & Yamaguchi, K. (1993). An experimental investigation on air-fuel mixture formation inside a low-pressure direct injection stratified charge rotary engine (SAE Paper 930678).

Kagawa, R., Okazaki, S., Somyo, N., & Akagi, Y. (1993). A study of a direct-injection stratified-charge rotary engine for motor vehicle application (SAE Paper 930677).

Muroki, T. (1984). Recent technology development of high-powered rotary engine at mazda (SAE Paper 841017).

Pariotis, E. G., Kosmadakis, G. M., & Rakopoulos, C. D. (2012). Comparative analysis of three simulation models applied on a motored internal combustion engine. *Energy Conversion and Management*, *60*, 45–55.

Shapovalov, V. (1999). The two-stroke rotary diesel engine (SAE Paper 1999–01–2888).

Shih, T. H., Liou, W. W., Shabbir, A., Yang, Z. G., & Zhu, J. (1995). A new $k-\varepsilon$ eddy viscosity model for high reynolds number turbulent flows. *Computers & Fluid*, *24*(3), 227–238.

Shih, T. I. P., Schock, H. J., Nguyen, H. L., & Stegeman, J. D. (1987). Numerical simulation of the flow field in a motored two-dimensional wankel engine. *Journal of Propulsion and Power*, *3*(3), 269–276.

Taghavifar, H., Khalilarya, S., Jafarmadar, S. (2014). Engine structure modifications effect on the flow behavior, combustion, and performance characteristics of DI diesel engine. *Energy Conversion and Management*, *85*, 20–32.

Tashima, S., Taqdokoro, T., Okimoto, H., & Niwa, Y. (1991). Development of sequential twin turbo system for rotary engine (SAE Paper 910624).

Wang, F., Reitz, R. D., Pera, C., Wang, Z., & Wang, J. X. (2013). Application of generalized RNG turbulence model to flow in motored single-cylinder PFI engine. *Engineering Applications of Computational Fluid Mechanics*, *7*(4), 486–495.

Warren, S., & Yang, D. C. H. (2013). Design of rotary engines from the apex seal profile (Abbr.: Rotary engine design by apex seal). *Mechanism and Machine Theory*, *64*, 200–209.

4

Investigation of turbulence model and numerical scheme combinations for practical finite-volume large eddy simulations

Adetokunbo A. Adedoyin[a,b], D. Keith Walters[a,b]* and Shanti Bhushan[b]

[a]*Department of Mechanical Engineering, Mississippi State University, Starkville, USA;* [b]*Center for Advanced Vehicular Systems, Mississippi State University, Starkville, USA*

Large eddy simulation (LES) is known to suffer from two primary error sources – the subgrid stress (SGS) model and the numerical discretization scheme. These cannot be separately quantified for finite-volume numerical simulations, but an appropriate combination can yield 'engineering accurate' prediction of turbulence dynamics. This paper seeks to evaluate combinations of commonly used second-order numerical schemes and Smagorinsky-type SGS models for practical LES. Error assessments are performed for isotropic decaying turbulence using both pseudospectral and finite-volume solvers, followed by validation for a complex turbulent flow of engineering interest. Error assessment using pseudospectral techniques is performed to isolate finite-differencing and modeling errors by explicitly adding numerical derivative error terms to the simulations. Error assessment using the finite-volume method focuses on identification of optimal model-discretization combinations for the best LES predictions using application solvers. The pseudospectral and finite-volume approaches show consistent predicted behavior of the interplay between numerical and modeling errors for different model-discretization combinations. Of those studied here, the two most optimal combinations are identified as: (1) standard or dynamic Smagorinsky SGS model with a bounded central difference scheme; and (2) Monotonic Integrated LES (MILES) with either a second-order upwind or QUICK scheme. These combinations were applied for an axisymmetric jet flow at Re $\sim 10^5$, using an 'engineering quality' mesh, where the MILES model with either an upwind or QUICK scheme showed the best predictive capability.

Keywords: turbulence modeling; large eddy simulation; eddy viscosity; discretization; finite-volume methods

1. Introduction

As available computing resources have continued to increase, large eddy simulation (LES) has grown in popularity among researchers and engineers for use in computational fluid dynamics (CFD) analysis of complex turbulent flows. The use of LES is motivated by a desire to compromise between the efficiency of Reynolds-averaged Navier-Stokes (RANS) and the theoretical accuracy of direct numerical simulation (DNS). While DNS for most high-Reynolds number engineering applications remains decades away, LES (or hybrid RANS/LES variants) are gradually becoming a common approach to problems with complex geometries and flow features in a number of application areas (Hanjalic, 2005). To date, most LES studies can be classified into one of two groups. The first is 'fundamental' studies, in which the subgrid stress (SGS) modeling aspects of LES are carefully studied using advanced simulation methods, such as spectral and pseudospectral techniques, for simple canonical flows. The second is 'application' studies, in which LES is used to solve problems of engineering relevance involving complex geometries and flow features, using readily available numerical methods and models. For example, Shalaby, Wozniak, & Wozniak (2008) performed LES for flow in a gas cyclone separator for industrial applications; Abdalla, Cook, & Hunt (2009) studied the dispersion characteristics of a buoyant plume above a point heat source using LES; and Su, Li, Li, Wei, & Zhao (2012) used LES to study complex vortical flow in a Francis turbine. In these cases, in order to validate the predictive capability of LES, results are usually analyzed by comparison with experimental data and/or other modeling methods (e.g., RANS). In many of these studies, the focus is on obtaining 'good' (i.e., engineering accurate) results using available general-purpose CFD tools. However, unlike their RANS counterparts, a detailed verification analysis is seldom performed to quantify numerical uncertainties and report on the accuracy of the solutions. One reason for the lack of such an analysis for LES is the strong coupling of SGS modeling and discretization errors, as both are functions of grid size.

Several authors have performed parametric studies to identify the best combination of modeling and numerical method, in terms of accuracy and computational cost. Baxevanou and Fidaros (2008) compared predictions using different combinations of k-ϵ and k-ω RANS turbulence models and numerical schemes for transient

* Corresponding author. Email: walters@me.msstate.edu

flow over an ONERA-A airfoil. The study concluded that total variation diminishing (TVD) schemes show faster convergence and more accurate predictions than upwind and central-difference schemes for both turbulence models, and recommended them for engineering applications. Robertson, Choudhury, Walters, & Bhushan (2014) performed a parametric study to identify the most stable and accurate RANS and hybrid RANS/LES turbulence model and numerical scheme combination available in OpenFOAM. The study concluded that a second-order upwind scheme with limiters is the most stable, has the smallest computational cost, and provides the most accurate results for RANS simulations. Among the hybrid RANS/LES models, an improved delayed Detached Eddy Simulation (DES) model with a second-order bounded central difference scheme was identified to be the most accurate. The interplay of SGS modeling and dissipative discretization errors have been intentionally utilized for LES modeling by Boris, Grinstein, Oran, & Kolbe (1992), which are referred to as the monotonically-integrated LES (MILES) approach.

The objective of this paper is to evaluate different combinations of turbulence (SGS) models and numerical schemes for complex turbulent flows. For this purpose, modeling and discretization errors are assessed using both a pseudospectral solver and a commercially-available, general-purpose finite-volume solver (Ansys FLUENT® version 6.3), focusing on errors associated with the most commonly used second-order discretization schemes and Smagorinsky, dynamic Smagorinsky and MILES SGS models. Error assessment using pseudospectral techniques focuses on the evaluation of the effect of finite-differencing and modeling errors. For this purpose the pseudospectral code was modified to explicitly include the effects of finite-differencing errors for several different numerical schemes. This allows modeling and numerical errors to be isolated from one another, which is not possible in actual finite-volume CFD simulations. Error assessment using the finite-volume solver focuses on the identification of optimal combinations of SGS model and numerical discretization scheme for the best LES predictions using application solvers. Error analysis is performed for high-Reynolds-number decaying isotropic turbulence, matching an experimental test case available in the literature (Kang, Chester, & Meneveau, 2003), and previously validated LES predictions (Bhushan & Warsi, 2005). The optimal combinations of SGS model and numerical discretization scheme are then applied to the LES of a high-Reynolds-number Re $\approx 10^5$, axisymmetric turbulent jet flow in order to validate their effectiveness.

2. Error analysis in LES

The governing equations for LES are obtained by filtering the Navier-Stokes equations, which for incompressible flow in divergence (conservative) form are (Pope, 2000, Eq. 2.94):

$$\frac{\partial \bar{u}_j}{\partial x_j} = 0, \tag{1}$$

$$\frac{\partial \bar{u}_i}{\partial t} + \frac{\partial(\bar{u}_i\bar{u}_j)}{\partial x_j} = -\frac{1}{\rho}\frac{\partial \bar{p}}{\partial x_i} + \nu\frac{\partial^2 \bar{u}_i}{\partial x_j \partial x_j} - \frac{\partial \tau_{ij}}{\partial x_j}; \quad i,j = 1,2,3, \tag{2}$$

where, $\bar{u}_i$ is the filtered velocity field, ρ is the fluid density, $\bar{p}$ is the filtered pressure field, ν is the kinetic energy, and τ_{ij} is the subgrid (or subfilter) stress tensor, and

$$\tau_{ij} = \overline{u_i u_j} - \bar{u}_i\bar{u}_j. \tag{3}$$

The subgrid stresses must be modeled for closure of the governing equations. In practice, for LES with finite resolution, filtering may be either explicit or implicit and consists of two operations: convolution of the velocity field with the filter kernel, and projection of the continuous velocity field onto a discrete representation (Carati, Winckelmans, & Jeanmart, 2001). We focus here on finite-volume numerical methods, in which the volume averaging procedure acts as an implicit filter, the net result of which is equivalent to convolution with a box filter kernel and the removal of all scales of motion beyond a cutoff wavenumber $\kappa_c = \pi/\Delta$, where Δ is the grid spacing. The box filter is smooth in spectral space, so that there is a subfilter component associated with all scales of motion, including the subgrid scales with a wavenumber less than κ_c. It has been pointed out that the use of smooth filtering results in some ambiguity regarding the strict definition of the subgrid stresses (De Stefano & Vasilyev, 2002). Herein, we adopt the point of view that the subgrid stresses represent the net effect of all unresolved motion on the resolved flow field. The modeling error may then be defined for any realization of the resolved flow field as the difference between the modeled SGS tensor and the exact SGS tensor, as shown in Figure 1.

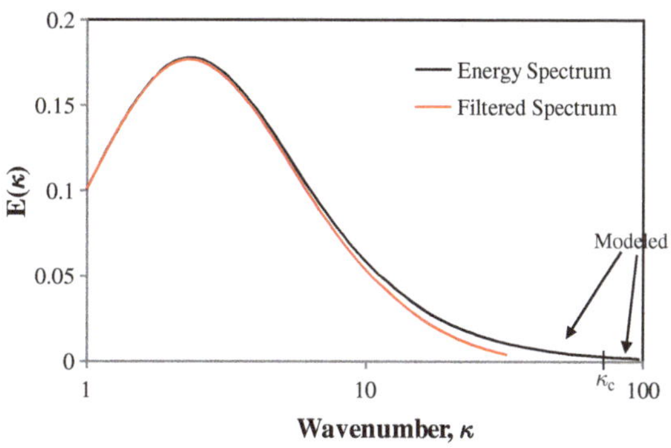

Figure 1. Implicit filtering due to volume averaging, indicating resolved and modeled (SGS) portions of the energy spectrum. The cutoff wavenumber κ_c is dictated by the grid size.

In addition, LES suffers from numerical error due to the discretization scheme used to approximate Eqs. (1) and (2). A formal analysis of discretization errors was presented by Ghosal (1996), in which two contributions were identified: aliasing error due to discrete computation of the nonlinear term in Eq. (2), and finite-differencing error due to the inability of the numerical differentiation operators to represent derivatives exactly, especially at high wavenumbers. The former is well understood for pseudospectral methods and can be minimized using dealiasing techniques (Canuto, Hussaini, Quarteroni, & Zang, 1988; Collis, 2001). In finite-volume LES with second-order discretization schemes, aliasing error is usually associated with the assumption that variable values are uniform over each face of a computational control volume. Finite-differencing errors are associated with the accuracy of the specific discretization scheme, and are principally due to the differentiation of the nonlinear terms (Chow & Moin, 2003; Kravchenko & Moin, 1997). Several studies have found that for LES with sharp cutoff filters, both aliasing and finite-differencing errors are generally larger than the total effect of the SGS model, especially in the high-wavenumber portion of the flowfield (Chow & Moin, 2003; Fedioun, Lardjane, & Gokalp, 2001; Ghosal, 1996). In general, numerical errors are dominated by finite-differencing errors for lower-order schemes and aliasing errors for higher-order schemes.

Park, Yoo, and Choi (2004) found that there exists an optimum degree of upwinding for a scheme of a given order in which the sum of the finite-differencing and aliasing errors is minimized. They also demonstrated that spectral analyses of static error contributions were misleading, and that a dynamic analysis was necessary to accurately assess the combined influence of errors on real LES simulations. Fedioun et al. (2001) pointed out that analysis of aliasing and finite-differencing errors obtained using pseudospectral simulations with modified wavenumbers is in fact not equivalent to the results obtained with actual finite-volume simulations. This is due to the fact that representations in the physical and spectral space do not transform exactly. They developed a modified approach to error analysis using pseudospectral simulation and showed that the two numerical error sources coupled with the modeling error combined to yield good results for the case of a second-order central difference scheme and a fourth-order compact scheme (Kravchenko & Moin, 1997). Their simulations were performed using both convective and divergence forms for the nonlinear term:

$$\frac{\partial(\bar{u}_i\bar{u}_j)}{\partial x_j}\ :\ \text{divergence form;} \qquad (4a)$$

$$\bar{u}_i\frac{\partial \bar{u}_i}{\partial x_j}\ :\ \text{convective form.} \qquad (4b)$$

It was demonstrated that the convective form yielded superior results. We note here that finite-volume codes almost exclusively adopt the divergence form for the nonlinear term in order to satisfy the conservation properties of the numerical scheme. It must also be noted that the results in Fedioun et al. (2001) were obtained using spectral eddy viscosity to model the SGS stresses, following the concept introduced by Kraichnan (1976). Such a model is difficult to implement into finite-volume simulations in physical space, which tend to rely on Smagorinsky-type eddy-viscosity models and are therefore likely to introduce greater amounts of modeling error, particularly in the higher wavenumbers.

This paper extends the previous analyses of modeling and discretization errors in LES to general-purpose finite-volume solvers using readily available numerical schemes and SGS models. The analysis proceeds in two parts. First, a pseudospectral code is used to evaluate the combined influence of SGS model and numerical flux formulation on the dynamical behavior of the LES of decaying isotropic turbulence. For this purpose, the numerical flux formulations use three different finite-differencing schemes: second-order central, second-order upwind (Barth & Jesperson, 1989), and the QUICK scheme (Leonard, 1979). Such relatively low-order formulations are used since these are most commonly employed in general-purpose CFD solvers. The evaluated SGS models include Smagorinsky (SM; Smagorinsky, 1963), dynamic Smagorinsky (DSM; Lilly, 1992) and no-model (denoted NM or MILES).[1] The unmodified pseudospectral results obtained for each SGS model represent reference LES cases. In the second part of the paper, results are obtained for the same cases as above in physical space using the general-purpose finite-volume code Ansys FLUENT®. The results are assessed in terms of the relative strength of the SGS and numerical error contribution to dissipation, and the predicted energy spectra. Finally, suitable combinations of SGS model and numerical discretization scheme are determined, and several combinations are tested for the LES of a high-Reynolds-number, momentum-conserving, axisymmetric turbulent jet.

It must be noted that in LES simulations, the grid size (or filter size) dictates the level of resolved turbulence prediction. Thus for DNS grid resolutions, LES is expected to resolve the entire turbulence scale similar to DNS. The dependence of SGS model on grid size Δ in Eq. (7) ensures that the effect of SGS model (and associated modeling error) decreases asymptotically with grid refinement. Numerical errors show a similar dependence on grid resolution and are expected to asymptotically decrease with grid refinement. Therefore, net error in a LES calculation for a given grid size is due to coupled SGS modeling and discretization errors.

3. Isotropic decaying turbulence test case

The first test case considered here is high-Reynolds-number decaying isotropic turbulence experimentally investigated by Kang et al. (2003). The experiments were performed for spatially decaying turbulence for up to $Re_\lambda = 720$, where λ is the Taylor length scale, using an inlet turbulence generator of size $M = 0.152$ m. In the experiment, the turbulent energy spectra were measured at downstream locations $x/M = 20$, 30, 40 and 48. The average streamwise velocity was $u_1(x) = 12.0$, 11.2, 11.0 and 10.8 m/s at $x/M = 20$, 30, 40 and 48, respectively. As summarized in Table 1, the turbulent characteristics at $x/M = 20$ included: turbulence intensity $u_{1,rms} = 1.85$ m/s; $Re_l = 3.06 \times 10^4$, where ℓ is the integral length scale; $\ell/\eta = 2272$, where η is the dissipation length scale; and $\ell/\lambda = 43$, where λ is the Taylor microscale. The turbulence decays downstream and the characteristics at $x/M = 48$ included: $\ell/\eta = 1844$ and $\ell/\lambda = 37.8$.

3.1. Numerical simulation setup

Temporal LES computations were performed in a periodic cubic box. The box size was selected to be 2π m per side for the pseudospectral simulations, and 2.56 m per side for the finite-volume simulations. Numerical simulations were initialized using the experimental energy spectrum at $x/M = 20$ and results compared against experimental data at $x/M = 30$, 40 and 48. The flow parameters were scaled from the experiment to the numerical simulation, using the length scale ratio

$$l^* = \frac{L_{Exp}}{2\pi} = 0.81487, \tag{5a}$$

where, $L_{Exp} = 5.12$ is the length of the decaying turbulence region reported in the experiment. The velocity scale is based on the turbulence intensity $u_{1,rms}$. The reference velocity in the simulation is chosen to be 2 m/s, which is the upper bound of the turbulence intensity in the experiment. The velocity ratio is:

$$u^* = \frac{u_{1,rms}^{Exp}}{2\ m/s} = 0.9285. \tag{5b}$$

The Reynolds number of the flow is computed based on the turbulence intensity and grid size of the turbulence generator:

$$Re = \frac{1}{u^* l^*} \frac{u_{1,rms}^{Exp} M}{\nu} = 2.47 \times 10^4. \tag{5c}$$

The measurement locations downstream of $x/M = 20$ were converted into simulation times by using Taylor's hypothesis:

$$t = \int_0^x \frac{dx}{u_1(x)}. \tag{5d}$$

Thus, $x/M = 30$, 40 and 48 correspond to $T = 0.1491, 0.3049$ and 0.4318 seconds, respectively.

The initial velocity field for the pseudospectral simulations was specified using random phases following Rogallo

Table 1. Summary of turbulent characteristics in the experiment, and grid resolution used in LES studies available in the literature and in the present study.

Case		Parameter	x/M = 20	x/M = 30	x/M = 40	x/M = 48	*Average*
Experiment (Kang et al. 2003)		ℓ (mm)	250	288	321	332	*298*
		λ (mm)	5.84	7.13	8.25	8.78	*7.5*
		η (mm)	0.11	0.14	0.16	0.18	*0.15*
LES studies in literature	Kang et al. (2003)	L = 5.12 m; $N^3 = 128^3$;	ℓ/h	6.25	7.2	8.025	8.3	*7.45*
		$h = 40$ mm	h/λ	6.84	5.61	4.85	4.56	*5.37*
			h/η	364	286	250	222	*267*
	Hickel et al. (2006) Margolin et al. (2006) Bhushan et al. (2006)	L = 2π m; $N^3 = 64^3$; $h = 98$ mm	ℓ/h	2.55	2.94	3.28	3.39	*3.04*
			h/λ	16.8	13.74	11.88	11.16	*13.07*
			h/η	891	700	612	545	*653*
Present study	Pseudospectral: L = 2π m; $N^3 = 64^3$; $h = 98$ mm		ℓ/h	2.55	2.94	3.28	3.39	*3.04*
			h/λ	16.8	13.74	11.88	11.16	*13.07*
			h/η	891	700	612	545	*653*
	Finite-volume: L = 2.56 m; $N^3 = 64^3$; $h = 40$ mm		ℓ/h	6.25	7.2	8.025	8.3	*7.45*
			h/λ	6.84	5.61	4.85	4.56	*5.37*
			h/η	364	286	250	222	*267*

(1981), with an energy spectrum reported by experiments (Kang et al., 2003) with Eq. (5e):

$$E(\kappa) = C_K \varepsilon^{2/3} \kappa^{-5/3} \left\{ \frac{\kappa \ell}{[(\kappa \ell)^{\alpha_2} + \alpha_1]^{1/\alpha_2}} \right\}^{\frac{5}{3}+\alpha_3} e^{-\alpha_4 \kappa \eta} \times F(\kappa \eta), \tag{5e}$$

where, ϵ, ℓ, and η are the experimentally reported values at the $x/M = 20$ measurement plane. The constants C_K and $\alpha_1 - \alpha_4$ are 1.613, 0.39, 1.2, 4.0 and 2.1, respectively. The factor F represents the bottleneck effect on the energy spectrum, but the simulations in the present work used a sufficiently large filter width such that accurate spectra were represented with $F = 1$. The initial flow field was run for a sufficient length of time until the skewness achieved a steady state and the initial random phases adjusted to a physically representative turbulent state. The velocity field was then rescaled in Fourier space to match the energy spectrum at $x/M = 20$, and the simulation was performed using the adjusted flow field (Kang et al., 2003; Walters & Bhushan, 2005).

For the finite-volume simulations using Ansys FLUENT®, the initial energy spectrum in Eq. (5e) was modified using the spectral box filter function $G(\kappa)$ to include the effect of the box-filtering operator implicit in the finite-volume formulation

$$E_{filt}(\kappa) = G^2(\kappa)E(\kappa) = \left[\frac{sin(\kappa\Delta)}{\kappa\Delta}\right]^2 E(\kappa). \tag{6}$$

The velocity field was specified using 30 discrete wavenumbers spanning length scales ranging from the Kolmogorov scale η to 10ℓ. The wavenumbers were logarithmically equispaced with a ratio of 1.122. The velocity field was specified with 16 modes per wavenumber.

Pseudospectral simulations were performed using 32^3 and 64^3 grid resolutions. The averaged grid resolution characteristics for the 32^3 case are: $\ell/h = 1.5$, $h/\lambda = 26$ and $h/\eta = 1306$. The grid characteristics of the 64^3 case are: $\ell/h = 3$, $h/\lambda = 13$ and $h/\eta = 653$, as summarized in Table 1. Both the grids have resolution $h/\eta < \ell/\eta$ – however, for the coarse grid, the grid resolution is barely fine enough to resolve the largest energy-containing length scales. Therefore, this study focuses mostly on the finer 64^3 grid. This grid is sufficiently fine such that the grid scale is located in the inertial subrange. Thus it is expected that the simulations will resolve the largest energy-containing eddies. However, considering large h/λ values, the grid is not fine enough to resolve all of the inertial subrange turbulent structures. Since the objective of the study is to investigate the best practices for practical LES applications for engineering problems, a grid resolution fine enough to resolve fine-scale turbulence is not warranted. Note that the 64^3 grid resolution has been successfully used by other pseudospectral studies for this particular test case with a variety of subgrid closures. For example, Hickel, Adams, and Domaradzki (2006) used the same grid for the validation of an adaptive local deconvolution method for LES. Margolin, Rider, and Grinstein (2006) used this grid for the validation of the implicit LES methodology, and reported that their results on the 64^3 grid are consistent with the 128^3 results reported in the literature. Bhushan, Warsi, and Walters (2006) used the grid for the validation of an algebraic model for LES. The study also included LES simulation using a dynamic Smagorinsky model.

Finite-volume simulations were performed using a 64^3 grid for which averaged $\ell/h = 7.45$, $h/\lambda = 5.4$ and $h/\eta = 267$. The grid resolution is same as that used by Kang et al. (2003) for their LES study. Note that we have considered a finer grid for finite-volume simulations compared to pseudospectral simulations, as the former is expected to have higher numerical error than the latter.

The time step for the simulation was $\Delta t = 0.00074$ and 0.00037 for the 32^3 and 64^3 grid cases, respectively. The time step size is an order of magnitude smaller than the time scale of the resolved turbulence eddies. It is expected that time discretization will have negligible impact on results.

The subgrid stresses were modeled as:

$$\tau_{ij} = -2C_S\Delta^2\sqrt{2S_{ij}S_{ij}}\,S_{ij}, \tag{7}$$

where, C_S is a model coefficient, Δ is the grid size, and $S_{ij} = \frac{1}{2}\left[\frac{\partial \bar{u}_i}{\partial x_j} + \frac{\partial \bar{u}_j}{\partial x_i}\right]$ is the rate-of-strain tensor. The study uses both SM and DSM models. For SM, $C_S = 0.03$ and 0.01 for isotropic decaying turbulence and free-shear jet flows, respectively (Bhushan & Warsi, 2007; Pope, 2000). For DSM, $C_S\Delta^2$ is computed dynamically at every time step for the domain, using explicit filtering and using the scale-similarity assumption (Lilly, 1992). Simulations were also performed without any turbulence modeling, i.e., $\tau_{ij} = 0$. For the pseudospectral simulations, these simulations are referred to as no-model (NM). The finite-volume simulations include dissipative finite-differencing errors, therefore such cases correspond to MILES (Boris et al., 1992), wherein the numerical dissipation replaces subgrid dissipation.

3.2. *Calculation of energy spectra*

To compute the energy spectra from the LES results, a fast Fourier transform (FFT) was performed for the velocity field (not required for the pseudospectral code) to obtain the Fourier components of the velocity ($\hat{u}_j$) in 3D wavenumber space, $\kappa_j = [-N/2 \ldots 0 \ldots N/2 - 1]$, where j = 1,2,3 correspond to the three spatial dimensions and N is the number of grid points in each direction of the domain. The data is then organized in the wavenumber space, such that wavenumbers $\kappa_1 = 0$, $\kappa_2 = 0$ and $\kappa_3 = 0$ lie at the center. A spherical domain is defined with the radius $\kappa = \sqrt{\kappa_j\kappa_j} = [0, N/2 - 1]$ and the velocity field is interpolated onto the surface of the domain. Note that repeated subscript notation represents summation. The energy spectrum is then obtained by numerical integration on the surface of the domain (ds) as below:

$$E(\kappa) = \frac{1}{4\pi\kappa^2} \iint \hat{u}_j \hat{u}_j^* ds, \tag{8}$$

where * denotes the conjugate component of the complex variable.

4. Pseudospectral simulation methodology

4.1. General methodology

The Fourier-Galerkin approximations of Eqs. (1) and (2) are (Bhushan & Warsi, 2007):

$$i\kappa_j \hat{u}_j(\boldsymbol{\kappa}) = 0, \tag{9a}$$

$$\left(\frac{d}{dt} + \nu\kappa^2\right) \hat{u}_i(\boldsymbol{\kappa}) = -i\kappa_i \frac{1}{\rho}\hat{p}(\boldsymbol{\kappa}) - \widehat{u_j u_{i,j}}(\boldsymbol{\kappa}) + \widehat{\tau_{ij,j}}(\boldsymbol{\kappa}), \tag{9b}$$

where $\hat{u}_i$ and $\hat{p}$ represent the Fourier coefficients of the velocity and pressure as functions of the discrete wavenumber vector $\boldsymbol{\kappa}$. The second term on the right in Eq. (9b),

$$-\widehat{u_j u_{i,j}}(\boldsymbol{\kappa}) = \widehat{\tau_{ij,j}}(\boldsymbol{\kappa}) = \hat{f}_i(\boldsymbol{\kappa}), \tag{10}$$

is the spectral representation of the nonlinear convection term. The pressure term in Eq. (7) can be eliminated by substituting Eq. (8) (the continuity equation) into Eq. (9) as

$$\hat{p}(\boldsymbol{\kappa}) = -i\kappa_j \hat{f}_j(\boldsymbol{\kappa})/\kappa^2. \tag{11}$$

The resulting ordinary differential equation that is integrated in time in the pseudospectral simulations is therefore

$$\frac{d}{dt}\hat{u}_i(\boldsymbol{\kappa}) = \hat{f}_i(\boldsymbol{\kappa}) - \kappa_i\kappa_j \hat{f}_j(\boldsymbol{\kappa})/\kappa^2 - \nu\kappa^2 \hat{u}_i(\boldsymbol{\kappa}). \tag{12}$$

The local time derivative is advanced by a second-order Runge-Kutta scheme. The nonlinear term and the SGS term on the right-hand side are determined by using the inverse FFT to compute the velocity field, evaluating the terms in physical space, and taking the FFT to yield their spectral representation. Since the governing equations are solved in wavenumber space, the solver involves (implicit) sharp spectral filtering. A 3/2-rule was employed to minimize aliasing errors (Collis, 2001), such that finite-difference errors could be isolated and studied.

4.2. Approximation of finite-differencing errors

Approximation of finite-differencing errors is accomplished by adding an error term to the spectral representation of the Navier-Stokes equations. In order to avoid any difficulties associated with the spectral representation of finite-difference errors, an alternative to the modified wavenumber method is employed, in which the error terms are evaluated using a Taylor series expansion in physical space and their spectral contribution is computed using the FFT.

The numerical errors that are investigated are those due to the numerical approximation of the derivative appearing in the convection term, which are expected to be the dominant contributing factor to the total numerical error (Fedioun et al., 2001). Defining the operator $\frac{\delta}{\delta x_j}$ as the difference between the numerical (denoted by ^{DISC}) and exact derivative (denoted by ^{EXACT}),

$$\frac{\delta}{\delta x_j} = \frac{\partial}{\partial x_j}^{DISC} - \frac{\partial}{\partial x_j}^{EXACT}, \tag{13}$$

the numerical error in the convective term is

$$e_i = \bar{u}_j \frac{\delta \bar{u}_i}{\delta x_j}. \tag{14a}$$

The differentiation error operator in Eq. (13) can be constructed from the truncated terms in the Taylor series expansion of the different finite-differencing schemes, which results in:

Central differencing(CD):

$$e_i = \bar{u}_j \frac{\Delta^2}{6} \frac{\partial^3 \bar{u}_i}{\partial x_j{}^3} + \bar{u}_j \frac{\Delta^4}{120} \frac{\partial^5 \bar{u}_i}{\partial x_j{}^5} + \bar{u}_j \frac{\Delta^6}{5040} \frac{\partial^7 \bar{u}_i}{\partial x_j{}^7} + \ldots \tag{14b}$$

Second – OrderUpwind(2U):

$$e_i = -\bar{u}_j \frac{\Delta^2}{3} \frac{\partial^3 \bar{u}_i}{\partial x_j{}^3} + |\bar{u}_j| \frac{\Delta^3}{4} \frac{\partial^4 \bar{u}_i}{\partial x_j{}^4} - \bar{u}_j \frac{7\Delta^4}{60} \frac{\partial^5 \bar{u}_i}{\partial x_j{}^5} + \ldots \tag{14c}$$

QUICK(Q):

$$e_i = -\bar{u}_j \frac{13\Delta^2}{48} \frac{\partial^3 \bar{u}_i}{\partial x_j{}^3} + |\bar{u}_j| \frac{7\Delta^3}{32} \frac{\partial^4 \bar{u}_i}{\partial x_j{}^4} - \bar{u}_j \frac{97\Delta^4}{760} \frac{\partial^5 \bar{u}_i}{\partial x_j{}^5} + \ldots \tag{14d}$$

It must be noted that in order to preserve the upwinding nature of the two 2U and Q schemes, the errors involving even derivative terms are multiplied by the absolute value of the velocity, $|\bar{u}_j|$, as derived in Appendix 1. An exact representation of the error would require the retention of all of the truncated terms, which is both impossible and unnecessary. As discussed later, numerical experiments were performed to determine the minimum number of terms that need to be retained in order to accurately approximate the numerical error.

The numerical errors computed using Eqs. (14) are added in Eq. (12) to obtain the modified LES equations:

$$\frac{d}{dt}\hat{u}_i(\boldsymbol{\kappa}) = \hat{f}_i(\boldsymbol{\kappa}) - \kappa_i\kappa_j \{\hat{f}_j(\boldsymbol{\kappa}) + \hat{e}_i(\boldsymbol{\kappa})\}/\kappa^2 - \nu\kappa^2 \hat{u}_i(\boldsymbol{\kappa}) + \hat{e}_i(\boldsymbol{\kappa}), \tag{15}$$

where $\hat{e}_i(\kappa)$ is the spectral representation of the numerical error in physical space given by $e_i(x)$. It must be noted that the calculation of higher-order derivative terms is simple in the pseudospectral solver, as it reduces to a simple arithmetic operation, i.e., FFT of $\left[\frac{\partial^n \bar{u}}{\partial x^n}\right] = (i\kappa)^n \hat{u}$.

5. Finite-volume simulation methodology

Finite-volume simulations were performed with the commercial flow solver Ansys FLUENT®. The solution algorithm was fully implicit and the time-dependent term was discretized using a second-order backward difference approximation. The simulations used the segregated solver, with which continuity and momentum equations were solved in a decoupled fashion during the outer iteration loop. The time advancement was carried out using non-iterative time advancement (NITA), i.e., a single outer iteration was performed during each time step, and the fractional step method was used for pressure-velocity coupling (Glaz, Bell, & Colella, 1993). The dimensionless time step size was the same as in the pseudospectral simulations. The efficacy of the NITA scheme was evaluated by comparison with an iterative (implicit) time advancement scheme that incorporated the SIMPLE algorithm (Patankar & Spalding, 1972) for pressure-velocity coupling and 25 outer iterations per time step. The results using both methods were found to be identical, but the NITA scheme reduced the computational time by an order of magnitude versus the iterative SIMPLE scheme.

Second-order extrapolation was used to evaluate the pressure at the control volume faces, and convective fluxes were evaluated at each face using several different spatial discretization schemes. These included the second-order central difference (CD), upwind (2U), QUICK (Q) and bounded central difference (BCD) schemes. The first-order upwind scheme (1U) was also used for comparison purposes. The BCD scheme is an adaptive scheme that represents a mixture of second-order central and upwind differencing, and first-order upwind differencing when the convection boundedness criterion is violated. This scheme potentially approximates a near-optimum scheme defined by Park et al. (2004), which includes just the right amount of dissipation to minimize the sum of aliasing and finite-differencing errors. SM, DSM and MILES were used for SGS modeling.

6. Isotropic decaying turbulence results

6.1. *Pseudospectral simulations*

The first necessary step in the error analysis was to determine the appropriate number of terms that need to be retained in the Taylor series expansion of the numerical discretization in Eqs. (14). To do this, simulations were conducted using a successively larger number of terms to represent the numerical differentiation error, i.e., SM-3 retains only 3rd derivative term, SM-34 retains 3rd and 4th derivative terms and SM-345 retains up to 5th derivative term. Figure 2 shows that retention of terms up to fifth-order error for the 2U scheme in Eq. (14c) does not show significant change in the computed energy density spectrum compared to retention of up to fourth-order error terms. It is expected that retention of even higher order terms will not lead to significant changes. Similar results were obtained for the CD and Q schemes, for which retention of seventh- and fifth-order error terms, respectively, did not show significant changes in the energy density spectrum. As a result, Eq. (14b) was truncated after the seventh-order derivative term and Eqs. (14b) and (14d) were truncated after the fifth-order derivative term. The naming convection used in the remainder of this section is to state the SGS model label first, followed by the (approximated) numerical discretization scheme label second. The absence of a second label indicates the unmodified pseudospectral result. For example, SM denotes pure pseudospectral LES with the constant-coefficient Smagorinsky model, while DSM-CD implies pseudospectral LES with the dynamic Smagorinsky model and added error contribution from the CD numerical flux formulation.

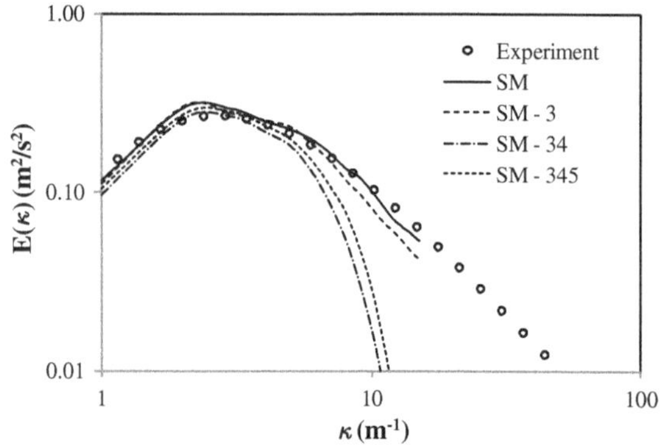

Figure 2. Pseudospectral results for isotropic decaying turbulence with successively higher-order approximations of the numerical error term for the 32^3 case, using the Smagorinsky model and second-order upwind scheme (2U), are compared with the experimental data at $x/M = 48$.

Figure 3(a) shows results on the 32^3 grid using the Smagorinsky SGS model in conjunction with numerical error from all three numerical schemes. The results highlight the dissipative nature of the upwind biased schemes. This effect is most apparent at higher wavenumbers, where the results clearly indicate that there is an excessive removal of energy and underprediction of the experimental data. SM-CD shows a pile up of energy at high wavenumbers, as the dispersive error of the CD scheme dominates the dissipative influence of the SGS model. It is clear from Eq. (14b) that the error term for the CD scheme is symmetric and does not remove kinetic energy from the system. The 64^3 grid predictions in Figure 3(b) are similar to those of the 32^3 grid for all of the schemes. The

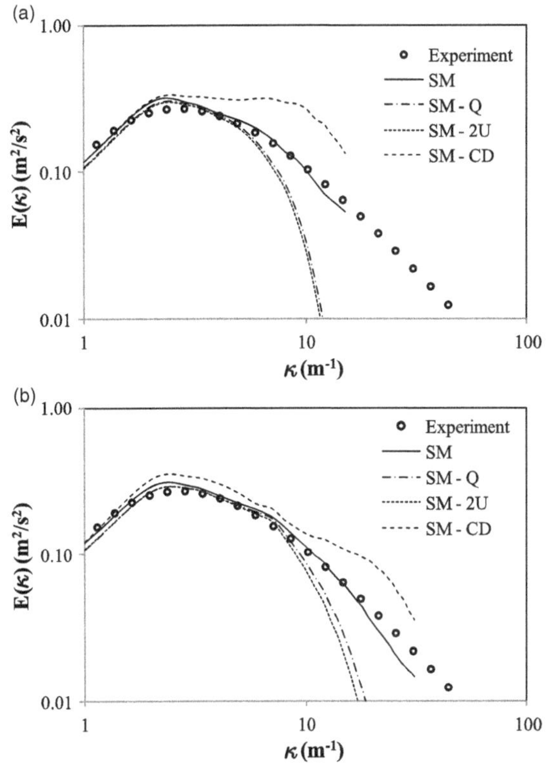

Figure 3. Pseudospectral results for isotropic decaying turbulence using the Smagorinsky model with numerical error from QUICK and second-order upwind and central-differencing schemes for (a) a 32^3 grid and (b) a 64^3 grid, are compared with pure LES and experimental data at $x/M = 48$.

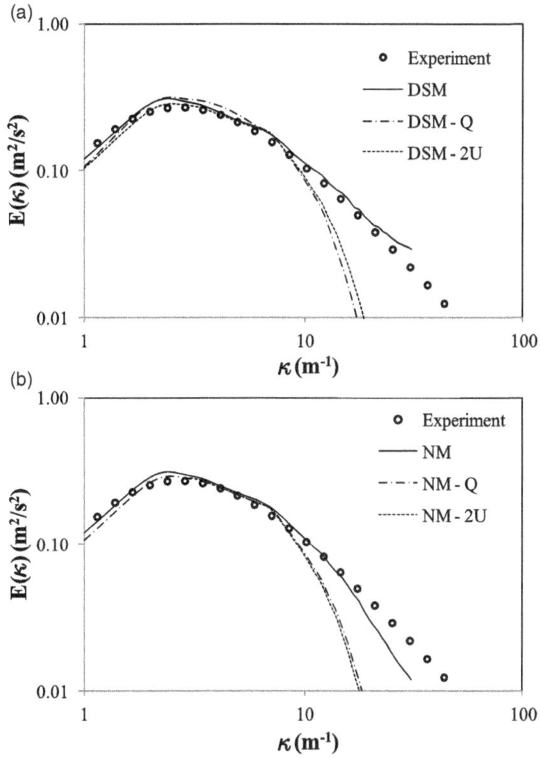

Figure 4. Pseudospectral results for isotropic decaying turbulence using (a) dynamic Smagorinsky and (b) no-model with numerical error from QUICK, second-order upwind and central-differencing schemes for 64^3 grid, are compared with pure LES and experimental data at $x/M = 48$.

DSM model with upwind biased schemes, as shown in Figure 4(a), shows a trend similar to the SM model. It should be noted that the DSM-CD solution diverged due to excess dispersion and non-physical energy pile up at high wavenumbers, which clearly demonstrates the nonlinear interaction between SGS models and numerical schemes. The eddy-viscosity coefficient in the DSM is dependent on the resolved strain-rate rather than on a fixed coefficient, hence the dissipative influence of this model is not sufficient to stabilize the simulation against the effects of the CD scheme. The NM simulations in Figure 4(b) showed predictions similar to SM and DSM for the upwind scheme, and diverged for the CD scheme, similar to DSM predictions.

Figures 3 and 4 show that NM simulations predict lower energy content than the SM and DSM cases, suggesting that NM has higher dissipation than SM or DSM. This is in contrast to the expectation that turbulence models will result in additional dissipation over NM. It must be noted that the results shown in Figures 3 and 4 are only at an instant, and do not show the turbulent kinetic energy (TKE) decay predicted by different models. Analysis of the TKE decay pattern reveals that NM simulations show excessive decay of TKE early on in the simulations, and show lower decay than SM and DSM later on in the simulations. The early excessive decay results in a lower TKE profile overall than SM and DSM. The early excessive dissipation in NM is likely due to inaccurate development of small-scale turbulence structures. However, it is apparent that the discretization scheme plays a more significant role than the SGS model in determining the predicted energy spectrum, particularly in the high-wavenumber regions. This suggests that the choice of SGS model may be of minor importance in 'application' LES simulations using general-purpose solvers with common second-order schemes. It also suggests that conclusions regarding model performance obtained using high-fidelity (e.g., pseudospectral) solvers in 'fundamental' LES studies will not necessarily translate to practical LES, and that care must be taken to ensure that any SGS model is not simply adding extra dissipation in an ad hoc fashion. Finally, it should be noted that these results do not reflect the influence of solid boundaries in wall-bounded flows. In such a case, it is expected that the combination of grid and SGS model will in fact play a critical role, but detailed investigation of such situations is beyond the scope of this work.

6.2. *Finite-volume simulations*

Results analogous to those above are presented in this section, obtained using the general-purpose finite-volume solver. Energy spectra corresponding to the downstream-most measurement station in the experiments are shown in Figures 5(a) and 5(b) for the SM and DSM simulations, respectively. As is evident, the BCD and CD scheme predictions are similar and show good agreement with experiments at the low wavenumbers. However, the former is underpredictive and the latter overpredictive at larger wavenumbers. Results with the 2U, Q, and 1U schemes show a substantial amount of dissipation, as expected. The 1U scheme is overly dissipative even at low wavenumbers, and the 2U and Q schemes provide similar energy spectra to each other. The energy spectrum obtained using MILES in Figure 5(c) shows the best agreement with experiments using either the BCD or Q schemes, although there is a slight overprediction of the energy in the lower wavenumbers with the Q scheme. The MILES-Q result is arguably as good as the DSM-BCD result in terms of the energy spectrum, although its formulation is much simpler both in terms of the numerical method and the SGS model. Interestingly, the MILES-BCD results show the best agreement with experiments over the entire energy spectrum. The MILES-CD result, not surprisingly, is highly dispersive and nonphysical. Overall, the finite-volume simulation results showed similar behavior to the pseudospectral simulations, including the dissipative nature of the upwind biased schemes and energy pile up in the high wavenumbers for the central difference scheme. In particular, the observation above, that numerical scheme plays a greater role in determining the predicted spectral behavior than the SGS model, is supported in the finite-volume results.

Results from the finite-volume simulations were analyzed to evaluate the net dissipative contribution from each of three sources – the discretization scheme, the SGS model, and the viscous stress terms. The total dissipation was estimated based on the time derivative of turbulent kinetic energy, dk/dt, at each time step. The SGS and viscous dissipation were estimated based on their volume averages within the domain at each time step, i.e.:

$$\text{SGS Dissipation} : \varepsilon_{SGS} = \oint \nu_T (2S_{ij}S_{ij})^2 dV; \quad (16a)$$

$$\text{Viscous Dissipation} : \varepsilon_{visc} = \oint \nu (2S_{ij}S_{ij})^2 dV. \quad (16b)$$

The numerical dissipation was calculated as the difference between the total and the SGS + viscous dissipation:

$$\text{Numerical Dissipation} : \varepsilon_{Num} = \frac{d}{dt} \oint k dV - \varepsilon_{SGS} - \varepsilon_{visc}. \quad (16c)$$

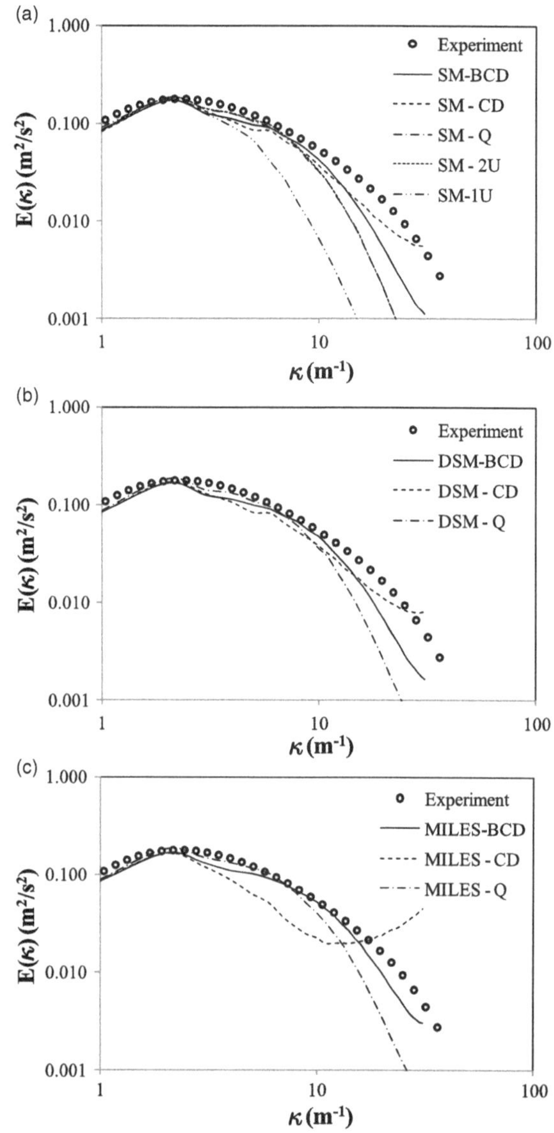

Figure 5. Finite-volume simulation energy spectra obtained for isotropic decaying turbulence using (a) Smagorinsky model, (b) dynamic Smagorinsky model and (c) MILES with different numerical schemes, are compared with experimental data at $x/M = 48$.

The computed viscous dissipation was found to be negligible for this case, which is expected since the mesh size is much larger than the Kolmogorov length scale.

The relative contribution of the numerical error from four of the discretization schemes, for simulations using the Smagorinsky model, is shown in Figure 6. It is clear that the more dissipative the scheme, the higher the contribution of numerical dissipation. Similar results were found for the cases using DSM. For all of the MILES simulations, of course, almost 100% of the dissipation arises due to the

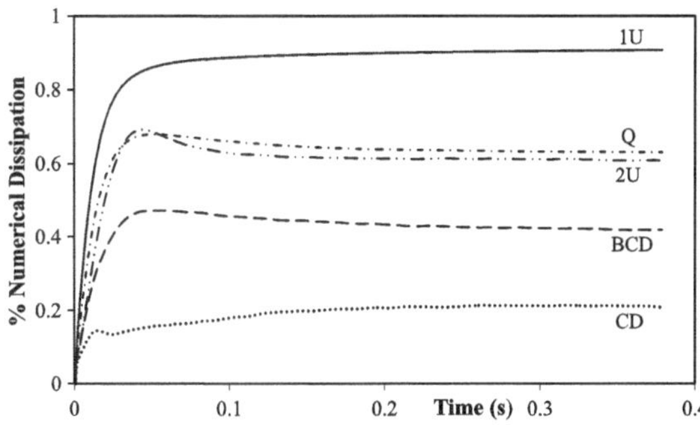

Figure 6. Relative contribution of numerical error to total dissipation for different numerical schemes in combination with the Smagorinsky model in finite-volume simulations.

numerical scheme. What is most relevant from Figure 6 is the fact that two of the schemes, CD and BCD, have smaller contributions to the overall dissipation from the numerical scheme than from the SGS model. For the Q scheme, approximately 65% of the dissipation arises due to numerical error, and only about 35% from the SGS model. The 2U scheme shows dissipation characteristics similar to the Q scheme, and results in only a net 3.5% higher numerical dissipation than the Q scheme. For the 1U scheme, not surprisingly, almost all of the dissipation is due to the numerical error. The results again confirm that for upwind-biased second-order schemes, the numerical error dominates over the contribution of the SGS model. Even for a less dissipative scheme such as BCD, the numerical contribution is almost equal to the SGS model contribution.

The predictive capability of the different combinations of SGS models and numerical discretization schemes are evaluated by estimating error in the prediction of energy spectra at T = 0.13, 0.27 and 0.38 (which corresponds to experimental data at x/M = 30, 40 and 48, respectively) using experimental data, and summarized in Table 2. For both SM and DSM models, the CD scheme shows the smallest error, due to overprediction of the energy spectrum at higher wavenumbers and underprediction for lower wavenumbers. BCD shows a 3% larger error than CD, whereas the 2U and Q schemes show up to 10% larger errors than the CD scheme. The 1U scheme shows a significantly larger error of 49%. MILES using BCD provides the best prediction among the model-scheme combinations, for which the error is 2.73%. MILES using CD predicts errors of 34%, whereas errors predicted using the 2U and Q models are nominally less than 10%.

Overall, the analysis indicates that the use of an upwind-biased discretization scheme (either 2U or Q) enables reasonably accurate resolution of low wavenumber velocity modes, but high wavenumber modes cannot be resolved due to high levels of numerical dissipation. This result is of course not surprising. The central-difference scheme shows non-physical behavior in the energy spectrum and significant overprediction of the decay rate regardless of the SGS model. The BCD model shows the most accurate spectral resolution for all of the models.

Table 2. Error in turbulent kinetic energy decay prediction for different combinations of turbulence models and numerical schemes for isotropic decaying turbulence simulations using the finite-volume solver Ansys FLUENT®.

	Turbulence model		
Numerical Scheme	SM (***E***)*	DSM (***E***)	MILES (***E***)
BCD	16.88%	11.64%	– 2.73%
CD	13.41%	8.32%	– 34.09%
QUICK	22.07%	19.06%	7.61%
2U	26.56%	22.38%	9.32%
1U	49.03%	–	–

Note: *Error (E) = (Experiment – CFD)/Experiment × 100.

Often, for engineering-level simulations, accurate resolution of high wavenumber velocity components is less critical than the accurate prediction of mean flow and large-eddy characteristics. Exceptions, for example acoustics simulations, generally require high-order/low-dissipation numerical schemes for accurate results. For more general flow analyses, these results indicate that the use of a second-order upwind scheme may in fact provide 'engineering accurate' solutions. Interestingly, for the decaying turbulence test case, the upwind-biased schemes apparently provide enough dissipation such that the addition of an SGS model is redundant, and may even decrease accuracy. This point underscores a more general observation, namely that conclusions drawn from fundamental LES studies regarding SGS model performance are often based on high-resolution (e.g., pseudospectral) numerical methods. Care must be taken when translating model performance characteristics to simulations using lower-resolution numerical methods, to ensure that SGS models are not merely adding extra dissipation to an already sufficiently dissipative solution.

7. Engineering application: round jet flow

To evaluate the above observations using a test problem more indicative of realistic engineering applications, simulations were performed using Ansys FLUENT® for a turbulent round jet flow issuing into a quiescent fluid. Three model-scheme combinations were selected for comparison, based on the results from the isotropic decaying turbulence simulations. These included two with the BCD scheme, which represents an advanced numerical method developed specifically for high-resolution simulations. The BCD scheme was combined with the DSM model – considered the most advanced of the eddy-viscosity SGS models investigated in this study – as well as with the SM model, which is a simpler and older model form but which showed better prediction of turbulence decay when paired with the BCD scheme. The third combination was

MILES-2U, which showed the best results in terms of the low wavenumber portion of the turbulence spectrum, and was found to be more stable than the MILES-BCD combination.

7.1. *Test case*

Simulations were performed to match the axisymmetric jet experiment of Hussein, Capp, & George (1994). The case included a relatively high Reynolds number (Re $\approx 10^5$), and a Mach number equal to approximately 0.2. The experiment was performed in a facility large enough to be considered an infinite environment, and recirculation flow entrainment was avoided. The data were collected using laser doppler anemometer (LDA), single hot wire (SHW), and flying hot wire (FHW) techniques, and included: (1) mean centerline velocity at several downstream locations; (2) profiles of the mean streamwise velocity, and streamwise, radial and azumithal turbulence intensities versus non-dimensional radial distance $\eta = \frac{r}{x-x_o}$, where x_o is the jet origin, from the jet centerline at $50D$, $70D$ and $90D$ downstream of the jet; and (3) estimate of energy balance, and pressure-velocity and pressure-strain rate correlations. Herein, the centerline velocity and turbulence intensity data are compared to simulation results for validation purposes.

7.2. *Simulation setup*

Initial tests were performed to obtain an economical grid for the simulations, which involved evaluation of the appropriate outer domain location such that the boundary

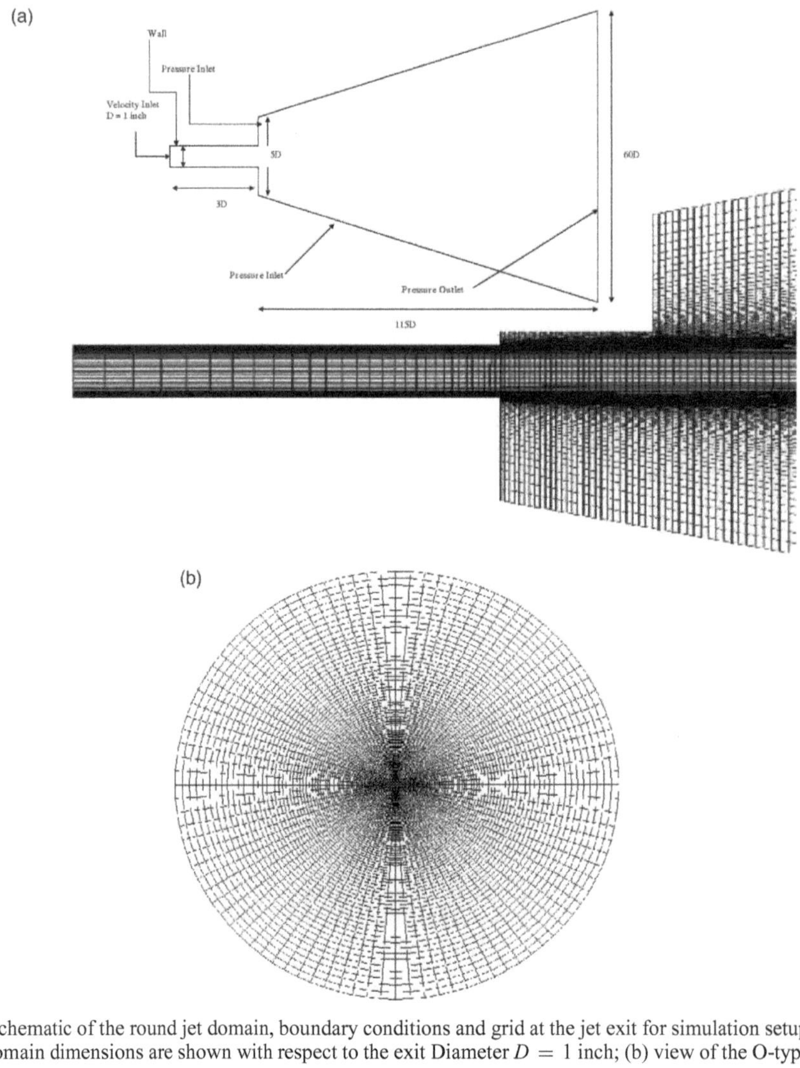

Figure 7. (a) 2D schematic of the round jet domain, boundary conditions and grid at the jet exit for simulation setup (following Hussein et al., 1994). The domain dimensions are shown with respect to the exit Diameter $D = 1$ inch; (b) view of the O-type grid used along the azimuthal direction.

condition does not affect the flow (Adedoyin, 2007). For this purpose, a cylindrical domain of radius 60D and a smaller conical domain as shown in Figure 7(a) were considered. Simulations were performed on these domains using a coarse 1M grid, which showed no significant effect of outer domain on the results. Thus, in order to minimize the numerical expense, the smaller conical domain – as shown in Figure 7(a) – was used for the study. Since the objective of this study is to evaluate the interplay of modeling-scheme errors for engineering applications, and not to obtain the best results for the jet flow, a reasonably fine grid resolution – comparable to that used in engineering applications – was sought. Therefore, the grid was refined to obtain a grid consisting of approximately 3M cells. The grid design for the domain is shown in Figure 7(b).

The simulation domain in Figure 7(a) was built with an initial (upstream) diameter 5 times larger than the jet

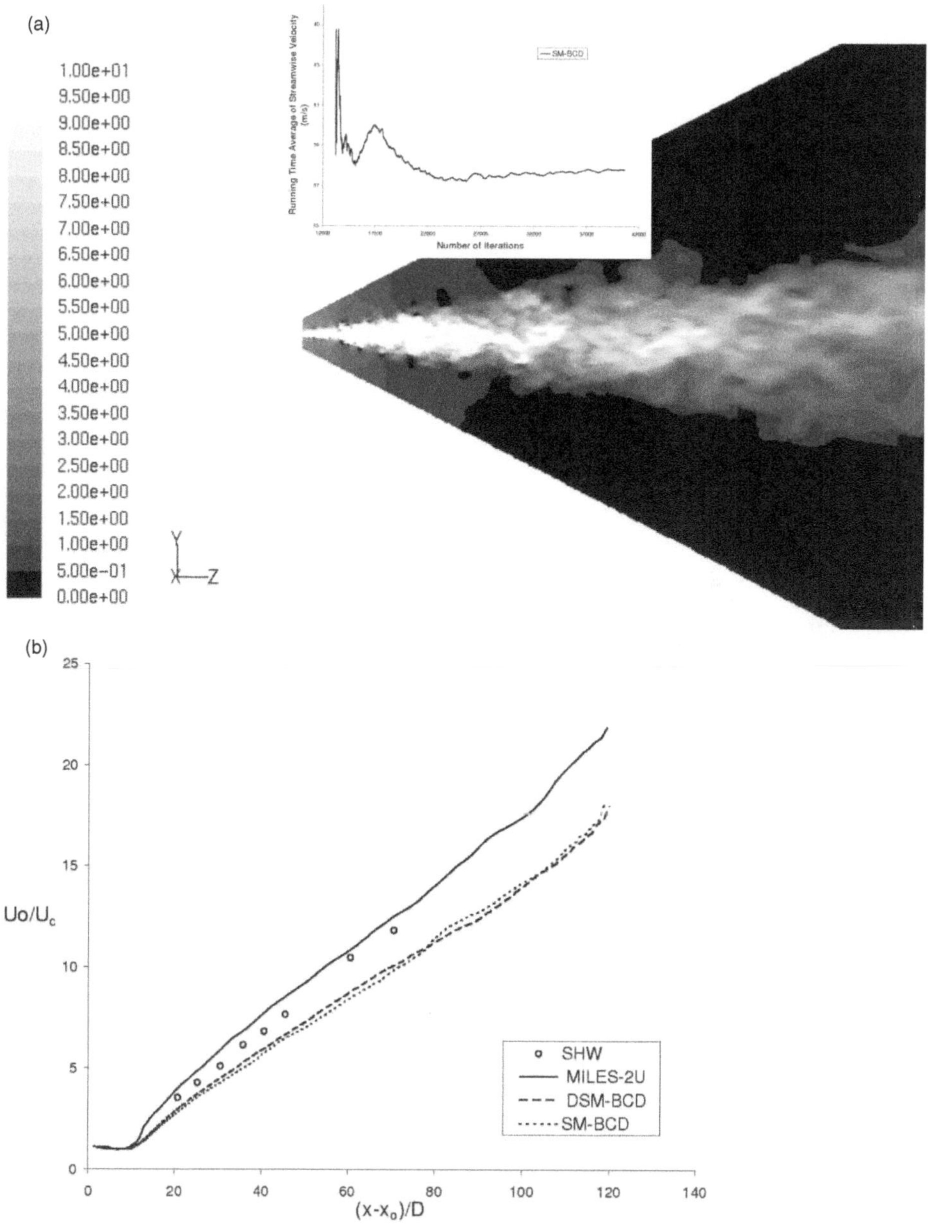

Figure 8. (a) Instantaneous streamwise velocity distribution in the jet. The inset figure shows the convergence history of the streamwise velocity component for SM-BCD; (b) Comparison of centerline velocity variation with distance from the jet exit for experimental data (Hussein et al., 1994) and MILES-2U, DSM-BCD and SM-BCD LES.

exit diameter ($D = 1$ inch), and the outer domain diameter grew at a rate of 0.5 inches per inch in the streamwise direction. The maximum streamwise domain extent was 115*D* downstream of the jet exit with a maximum outer diameter of 60*D*. All of the outer boundaries external to the jet were specified as having a constant (ambient) pressure. This choice of boundary condition allows the low velocity entrainment of ambient fluid into the jet and accurately reproduces the effects of an infinite domain. The experimental study reported that the jet exit velocity closely resembled a top-hat profile with a laminar boundary layer thickness of $\delta_{95} = 0.7$ mm at the jet exit. To achieve the required boundary layer thickness at the jet exit, an inlet supply pipe was added for the simulation as shown. The length of the supply pipe (x) was estimated from the flat plate laminar boundary layer equation:

$$\frac{\delta}{x} = \frac{5}{Re^{0.5}}. \tag{17}$$

For jet exit velocity $U = 56.2$ m/s, this yields an inlet length of 3 inches. This supply tube length was used in the simulation, and a uniform velocity was applied at the supply tube inlet. Results showed that the predicted exit velocity profile closely matched the experimental profile. The time step size for the simulations was specified to have a maximum (convective) Courant–Friedrichs–Lewy (CFL) number of 1.0, which corresponded to $\Delta t \approx 10^{-6}$ s.

7.3. Results

Figure 8(a) shows a representative plot of the convergence history of the solution and the instantaneous contours of the streamwise velocity distribution in the jet at a statistically stationary state for the SM-BCD combination. Temporal convergence was assessed using the running time average of the streamwise velocity at a location 100*D* downstream of the jet exit. It took approximately 40 K time steps to achieve convergence for SM-BCD and MILES-2U, and 50 K time steps for DSM-BCD.

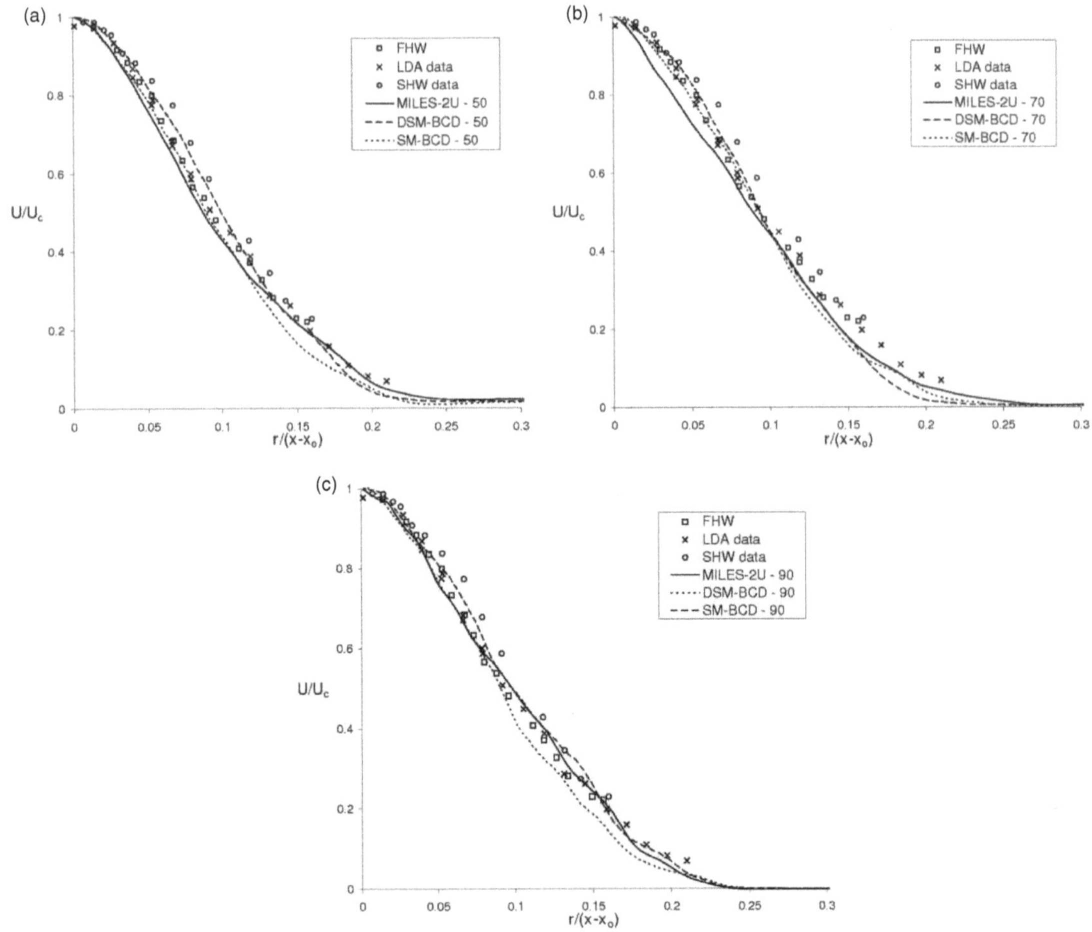

Figure 9. Comparison of mean velocity profile obtained using MILES-2U, DSM-BCD and SM-BCD LES with experimental data (Hussein et al., 1994) at downstream distance (a) 50*D*, (b) 70*D* and (c) 90*D*.

The instantaneous streamwise velocity contours in Figure 8(a) highlight the fundamental aspect of LES, namely that the large turbulent flow structures are well resolved in the simulation. Also apparent is the low speed entrainment of ambient fluid into the jet. Figure 8(b) compares the inverse of the centerline mean streamwise velocity versus the downstream distance with the experimental data. The virtual origin x_0 for the simulations was determined to be $4.9D \pm 0.6D$, upstream of the jet exit. All of the simulations show a linear increase, similar to what was observed in the experiment. In terms of centerline velocity decay, the MILES-2U slope predictions compare within 8% of the experimental data, while the SM-BCD and DSM-BCD slopes were underpredicted, indicating that turbulent mixing of mean momentum using both BCD combinations was less than that observed in the experiments. This suggests a complex interaction between numerical and model dissipation, and energy transfer mechanisms from large resolved to small resolved turbulent scales. Consistent with the isotropic turbulence results above, the use of a more dissipative (upwind-biased) discretization scheme with no SGS model yields more accurate results than a less dissipative scheme with common eddy-viscosity SGS models.

In Figure 9, the radial profile of the normalized mean streamwise velocity is compared with the experimental data at the three downstream locations, for each of the three tested model-scheme combinations. The predicted profiles demonstrate self-similar behavior downstream, which is expected for the mean quantities. SM-BCD performs slightly better than the other models – however, the averaged error is reasonable ($< 6\%$) for all the combinations.

Figures 10–12 compare the resolved turbulence intensity components with the experimental data at each downstream location. The turbulence intensity components show self-similar behavior, similar to the mean velocity profile. All combinations underpredict the turbulence

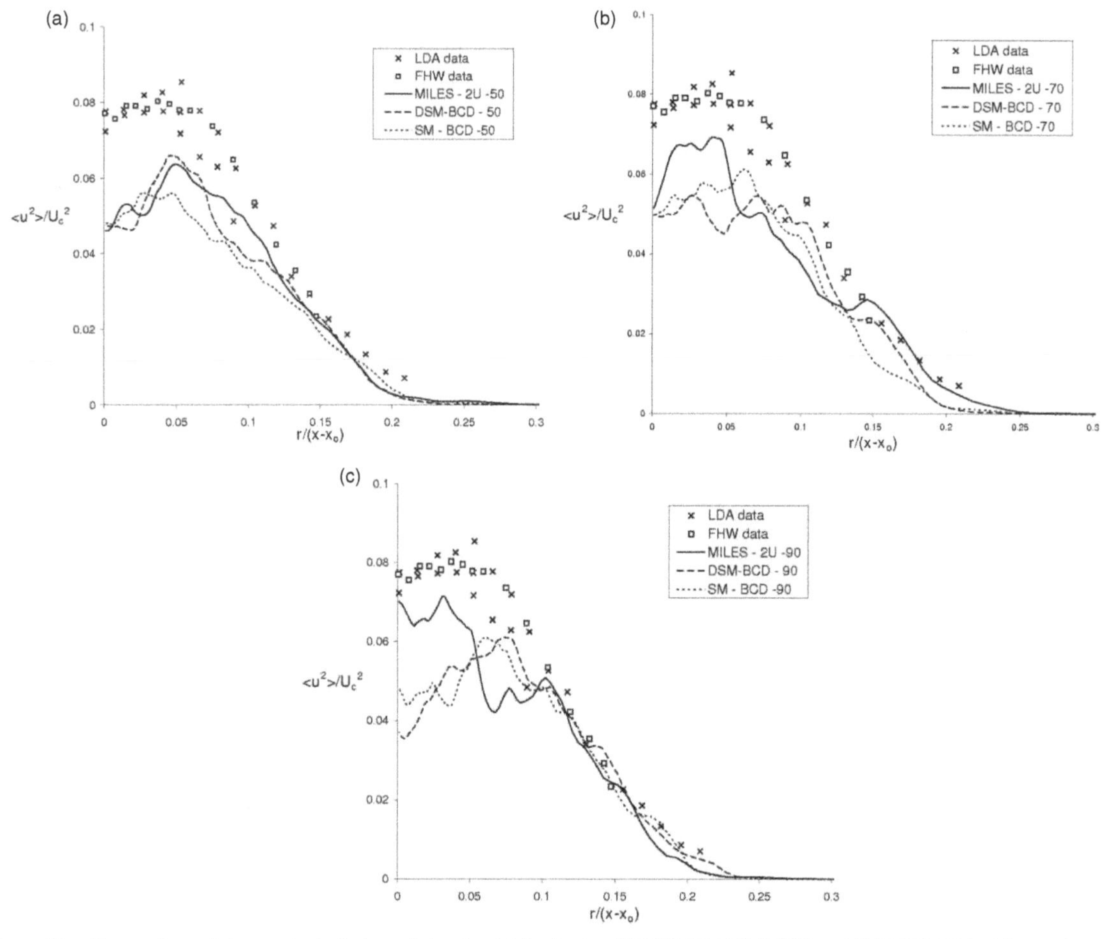

Figure 10. Comparison of streamwise turbulence intensity obtained using MILES-2U, DSM-BCD and SM-BCD LES with experimental data (Hussein et al., 1994) at downstream distance (a) 50*D*, (b) 70*D* and (c) 90*D*.

intensities close to the jet center, i.e., $0 \leq \eta \leq 0.06$, by up to 25–30%. An underprediction of the resolved turbulence intensity is expected in LES because the simulation does not resolve the entire turbulence scale, as in the experiment. LES comparison with experimental data can be improved by adding modeled turbulence intensity, which is unfortunately not available in the solutions, i.e., modeled turbulent kinetic energy is not computed in the SM, DSM and MILES simulations. Among the combinations considered, MILES-2U shows the highest resolved turbulence level, whereas no significant differences are observed between the SM and DSM predictions.

Overall, the results for the jet flow case show reasonable agreement with experiments, including self-similarity, good agreement for radial profiles of mean velocity, and reasonable agreement of radial profiles of second moments. The differences between the different model-scheme combinations manifest primarily in the prediction of centerline mean velocity. Surprisingly, but consistent with the results for decaying isotropic turbulence, the best agreement was found for the case of no-model (MILES) in combination with a second-order upwind discretization scheme.

8. Summary and conclusions

The goal of this research effort was to evaluate combinations of SGS model and discretization schemes for the LES of complex turbulent flows using general-purpose CFD algorithms, represented here by the commercial code Ansys FLUENT®. To achieve this goal, a formal quantification of coupled modeling and discretization errors present in finite-volume/finite-difference simulations was first conducted for isotropic decaying turbulence. The study focused on commonly used second-order finite-differencing schemes, as well as the Samagorinsky and dynamic Smagorinsky SGS models, and MILES (i.e., no SGS model). The errors were assessed using a pseudospectral solver, in which finite-differencing errors were added as explicit terms on the right-hand side of the spectral

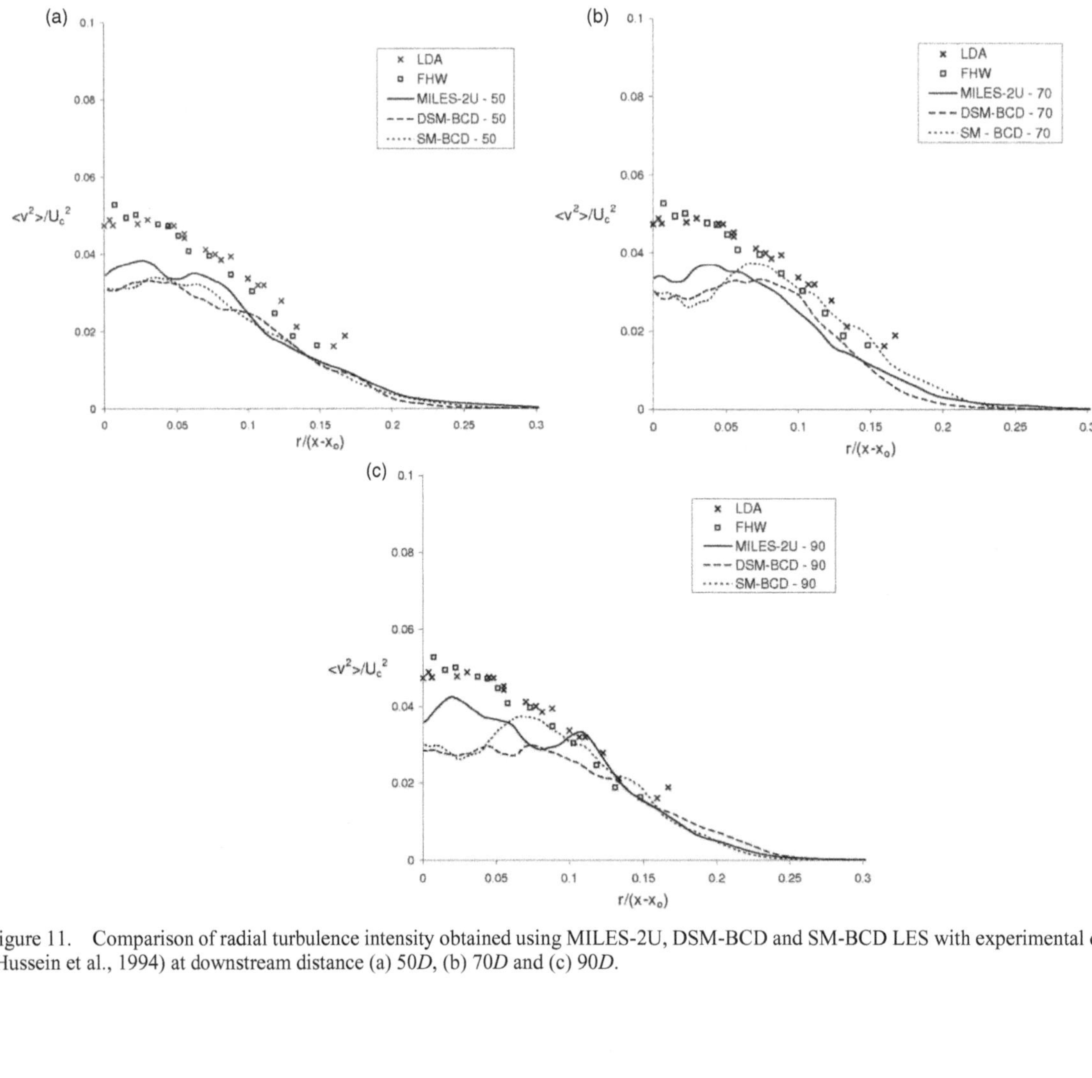

Figure 11. Comparison of radial turbulence intensity obtained using MILES-2U, DSM-BCD and SM-BCD LES with experimental data (Hussein et al., 1994) at downstream distance (a) 50*D*, (b) 70*D* and (c) 90*D*.

representation of the governing momentum equations. This allowed the isolation of the two error sources due to modeling and numerical scheme. Following this, net modeling and discretization errors in the application solver were studied, and the optimal combinations of SGS model and numerical discretization scheme for the best LES predictions were identified. The effectiveness of the optimal combinations of model/scheme identified from the isotropic decaying case were applied for the LES of a high $Re \approx 10^5$ axisymmetric turbulent jet flow.

The interplay of the modeling and discretization errors shows a consistent behavior in both the pseudospectral and finite-volume simulations of isotropic decaying turbulence. Conclusions are summarized as:

(1) the upwind and QUICK schemes are dissipative and appear to provide an appropriate level of dissipation when combined with the MILES approach;
(2) the central-difference scheme shows dispersion errors, which helps in counteracting SGS errors in selected cases, but is not reliable as it leads to energy pile-up at higher wavenumbers;
(3) the bounded central-difference scheme, which combines the effects of a central differencing scheme and an upwind scheme, helps to eliminate the energy pile-up issue of the central-difference scheme, and is found to be appropriate for SM and DSM simulations; and
(4) MILES with second-order upwind or QUICK schemes performed better than MILES with BCD for the prediction of large-scale turbulent structures, and provided the best results for the engineering applications.

Taken in their entirety, the results from this study suggest that useful predictions may indeed be obtained from LES of engineering flows using commonly available discretization schemes and SGS models found in general-purpose flow solvers. Important guiding principles are that the discretization scheme should be non-dispersive, and

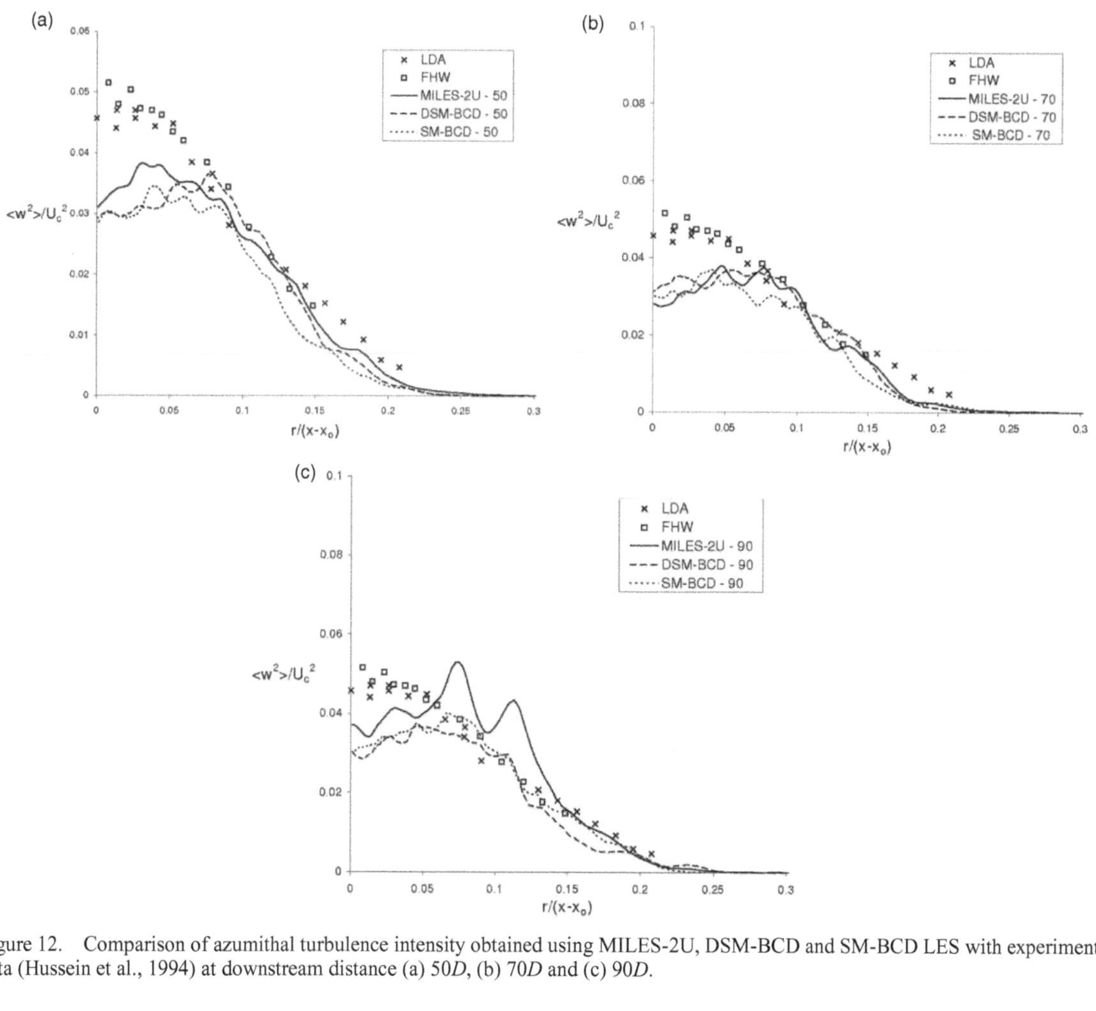

Figure 12. Comparison of azumithal turbulence intensity obtained using MILES-2U, DSM-BCD and SM-BCD LES with experimental data (Hussein et al., 1994) at downstream distance (a) 50*D*, (b) 70*D* and (c) 90*D*.

that the combined effects of the discretization scheme and SGS model should provide a sufficient but not overly large level of dissipation. Satisfying these criteria has been found to result in a good prediction of the large-scale flow behavior, including the energy spectrum (except near the cutoff wavenumber), turbulent energy decay, and turbulent mixing dynamics. The four combinations investigated here that most closely meet this requirement are SM or DSM with bounded central-difference, and MILES with the second-order upwind or QUICK scheme. Perhaps surprisingly, it is the latter two, which rely on no explicit SGS model and on strongly upwind discretization schemes, that yield the best results overall for the test cases considered here.

More generally, the results suggest that the choice of SGS model may be of minor importance in 'application' LES simulations using general-purpose solvers with common second-order schemes, and that upwind-biased schemes apparently provide enough dissipation such that the addition of an SGS model is redundant, and may even decrease accuracy. It must again be pointed out that these conclusions are only applicable to the non-wall-bounded turbulent flows considered here, and it is expected that grid anisotropy and SGS model details will play a critical role in near-wall turbulence simulations. Although such an analysis is beyond the scope of this study, it should be considered as a future research effort.

Ongoing and future work efforts are focusing on the identification of the best model-discretization combination for hybrid RANS/LES models, since such models are gaining popularity in the CFD application community due to having lower grid resolution requirements than LES. Some initial results have shown that the choice of MILES for the LES mode with the BCD scheme shows improved resolved turbulence level predictions in comparison to the eddy viscosity type LES model (Walters, Bhushan, Alam, & Thompson, 2013). Future work will focus on a systematic study, similar to the present study, for hybrid RANS/LES models.

Disclosure statement

No potential conflict of interest was reported by the authors.

Funding

This work was supported by the United States (US) National Science Foundation [grant number CBET-0645138].

Note

1. The no-model simulations are only designated as MILES for the finite-volume simulations, since the unmodified pseudospectral simulations do not include any dissipative numerical operators in their formulation, which are integral to the MILES approach.

References

Abdalla, I.E., Cook, M.J., & Hunt, G.R. (2009). Numerical study of thermal plume characteristics and entrainment in an enclosure with a point heat source. *Engineering Application of Computational Fluid Mechanics, 3*, 608–630.

Adedoyin, A.A. (2007). *Determination of best practice guidelines for performing large eddy simulation of flows in configurations of engineering interest* (Unpublished master's thesis). Mississippi State University, Starkville, USA.

Barth, T.J., & Jesperson, D. (1989). *The design and application of upwind schemes on unstructured meshes* (AIAA Paper No. AIAA-89-0366). Paper presented at the 27th AIAA Aerospace Sciences Meeting and Exhibit, January 1989, Reno, Nevada, USA.

Baxevanou, C.A., & Fidaros, D.K. (2008). Validation of numerical schemes and turbulence models combinations for transient flow around airfoil. *Engineering Application of Computational Fluid Mechanics, 2*, 208–221.

Bhushan, S., & Warsi, Z.U.A. (2005). Large eddy simulation of turbulent channel flow using an algebraic model. *International Journal for Numerical Methods in Fluids, 49*, 489–519.

Bhushan, S., & Warsi, Z.U.A. (2007). Large eddy simulation of free shear flow using algebraic model. *Computers and Fluids, 36*, 1384–1397.

Bhushan, S., Warsi, Z.U.A., & Walters, D.K. (2006). Estimating backscatter in subgrid scale turbulence through algebraic modeling. *AIAA Journal, 44*, 837–847.

Boris, J.P., Grinstein, F.F., Oran, E.S., & Kolbe, R.J. (1992). New insights into large-eddy simulation. *Fluid Dynamics Research, 10*, 199–227.

Canuto, C., Hussaini, M.Y., Quarteroni, A., & Zang, T.A. (1988). *Spectral methods in fluid dynamics*. New York, NY: Springer.

Carati, D., Winckelmans, G.S., & Jeanmart, H. (2001). On the modelling of the subgrid-scale and filtered-scale stress tensors in large-eddy simulation. *Journal of Fluid Mechanics, 441*, 119–138.

Chow, F.K., & Moin, P. (2003). A further study of numerical errors in large-eddy simulations. *Journal of Computational Physics, 184*, 366–380.

Collis, S.S. (2001). Monitoring unresolved scales in multiscale turbulence modeling. *Physics of Fluids, 13*, 1800–1806.

De Stefano, G., & Vasilyev, O.V. (2002). Sharp cutoff versus smooth filtering in large eddy simulation. *Physics of Fluids, 14*, 362–369.

Fedioun, I., Lardjane, N., & Gokalp, I. (2001). Revisiting numerical errors in direct and large eddy simulations of turbulence: Physical and spectral space analysis. *Journal of Computational Physics, 174*, 816–851.

Ghosal, S. (1996). An analysis of numerical errors in large-eddy simulations of turbulence. *Journal of Computational Physics, 125*, 187–206.

Glaz, H.M., Bell, J.B., & Colella, P. (1993). An analysis of the fractional-step method. *Journal of Computational Physics, 108*, 51–58.

Hanjalic, K. (2005). Will RANS survive LES? A view of perspectives. *Journal of Fluid Engineering, 127*, 831–839.

Hickel, S., Adams, N.A., & Domaradzki, J.A. (2006). An adaptive local deconvolution method for implicit LES. *Journal of Computational Physics, 213*, 413–436.

Hussein, H.J., Capp, S.P., & George, W.K. (1994). Velocity measurements in a high-Reynolds-number, momentum-conserving, axisymmetric, turbulent jet. *Journal of Fluid Mechanics, 258*, 31–75.

Kang, H.S., Chester, S., & Meneveau, C. (2003). Decaying turbulence in an active-grid-generated flow and comparisons with large-eddy simulation. *Journal of Fluid Mechanics*, *480*, 129–160.

Kraichnan, R.H. (1976). Eddy viscosity in two and three dimensions. *Journal of Atmospheric Science*, *33*, 1521–1536.

Kravchenko, G., & Moin, P. (1997). On the effect of numerical errors in large-eddy simulations of turbulent flows. *Journal of Computational Physics*, *131*, 310–322.

Leonard, B.P. (1979). A stable and accurate convective modelling procedure based on quadratic upstream interpolation. *Computer Methods in Applied Mechanics and Engineering*, *19*, 59–98.

Lilly, D.K. (1992). A proposed modification of the Germano subgrid-scale closure method. *Physics of Fluids A*, *4*, 633–634.

Margolin, L.G., Rider, W.J., & Grinstein, F.F. (2006). Modeling turbulent flow with implicit LES. *Journal of Turbulence*, *7*, N15.

Park, N., Yoo, J.Y., & Choi, H. (2004). Discretization errors in large eddy simulation: On the suitability of centered and upwind-biased compact difference schemes. *Journal of Computational Physics*, *198*, 580–616.

Patankar, S.V., & Spalding, D.B. (1972). A calculation procedure for heat, mass, and Momentum transfer in three-dimensional parabolic flows. *International Journal of Heat and Mass Transfer*, *15*, 1787–1806.

Pope, S.B. (2000). *Turbulent flows*. Cambridge: Cambridge University Press.

Robertson, E., Chitta, V., Walters, D.K., & Bhushan, S. (2014). On the vortex breakdown phenomenon in high angle of attack flows over delta wing geometries. Proceedings of the ASME 2014 International Mechanical Engineering Congress & Exposition, IMECE2014 November 14–20, 2014, Montreal, Quebec, Canada.

Rogallo, R.S. (1981). *Numerical experiments in homogeneous turbulence* (NASA Technical Memorandum 81315). Washington, DC: NASA.

Shalaby, H., Wozniak, K., & Wozniak, G. (2008). Numerical calculation of particle-laden cyclone separator flow using LES. *Engineering Application of Computational Fluid Mechanics*, *2*, 382–392.

Smagorinsky, J. (1963). General circulation experiments with the primitive equations. *Monthly Weather Review*, *91*, 99–164.

Su, W.-T., Li, F.-C., Li, X.-B., Wei, X.-Z., & Zhao, Y. (2012). Assessment of LES performance in simulating complex 3D flows in turbo-machines. *Engineering Application of Computational Fluid Mechanics*, *6*, 356–365.

Walters, D.K., & Bhushan, S. (2005). *Specification of time-dependent inlet boundary conditions for LES, VLES, and DES of turbulent flow (AIAA Paper No. AIAA-2005-1284)*. Paper presented at the 43rd AIAA Aerospace Sciences Meeting, January 2005, Reno, Nevada, USA.

Walters, D.K., Bhushan, S., Alam, F., & Thompson, D. (2013). Investigation of a dynamic hybrid RANS/LES modeling methodology for finite-volume CFD simulations. *Flow, Turbulence and Combustion*, *91*, 643–667.

Appendix 1. Numerical error for second-order upwind scheme

For a 1D system, the error in the convection term is:

$$e = u\frac{\delta u}{\delta x} = \begin{cases} u\frac{\delta u}{\delta x}^{BD} & : u \geq 0 \\ u\frac{\delta u}{\delta x}^{FD} & : u < 0 \end{cases}, \tag{A1}$$

where, ^{FD} and ^{BD} are for forward and backward difference, respectively. The second-order derivative approximations using forward and backward difference schemes are:

$$\begin{aligned} \frac{\delta u}{\delta x}^{FD} &= \frac{-u(x+2\Delta)+4u(x+\Delta)-3u(x)}{2\Delta} - \frac{\partial u}{\partial x} \\ &= -\frac{\Delta^2}{3}\frac{\partial^3 \bar{u}}{\partial x^3} - \frac{\Delta^3}{4}\frac{\partial^4 \bar{u}}{\partial x^4} - \frac{7\Delta^4}{60}\frac{\partial^5 \bar{u}}{\partial x^5} + \ldots \end{aligned} \tag{A2}$$

$$\begin{aligned} \frac{\delta u}{\delta x}^{BD} &= \frac{3u(x)-4u(x-\Delta)+u(x-2\Delta)}{2\Delta} - \frac{\partial u}{\partial x} \\ &= -\frac{\Delta^2}{3}\frac{\partial^3 \bar{u}}{\partial x^3} + \frac{\Delta^3}{4}\frac{\partial^4 \bar{u}}{\partial x^4} - \frac{7\Delta^4}{60}\frac{\partial^5 \bar{u}}{\partial x^5} + \ldots \end{aligned} \tag{A3}$$

Therefore, the error for the upwind scheme can be expressed in a single equation as given below:

$$e = u\frac{\delta u}{\delta x} = -u\frac{\Delta^2}{3}\frac{\partial^3 \bar{u}}{\partial x^3} + |u|\frac{\Delta^3}{4}\frac{\partial^4 \bar{u}}{\partial x^4} - u\frac{7\Delta^4}{60}\frac{\partial^5 \bar{u}}{\partial x^5} + \ldots \tag{A4}$$

A.1. Fourier analysis of numerical error

Fourier expansion of the velocity field and associated derivative is:

$$u(x) = \sum \grave{u} e^{ikx}, \tag{A5}$$

$$\frac{\partial u}{\partial x}^{e} = \sum ik\grave{u}\, e^{ikx} = \sum \grave{D}_e\, e^{ikx}, \tag{A6}$$

where, $\grave{D}_e = ik\grave{u}$ are the Fourier components of the exact velocity derivative. The Fourier representation of error in the convection term is:

$$\grave{e} = \grave{u} * \frac{\delta \grave{u}}{\delta x}, \tag{A7}$$

where $*$ represents convolution. The Fourier component of the forward and backward difference schemes are computed as follows:

$$\begin{aligned} \frac{\partial u}{\partial x}^{FD} &= \frac{1}{2\Delta}\sum \grave{u} e^{ikx}[-e^{2ik\Delta} + 4e^{ik\Delta} - 3] \\ &= \sum ik\grave{u}\left[\frac{\sin(k\Delta)\,\{2-\cos(k\Delta)\}}{k\Delta} \right. \\ &\quad \left. + i\frac{\{\cos(k\Delta)-1\}^2}{k\Delta}\right] e^{ikx} \end{aligned} \tag{A8}$$

$$\Rightarrow \grave{D}_{FD} = \grave{D}_e\left[\frac{\sin(k\Delta)}{k\Delta}\{2-\cos(k\Delta)\} + i\frac{\{\cos(k\Delta)-1\}^2}{k\Delta}\right]; \tag{A9}$$

$$\begin{aligned} \frac{\partial u}{\partial x}^{BD} &= \frac{1}{2\Delta}\sum \grave{u} e^{ikx}[3 + e^{-2ik\Delta} - 4e^{-ik\Delta}] \\ &= \sum ik\grave{u}\left[\frac{\sin(k\Delta)\{2-\cos(k\Delta)\}}{k\Delta} \right. \\ &\quad \left. - i\frac{\{\cos(k\Delta)-1\}^2}{k\Delta}\right] e^{ikx} \end{aligned}$$

$$\Rightarrow \grave{D}_{BD} = \grave{D}_e\left[\frac{\sin(k\Delta)}{k\Delta}\{2-\cos(k\Delta)\} - i\frac{\{\cos(k\Delta)-1\}^2}{k\Delta}\right]. \tag{A10}$$

The Fourier components of the error in derivative calculations are:

$$\frac{\delta\grave{u}}{\delta x}^{FD}(k) = \grave{D}_e - \grave{D}_{FD} = -\frac{\{\cos(k\Delta) - 1\}^2}{\Delta} + ik\left[1 - \frac{\sin(k\Delta)}{k\Delta}\{2 - \cos(k\Delta)\}\right]; \quad \text{(A11)}$$

$$\frac{\delta\grave{u}}{\delta x}^{BD}(k) = \grave{D}_e - \grave{D}_{BD} = \frac{\{\cos(k\Delta) - 1\}^2}{\Delta} + ik\left[1 - \frac{\sin(k\Delta)}{k\Delta}\{2 - \cos(k\Delta)\}\right]. \quad \text{(A12)}$$

Thus, the error equation for both $u \geq 0$ and $u < 0$ for the upwind scheme is:

$$\grave{e} = -|\grave{u}| * \frac{\{\cos(k\Delta) - 1\}^2}{\Delta} + \grave{u} * ik\left[1 - \frac{\sin(k\Delta)}{k\Delta}\{2 - \cos(k\Delta)\}\right]. \quad \text{(A13)}$$

It must be noted that the real component of the error in the derivative calculation is due to the even derivative terms. Thus, even derivative terms need to be multiplied by the absolute value of the velocity, consistent with the Taylor series expansion analysis above.

5

The effect of vent openings on the microclimate inside multi-span greenhouses during summer and winter seasons

Ke-shi He[a], Da-yue Chen[a], Li-juan Sun[b], Zheng-lu Liu[c] and Zhen-yu Huang[a]*

[a]*Laboratory of Complex Dynamics Simulation, Department of Instrument Science and Engineering, School of Electronic, Information and Electrical Engineering, Shanghai Jiao Tong University, Shanghai, China;* [b]*Institute of Crop Science, Chinese Academy of Agricultural Sciences, Beijing, China;* [c]*Shanghai Key Laboratory of Protected Horticultural Technology, Sunqiao Modern Agriculture Development Zone, Shanghai, China*

In research literature, little attention has been applied to the effect of vent openings on greenhouse climates during cooling and dehumidification processes with natural ventilation, which would provide guidelines for greenhouse management. To address this problem, a 3D CFD model was successfully designed based on 11-span plastic greenhouses. The model was validated with the measured air temperature and relative humidity, and then used to investigate the effects of vent configuration and opening size on the greenhouse climate. The simulations show that the vent configuration affects greenhouse microclimate patterns and the spatial and temporal variations of the internal climate. Meanwhile, the vent opening size affects greenhouse dehumidification time, air temperature and relative humidity during the dehumidification process. Finally, the assessments of ventilation performance highlight that the roof plus side opening is most suitable for summer cooling, while the roof opening is most suitable for winter dehumidification.

Keywords: vent configuration; greenhouse microclimate; ventilation performance; natural ventilation; CFD model

1. Introduction

Greenhouse cultivation can enable vegetable production and thus remove the constraints of season and climate, which is becoming an important factor in traditional vegetable production in fields in China. China is a huge country, covering a wide range of latitude and longitude. The extreme differences of climate and geographical environment give rise to the existence of many different types of representative greenhouses in different regions. For example, solar greenhouses are common in northern China (Tong, Christopher, & Li, 2009), but this type of greenhouse is hardly ever used in eastern China, where the tunnel greenhouse is popular.

Greenhouse climate management depends on its natural ventilation system. Through promoting the exchange between indoor and outdoor air, natural ventilation can dissipate surplus heat and vapor in summer, while it can exclude excessive vapor and provide a suitable thermal climate in winter (Baptista, Bailey, Randall, & Meneses, 1999). However, the ventilation mechanism is complex and poorly understood, since it depends on many factors.

In eastern China, the usage of multi-span plastic greenhouses has rapidly expanded in the past decade, traditional, simple structures covering a large cultivated area. There are economic advantages to building this type of greenhouse, so these greenhouses play an important role in horticultural production. However, for the management of large greenhouses, it is very difficult to maintain uniform, suitable environmental factors such as internal airflow, air temperature and humidity (Hong et al., 2008). Moreover, greenhouse management faces the major challenge of controlling excess temperature in summer and high humidity in winter.

In recent years, the computational fluid dynamics (CFD) technique has become an important tool which is widely used in many fields, such as waterfront planning (Zhang, You, & Zhao, 2014), aeronautical engineering (Yu, Xu, & Gan, 2014), and fuel-cell design (Bao et al., 2014). In the greenhouse engineering community, many researchers have employed CFD to study the effects of greenhouse design (Bartzanas, Boulard, & Kittas, 2004; Bournet, Ould Khaoua, & Boulard, 2007, Bournet, Ould Khaoua, Boulard, Migeon, & Chassériaux, 2007; Hong et al., 2008; Kacira, Sase, & Okushima, 2004; Kacira, Short, & Stowell, 1998; Kittas and Bartzanas, 2007; Lee, Short, Sase, Okushima, & Qiu, 2000; Ould Khaoua, Bournet, Migeon, Boulard, & Chassériaux, 2006), specific equipment (Baeza et al., 2009; Fatnassi, Boulard, & Bouirden, 2003; Fatnassi, Boulard, Poncet, & Chave, 2006; Montero et al., 2013; Tamimi, Kacira, Choi, & An, 2013) and external weather conditions (Hong et al., 2008; Lee & Short, 2000; Nebbali, Roy, & Boulard, 2012; Ould

*Corresponding author. Email: bighuang@sjtu.edu.cn

Khaoua et al., 2006; Teitel, Ziskind, Liran, Dubovsky, & Letan, 2008; Tong et al., 2009) on greenhouse climates. Their results provide valuable guidance for ventilation system management; however, they are limited to specific greenhouse structures and local climate conditions.

So far, most numerical studies of natural ventilation have focused on greenhouse climate and ventilation performance during the summer (Baeza et al., 2009; Bartzanas et al., 2004; Bournet, Ould Khaoua, & Boulard, 2007, Bournet, Ould Khaoua, Boulard, et al., 2007; Fatnassi et al., 2003, 2006; Hong et al., 2008; Kacira et al., 1998, 2004; Kichah, Bournet, Migeon, & Boulard, 2012; Lee et al., 2000; Lee & Short, 2000; Nebbali et al., 2012; Ould Khaoua et al., 2006; Teitel et al., 2008). Few have examined the effect of vent opening on greenhouse climates during the different seasons of summer and winter when natural ventilation is respectively used for cooling and dehumidification. Therefore, a CFD model is designed for the type of multi-span plastic greenhouses employed in eastern China to investigate the internal climate mechanism associated with the ventilation processes. CFD-assisted exploration can help horticulturists to improve the efficiency of ventilation system management according to the seasonal climate variations.

In this paper, a 3D CFD model was designed to investigate greenhouse climates during the cooling and dehumidification processes with natural ventilation. First, the effect of vent configuration on the greenhouse microclimate patterns was investigated. Then, the spatial and temporal variations of the greenhouse climate were revealed. In addition, the effect of the vent opening size on the greenhouse dehumidification time, air temperature and relative humidity was studied. Finally, the ventilation process was assessed using different vent configurations in different seasons.

2. Materials and methods

2.1. *Greenhouse and plant material*

The studied multi-span plastic greenhouse is north-south (N-S) oriented and located in the Sunqiao Modern Agriculture Development Zone, Shanghai (31°18′ N, 121°63′ E). The greenhouse is divided into two independent compartments by a central partition: the experimental section (Compartment I) composed of five spans and the adjacent section (Compartment II) composed of six spans. Each span has two roof vents (roll-up type, 1 m × 30 m each) except for the first and last spans. This type of vent is also installed in the sidewalls. A sketch of the greenhouses is given in Figure 1.

In the greenhouse, lettuces (*Lactuca Satira L*, 'Yang Shan') are laid out with an N-S orientation in 22 rows with a row spacing of 0.4 m and an in-row spacing of 0.2 m. They are grown in bare soil and cultivated by using drip irrigation during the greenhouse crop management.

2.2. *Experimental design*

Two measurements were conducted in Compartment I during summer and winter days, when the wind was almost perpendicular to the greenhouse orientation. The first measurement (EXP1) was performed on 11:00–12:20, 11 July 2011. The roof vents were fully open for the entire of the experimental period. The second measurement (EXP2) was performed on 11:20–12:20, 16 January 2012. The greenhouse was initially kept closed. The internal air temperature and relative humidity were 20.6°C and 93%. After 20 minutes, the roof vents were fully opened.

The external climate parameters were monitored by a meteorological station (WE800, Global Water Instrumentation Inc, USA) which was set at a height of 2.0 m and positioned far away from the greenhouse. The air conditions inside the greenhouse were measured with five temperature and humidity sensors (PH100TMPA/PH100HUMA, XPH Inc, China) placed uniformly at a height of 1.6 m along the greenhouse width in the intermediate cross-section of the greenhouse. The ground temperature was recorded with an infrared radiation thermometer (CA380, Cason Inc, Hong Kong). The crop and cover temperatures were measured by means of Pt100 probes that were fixed on the lower surface of the leaves and on the inner surface of the covers. The inside radiation was monitored by a net radiometer (Q-7.1, Campbell Scientific Inc, USA) located at a height of 0.5 m (just above the crop canopy). The leaf area index was recorded by a crop canopy analyzer (LAI-2200, LI-COR Inc, USA). All experiments were performed under almost steady-state weather conditions. Outside and inside climate parameters

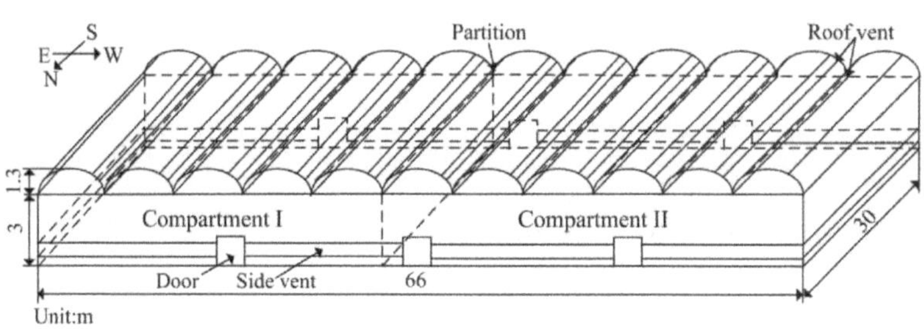

Figure 1. Sketches of multi-span plastic greenhouses.

were recorded at a sample rate of 10 Hz and then averaged every 5 minutes with two data acquisition cards (PCI9112 and PCI9118, ADLINK Technology Inc, Taiwan).

3. CFD Model design

3.1. Governing equations

In this study, a 3D CFD model was designed with reference to the 11-span plastic greenhouses cultivated with a mature lettuce. The model is based on the resolution of the governing equations (Patankar, 1980) with the finite volume method. These equations can be expressed as

$$\frac{\partial(U\phi)}{\partial x} + \frac{\partial(V\phi)}{\partial y} + \frac{\partial(W\phi)}{\partial z} = \Gamma \nabla^2 \phi + S_\phi, \quad (1)$$

where ϕ represents the concentration of the transported quantity in a dimensional form (the momentum, the scalars mass and energy conservation equations), x, y and z are the Cartesian space coordinates, Γ is the diffusion coefficient, U, V and W are the components of velocity vector, S_ϕ is the source term, and ∇^2 is the Laplace operator.

As pointed out by Roy, Boulard, Kittas, and Wang (2002), the fluid flow becomes turbulent once the Reynolds number exceeds 5×10^4. In our studies, the Reynolds number was estimated to range between 2.4×10^5 and 4.2×10^5 with different vent openings and inlet velocities. Therefore, it can be concluded that the airflows in the naturally ventilated multi-span greenhouse were turbulent for all the cases studied. In the greenhouse engineering community, the standard k-ϵ model is widely used for describing the phenomenon of turbulence. For the simulated cases (with a Reynolds number value of $1.9 \times 10^5 \sim 4.7 \times 10^5$) of Boulard and Wang (2002), Bartzanas et al. (2004) and Bournet, Ould Khaoua, and Boulard (2007), it has been proved that the model can provide not only the easiest convergence but also acceptable accuracy. Therefore, the standard k-ϵ model was adopted to describe the turbulence (for the complete set of equations for the k-ϵ model, see Launder & Spalding, 1974). The Boussinesq model was activated to treat the density as the constant value in all solved equations except for the buoyancy term. The standard wall function was used for the near-wall treatment (Launder & Spalding, 1974). The species model was activated to calculate the transport equation.

The discrete ordinate model (DOM) was adopted to account for the radiative transfer between the greenhouse and the atmospheric environment. The no-gray model was activated by dividing the radiative spectrum into the solar radiation band [0.4–2.4 μm] and the long wave band [2.4–180 μm]. The crop radiation parameters were only valid for the solar radiation band. However, as for other materials, the radiation parameters were considered as the constants independent of the wavelength band. The beam direction and irradiation were computed by a solar calculator according to a given position, date and time. The flow iteration was set to 10 per radiation iteration. The DOM solution required material optical parameters, as summarized in Table 1.

Table 1. Thermal and optical properties of the materials involved in the simulation.

Properties	Sidewall/ partition	Roof	Crop	Soil
Density (kg m^{-3})	920	920	1,001	2,000
Specific heat (J kg^{-1} °C^{-1})	2,600	2,600	3,300	1,550
Thermal conductivity (W m^{-1} °C^{-1})	0.33	0.33	0.4	1.58
Absorptivity	0.92	0.06	0.95	0.9
Scattering coefficient	0	0	0	− 15
Refractive index	1.53	1.53	1.51	3
Emissivity	0.70	0.70	0.95	0.92

3.2. Computational domain and mesh

The pre-processing software Gambit 2.3.16 was utilized to generate a large computational domain (90 m × 132 m × 30 m) composed of a greenhouse and its surroundings. A coupled approach was used to model the internal and external greenhouse airflows, since the decoupled approach may introduce significant errors in the case of large ventilation openings (Ramponi & Blocken, 2012). Therefore, the greenhouse ventilation openings were considered to be open in the simulation and both the external and internal airflows were solved simultaneously by using the governing equations within the same computational domain.

The indoor domain was meshed with an unstructured 0.4 m grid, whereas the outdoor domain was meshed with an unstructured grid with a size varying from 1.0 m to 2.0 m. Meanwhile, the crop canopy was meshed with a structured, 0.05 m grid. The grid density in the vicinities of cover, ground and ventilator was increased. A check for grid dependency was made with three different mesh numbers (4,289,292 cells, 5,215,444 cells, and 7,125,810 cells) to ensure that the grid resolution would not have a noticeable influence on the numerical solution. The details of the mesh created in different regions with three different mesh numbers are presented in Table 2. Table 3 exhibits a substantial increase in the ventilation rate between the first and second grids, indicating that the grid resolution does have an influence on the solution. A small increase in the ventilation rate between the second and third grids indicates that the grid resolution has almost no influence on the solution. Finally, the grid with 5,215,444 cells was created in the domain. The choice resulted from a good compromise between a refined grid which required excessive computation time and a coarser one which reduced the accuracy

Table 2. Statistics of the mesh created in the domain with three different mesh strategies.

Domain	Mesh type	Mesh size (m)		
		Mesh strategy 1	Mesh strategy 2	Mesh strategy 3
Greenhouse	TGrid, Tet/Hybrid	0.4	0.4	0.4
Crop canopy	Map, Hex	0.05	0.05	0.05
Surroundings A	Cooper, Hex	1.5	1.5	0.5
Surroundings B	Cooper, Wedge	2.0	1.0	1.0
Surroundings C	Map, Hex	2.0	2.0	2.0

Table 3. Calculated ventilation rates for the numerical model with three different mesh strategies.

Mesh strategy	Cell number	Ventilation rate ($m^3\ m^{-2}\ s^{-1}$)
Mesh strategy 1	4,289,292	0.112
Mesh strategy 2	5,215,444	0.125
Mesh strategy 3	7,125,810	0.127

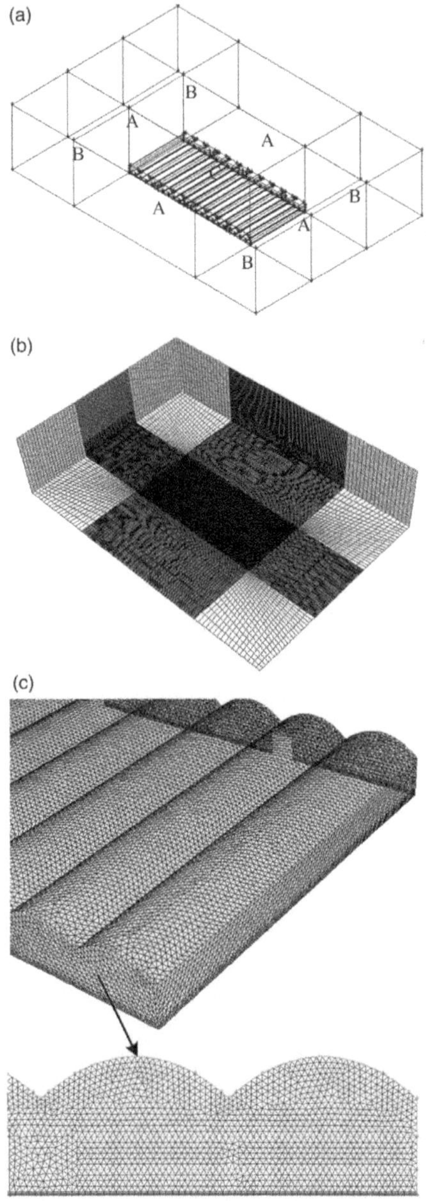

Figure 2. The computational domain and mesh.

of the numerical calculations. The maximum value of the EquiAngle skew along with the detail of the mesh in the ventilator region and near the wall is about 0.4, which falls within the acceptable range. The generated boundary-layer mesh for the ground consists of five layers. The growth factor of the boundary-layer mesh was set to 1.2. The value of the y^+ criterion is controlled within the region of the validity of the logarithmic distribution (i.e., $20 < y^+ < 200$). Likewise, for the other cases, the y^+ criterion can be also satisfied. The computational domain and mesh are shown in Figure 2.

3.3. Boundary condition

The velocity-inlet boundary condition was imposed for the domain inlet with a logarithmic velocity profile (Richards & Hoxey, 1993). The profile was linked to the CFD main module using the user-defined function (UDF). The surface roughness length was set to 0.015 m, given for the local terrain according to the responding standard (Wieringa, 1992).

The pressure-outlet boundary condition was selected for the domain outlet. The symmetry boundary condition was applied at the top of the domain. The wall boundary condition was used along the solid regions (ground, walls and roofs). The fixed temperature condition was imposed at the limits of the domain and all wall boundaries. The ground was treated as the opaque and diffuse wall. The temperature of the ground of each span was taken as the constant. The roof and sidewall were considered as the semi-transparent walls with finite thicknesses. The diffuse fraction of the irradiation was set to 20% for the roof. The sidewall was treated as the opaque and adiabatic material. The boundary condition settings of the wall stem from the material properties, as summarized in Table 1. The input values in the boundary conditions are derived from experimental data, as shown in Table 4.

In the simulation, the lettuce crop was treated as a porous medium to incorporate its resistance on the ambient airflow. The drag force generated by the airflow through the porous medium can be described by the

Table 4. Boundary conditions used in the CFD simulations.

Parameters	Values Case 1	Case 2
Air speed (m s^{-1})	1.3	2.0
Air temperature (°C)	35.0	10.5
Air relative humidity (%)	60	47
Air density (kg m^{-3})	1.15	1.25
Air specific heat (J kg^{-1} $°C^{-1}$)	1,006.7	1,005.8
Air thermal expansion coefficient (°C $^{-1}$)	0.00325	0.00353
Air thermal conductivity (W m^{-1} $°C^{-1}$)	0.0267	0.0249
Air dynamics viscosity (kg m^{-1} s^{-1})	1.89×10^{-5}	1.77×10^{-5}
Air gravitational acceleration (m s^{-2})	9.81	9.81
Solar radiation (W m^{-2})	688	340
Net radiation (W m^{-2})	490	228
Leaf surface temperature (°C)	37.2	14.0
Outside ground temperature (°C)	42.0	12.6
Inside ground temperature (°C)	41.0	21.5

Note: Case 1: 11:00–12:20, 11 July 2011; Case 2: 11:20–12:20, 16 January 2012.

Darcy-Forchheimer equation,

$$S_\phi = -\left(\frac{\mu}{K_p}U + \rho\frac{C_F}{\sqrt{K_p}}U^2\right), \quad (2)$$

where ρ is the air density (kg m^{-3}), U is the air velocity (m s^1), K_p is the permeability (m^2), μ is the dynamic viscosity (kg m^{-1} s^{-1}), and C_F is the non-linear momentum loss coefficient.

For simplification, it is assumed that the pressure force constitutes the major portion of the total drag of the crop canopy (Wilson & Shaw, 1977). The sink of the momentum per the crop volume can be expressed as

$$S_\phi = -\rho I_{\text{LAD}} C_d U^2, \quad (3)$$

where ρ is the air density (kg m^{-3}), I_{LAD} is the leaf area density (m^2 m^{-3}), U is the air velocity (m s^{-1}), and C_d is the drag coefficient of the crop canopy.

For the crop canopy and the range of the air speed, the term in U of Eq. 2 is much smaller than the quadratic term, and thus can be ignored. Consequently, combining Eq. 2 and Eq. 3, one can deduce the permeability and the non-linear momentum loss coefficient of the porous medium:

$$\frac{C_F}{\sqrt{K_p}} = I_{\text{LAD}} C_d. \quad (4)$$

In the study, I_{LAD} was defined as I_{LAI}/H (the leaf area index, I_{LAI}: 4.5 m^2 m^{-2} and 3.8 m^2 m^{-2}; the crop height, H: 0.20 m and 0.18 m). The drag coefficient C_d was determined to be 0.32. Boulard and Wang (2002) successfully used this value (0.32) for the drag coefficient for a mature lettuce in their simulation.

The crop-air interactions of thermal and water vapor were implemented with a UDF programmed in C+ + (for details, see Boulard & Wang, 2002).

3.4. *Numerical procedure*

The CFD software package ANSYS FLUENT 12.0 was used to perform two sets of simulations. The first set is the steady simulation employed for summer ventilation to reveal the greenhouse climate mechanism during the cooling process with natural ventilation. The second set is the unsteady simulation employed for winter ventilation to investigate the time-dependent greenhouse climate parameters during the dehumidification process with natural ventilation.

The semi-implicit method for the pressured-linked equation (SIMPLE) and the pressure-implicit splitting of operators (PISO) algorithms were applied to solve the flow field for the steady and unsteady flows, respectively. The body-force weighted scheme was used for the pressure discretization. The second-order upwind discretization scheme was selected for the momentum and turbulence equations. The first-order implicit scheme was employed for the time discretization. The unsteady simulation was continuously executed until the initial humidity (i.e., 93%) in the crop canopy dropped to a value below 75%, which indicates the end of the greenhouse dehumidification. Considering the computational efficiency and the temporal accuracy, the time step size was set to 0.05 s based on the preliminary grid-dependence test. The maximum iterations was set to 20 per time step. The convergence criterion was set to 10^{-6} for all variables. The under-relaxation factors used for the CFD simulation are shown in Table 5. These factors are appropriately reduced when the scaled residuals become unstable.

Table 5. Under-relaxation factors used in the CFD simulation.

Under-relaxation Factors	Values
Pressure	0.3
Density	1
Body Forces	1
Momentum	0.7
Turbulent Kinetic Energy	0.8
Turbulent Dissipation Rate	0.8
Turbulent Viscosity	1
H_2O	1
O_2	1
Energy	1
Discrete Ordinate	1

4. Results and discussion

4.1. Validation of the CFD model

The experimental and simulated temperature profile (Figure 3) at a height of 1.6 m during the cooling process shows good agreement, with a difference of less than 1 °C. The reasons for the difference are: a) any heat loss through the covers was ignored in the simulation; b) the crop parameters (such as the crop drag coefficient) were roughly estimated in the model. The results of the field measurements indicate that the model is able to provide an acceptable prediction of inside air temperature during the cooling process.

The experimental and simulated relative humidity profile (Figure 4) at a height of 1.6 m after the four-minute dehumidification process with natural ventilation gradually decreases with the increase of the distance from the side opening, with a difference of less than 6%. The explanations for this difference are: a) the leakage of air vapor from the internal to external environments is unavoidable in practice, whereas this leakage was ignored in the simulation; b) the heterogeneity of the internal humidity is inevitable in practice, whereas a perfect homogenization was assumed in the simulation; and c) the fluctuations of the outside climate are inevitable in practice, whereas the steady-state climate condition was assumed in the simulation. Nevertheless, the close agreement between the experimental and simulated humidity indicates that the model can be suitable for exploring the dynamic response of the indoor humidity to the external climate during the dehumidification process.

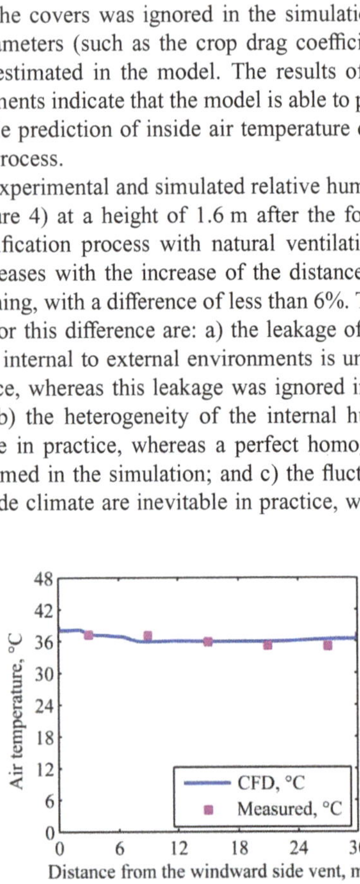

Figure 3. Experimental and numerical air temperature at a height of 1.6 m inside the greenhouse during the cooling process in summer.

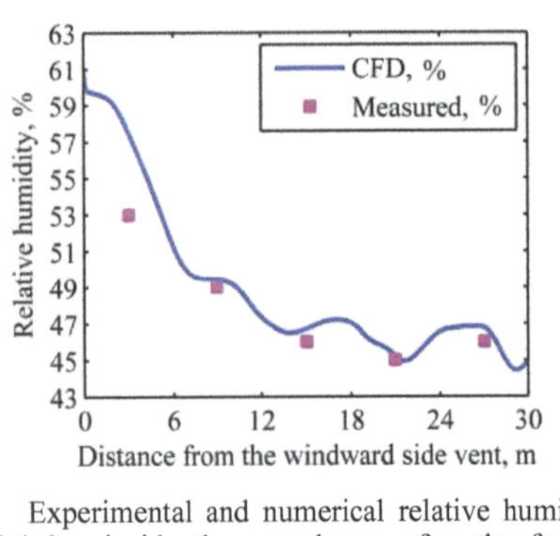

Figure 4. Experimental and numerical relative humidity at a height of 1.6 m inside the greenhouse after the four-minute dehumidification process in winter.

4.2. Effect of the vent configuration on the greenhouse microclimate patterns and the turbulent kinetic energy distribution

Figures 5–7 show that, during the cooling process, the external colder airflow enters into the greenhouse and excludes the hotter internal air. The greater the airflow velocity (Figure 5), the lower the air temperature (Figure 6). During the dehumidification process, drier airflow from outside penetrates into the greenhouse and replaces the wetter internal air. The humidity pattern (Figure 7) at the dehumidification end is the consequence of the airflow pattern (Figure 5) in the greenhouse.

With the side opening, a recirculation loop occurs in the first span with a good mixing and a relatively higher velocity. However, stiller air conditions prevail in the other span. The temperature pattern (Figure 6(a)) shows that the air temperature in most of the regions of the first three spans is over 1 °C lower than in the other span. The humidity pattern (Figure 7(a)) shows that the relative humidity in the vicinity of the crop canopy is lowered to reach the level of 40–50% at the dehumidification end. However, a significant amount of wet air is still confined adjacent to the roofs of the leeward area.

With the roof opening, most of the airflow enters the greenhouse through the windward roof vents, travels along the interior and finally towards the last two roof vents. The temperature pattern (Figure 6(b)) shows that the air temperature in the first span where the airflow is stagnant is approximately 2–3 °C warmer than in the other span. The air temperature reaches 35.0 °C in the vicinity of the roofs. The humidity pattern (Figure 7(b)) shows that the relative humidity is approximately 48% in the center. As a result of low velocities in the first and last spans, the relative humidity ranges between 60% and 75%.

With the roof plus side opening, the external airflow penetrates into the windward side vent and is then shunted. A part of the inlet airflow combined with inside hot air produces a circulation in the first span. The remaining airflow travels through the center and moves towards the last two roof vents and the leeward side vent. Compared to the other configurations, this one induces larger reductions in the indoor temperature and humidity level. The temperature pattern (Figure 6(c)) shows that the air temperature varies from 35.8 °C to 37.2 °C in the lower part of the side openings, while, in the upper part of the side openings, with the air speed being greater, the air temperature ranges between 34.4 °C and 35.8 °C. The humidity pattern (Figure 7(c)) shows that the relative humidity in most regions is lowered

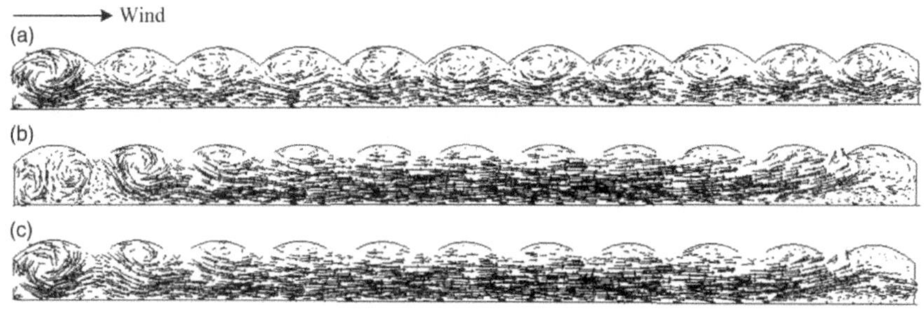

Figure 5. Airflow pattern inside the greenhouse with three different vent configurations: (a) side opening; (b) roof opening; and (c) roof plus side opening.

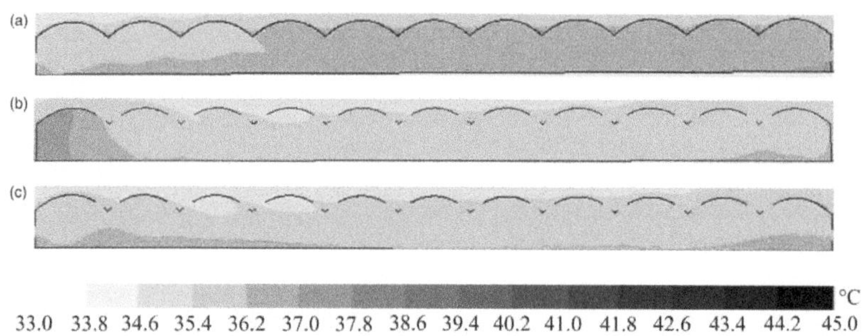

Figure 6. Temperature pattern inside the greenhouse during the cooling process with natural ventilation in summer: (a) side opening; (b) roof opening; and (c) roof plus side opening.

to a value below 55% except in the vicinity of the ground of the last span, where the humidity ranges between 63% and 73%.

Figure 8 shows the different turbulent kinetic energy (referred to as turbulence level) distributions in the greenhouse with three different vent configurations. It is clear that, in the greenhouse, the higher the airflow velocity, the higher the level of turbulence. A high level of turbulence may occur in the first span or in the vicinities of the first two windward roof ventilators and the last roof ventilator. Meanwhile, an increase in turbulence level tends to induce more homogenous temperature and humidity distributions in the greenhouse.

4.3. *Effect of the vent configuration on the spatial and temporal variations of the greenhouse climate*

The horizontal profiles of the differences between inside and outside air temperature and humidity at a height of 1.0 m during the cooling and dehumidification processes are shown (Figures 9 and 10). With the side opening, internal air temperature and humidity elevations are increased

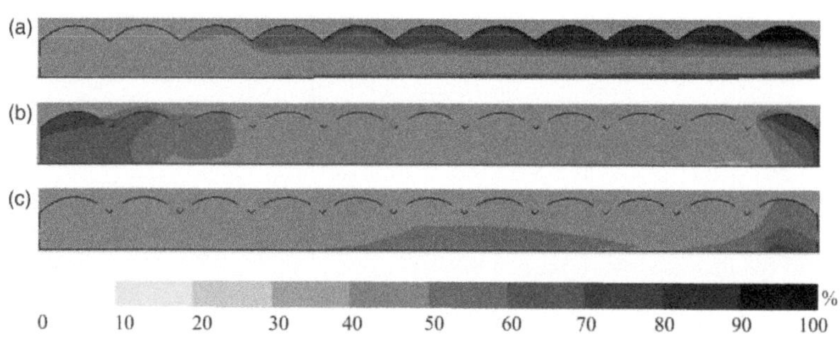

Figure 7. Humidity pattern inside the greenhouse at the end of the dehumidification with natural ventilation in winter: (a) side opening; (b) roof opening; and (c) roof plus side opening.

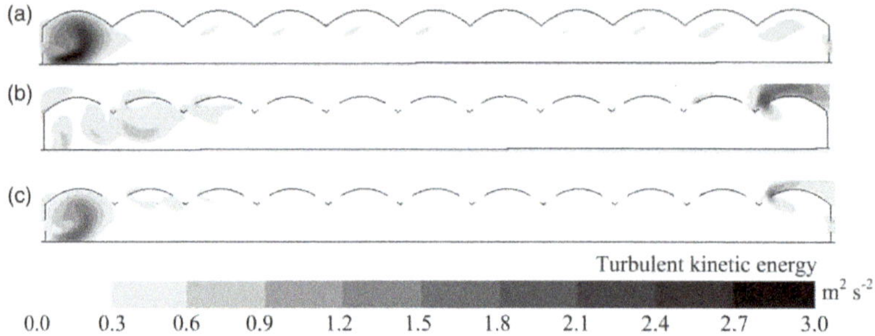

Figure 8. Turbulent kinetic energy distribution inside the greenhouse with three different vent configurations: (a) side opening; (b) roof opening; and (c) roof plus side opening.

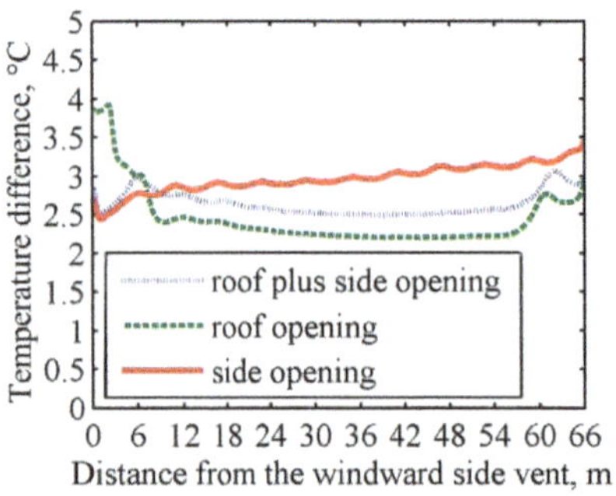

Figure 9. The difference between inside and outside air temperature at a height of 1.0 m during the cooling process in summer.

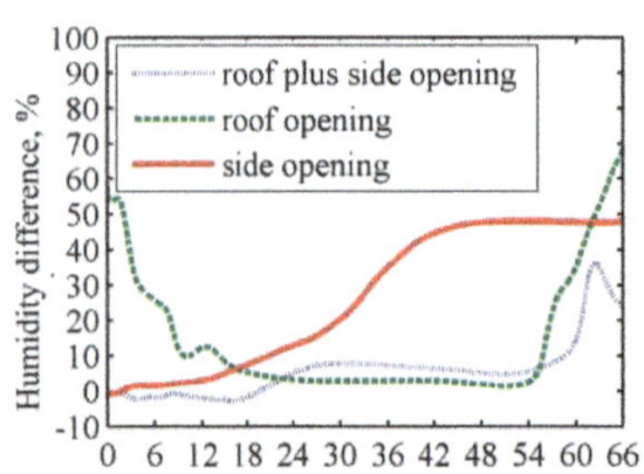

Figure 10. The difference between inside and outside air humidity at a height of 1.0 m after the one-minute dehumidification process in winter.

from 2.5 °C and 0% at the windward side to 3.5 °C and 50% at the leeward side more than the external conditions, respectively. With the roof opening, the air temperature and humidity drop sharply in the first span, vary little in the center and are elevated noticeably in the last span. With the roof plus side opening, the temperature distribution is relatively homogeneous. However, a clear humidity gradient occurs between the two sides of the greenhouse. It is clear that the humidity is close to the external conditions in the windward side, whereas it increases up to 24% more than the external conditions in the leeward side.

It is well known that the prerequisite of condensation formation is that the indoor air temperature is lower than the air dew-point temperature. The horizontal profiles of the difference between the air temperature and the air dew-point temperature at a height of 1.0 m in the middle of the greenhouse after the one-minute dehumidification process (Figure 11) show that the air in the leeward area is prone to condensation. Moreover, compared to other configurations, the side opening more easily induces condensation formation.

The time courses of the indoor mean temperature and relative humidity during the dehumidification process with natural ventilation (Figure 12) show that, in the first two minutes, the side opening induces a slower decrease in the indoor air temperature and relative humidity than the roof opening and the roof plus side opening. The latter two configurations result in similar variation trends in the air temperature and relative humidity. With roof, side and roof plus side opening, the greenhouse dehumidification

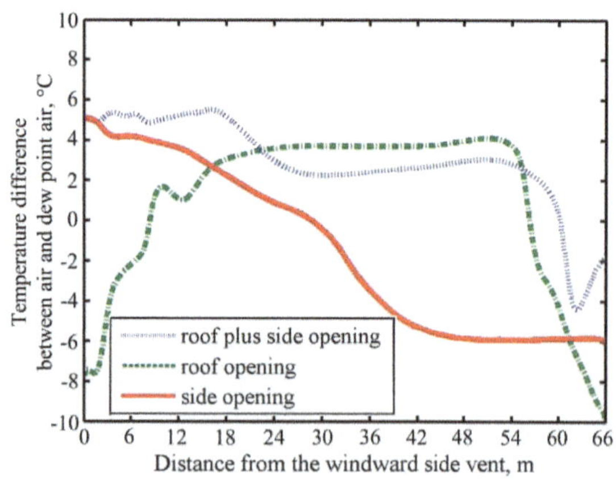

Figure 11. The difference between the air temperature and the air dew point temperature at a height of 1.0 m in the greenhouse after the one-minute dehumidification process in winter.

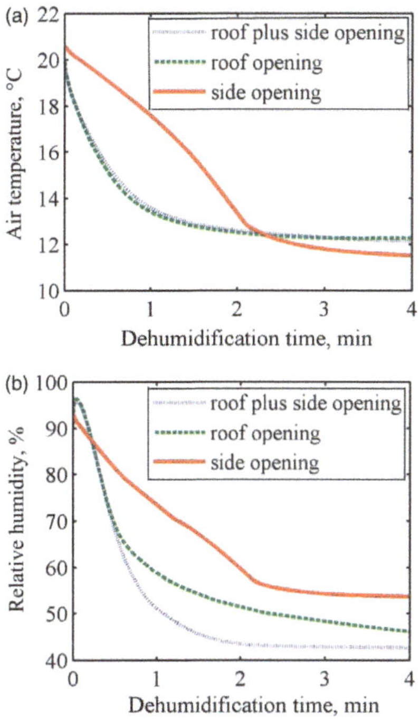

Figure 12. Time courses of the internal (a) mean temperature and (b) relative humidity during the dehumidification process in winter.

time is 2.5 minutes, 3.7 minutes and 1.3 minutes, respectively. Correspondingly, the mean values at the dehumidification end are 12.4 °C, 11.6 °C and 13.2 °C for the air temperature, and 50%, 54% and 48% for the relative humidity.

4.4. *Effect of the vent opening size on the greenhouse dehumidification time and inside climate*

The relationship between the greenhouse dehumidification time and vent opening size is illustrated (Figure 13). The data can be fitted to a function $y = a + b \exp(-c\ x)$ where y is the dehumidification time and x is the vent opening size. Under the roof opening sizes of 0.2 m, 0.4 m, 0.6 m, 0.8 m and 1.0 m, the dehumidification time is 6.5 minutes, 4.0 minutes, 2.8 minutes, 2.6 minutes and 2.5 minutes, respectively. By using this equation, the dehumidification time can be estimated across the whole range. It is clear that as the vent opening is increased from 0.2 m to 1.0 m, the dehumidification time decreases exponentially from 6.5 minutes to 2.5 minutes. As shown in Figure 14, at the dehumidification end, the mean values of inside temperature and humidity drop from 13.9 °C to 12.4 °C and from 66% to the value around 50%, respectively. The corresponding standard deviations vary in the range of 0.85–1.3°C and 7.4–9.0%.

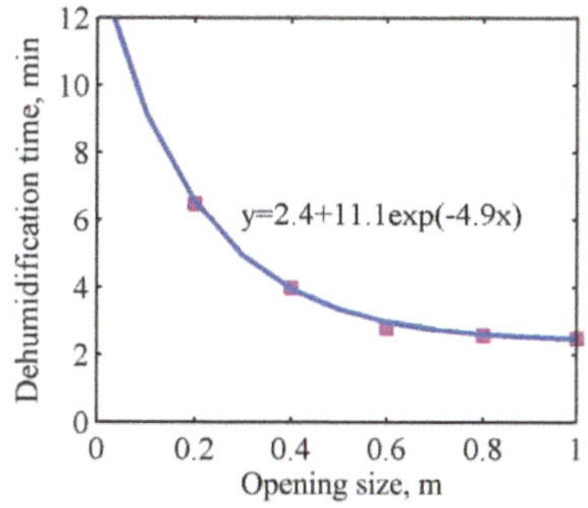

Figure 13. Evaluation of the greenhouse dehumidification time as the vent opening size increases in winter.

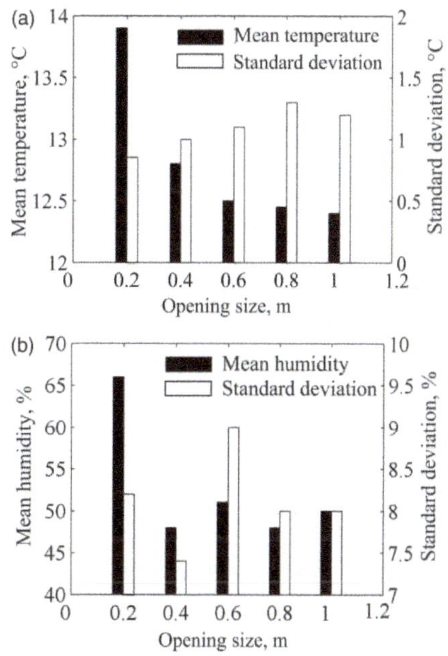

Figure 14. Mean and standard deviation values of the internal (a) air temperature and (b) humidity with five different roof opening sizes at the dehumidification end in winter.

4.5. *Optimization of vent opening during the cooling and dehumidification process with natural ventilation*

From the growers' viewpoint, the primary aim of a ventilation management system is to fix the optimal vent configuration according to the prevailing weather conditions. Therefore, in this study, the summer ventilation process was assessed based on cooling efficiency and homogeneity performance, while the winter ventilation process was assessed based on heat loss, dehumidification efficiency and homogeneity performance. The natural ventilation performance corresponding to different vent configurations is summarized in Tables 6 and 7.

During the summer ventilation process, the side opening has the poorest cooling capacity. Moreover, it offers

Table 6. Evaluation of three different vent configurations during the cooling processes.

	Cooling efficiency	Climate parameters		Homogeneity	
Case	Y_c	T_m	RH_m	η_T	η_{RH}
Roof opening	1.5	35.9	57	1.75	24.8
Side opening	8.3	36.5	56	1.40	19.7
Roof plus side	1.3	36.0	57	1.58	21.7

Note: $Y_c = [(T_m - T_o)/A] \times 10^3$, where Y_c is the cooling efficiency in °C m^{-2}, T_m is the mean indoor temperature after the cooling process in °C, T_o is the outside air temperature in °C, and A is the ventilation area in m^2; RH_m = mean indoor relative humidity at the dehumidification end in %; $\eta_T = (\Delta T/T_o) \times 10^2$, where η_T is the temperature homogeneity in the crop canopy, and ΔT is the difference of the maximum and minimum values of the air temperature in the crop canopy in °C; $\eta_{RH} = (\Delta RH/RH_o) \times 10^2$, where η_{RH} is the humidity homogeneity in the crop canopy, ΔRH is the difference of the maximum and minimum values of the relative humidity in the crop canopy in %, and RH_o is the outside air humidity in %.

Table 7. Evaluation of three different vent configurations during the dehumidification processes.

	Heat loss		Dehumidification efficiency	Climate parameters		Homogeneity	
Case	Q_{sen}	Q_{lat}	Y_d	T_m	RH_m	η_T	η_{RH}
Roof opening	354	767	4.8	12.4	50	2.54	96.2
Side opening	230	498	9.7	11.6	54	3.56	110.0
Roof plus side	689	1,489	7.7	13.2	48	2.93	93.0

Note: $Q_{sen} = \rho C_p G(T_i - T_o)$, where Q_{sen} is the sensible heat flux in W m^{-2}, ρ is the air density in kg m^{-3}, C_p is the volumetric specific heat of air in J m^{-3} °C^{-1}, G is the air exchange through the window in m^{-3} s^{-1} m^{-2}, and T_i and T_o are the temperatures of the air inside and outside the greenhouse in °C, respectively; $Q_{lat} = \lambda G(x_i - x_o)$, where Q_{lat} is the latent heat flux in W m^{-2}, λ is the latent heat of evaporation in J kg^{-1}, and x_i and x_o are the concentration of vapor in the air inside and outside the greenhouse in kg m^{-3}, respectively; $Y_d = [(RH_{in} - RH_m)/(A \times t)] \times 10^6$, where Y_d is the dehumidification efficiency in % m^{-2} s^{-1}, RH_{in} is the initial indoor humidity before the dehumidification begins in %, RH_m is the mean indoor relative humidity at the dehumidification end in %, A is the ventilation area in m^2, and t is the greenhouse dehumidification time in s; T_m = mean indoor temperature after the cooling process in °C; $\eta_T = (\Delta T/T_o) \times 10^2$, where η_T is the temperature homogeneity in the crop canopy and ΔT is the difference of the maximum and minimum values of the air temperature in the crop canopy in °C; $\eta_{RH} = (\Delta RH/RH_o) \times 10^2$, where η_{RH} is the humidity homogeneity in the crop canopy, ΔRH is the difference of the maximum and minimum values of the relative humidity in the crop canopy in %, and RH_o is the outside air humidity in %.

the worst cooling efficiency ($Y_c = 8.3$ °C m^{-2}). Therefore, this configuration is not recommended. Roof opening offers a better cooling performance than side opening, but slightly increases the indoor climate heterogeneity. This finding differs from the previous results (Bartzanas et al., 2004; Bournet, Ould Khaoua, & Boulard, 2007; Ishii et al., 2008; Kittas, Katsoulas, Bartzanas, Mermier, & Boulard, 2008). In those studies, roof opening offers the best homogeneity performance. The reason for this difference is that the non-symmetric design of the roof vents of the studied greenhouse results in a relatively worse homogeneity. Roof plus side opening provides the best ventilation efficiency of approximately 6.4 times more than that of side opening alone and approximately 1.2 times more than that of roof opening alone. Moreover, it offers a good homogeneity performance ($\eta_T = 1.58$, $\eta_{RH} = 21.7$). Therefore, from the considerations of cooling efficiency and homogeneity performance, the roof plus side opening configuration tends to be recommended.

Presently, a small number of research studies have been conducted concerning the ventilation performance during the winter period. In this study, the greenhouse climate has characteristics with a more obvious heterogeneity at the dehumidification end. Roof plus side opening provides the best dehumidification performance, whereas this configuration results in significantly higher heat loss than the other configurations. The side opening is more appropriate for reducing the heat loss and offers the best dehumidification efficiency ($Y_d = 9.7\%$ m^{-2} s^{-1}), but it provides the worst homogeneity of the air temperature and humidity in the crop canopy ($\eta_T = 3.56$, $\eta_{RH} = 110$). Therefore, from the considerations of dehumidification efficiency, heat loss and homogeneity performance, the roof opening configuration makes a good compromise and tends to be suggested. Moreover, from a practical point of view, this configuration can prevent very cold external air from flowing into the crop canopy directly.

5. Conclusion

In this paper, the comparative effect of vent opening on the greenhouse microclimate under different ventilation processes was investigated. The results reveal that vent configuration strongly affects not only greenhouse microclimate patterns but also the spatial distribution and dynamics behavior of the internal temperature and relative humidity. Moreover, during the dehumidification process, increasing the vent opening size induces an exponential decrease in the greenhouse dehumidification time and further affects the inside climate. The assessment of ventilation performance demonstrates that the roof plus side opening is the most suitable for greenhouse cooling in summer, while the roof opening is the most suitable for greenhouse dehumidification in winter. These results can help the horticulturists not only to understand the ventilation mechanism but also to have better control of the ventilation system in different seasons. One limitation of this study is that some physical mechanisms (e.g., condensation) were not taken into consideration in the CFD model. Meanwhile, another limitation is that the grid density across the sidewall vents should be further increased. Therefore, the suggested improvement is to enhance the accuracy of the CFD model. In the future, the main direction of this work will be to investigate the effects of external climate conditions (e.g., the wind regime) and internal elements (e.g., insect screens) on the greenhouse microclimate.

Disclosure statement

No potential conflict of interest was reported by the authors.

Funding

This work was supported by the Shanghai Municipal Agriculture Commission [Project No. 2009(8-1)].

Nomenclature

A	ventilation area (m^2)
C_d	drag coefficient
C_F	non-linear momentum loss coefficient
C_p	volumetric specific heat of air (J m^{-3} °C^{-1})
G	air exchange through the window (m^{-3} s^{-1} m^{-2})
H	crop height (m)
I_{LAI}	leaf area index (m^2 m^{-2})
I_{LAD}	leaf area density (m^2 m^{-3})
K_p	permeability of the porous medium (m^2)
Q_{sen}, Q_{lat}	sensible and latent heat flux (W m^{-2})
RH_{in}	initial indoor humidity before the dehumidification begins (%)
RH_o	outside air humidity (%)
RH_m	mean indoor relative humidity at the dehumidification end (%)
S_ϕ	source term
T_i, T_o	inside and outside air temperature (°C).
T_m	mean indoor temperature after the cooling process (°C)
t	ventilation time (s)
U, V, W	components of velocity vector
u_h	air velocity at the reference height (m s^{-1})
x_i, x_o	concentration of vapor in the air inside and outside the greenhouse (kg m^{-3})
x, y, z	Cartesian space coordinates
Y_c	cooling efficiency (°C m^{-2})
Y_d	dehumidification efficiency (% m^{-2} s^{-1})

Greek letters

Γ	diffusion coefficient (m^2 s^{-1})
ΔRH	difference of the maximum and minimum value of the relative humidity in the crop canopy (%)
ΔT	difference of the maximum and minimum value of the air temperature in the crop canopy (°C)
η_{RH}	humidity homogeneity in the crop canopy
η_T	temperature homogeneity in the crop canopy
λ	latent heat of evaporation (J kg^{-1})
μ	dynamic viscosity (kg m^{-1} s^{-1})
ρ	air density (kg m^{-3})
ϕ	concentration of the transported quantity
∇^2	Laplace operator

References

Baeza, E. J., Pérez-Parra, J. J., Bailey, B. J., Montero, J. I., Lopez, J. C., & Gazquez, J. C. (2009). Analysis of the role of sidewall vents on buoyancy-driven natural ventilation in parral-type greenhouses with and without insect screens using computational fluid dynamics. *Biosystems Engineering*, *104*(1), 86–96.

Bao, N., Zhou, Y. B., Jiao, K., Yin, Y., Du, Q., & Chen, J. (2014). Effect of gas diffusion layer deformation on liquid water transport in proton exchange membrane fuel cell. *Engineering Applications of Computational Fluid Mechanics*, *8*(1), 26–43.

Baptista, F. J., Bailey, B. J., Randall, J. M., & Meneses, J. F. (1999). Greenhouse Ventilation Rate: Theory and Measurement with Tracer Gas Techniques. *Journal of Agricultural Engineering Research*, *72*(4), 363–374.

Bartzanas, T., Boulard, T., & Kittas, C. (2004). Effect of vent arrangement on windward ventilation of a tunnel greenhouse. *Biosystems Engineering*, *88*(4), 479–490.

Boulard, T., & Wang, S. (2002). Experimental and numerical studies on the heterogeneity of crop transpiration in a plastic tunnel. *Computers and Electronics in Agriculture*, *34*(1–3), 173–190.

Bournet, P. E., Ould Khaoua, S. A., & Boulard, T. (2007). Numerical prediction of the effect of vent arrangements on the ventilation and energy transfer in a multi-span glasshouse using a bi-band radiation model. *Biosystems Engineering*, *98*(2), 224–234.

Bournet, P. E., Ould Khaoua, S. A., Boulard, T., Migeon, C., & Chassériaux, G. (2007). Effect of roof and side opening combinations on the ventilation of a glasshouse using computer simulations. *Transactions of the ASABE*, *50*(1), 201–212.

Fatnassi, H., Boulard, T., & Bouirden, L. (2003). Simulation of climatic conditions in full-scale greenhouse fitted with insect-proof screens. *Agricultural and Forest Meteorology*, *118*(11–2), 97–111.

Fatnassi, H., Boulard, T., Poncet, C., & Chave, M. (2006). Optimisation of greenhouse insect screening with computational fluid dynamics. *Biosystems Engineering*, *93*(3), 301–312.

Hong, S. W., Lee, I. B., Hwang, H. S., Seo, I. H., Bitog, P., Yoo, J. I., ... Yoon, N. K. (2008). Numerical simulation of ventilation efficiencies of naturally ventilated multi-span greenhouses in Korea. *Transaction of the ASABE*, *51*(4), 1417–1432.

Ishii, M., Sase, S., Okushima, L., Moriyama, H., Hayash, M., & Yamamoto, Y. (2008). Comparison of field measurement and CFD simulation for a natural ventilated multi-span greenhouse in summer. *Acta Horticulturae*, *801*, 941–948.

Kacira, M., Sase, S., & Okushima, L. (2004). Effects of Side Vents and Span Numbers on Wind-Induced Natural Ventilation of a Gothic Multi-Span Greenhouse. *Japan Agricultural Research Quarterly*, *38*(4), 227–233.

Kacira, M., Short, T. H., & Stowell, R. R. (1998). A CFD evaluation of naturally ventilated, multi-span, sawtooth greenhouses. *Transactions of the ASAE*, *41*(3), 833–836.

Kichah, A., Bournet, P. E., Migeon, C., & Boulard, T. (2012). Measurement and CFD simulation of microclimate characteristics and transpiration of an Impatiens pot plant crop in a greenhouse. *Biosystems Engineering*, *112*(1), 22–34.

Kittas, C., & Bartzanas, T. (2007). Greenhouse microclimate and dehumidification effectiveness under different ventilator configurations. *Biosystems Engineering*, *42*(10), 3774–3784.

Kittas, C., Katsoulas, N., Bartzanas, T., Mermier, M., & Boulard, T. (2008). The impact of insect screens and ventilation openings on the greenhouse microclimate. *Transactions of the ASABE*, *51*(6), 2151–2165.

Launder, B. E., & Spalding, D. B. (1974). The numerical computation of turbulent flows. *Computer Methods in Applied Mechanics and Engineering*, *3*(2), 269–289.

Lee, I. B., & Short, T. H. (2000). Two-dimensional numerical simulation of natural ventilation in a multi-span greenhouse. *Transactions of the ASAE*, *43*(3), 745–753.

Lee, I. B., Short, T. H., Sase, S., Okushima, L., & Qiu, G. Y. (2000). Evaluation of structural characteristics of naturally ventilated multi-span greenhouses using computer simulation. *Japan Agricultural Research Quarterly*, *34*(4), 247–256.

Montero, J. I., Munoz, P., Sanchez-Guerrero, M. C., Medrano, E., Piscia, D., & Lorenzo, P. (2013). Shading screens for the improvement of the night-time climate of unheated greenhouses. *Spanish Journal of Agriculture Research*, *11*(1), 32–46.

Nebbali, R., Roy, J. C., & Boulard, T. (2012). Dynamic simulation of the distributed radiative and convective climate within a cropped greenhouse. *Renewable Energy*, *43*, 111–129.

Ould Khaoua, S. A., Bournet, P. E., Migeon, C., Boulard, T., & Chassériaux, G. (2006). Analysis of Greenhouse Ventilation Efficiency based on Computational Fluid Dynamics. *Biosystems Engineering*, *95*(1), 83–98.

Patankar, S. V. (1980). *Numerical heat transfer and fluid flow*. New York, USA: Hemisphere Publishing Corporation.

Ramponi, R., & Blocken, B. (2012). CFD simulation of cross-ventilation for a generic isolated building: Impact of computational parameters. *Building and Environment*, *53*, 34–38.

Richards, P. J., & Hoxey, R. P. (1993). Appropriate boundary conditions for computational wind engineering models using the k-ϵ turbulence model. *Journal of Wind Engineering and Industrial Aerodynamics*, *46–47*, 145–153.

Roy, J. C., Boulard, T., Kittas, C., & Wang, S. (2002). Convective and Ventilation Transfers in Greenhouses, Part 1: The Greenhouse considered as a Perfectly Stirred Tank. *Biosystems Engineering*, *83*(1), 1–20.

Tamimi, E., Kacira, M., Choi, C. Y., & An, L. (2013). Analysis of microclimate uniformity in a naturally ventilated greenhouse with a high-pressure fogging system. *Transactions of the ASABE*, *56*(3), 1241–1254.

Teitel, M., Ziskind, G., Liran, O., Dubovsky, V., & Letan, R. (2008). Effect of wind direction on greenhouse ventilation rate, airflow patterns and temperature distributions. *Biosystems Engineering*, *101*(3), 351–369.

Tong, G., Christopher, D. M., & Li, B. (2009). Numerical modelling of temperature variations in a Chinese solar greenhouse. *Computers and Electronics in Agriculture*, *68*(1), 129–139.

Wieringa, J. (1992). Updating the davenport roughness classification. *Journal of Wind Engineering and Industrial Aerodynamics*, *41*(1–3), 357–368.

Wilson, N. R., & Shaw, R. H. (1977). A higher order closure model for canopy flow. *Journal of Applied Meteorology*, *16*, 1197–1250.

Yu, Y., Xu, J. L., & Gan, N. G. (2014). Effects of parameters on continuously working plasma synthetic jet. *Engineering Applications of Computational Fluid Mechanics*, *8*(1), 55–69.

Zhang, C. X., You, X. Y., & Zhao, S. M. (2014). Application of EFDC model to waterfront planning: Case study of tianjin harbor economic zone, China. *Engineering Applications of Computational Fluid Mechanics*, *8*(1), 1–13.

6

An extrapolation-based boundary treatment for using the lattice Boltzmann method to simulate fluid-particle interaction near a wall

Cheng-Hsien Lee[a], Zhenhua Huang[b]* and Yee-Meng Chiew[c]

[a]*Department of Civil and Environmental Engineering, National University of Singapore, Singapore;* [b]*Department of Ocean and Resources Engineering, School of Ocean and Earth Science and Technology, University of Hawaii at Manoa, Honolulu, USA;* [c]*School of Civil and Environmental Engineering, Nanyang Technological University, Singapore*

Two important issues need to be addressed when using lattice Boltzmann methods to simulate particulate flows: complex moving boundaries and near contact of two solid objects. This study introduces a new boundary treatment method for using lattice Boltzmann methods to study fluid-particle interaction problems in which solid objects are close to a wall or two solid objects are near contact. The new method does not need to use any additional lubrication correction to total hydrodynamic force, instead implementing the no-slip boundary condition by introducing an appropriate fictitious flow field outside the fluid domain. The new method was verified by simulating a spherical particle falling down in fluids with and without rebound. It is concluded that the proposed boundary treatment method can accurately simulate the fluid-particle interaction near a rigid wall and collision in fluid.

Keywords: lattice Boltzmann method; solid-fluid interactions; fluid-particle interaction; boundary treatment

1. Introduction

Particulate flows are ubiquitous in industrial and natural processes, but they are still not well understood. Numerical simulations can further our understanding of particulate flows. Variations of the lattice Boltzmann method (LBM), a class of kinetic-theory-based method (Higuera & Succi, 1989; Higuera, Succi, & Benzi, 1989; McNamara & Zanetti, 1988; Succi, Foti, & Higuera, 1989), have been applied to the simulation of particulate flows (Derksen, 2011; Z. Feng & Michaelides, 2005, 2009; Jafari, Yamamoto, & Rahnama, 2011; Ladd, 1994a, 1994b; Lin, Lin, Chin, & Tai, 2011; Nguyen & Ladd, 2002; Suzuki & Inamuro, 2013; ten Cate, Nieuwstad, Derksen, & Van den Akker, 2002), and a key component in LBMs is the so-called particle distribution function (PDF), which represents a number of fluid particles with certain velocities. When applying an LBM scheme, it is necessary to transform the velocities into the values of PDFs at boundaries. Several methods are available to perform this transformation, including immersed-boundary methods, bounce-back methods, and extrapolation methods.

In most of the immersed-boundary lattice Boltzmann methods (IBLBMs; Dash & Lee, 2014; Y. Feng, Han, & Owen, 2010; Z. Feng & Michaelides, 2005; ten Cate et al., 2002), fluid is assumed to be filled within the solid domain and Lagrangian markers are distributed around the boundaries. Appropriate extra body-forces are applied to the makers to ensure the no-slip boundary condition. The advantage of Eulerian-Lagrangian-based IBLBMs is that it is easy to trace the motion of a boundary. However, Suzuki and Inamuro (2013) indicated that Eulerian-Lagrangian-based IBLBMs introduce a discontinuity in the velocity gradient at boundary, reducing their numerical accuracy. Suzuki and Inamuro further proposed a high-order Eulerian-Lagrangian-based IBLBM to improve the numerical accuracy by distributing the Lagrangian makers inside the solid, but the implementation of their model become complicated.

Bounce-back methods (BBMs; Cornubert, d'Humières, & Levermore, 1991; Frisch et al., 1987) assume that the fluid particles bounce back in their reversed directions when they encounter a solid surface, and are applicable only to stationary solid boundaries. To apply BBMs to moving boundaries, Ladd (1994a) introduced an adjustment term to account for the velocity of a moving boundary. However, this method assumes that the boundary always coincides with the lattice points or lattice midpoints, leading to low numerical resolution for general surface geometries. The numerical resolution at boundary can be improved by the following methods: considering the volume fraction for each boundary fluid node (Verberg & Ladd, 2001), interpolating the PDFs at the boundary (Bouzidi, Firdaouss, & Lallemand, 2001; Filippova & Hänel, 1998; Mei, Luo, & Shyy, 1999) and extrapolating/interpolating the velocity at the lattice midpoints of the boundary lattice links (Yin & Zhang, 2012). Each of these

*Corresponding author. Email: zhenhua@hawaii.edu

methods can improve the accuracy of BBMs to second order.

Extrapolation methods introduce ghost nodes outside the fluid domain but adjacent to boundaries. The PDFs in the ghost nodes are decomposed into two parts: equilibrium and non-equilibrium. Equilibrium parts are computed using the velocity and density extrapolated from the fluid domain, and non-equilibrium parts are directly extrapolated from the fluid domain. An extrapolation can be link-based (Guo, Zheng, & Shi, 2002) or normal-direction-based (Tiwari & Vanka, 2012). Extrapolation methods are also of second-order accuracy.

In applying the improved BBMs and extrapolation methods, the total hydrodynamic force exerting on a particle is the sum of the momentum changes occurring on the particle surface. If the distance between two particles or between a particle and a wall is less than a grid size, momentum exchanges at some boundary links crossing two solid boundaries cannot be calculated, and the hydrodynamic force exerting on the particles will be underestimated. Lubrication corrections (Ding & Aidun, 2003; Nguyen & Ladd, 2002) have been proposed, but few studies have checked the accuracy of lubrication corrections by comparing the simulation results with experimental data. The lubrication corrections may significantly influence the behavior of particles near a boundary, as well as the particle-particle interaction and the particle-wall interaction.

This study aims to develop an easy-to-implement and accurate method to simulate fluid-particle interaction near a wall by mirroring flow field near a boundary anti-symmetrically into the non-fluid domain. This method can not only treat complex moving boundaries, but it can also accurately determine the hydrodynamic forces exerting on a particle that is very close to or in contact with a solid boundary without any additional lubrication correction.

This study introduces a method to accurately model forces exerting on moving particles near or in contact with a rigid boundary. To check its robustness in dealing with complex boundaries, the present method is applied to three problems: cylindrical Couette flows, flows past a sphere with Reynolds numbers ranging from 25 to 200, and sedimentation of a spherical particle in a viscous fluid with and without rebound. The accuracy of the proposed method is validated by comparisons of the sedimentation trajectories and velocities with those reported in published literature (Li, Hunt, & Colonius, 2012; ten Cate et al., 2002).

2. Numerical scheme

2.1. *Lattice Boltzmann equation*

The core of the lattice Boltzmann method is the so-called particle distribution function, $f_i(\mathbf{x}, t)$, which represents the probability of a particle appearing at location $\mathbf{x}$ at time t. The particle moves at a velocity $\mathbf{c}_i$, which represents the

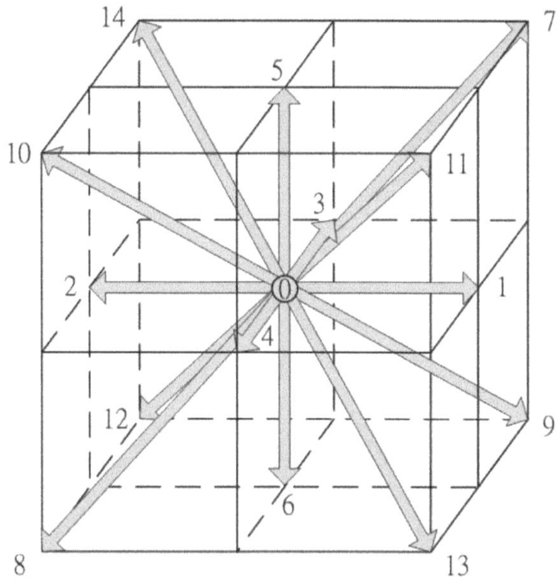

Figure 1. D3Q15 lattice.

ith component of a discrete velocity set. For simplicity, this study adopts a three-dimensional discrete velocity set with 15 components (D3Q15; Qian, Succi, & Orszag, 1995) for three-dimensional simulations. The velocities in D3Q15, as shown in Figure 1, are given by

$$\mathbf{c}_i = c \times \begin{cases} (0,0,0) & i=0 \\ (\pm 1,0,0),(0,\pm 1,0), & i=1,2,\ldots,6, \\ (0,0,\pm 1) & i=7,8,\ldots,14 \\ (\pm 1,\pm 1,\pm 1) & \end{cases} \quad (1)$$

where $c \equiv \Delta x/\Delta t$ with Δx being the grid size and Δt the time step. It is required that the simulated maximum velocity should be much smaller than the basic speed on the lattice $c = \Delta x/\Delta t$. After having found the particle distribution function $f_i(\mathbf{x}, t)$, the fluid density (ρ), and the fluid momentum ($\rho\mathbf{u}$; He & Luo, 1997a, 1997c) are calculated by

$$\rho = \sum_{i=0}^{14} f_i \quad \text{and} \quad \rho\mathbf{u} = \sum_{i=0}^{14} \mathbf{c}_i f_i. \quad (2)$$

The lattice Boltzmann equation, which describes the evolution of $f_i(\mathbf{x}, t)$, is a discrete form of the following single-relaxation-time Boltzmann equation (He & Luo, 1997c):

$$f_i(\mathbf{x} + \mathbf{c}_i\Delta t, t + \Delta t) - f_i(\mathbf{x}, t) = -\frac{1}{\tau}[f_i(\mathbf{x}, t) - f_i^{\text{eq}}(\mathbf{x}, t)], \quad (3)$$

where τ is the relaxation time and f_i^{eq} is the equilibrium distribution function to be specified. Eq. (3) recovers to the Navier-Stokes equation to the second order in space and time by applying the Chapman-Enskog expansion with low-Mach-number and with τ being given by

$$\tau = \frac{3\nu}{c^2 \Delta t} + 0.5, \quad (4)$$

where ν is the kinematic viscosity of the fluid (He & Luo, 1997b). Since ν, c, and Δt are all positive, Eq. (4) implies that $\tau \geq 0.5$. The equilibrium distribution functions depend on the local fluid state and are given by

$$f_i^{\mathrm{eq}} = \omega_i \rho \left[1 + \frac{3}{c^2}\mathbf{c}_i \cdot \mathbf{u} + \frac{9}{2c^4}(\mathbf{c}_i \cdot \mathbf{u})^2 - \frac{3}{2c^4}\mathbf{u} \cdot \mathbf{u} \right] \quad (5)$$

where ω_i are weighting factors and are given by (Aidun & Clausen, 2010; Chen & Doolen, 2003; He & Luo, 1997c)

$$\omega_i = \begin{cases} \frac{2}{9} & i = 0 \\ \frac{1}{9} & i = 1, 2, \ldots, 6, \\ \frac{1}{72} & i = 7, 8, \ldots, 14 \end{cases} \quad (6)$$

Implementation of a lattice Boltzmann method has two steps: a collision step and a streaming step.

At the collision step

$$\tilde{f}_i(\mathbf{x}, t) - f_i(\mathbf{x}, t) = -\frac{1}{\tau}[f_i(\mathbf{x}, t) - f_i^{\mathrm{eq}}(\mathbf{x}, t)] \quad (7)$$

and at the streaming step

$$f_i(\mathbf{x} + \mathbf{c}_i \Delta t, t + \Delta t) = \tilde{f}_i(\mathbf{x}, t) \quad (8)$$

where $\tilde{f}_i$ represents the post-collision state of the PDF. The collision step simulates fluid particle collisions and their scatters into different directions. The streaming step advances the fluid particles to the next lattice node along their directions of motion. Since both steps take little computational effort, the lattice Boltzmann method is highly efficient and simple to implement.

As aforementioned, the lattice Boltzmann equation (Eq. 3) can recover to the Navier-Stokes equation only for small $\mathrm{Ma} = u_{\max}/c$ with $u_{\max}$ being the maximum velocity. Therefore, it is desirable to maintain Ma as close to zero as possible; in this regard, He and Luo (1997b) suggest $\mathrm{Ma} < 0.15$, and Y. Feng, Han, and Owen (2007) suggest $\mathrm{Ma} < 0.1$. Y. Feng et al. also mention that numerical instability will occur when $\tau \to 0.5$. Mathematically, the computational Mach number (Ma) and the relaxation time (τ) represent dimensionless grid size (Δx) and time step (Δt), thus both can affect the numerical accuracy and stability (Chen & Doolen, 2003; Y. Feng et al., 2007). Therefore, Δx and Δt should be chosen such that τ should be larger than 0.5 and Ma should be as small as possible. However, the spatial resolution of a flow field is controlled by grid size Δx. For a given grid size, both Ma and τ decrease with decreasing Δt. For a high value of Re, a small Δt is required to ensure a small Ma. Therefore this method may not be suitable for high Reynolds number flows.

This study applies the LBM to solving moving-particle problems. The previous study (Ladd, 1994b) suggested using an LBM in both the fluid and the solid domains by assuming that the solid domain is also filled with fluid, but that the fluid in the solid domain will reduce the numerical stability (Nguyen & Ladd, 2002). In this study, an LBM is performed only in the fluid domain as suggested by previous studies (Aidun & Lu, 1995; Nguyen & Ladd, 2002). A moving particle leads to a time-varying fluid domain. The density of the new fluid node which is occupied by the particle at the previous time step is given by the initial fluid density for simplicity, and the velocity of the new fluid node is given by the solid velocity at the previous time step. Finally, f_i of the new fluid node is computed using Eq. (5) as in Aidun and Lu (1995).

2.2. *Boundary conditions*

We propose a new extrapolation method for treating the no-slip boundary condition, which is similar to the method of images (Graebel, 2001). At first, we consider a simple case in which a boundary is static and located at the midpoint of a lattice link, as shown in Figure 2, where B represents the boundary; the subscripts S and F represent, respectively, the ghost node in the solid domain and the fluid node adjacent to the boundary. We assume that a fictitious flow field exists outside the fluid domain, and the velocity of the fictitious flow field is anti-symmetric to that in the fluid domain. This assumption yields $\mathbf{u}_S = -\mathbf{u}_F$, $\partial(-\mathbf{u}_S)/\partial(-\mathbf{x}) = \partial \mathbf{u}_F/\partial \mathbf{x}$, and a zero velocity at the boundary. Based on this assumption, the amount of fluid particles leaving the fluid domain, $\tilde{f}_\alpha(\mathbf{x}_F, t)$, equals that of fictitious fluid particles entering the fluid domain, $\tilde{f}_{-\alpha}(\mathbf{x}_S, t)$, i.e.,

$$\tilde{f}_{-\alpha}(\mathbf{x}_S, t) = \tilde{f}_\alpha(\mathbf{x}_F, t) \quad (9)$$

Eq. (9) suggests $f_{-\alpha}(\mathbf{x}_F, t + \Delta t) = \tilde{f}_\alpha(\mathbf{x}_F, t)$, which is the same as the BBM. Now, we further divide $\tilde{f}_i$ into a component for the equilibrium state, $\tilde{f}_i^{\mathrm{eq}}$, and a component for the non-equilibrium state, $\tilde{f}_i^{neq}$. The component for the equilibrium state can be calculated using Eq. (5). Because

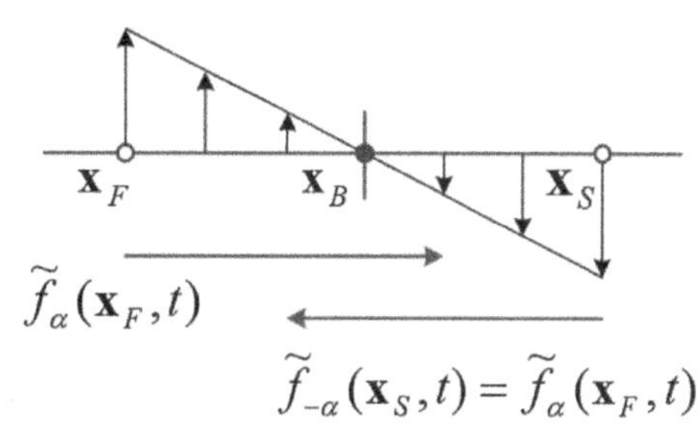

Figure 2. Illustration of the method of images for boundary treatment in the LBM when the stationary wall is located at the midpoint of a boundary link.

$\mathbf{u}_S = -\mathbf{u}_F$ and $\mathbf{c}_\alpha = -\mathbf{c}_{-\alpha}$,

$$\tilde{f}^{\,\mathrm{eq}}_{-\alpha}(\mathbf{x}_S, t) = \tilde{f}^{\,\mathrm{eq}}_{\alpha}(\mathbf{x}_F, t), \tag{10}$$

and consequently

$$\tilde{f}^{\,neq}_{-\alpha}(\mathbf{x}_S, t) = \tilde{f}^{\,neq}_{\alpha}(\mathbf{x}_F, t), \tag{11}$$

$\tilde{f}^{\,neq}_i$ is associated with the shear stress of the flow (Mei, Yu, Shyy, & Luo, 2002). Since a shear stress of an incompressible Newtonian fluid depends only on the deformation rate of the fluid element, $\tilde{f}^{\,neq}_{-\alpha}(\mathbf{x}_S, t) = \tilde{f}^{\,neq}_{\alpha}(\mathbf{x}_F, t)$ holds when $\partial(-\mathbf{u}_S)/\partial(-\mathbf{x}) = \partial \mathbf{u}_F/\partial \mathbf{x}$.

Now, we generalize the above-mentioned boundary treatment method to a moving boundary. For the case in which a boundary moves and does not locate at the midpoint of a link, we assume that the fictitious velocity field relative to the velocity of the boundary is anti-symmetric to the relative velocity field in the fluid domain for each link, as shown in Figure 3. Consequently, one can extrapolate $\mathbf{u}_S$ from the velocity in the fluid domain near the boundary. For numerical stability, as the fraction of the boundary lattice link, $\Delta \equiv |\mathbf{x}_B - \mathbf{x}_F|/|\mathbf{x}_S - \mathbf{x}_F|$, exceeds 0.5, $\mathbf{u}_S$ is calculated using

$$\mathbf{u}_S = \frac{(\mathbf{u}_B - \mathbf{u}_F)}{\Delta} + \mathbf{u}_F. \tag{12}$$

For $\Delta < 0.5$,

$$\mathbf{u}_S = \frac{2(\mathbf{u}_B - \mathbf{u}_{FF})}{1 + \Delta} + \mathbf{u}_{FF}, \tag{13}$$

where $\mathbf{u}_{\mathrm{FF}}$ represents the velocity at $\mathbf{x}_F + \mathbf{c}_{-\alpha}\Delta t$. This extrapolation method is also adopted in calculating the velocity at a midpoint of a boundary link (Yin, Le, & Zhang, 2012). We use $\mathbf{u} = \mathbf{u}_S$ in Eq. (5) to calculate $\tilde{f}^{\,\mathrm{eq}}_{-\alpha}(\mathbf{x}_S, t)$. As a particle is close to a wall or two particles are near contact, $\mathbf{u}_{\mathrm{FF}}$ in Eq. (13) may not be available; in this situation we approximate $\mathbf{u}_{\mathrm{FF}}$ by the velocity of the solid instead.

For the non-equilibrium component, $\tilde{f}^{\,neq}_{-\alpha}(\mathbf{x}_S, t)$ should be given by $\tilde{f}^{\,neq}_{\alpha}(\mathbf{x}_B - \Delta t \mathbf{c}_\alpha[1 - \Delta], t)$ strictly, and $\tilde{f}^{\,neq}_{\alpha}(\mathbf{x}_B - \Delta t \mathbf{c}_\alpha[1 - \Delta], t)$ can be interpolated/extrapolated

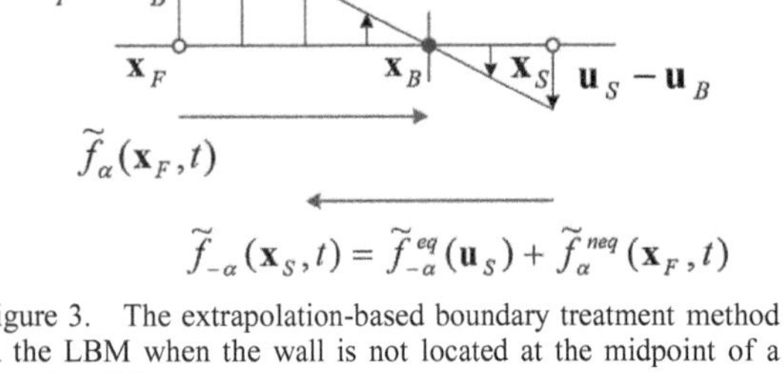

Figure 3. The extrapolation-based boundary treatment method in the LBM when the wall is not located at the midpoint of a boundary link.

from $\tilde{f}^{\,neq}_{\alpha}(\mathbf{x}_F, t)$ and $\tilde{f}^{\,neq}_{\alpha}(\mathbf{x}_{FF}, t)$. Since the non-equilibrium component of the PDF is much smaller than the equilibrium component (He & Luo, 1997b), applying a high-order interpolation/extrapolation method may not be necessary. For simplicity, we approximate $\tilde{f}^{\,neq}_{-\alpha}(\mathbf{x}_S, t)$ by $\tilde{f}^{\,neq}_{\alpha}(\mathbf{x}_F, t)$. Therefore, $\tilde{f}_{-\alpha}(\mathbf{x}_S, t)$ is given by

$$\tilde{f}_{-\alpha}(\mathbf{x}_S, t) = \tilde{f}^{\,\mathrm{eq}}_{-\alpha}(\mathbf{u}_S) + \tilde{f}^{\,neq}_{\alpha}(\mathbf{x}_F, t). \tag{14}$$

As mentioned in the introduction, the existing extrapolation methods can be further classified into two groups: link-based methods (Guo et al., 2002) and normal-direction-based methods (Tiwari & Vanka, 2012). The advantage of the normal-direction-based method over the link-based method is its generality, but its extrapolation process is complicated. Before extrapolating $\mathbf{u}_S$ along a direction normal to a boundary, the velocity at a point in the normal direction needs to be interpolated in multiple directions. As a result, the normal-direction-based method is numerically expensive for three-dimensional problems with moving boundaries. The advantage of the link-based method is that it is easy to implement and numerically inexpensive due $\mathbf{u}_S$ only being extrapolated along the lattice link. The present boundary treatment method is in essence a link-based method. However, the proposed method differs from the existing link-based extrapolation method proposed by Guo et al. (2002) with respect to how $\tilde{f}^{\,neq}_{-\alpha}(\mathbf{x}_S, t)$ is determined. The extrapolation method of Guo et al. (2002) extrapolates $\tilde{f}^{\,neq}_{-\alpha}(\mathbf{x}_S, t)$ from $\tilde{f}^{\,neq}_{-\alpha}(\mathbf{x}_F, t)$ and $\tilde{f}^{\,neq}_{-\alpha}(\mathbf{x}_{\mathrm{FF}}, t)$. The present extrapolation method is based on the method of images (Graebel, 2001) with $\tilde{f}^{\,neq}_{-\alpha}(\mathbf{x}_S, t) = \tilde{f}^{\,neq}_{\alpha}(\mathbf{x}_F, t)$. The boundary treatment is performed after each streaming step (see Eq. 8). At the streaming step, $\tilde{f}_\alpha(\mathbf{x}_F, t) = \tilde{f}^{\,\mathrm{eq}}_{\alpha}(\mathbf{x}_F, t) + \tilde{f}^{\,neq}_{\alpha}(\mathbf{x}_F, t)$ is advanced from $\mathbf{x}_F$ to $\mathbf{x}_S$; thus, determining $\tilde{f}^{\,neq}_{-\alpha}(\mathbf{x}_S, t)$ is straightforward.

2.3. *Particle dynamics*

When a solid object is close to a wall or two solid objects are near contact, the dynamic coupling between fluid and particles become important, but was not addressed by the previous methods (Guo et al., 2002; Tiwari & Vanka, 2012). The motion of a sphere in a viscous fluid is governed by

$$m\frac{\mathrm{d}^2\mathbf{x}_p}{\mathrm{d}t^2} = \mathbf{F}^h + \mathbf{F}^c + m\mathbf{g}_{\mathrm{eff}}, \tag{15}$$

where $\mathbf{x}_p$ is the location of the sphere, $\mathbf{F}^h$ is the hydrodynamic force exerting on the sphere, $\mathbf{F}^c$ is the contact force, m is the mass of the sphere, and $\mathbf{g}_{\mathrm{eff}}$ represents the effective acceleration due to gravity. A linear spring-dashpot model (Silbert et al., 2001) is adopted to calculate the normal contact force for simplicity and the tangential contact force is ignored here. $\mathbf{g}_{\mathrm{eff}}$ is defined as $\mathbf{g}_{\mathrm{eff}} = (1 -$

$\rho/\rho_p)\mathbf{g}$, which incorporates both the buoyant and gravity forces. The angular velocity of a sphere in viscous fluid is governed by

$$\mathbf{I}\cdot\frac{d\omega_p}{dt}=\mathbf{T}^h, \qquad (16)$$

where ω_p is the angular velocity, $\mathbf{I}$ is the moment of inertia, and $\mathbf{T}^h$ represents the hydrodynamic torque exerting on the particle. Because the tangential force is not considered in $\mathbf{F}^c$, the torque resulting from inter-particle contact is zero. In this study, Eq. (15), Eq. (16) and the contact model (Silbert et al., 2001) are solved using an open source code Large-scale Atomin/Molecular Massively Parallel Simulator (LAMMPS; Plimpton, 1995) which adopts the second-order Verlet integration.

The hydrodynamic force, $\mathbf{F}^h$, on a solid surface can be evaluated by the net momentum flux through the solid surface (Ladd, 1994b), which is given for each boundary lattice link, l, by

$$\mathbf{F}_l(\mathbf{x}_B,t)=\mathbf{c}_\alpha[f_\alpha(\mathbf{x}_S,t)-f_{-\alpha}(\mathbf{x}_F,t)]\frac{\Delta x^2}{\Delta t}. \qquad (17)$$

The total force $\mathbf{F}^h$ is simply the summation of $\mathbf{F}_l(\mathbf{x}_B,t)$ for all boundary links

$$\mathbf{F}^h=\sum\mathbf{F}_l(\mathbf{x}_B,t). \qquad (18)$$

Similarly, the total torque exerting on a particle, $\mathbf{T}^h$, can be computed by

$$T^h=\sum(\mathbf{x}_B-\mathbf{x}_p)\times\mathbf{F}_l(\mathbf{x}_B,t). \qquad (19)$$

When the gap between two particles or between a particle and a wall becomes less than Δx, as shown in Figure 4, we modify Eq. (17) to calculate $\mathbf{F}_l(\mathbf{x}_{B_1},t)$. Due to lacking the information for the non-equilibrium component of PDF, the non-equilibrium component of PDF at $\mathbf{x}_{B_1}$ and $\mathbf{x}_{B_2}$ is neglected. Consequently, Eq. (17) becomes

$$\mathbf{F}_l(\mathbf{x}_{B_1},t)=\mathbf{c}_\alpha[\tilde{f}_\alpha^{\mathrm{eq}}(\mathbf{u}_{S_1})-\tilde{f}_{-\alpha}^{\mathrm{eq}}(\mathbf{u}_{S_2})]\frac{\Delta x^2}{\Delta t}. \qquad (20)$$

$\mathbf{u}_{S_1}$ and $\mathbf{u}_{S_2}$ in Eq. (20) can be extrapolated from $\mathbf{u}_{B_1}$ and $\mathbf{u}_{B_2}$. However, when $|\mathbf{x}_{B_1}-\mathbf{x}_{B_2}|$ is too small or $|\mathbf{u}_{B_1}-\mathbf{u}_{B_2}|$ is too large, the extrapolation may lead to numerical instability. Consequently, $\mathbf{u}_{S_1}$ and $\mathbf{u}_{S_2}$ are given by the velocities

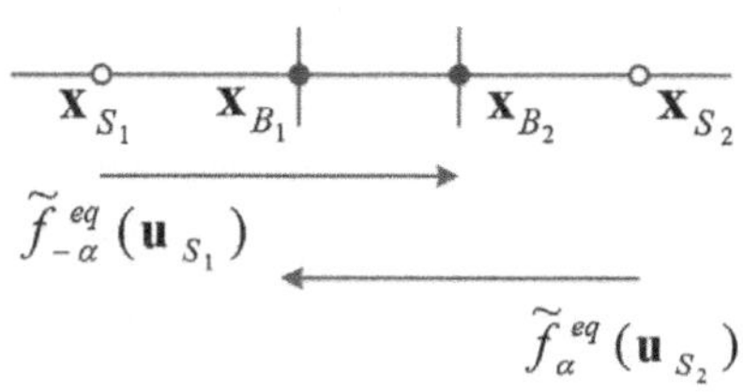

Figure 4. The extrapolation-based boundary treatment method for the near-contact problem.

of the solid at $\mathbf{x}_{S_1}$ and $\mathbf{x}_{S_2}$ for simplicity and numerical stability.

We stress here that using Eq. (20) to calculate $\mathbf{F}_l(\mathbf{x}_{B_1},t)$ is one of the novelties of the present boundary treatment method. In previous studies, empirical formulas for lubrication force (Ding & Aidun, 2003; Nguyen & Ladd, 2002) were adopted to correct the total hydrodynamic forces exerting on particles in regions where $\mathbf{F}_l(\mathbf{x}_{B_1},t)$ cannot be calculated. A detailed comparison between the present boundary treatment method and the lubrication-correction method is presented below.

3. Validation and discussion

To examine its capabilities, the proposed method is applied to four problems in this section: a cylindrical Couette flow, flow past a sphere, sedimentation of a sphere in a viscous fluid, and a sphere rebounding in a viscous fluid.

3.1. *Cylindrical Couette flow*

To study the capacity of the proposed method to treat a fixed complex boundary, a cylindrical Couette flow, which is confined by two co-axial cylinders, is simulated. An inner cylinder of radius R_1 is rotated at the angular velocity ω, and the outer cylinder of radius R_2 remains stationary. The analytical solution of the tangential velocity is given by

$$u_\theta(r)=\frac{\omega R_1}{R_2^2-R_1^2}\frac{R_2}{R_1}\left(\frac{R_2}{r}-\frac{r}{R_2}\right), \qquad (21)$$

where r is the distance to the cylinder center (Guo & Zhao, 2002). The discrete velocity set of D2Q9 (He & Luo, 1997b), $\tau=0.6$, and $\Delta x/R_1=1/8$ were used to simulate the cylindrical Couette flow. The computed and analytical velocity profiles for $R_1/R_2=0.1$, 0.2, and 0.5 are shown in Figure 5 with $\mathrm{Re}=10$, where the Reynolds number is defined as $Re\equiv R_1^2\omega/\nu$; good agreements between the simulated and analytical results can be observed. To study the convergence rates of the present boundary treatment method, an error analysis was performed by varying $\Delta x/R_1$. The variations of the relative global L_2 norm errors (He & Luo, 1997b), E, are shown in Figure 6 for $Re=5$, 10, 20 and 30 with $R_1/R_2=0.5$. The computational Mach number, defined by $\mathrm{Ma}\equiv u_{\max}/c$, where $u_{\max}$ is the maximum flow velocity, is less than 0.02 for these cases, satisfying the assumption of a low Mach number required by the LBM. As shown in Figure 6, the slopes of the curves for the different values of Re are all close to 2, indicating that the boundary treatment is of second-order accuracy. The two previous extrapolation methods (Guo et al., 2002; Tiwari & Vanka, 2012) were also applied to the same problem, and the numerical results showed that they are of second-order accuracy too. For the value of E, we remark that the present boundary treatment method gives numerical errors similar to the method of Guo et al. (2002). As an

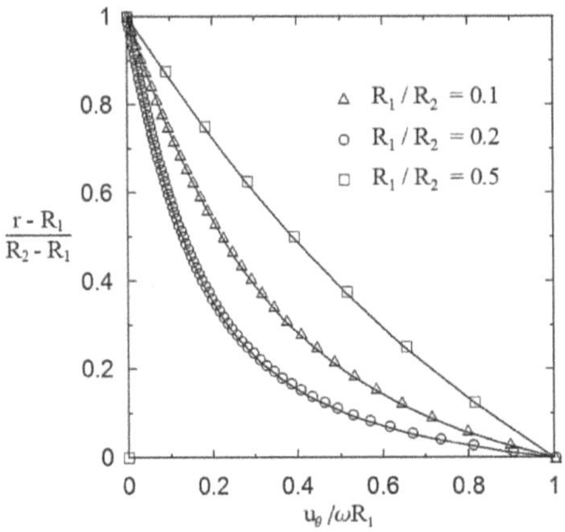

Figure 5. Velocity profiles of cylindrical Couette flows.
Note: The symbols represent the simulation results and the solid lines represent the analytical results.

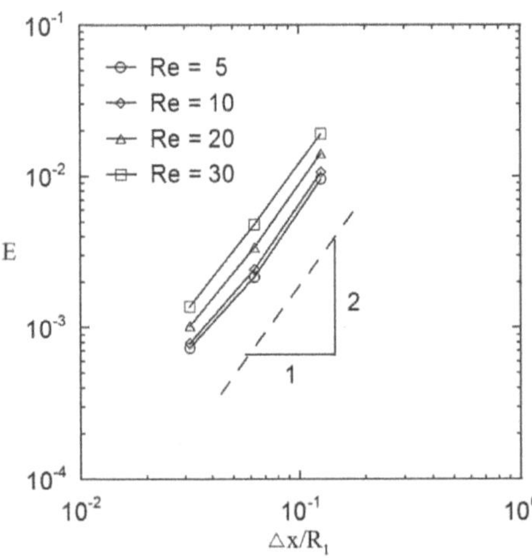

Figure 6. Relative global L_2 norm errors against grid sizes ($\tau = 0.6$).

example, using Re $= 30$ and $\Delta x/R_1 = 0.625$, the values of E obtained by the method of Guo et al. (2002) and the present method are 7×10^{-3} and 5×10^{-3}, respectively. Since the numerical error given in the study of Tiwari and Vanka (2012) is the difference of the two numerical solutions using different grid sizes, no comparison with it is made here.

3.2. *Flow past a sphere*

A schematic diagram of a flow past a sphere of diameter d is shown in Figure 7, where the sphere is fixed at the location $(2.5d, 2.5d, 2.5d)$ in a domain of dimensions $15d \times 5d \times 5d$. A uniform flow field with velocity $= U_1$ in the entire simulation domain is specified as the initial condition, and a uniform flow with velocity $= U_1$ is given

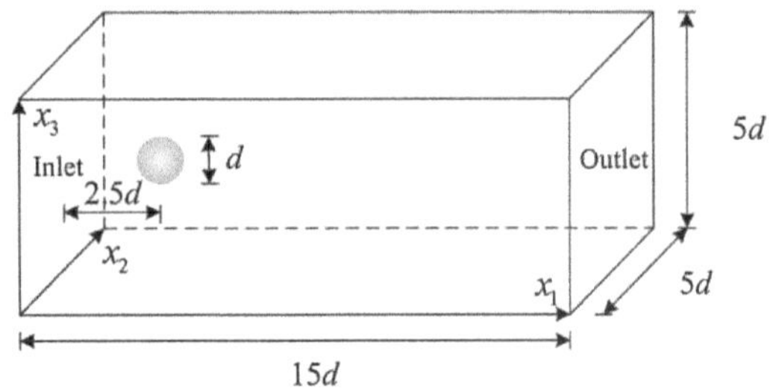

Figure 7. Flow over a three-dimensional stationary sphere.

at the inlet $U_1 = 0$, and the gradient-free PDF in the U_1-direction (U_1) is imposed at the outlet $U_1 = 15d$. For this problem, a drag coefficient for the sphere can be defined by

$$\mathrm{Cd} \equiv \frac{8F_D}{\pi \rho d^2 U_1^2}, \tag{22}$$

where F_D is the drag force (the projection of $\mathbf{F}^h$ in the x_1-direction). Previous studies (Abraham, 1970; Johnson & Patel, 1999) have shown that the drag coefficients only vary with a Reynolds number defined by $Re = U_1 d/\nu$, and using the following empirical formula for Cd:

$$\mathrm{Cd} = \frac{64}{Re} + \frac{6}{1 + \sqrt{Re}} + 0.4, \tag{23}$$

To simulate the problem using the present boundary treatment method, the following parameters are used: $\nu = 1 \times 10^{-4} \mathrm{m}^2/\mathrm{sec}$, $d = 0.2\,\mathrm{m}$, the grid size $\Delta x = 0.01\mathrm{m}$, and $\Delta t = 7.5 \times 10^{-5}$ sec, which give $\tau = 0.59$. The stationary state is achieved when the relative error in the calculated drag coefficient is less than $1.e-5$ between two consecutive time steps.

Figure 8 shows the streamlines computed for Re $= 50$, 100, 150, and 200, indicating that the flow patterns are similar in the range of $50 <$ Re < 200: all have two symmetric vortexes in the leeside of the sphere.

The values of Cd computed for $25 <$ Re < 200 are shown in Figure 9, together with the numerical results reported in the previous study (Johnson & Patel, 1999), and the empirical relationship Eq. (23). The good agreement suggests that the proposed method can adequately treat stationary complex boundaries.

Figure 10 shows the effects of relaxation time τ on the values of Cd. It seems that the drag coefficient is not sensitive to the values of relaxation time: the maximum relative difference in the computed drag coefficients is only about 1% in the range of $0.53 < \tau \leq 0.68$.

When choosing the grid size (Δx) and the time step (Δt) for an LBM simulation, in addition to the requirements on the dimensionless parameters Ma and τ there is also a requirement on the maximum grid size. For this problem, if $d/\Delta x$ is too small, the flow separation cannot be resolved well, leading to underestimated or overestimated values of Cd. As shown in Figure 11, the grid

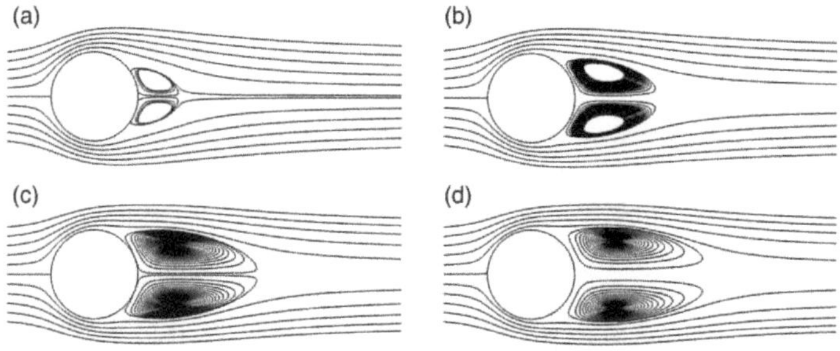

Figure 8. Computed axisymmetric streamlines for steady flow past the sphere: (a) Re = 50; (b) Re = 100; (c) Re = 150; (d) Re = 200.

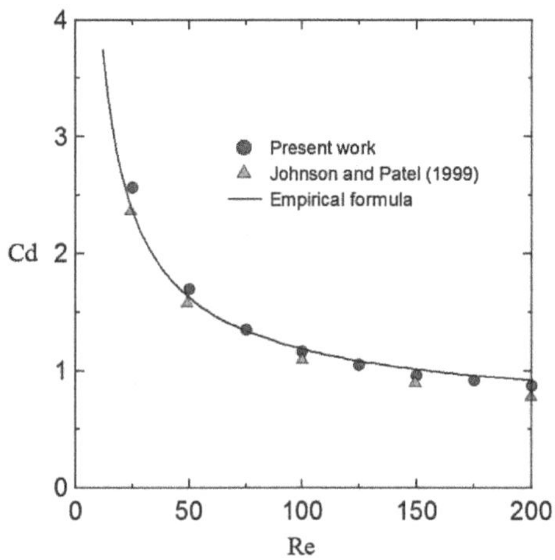

Figure 9. Drag coefficient against Reynolds number for $\tau = 0.59$ and $d/\Delta x = 20$.

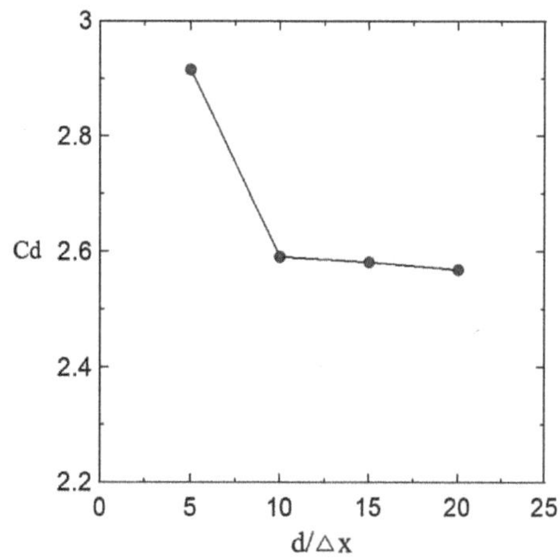

Figure 11. Drag coefficient against grid size for $\tau = 0.59$ and $Re = 25$.

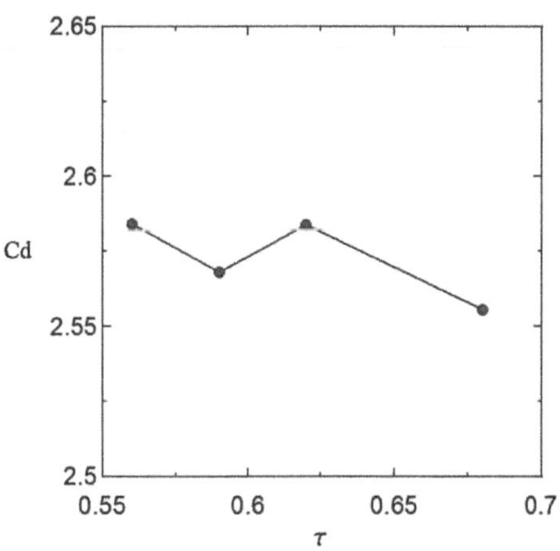

Figure 10. Drag coefficient against τ for $d/\Delta x = 20$ and $Re = 25$.

size has negligible effects on the values of Cd as long as $d/\Delta x \geq 10$.

We remark that the values of Ma for $Re =$ 50, 100, 150, and 200 using $\tau = 0.59$ are 0.0848, 0.182, 0.266 and 0.358, respectively. Even though large values of Ma ($0.15 \leq \text{Ma} \leq 0.358$) can give acceptable results for flow past a sphere, we still adopt $\text{Ma} \leq 0.15$ as a constraint in this study. Therefore, in this study, the grid size (Δx) and the time step (Δt) are chosen such that$d/\Delta x \geq 10$, $0.53 < \tau \leq 0.68$, and $\text{Ma} \leq 0.15$ to ensure numerical accuracy and stability.

3.3. *Sedimentation of a spherical particle in a viscous fluid*

Sedimentation of a sphere in a tank has been studied experimentally by ten Cate et al. (2002). This is a good example for illustrating how well the proposed method can deal with moving-particle problems. Figure 12 shows a schematic diagram of the experimental setup of ten Cate et al., in which the dimensions of the tank are $0.1\,\text{m} \times 0.1\,\text{m} \times 0.16\,\text{m}$. A sphere of $d = 1.5 \times 10^{-2}\,\text{m}$ and $\rho_p = 1120\,\text{kg/m}^3$ was released from a specified location $(0.05\,\text{m}, 0.05\,\text{m}, 0.1275\,\text{m})$ in a fluid, and four fluids were examined in the experiments. In Figure 12, h represents the length of the vertical distance between the sphere and the bottom, and $\mathbf{g}$ = gravitational acceleration.

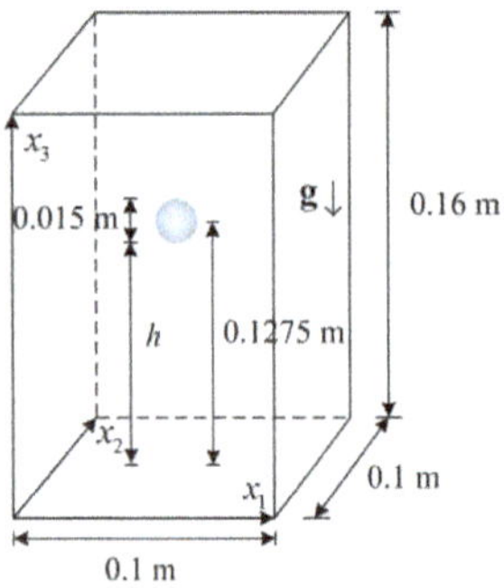

Figure 12. Experimental setup for a single sphere settling under gravity.

The details of the fluid properties and the corresponding numerical setup are given in Table 1. A fixed Δx is used for all four fluids. The boundaries of the tank lie in the midpoints of the boundary links. The values of Δt were selected to ensure that τ lies within 0.6 and 0.65. The Stoke number, St, in Table 1 is defined as $\mathrm{St} \equiv 1/9(\rho_p/\rho)w_{max}d/\nu$, with $w_{\max}$ being the maximum falling velocity of the sphere, and the rebound of a sphere occurs only for large Stoke numbers (St > 10; Ardekani & Rangel, 2008; Joseph, Zenit, Hunt, & Rosenwinkel, 2001; Li et al., 2012). The Stoke numbers for the cases in this section ranged from 0.17 to 3.89, and no rebound occurred (ten Cate et al., 2002).

The computed elevation (h) and falling velocity (w_p) of the sphere are plotted in Figure 13, where the experimental results of ten Cate et al. (2002) are also shown for comparison. Good agreement is found between the simulated and experimental results for all four fluids. To compare the sensitivity of the numerical results with the relaxation time, Case 4 was also simulated using $\tau = 0.68$ ($\Delta t = 6 \times 10^{-4}$ sec and $\Delta x = 1 \times 10^{-3}$ m), and nearly identical results were obtained.

As pointed out in the introduction, when the sphere is so close to the bottom that the distance between the particle and a wall is less than a grid size, the hydrodynamic forces exerting on particles for some boundary links cannot be calculated due to the resolution issue. One remedy is to use the lubrication force F_{lub} proposed by Nguyen and Ladd (2002) to correct the force acting on the sphere when h (the distance between the sphere and the bottom) is smaller than a threshold distance h_Δ:

$$F_{\text{lub}} = -3\pi\rho\nu d u_\perp \left(\frac{d}{h} - \frac{d}{h_\Delta}\right) \tag{24}$$

where $u_\perp$ is the velocity of the sphere perpendicular to the bottom. To examine the ability of the present boundary treatment method to simulate particle movement near a wall, the computed velocity of the sphere near the bottom is shown in Figure 14 for Case 1 in Table 1. Both the experimental and computational results of ten Cate et al. (2002) are given in Figure 14 for comparison. An IBLBM was used by ten Cate et al., and Eq. (24) was adopted to correct the total hydrodynamic force exerting on the sphere. Some previous studies (Z. Feng & Michaelides, 2009; Lin et al.,

Table 1. Fluid properties and numerical setup for sedimentation of a sphere in different viscous fluids.

Case	ρ(kg/m^3)	ν(m^2/sec)	Δx(m)	Δt(sec)	$d/\Delta x$	τ	Ma	St
1	970	3.845×10^{-4}	1×10^{-3}	1×10^{-4}	15	0.615	3.4×10^{-3}	0.17
2	965	2.197×10^{-4}	1×10^{-3}	2×10^{-4}	15	0.632	1.1×10^{-2}	0.48
3	962	1.175×10^{-4}	1×10^{-3}	4×10^{-4}	15	0.641	3.5×10^{-2}	1.45
4	960	6.042×10^{-5}	1×10^{-3}	6×10^{-4}	15	0.608	7.2×10^{-2}	3.89

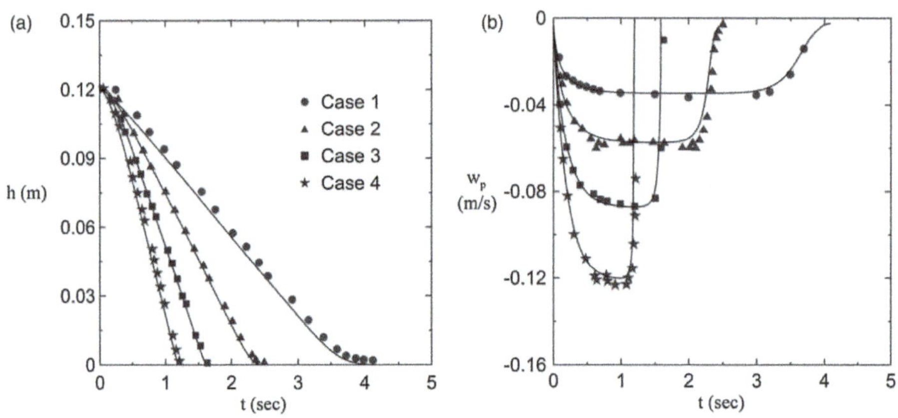

Figure 13. The sphere settling trajectory z_p (a) and sedimentation velocity w_p (b).
Note: The solid lines represent simulation results and the symbols represent experimental results.

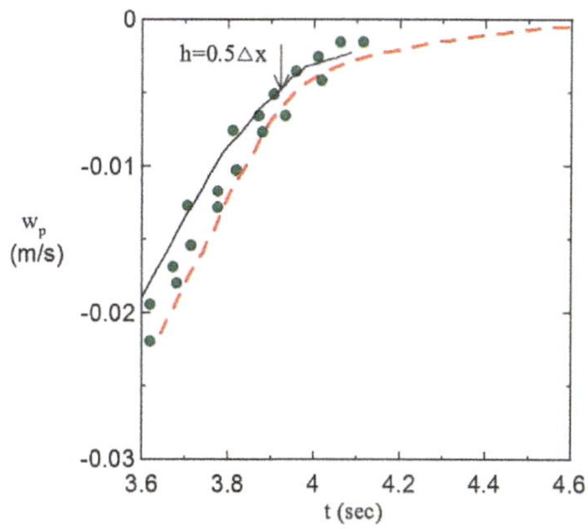

Figure 14. The sedimentation velocity of a sphere near the bottom for Case 1.
Note: The solid lines represent our simulation results, and the dashed lines and symbols represent the numerical results and experimental results of ten Cate et al. (2002).

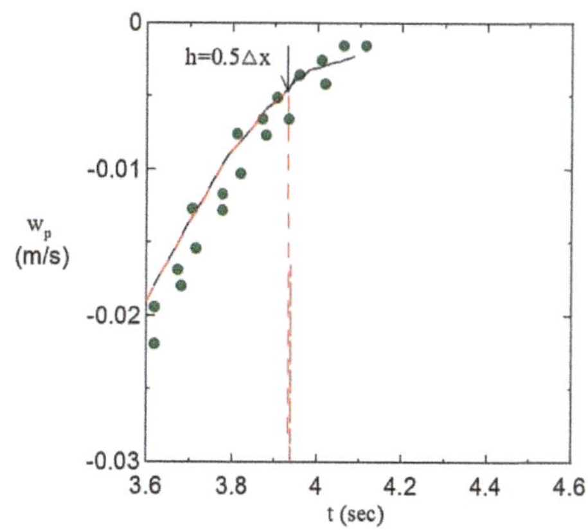

Figure 15. The sedimentation velocity of the sphere near the bottom for Case 1.
Note: The solid lines were obtained using Eq. (20) and the dashed lines were obtained using Eq. (24) to correct the hydrodynamic force. The symbols represent the experimental results of ten Cate et al. (2002).

2011; Suzuki & Inamuro, 2013; ten Cate et al., 2002) also applied various boundary treatment methods using LBMs to reproduce the experiments (ten Cate et al., 2002), but only ten Cate et al. examined the computed velocity of the sphere near the bottom. The good agreement among these three results can be found in Figure 14 even when the gap between the sphere and the bottom is smaller than one half of the grid size. Nevertheless, while the computed results using the present boundary treatment method are within the range of the experimental results of ten Cate et al., the computed results of ten Cate et al. are outside the range of the experimental results when $t < 3.9$ sec. Therefore, we believe that the present boundary treatment method is more accurate than that of ten Cate et al. for this case. Additionally, ten Cate et al. indicated that using Eq. (24) might yield an unrealistically long sedimentation time; however, our method does not have this shortcoming.

We also adopted Eq. (24) rather than Eq. (20) to correct the force exerting on the sphere for Case 1. The computed results using Eq. (24) with $h_\Delta = 0.5\Delta x$ for Case 1 are shown in Figure 15, together with the computed results using Eq. (20) and the experimental results of ten Cate et al. (2002). The computed falling velocity ($-w_p$) using Eq. (24) increases sharply as $h < 0.5\Delta x$, which is physically unreasonable (Eq. (24) with $h_\Delta = 0.5\Delta x$ suggests that $F_{\text{lub}} = 0$ when $h \cong 0.5\Delta x$); this suggests that the success of the lubrication correction model proposed by Nguyen and Ladd (2002) strongly depends on the choice of the value of h_Δ: an inappropriate value of h_Δ may not be able to simulate the lubrication force correctly. The lubrication correction model proposed by Ding and Aidun (2000) is a link-based model; as with Eq. (24), this method also depends on $(1/h - 1/h_\Delta)$. Therefore, this lubrication correction model has the same shortcoming as that of Nguyen and Ladd.

3.4. *Rebound of a spherical particle in a viscous fluid*

When St > 10, the spherical particle will rebound from a wall (Ardekani & Rangel, 2008; Joseph et al., 2001; Li et al., 2012). The rebound of a spherical particle has been studied experimentally (Li et al., 2012). Even though some computational fluid dynamics simulations of rebound problems can be found in the literature (Ardekani & Rangel, 2008; Li et al., 2012), no attempt has been made at the present to simulate rebound problems using LBMs.

We now use the present boundary treatment method to reproduce the experiments of Li et al. (2012). The experiments were performed in a glass rectangular tank filled with a liquid of density 1203 (kg/m^3) and kinematic viscosity 4.173×10^{-5} m^2/sec. A steel sphere of $d = 9.5 \times 10^{-3}$ m and $\rho_p = 7780$ kg/m^3 was released from an electromagnetic plate (a release mechanism) under the fluid surface. Seven release heights (h_r) were examined in the experiments; however, only release heights where the computational Mach numbers are smaller than 0.18 and the relaxation times are greater than 0.53 can be simulated efficiently using LBMs with acceptable accuracy and without the numerical stability problem: $h_r = 5.5 \times 10^{-3}$ m and $h_r = 1.05 \times 10^{-2}$ m.

The computational domain is similar to the one shown in Figure 12, but with the width of the tank set to 3.8×10^{-2} m. The bottom of the electromagnetic plate was located at the midpoints of numerical boundary links. The elastic constant in the linear spring-dashpot contact model (Silbert et al., 2001) was set to be 3.54×10^{-7} N/m, and the coefficient of restitution was set to 0.97 for the collision between a steel sphere and a zerodur wall (Joseph et al., 2001). Since the timescale for the collision between the particle and the wall is much smaller than the timescale for flows, the time step for solving the equations for the

Table 2. Numerical setup for simulating the rebound of a sphere released from various heights.

Case	h_r(m)	Δx(m)	Δt(sec)	$d/\Delta x$	τ	Ma	St
1	5.5×10^{-3}	9.5×10^{-4}	5×10^{-4}	10	0.569	0.12	37.91
2	5.5×10^{-3}	9.5×10^{-4}	2.5×10^{-4}	10	0.535	0.06	37.91
3	5.5×10^{-3}	6.33×10^{-4}	2.21×10^{-4}	15	0.569	0.08	39.37
4	1.05×10^{-2}	9.5×10^{-4}	5×10^{-4}	10	0.569	0.17	52.84
5	1.05×10^{-2}	9.5×10^{-4}	2.5×10^{-4}	10	0.535	0.08	54.08
6	1.05×10^{-2}	6.33×10^{-4}	2.21×10^{-4}	15	0.569	0.11	53.76

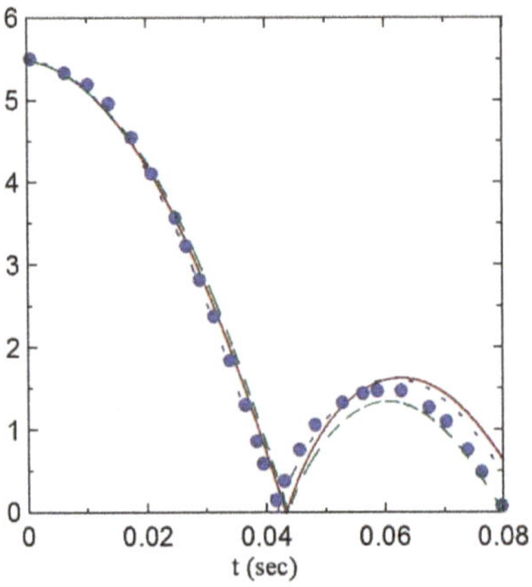

Figure 16. Comparison between the simulated (lines) and measured (symbols) trajectories for $h_r = 5.5 \times 10^{-3}$m.
Note: The solid line is for Case 1, the dashed line for Case 2, and the dotted line for Case 3.

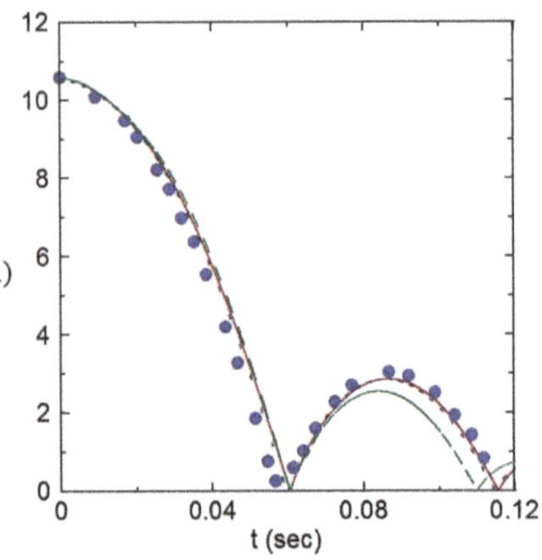

Figure 17. Comparison between the simulated (lines) and measured (symbols) trajectories for $h_r = 1.05 \times 10^{-2}$ m.
Note: The solid line is for Case 4, the dashed line for Case 5, and the dotted line for Case 6.

motion of the particle, Eqs. (15) and (16), Δt_p, was set to be $0.01\Delta t$ (the hydrodynamic force was assumed to be constant within Δt).

Table 2 summarizes all the model parameters used in our numerical experiments, and Figures 16 and 17 show the computed trajectories for two released heights: $h_r = 5.5 \times 10^{-3}$ m and $h_r = 1.05 \times 10^{-2}$ m Before collision, the numerical results are not sensitive to the relaxation time as long as it is greater than 0.535. However, the relaxation time still can slightly affect the simulated trajectory of the particle after collision. Since the effects of surface roughness (Ardekani, Dabiri, & Rangel, 2008; Li et al., 2012) were not considered in our simulations, the agreement between the simulated and experimental results can be regarded very good, indicating that the present boundary treatment method is capable of modeling collision processes. We remark that reasonable results can be obtained even for Ma as large as 0.17, which is higher than the upper limit suggested by He and Luo (1997b).

4. Conclusion

There are two important issues that need to be addressed when using lattice Boltzmann methods to simulate fluid-particle interaction near a wall: (1) the treatment of moving boundaries and (2) the determination of the hydrodynamic force exerting on a particle that is close to or in contact with a wall. This study introduces a new extrapolation-based boundary treatment method to address these two important issues: the first issue is addressed by introducing an appropriate fictitious flow field outside the fluid domain, and the second issue is addressed by adopting the velocity of the solid to calculate the equilibrium component of the PDF at the ghost node and evaluating the hydrodynamic force for the boundary link using the net exchange of the equilibrium PDFs at two linked ghost nodes. One feature of the present boundary treatment method is that it is easier to implement than the existing extrapolation methods without sacrificing numerical accuracy. Another important feature of the present boundary treatment method is that it does not need to use any empirical lubrication correction in regions close to rigid boundary.

The present boundary treatment method was applied to four problems: a cylindrical Couette flow, flow past a sphere, sedimentation of a sphere in a viscous fluid, and a sphere rebounding from a wall in a viscous fluid. The present boundary treatment is of second-order accuracy when applied to the study of a cylindrical Couette flow. For flow past a sphere, the present boundary treatment method can compute the drag force accurately. For sedimentation of a sphere in a viscous fluid, the present boundary treatment method can capture the fluid-particle interaction very well, even when the gap between two solid objects is smaller than one half of the grid size. For a sphere rebounding from a wall in a viscous fluid, the present boundary treatment method can accurately model the entire collision process in fluid. It is concluded that the proposed boundary treatment method can adequately handle problems involving particles in near contact and collision of a particle with a wall. The method can easily be applied to the simulation of particulate flows using LBMs.

Acknowledgments

The second author acknowledges the support from the University of Hawaii at Manoa (UHM) and Nanyang Technological University. This is UHM SOEST contribution number 9442.

Disclosure statement

No potential conflict of interest was reported by the authors.

References

Abraham, F. F. (1970). Functional dependence of drag coefficient of a sphere on Reynolds number. *Physics of Fluids*, *13*(8), 2194–2195.

Aidun, C. K., & Clausen, J. R. (2010). Lattice-boltzmann method for complex flows. *Annual Review of Fluid Mechanics*, *42*, 439–472.

Aidun, C. K., & Lu, Y. (1995). Lattice Boltzmann simulation of solid particles suspended in fluid. *Journal of Statistical Physics*, *81*(1), 49–61.

Ardekani, A. M., Dabiri, S., & Rangel, R. H. (2008). Collision of multi-particle and general shape objects in a viscous fluid. *Journal of Computational Physics*, *227*(24), 10094–10107.

Ardekani, A. M., & Rangel, R. H. (2008). Numerical investigation of particle-particle and particle-wall collisions in a viscous fluid. *Journal of Fluid Mechanics*, *596*, 437–466.

Bouzidi, M. h., Firdaouss, M., & Lallemand, P. (2001). Momentum transfer of a Boltzmann-lattice fluid with boundaries. *Physics of Fluids*, *13*, 3452–3459.

Chen, S., & Doolen, G. D. (2003). Lattice Boltzmann method for fluid flows. *Annual Review of Fluid Mechanics*, *30*(1), 329–364.

Cornubert, R., d'Humières, D., & Levermore, D. (1991). A Knudsen layer theory for lattice gases. *Physica D: Nonlinear Phenomena*, *47*(1), 241–259.

Dash, S. M., & Lee, T. S. (2014). Natural convection from inclined square cylinder using novel flexible forcing IB-LBM approach. *Engineering Applications of Computational Fluid Mechanics*, *8*(1), 91–103.

Derksen, J. (2011). Simulations of granular bed erosion due to laminar shear flow near the critical Shields number. *Physics of Fluids*, *23*(11), 113303.

Ding, E.-J., & Aidun, C. K. (2000). The dynamics and scaling law for particles suspended in shear flow with inertia. *Journal of Fluid Mechanics*, *423*, 317–344.

Ding, E.-J., & Aidun, C. K. (2003). Extension of the lattice-Boltzmann method for direct simulation of suspended particles near contact. *Journal of Statistical Physics*, *112*(3–4), 685–708.

Feng, Y., Han, K., & Owen, D. (2007). Coupled lattice Boltzmann method and discrete element modelling of particle transport in turbulent fluid flows: Computational issues. *International Journal for Numerical Methods in Engineering*, *72*(9), 1111–1134.

Feng, Y., Han, K., & Owen, D. (2010). Combined three-dimensional lattice Boltzmann method and discrete element method for modelling fluid–particle interactions with experimental assessment. *International Journal for Numerical Methods in Engineering*, *81*(2), 229–245.

Feng, Z. G., & Michaelides, E. E. (2005). Proteus: a direct forcing method in the simulations of particulate flows. *Journal of Computational Physics*, *202*(1), 20–51.

Feng, Z. G., & Michaelides, E. E. (2009). Robust treatment of no-slip boundary condition and velocity updating for the lattice-Boltzmann simulation of particulate flows. *Computers & Fluids*, *38*(2), 370–381.

Filippova, O., & Hänel, D. (1998). Grid refinement for lattice-BGK models. *Journal of Computational Physics*, *147*(1), 219–228.

Frisch, U., d'Humieres, D., Hasslacher, B., Lallemand, P., Pomeau, Y., & Rivet, J. P. (1987). Lattice gas hydrodynamics in two and three dimensions. *Complex Systems*, *1*(4), 649–707.

Graebel, W. P. (2001). *Engineering fluid mechanics*. New York: Taylor & Francis.

Guo, Z., & Zhao, T. (2002). Lattice Boltzmann model for incompressible flows through porous media. *Physical Review E*, *66*(3), 036304.

Guo, Z. L., Zheng, C. G., & Shi, B. C. (2002). An extrapolation method for boundary conditions in lattice Boltzmann method. *Physics of Fluids*, *14*(6), 2007–2010.

He, X., & Luo, L.-S. (1997a). A priori derivation of the lattice Boltzmann equation. *Physical Review E*, *55*(6), R6333–R6336.

He, X., & Luo, L.-S. (1997b). Lattice Boltzmann model for the incompressible Navier–Stokes equation. *Journal of Statistical Physics*, *88*(3), 927–944.

He, X., & Luo, L.-S. (1997c). Theory of the lattice Boltzmann method: From the Boltzmann equation to the lattice Boltzmann equation. *Physical Review E*, *56*(6), 6811–6817.

Higuera, F., & Succi, S. (1989). Simulating the flow around a circular cylinder with a lattice Boltzmann equation. *Europhysics Letters*, *8*(6), 517–521.

Higuera, F., Succi, S., & Benzi, R. (1989). Lattice gas dynamics with enhanced collisions. *Europhysics Letters*, *9*(4), 345–349.

Jafari, S., Yamamoto, R., & Rahnama, M. (2011). Lattice-Boltzmann method combined with smoothed-profile method for particulate suspensions. *Physical Review E*, *83*(2), 026702.

Johnson, T. A., & Patel, V. C. (1999). Flow past a sphere up to a Reynolds number of 300. *Journal of Fluid Mechanics*, *378*, 19–70.

Joseph, G. G., Zenit, R., Hunt, M. L., & Rosenwinkel, A. M. (2001). Particle-wall collisions in a viscous fluid. *Journal of Fluid Mechanics*, *433*, 329–346.

Ladd, A. J. C. (1994a). Numerical simulations of particulate suspensions via a discretized Boltzmann equation. Part 1. Theoretical foundation. *Journal of Fluid Mechanics*, *271*, 285–309.

Ladd, A. J. C. (1994b). Numerical simulations of particulate suspensions via a discretized Boltzmann equation. Part 2. Numerical results. *Journal of Fluid Mechanics*, *271*, 311–339.

Lin, S. Y., Lin, C. T., Chin, Y. H., & Tai, Y. H. (2011). A direct-forcing pressure-based lattice Boltzmann method for solving

fluid-particle interaction problems. *International Journal for Numerical Methods in Fluids*, *66*(5), 648–670.
Li, X. B., Hunt, M. L., & Colonius, T. (2012). A contact model for normal immersed collisions between a particle and a wall. *Journal of Fluid Mechanics*, *691*, 123–145.
McNamara, G. R., & Zanetti, G. (1988). Use of the Boltzmann equation to simulate lattice-gas automata. *Physical Review Letters*, *61*(20), 2332–2335.
Mei, R., Luo, L- S., & Shyy, W. (1999). An accurate curved boundary treatment in the lattice Boltzmann method. *Journal of Computational Physics*, *155*(2), 307–330.
Mei, R., Yu, D., Shyy, W., & Luo, L.-S. (2002). Force evaluation in the lattice Boltzmann method involving curved geometry. *Physical Review E*, *65*(4), 041203.
Nguyen, N. Q., & Ladd, A. (2002). Lubrication corrections for lattice-Boltzmann simulations of particle suspensions. *Physical Review E*, *66*(4), 046708.
Plimpton, S. (1995). Fast parallel algorithms for short-range molecular dynamics. *Journal of Computational Physics*, *117*(1), 1–19.
Qian, Y. H., Succi, S., & Orszag, S. (1995). Recent advances in lattice Boltzmann computing. *Annual Reviews of Computational Physics*, *3*, 195–242.
Silbert, L. E., Ertas, D., Grest, G. S., Halsey, T. C., Levine, D., & Plimpton, S. J. (2001). Granular flow down an inclined plane: Bagnold scaling and rheology. *Phys. Rev. E*, *64*(5), 051302.
Succi, S., Foti, E., & Higuera, F. (1989). Three-dimensional flows in complex geometries with the lattice Boltzmann method. *Europhysics Letters*, *10*(5), 433–438.
Suzuki, K., & Inamuro, T. (2013). A higher-order immersed boundary–lattice Boltzmann method using a smooth velocity field near boundaries. *Computers & Fluids*, 105–115.
ten Cate, A., Nieuwstad, C. H., Derksen, J. J., & Van den Akker, H. E. A. (2002). Particle imaging velocimetry experiments and lattice-Boltzmann simulations on a single sphere settling under gravity. *Physics of Fluids*, *14*(11), 4012–4025.
Tiwari, A., & Vanka, S. P. (2012). A ghost fluid Lattice Boltzmann method for complex geometries. *International Journal for Numerical Methods in Fluids*, *69*(2), 481–498.
Verberg, R., & Ladd, A. J. C. (2001). Accuracy and stability of a lattice-Boltzmann model with subgrid scale boundary conditions. *Physical Review E*, *65*(1), 016701.
Yin, X. W., Le, G. G., & Zhang, J. F. (2012). Mass and momentum transfer across solid-fluid boundaries in the lattice-Boltzmann method. *Physical Review E*, *86*(2), 026701.
Yin, X. W., & Zhang, J. F. (2012). An improved bounce-back scheme for complex boundary conditions in lattice Boltzmann method. *Journal of Computational Physics*, *231*(11), 4295–4303.

7

Computational study of a complex three-dimensional shock boundary-layer interaction

Joshua Katzenberg and David MacManus*

Department of Power and Propulsion, School of Engineering, Cranfield University, Cranfield, MK42 0AL, Bedfordshire, UK

Shock boundary–layer interactions occur in many high-speed aerodynamic flows and they can have a notable impact on design considerations due to the aerodynamic and heat transfer effects. Consequently there is a notable interest in understanding the ability of computational tools to calculate the complex flow fields that can arise in a range of engineering applications. Three-dimensional complex shock boundary layer interaction studies are expensive in both time and computational resources. Although recent studies have begun to focus on the use of more complex computational methods such as large eddy simulations, the aim of this research is to assess the ability of steady Reynolds averaged Navier Stokes turbulence models to simulate the interaction of a planar shock impinging on a cylindrical body under supersonic conditions and to determine if these models have a role to play in engineering design applications. The performance of both eddy viscosity and Reynolds stress models are evaluated relative to an established experimental test case. The impact of Reynolds number and impinging shock strength are also considered. Of the eddy viscosity models it was shown that the Spalart-Allmaras model is unsuitable for this complex interaction and that the k-ε and Reynolds stress methods both gave notably better agreement with the measured surface static pressures. Overall it was considered that the Reynolds stress method was the best model as it also provided better agreement with the measured surface flow topology. It was concluded that, although a steady Reynolds averaged Navier Stokes approach has known limitations for this type of complex interaction, within an engineering context it can also provide useful results when applied appropriately.

Keywords: shock boundary layer interaction; boundary layer separation; Reynolds stress model; oblique shock interaction; multibody; supersonic

1. Introduction

The topic of shock boundary layer interactions (SBLI) is a broad subject which has been the focus of extensive research over many years. It is a complex, multifaceted topic, which is not yet fully understood and still causes significant difficulties in the design of a wide range of high-speed vehicles and aerospace components (Allen, Heaslet, & Nitzberg, 1947; Borovoy et al., 2013; Brosh & Kussoy, 1983; Chaplin et al., 2011; Délery, 1999; Donaldson, 1944; Holden, Wadhams, MacLean, & Mundy, 2010). Initial investigations of nominally two-dimensional SBLIs were previously undertaken for aerofoils under supersonic conditions. For an aerofoil under high transonic conditions, a normal shock wave can form and interact with the boundary layer, typically on the suction side (Babinsky & Harvey, 2011). In this interaction, compression waves form near the wall as the boundary layer is unable to support the rise in static pressure across the shock. As the shock strength increases, along with the concomitant adverse pressure gradient, boundary layer separation can occur. Further instances of nominally two-dimensional SBLIs have been investigated extensively, most notably for compression ramps and oblique shocks impinging on flat plates where it has been noted that the interaction depends on the nominal pressure rise across the shock as well as the state and characteristics of the approaching boundary layer (Arnal & Délery, 2004; Délery, Marvin, & Reshotko, 1986; Délery, & Coet, 1990; Délery, 1996). For example, an increase in the shock strength, by changing the ramp angle, can lead to separation of the boundary layer if the pressure rise across the shock is high enough. Separation results in an influence from the SBLI which extends further upstream than with an attached SBLI and an additional shock system of separation and reattachment shocks can arise. Similar effects arise with the impingement of a two-dimensional oblique shock onto a flat plate where the induced increase in the boundary layer thickness generates an additional shock wave ahead of the primary reflected shock which in turn crosses the impinging shock wave (Arnal & Délery, 2004). If the shock is of significant strength, an increase in the thickness of the boundary layer occurs and a separation bubble can also arise. Thus, in nominally two-dimensional SBLIs, shock strength and angle, dependant on Mach number and geometry, along

*Corresponding author. Email: d.g.macmanus@cranfield.ac.uk

with the Reynolds number, state and shape factor of the approaching boundary layer are key parameters as to the nature of the shock interaction.

Nominally three-dimensional SBLIs are typically more complex and occur in many high-speed aerodynamic flows (Brosh & Kussoy, 1983; Pamadi et al., 2005). Examples include planar shocks impinging onto axisymmetric bodies (Brosh & Kussoy, 1983) and conical shock waves impinging onto flat plates (Gai & Teh, 2000). For these configurations, along the central symmetry plane there are broad similarities with the nominally two-dimensional SBLIs such as the local boundary layer thickening which can lead to separation. Critical factors which affect the likelihood and magnitude of separation are shock strength and shock angle which are typically controlled by the distance of the shock generator from the boundary layer, approaching Mach number and the characteristics of the approaching boundary layer. For these configurations the increase in surface static pressure from the SBLI typically arises several boundary layer thicknesses upstream relative to the nominal ideal shock impingement point and locally strong cross-flows can occur due to the SBLI. Distinct regions arise within the flow field which depend on the initial SBLI and the subsequent evolution of the boundary layer, shock waves, expansion systems, crossflow and indeed the main flow (Chaplin et al., 2011). Adverse pressure gradients at the initial interaction point can be affected by lateral, or azimuthal, flow migration away from the position of peak shock intensity which results in potent cross-flows (Brosh & Kussoy, 1983). In the case of a planar oblique shock impinging onto a cylindrical body (Brosh & Kussoy, 1983), the coalescing of the diffracted shocks on the leeward side resulted in an additional separation region and thereby deflected the flow in a similar manner to an obstacle in the boundary layer (Peake & Tobak, 1982; Sedney & Kitchens, 1975). Further detailed studies of three-dimensional SBLIs are reported by other researchers (Arnal & Délery, 2004; Babinsky & Harvey, 2011; Peake & Tobak, 1982; Saric et al., 1996).

1.1. Computational studies of SBLIs

SBLIs comprise a very wide range of configurations and flow regimes and the aerodynamic characteristics of the interaction is highly dependent on the defining conditions. The interactions can encompass aspects such as transition, separation, highly skewed boundary layers, reflected shocks and expansion fans, vortical flows and strongly three-dimensional flow gradients (Babinsky & Harvey, 2011). Furthermore, it is considered that all SBLIs are fundamentally unsteady to some degree (Oliver, Lillard, Schwing, Blaisdell, & Lyrintzis, 2007). Overall these features present a very challenging problem for computational fluid dynamics (CFD). Nevertheless, a broad variety of SBLI problems have been tackled using a range of computational methods with varying degrees of success (Bhagwandin & DeSpirito, 2011; Dolling, 2001; Oliver et al., 2007). These methods have included various forms of Reynolds Averaged Navier Stokes (RANS) approaches (DeBonis et al., 2010; Thivet, 2002; Vallet, 2007), unsteady RANS (URANS) (Barakos, Doerffer, Hirsch, Dussauge, & Babinsky, 2010; Hirsch, 2010a; Sandham, 2010), Detached Eddy Simulations (DES) (Garnier, 2009; Shams & Comte, 2010), Large Eddy Simulations (LES) (DeBonis et al., 2010; Eagle, Driscoll, & Benek, 2012; Hadjadj, 2012) and Direct Numerical Simulations (DNS) (Adams, 2000; Tokura & Maekwa, 2011). Clearly these methods provide different levels of modeling fidelity and sophistication along with concomitant resource demands.

Oliver et al. (2007) examined the ability of some steady RANS methods, Spalart-Allmaras (SA), Menter's Shear Stress Transport (SST) and Olsey and Coakley's Lag model, to calculate SBLIs occurring on flat plates and a compression ramp at Mach numbers ranging between 2.87 and 6.0. Simulations of relatively benign SBLIs produce results that are in reasonable agreement with experimental results. However, as shock strength increases, the SBLI becomes more difficult to determine computationally and the difference in accuracy of various RANS models becomes more apparent. It was found that the calculation of too large a separation bubble as well as a region of upstream influence which arises in two-dimensional calculations can be improved with a three-dimensional computational model as this includes relief effects as well as the impact of the wind tunnel walls which arises in many experimental configurations. For these cases, RANS models generally calculate the inviscid flow field as well as the overall effects such as separation and the region of upstream influence to a level deemed adequate for most engineering design work. However, Oliver et al. (2007) showed that the thermal environment and interactions proved more difficult to compute. In addition, the detailed flow visualization and skin-friction-related distributions were adversely affected even if surface pressures correlated closely with experimental data. It was noted that the low-frequency shock unsteadiness, a result of large streamwise turbulent fluctuations observed in the experiment, was not captured by RANS models due to the averaging of turbulent fluctuations, and thus, there was no physical source for such effects in RANS models. This problem was partially addressed by reducing the grid aspect ratio near the separation region and of the three models tested, the simple formulation of the one-equation SA model providing the lowest fidelity, with the lag model producing results in-between the SA and SST models (Oliver et al., 2007).

Subsequent studies on the ability of steady RANS methods to calculate the SBLI flow field showed that the accuracy depends greatly on the turbulence model (DeBonis et al., 2010). It was reported that for weak SBLIs of planar shocks impinging on a flat plate, differences in the calculated streamwise velocity distributions were within 0.5% of each RANS turbulence model considered for the Menter SST, k-ω variants and Spalart-Allmaras models. However, as the shock intensity increased, the error in all the solutions also increased, although the relative error between each turbulence model was relatively consistent. Grid construction, either structured or unstructured, showed no discernible difference on the calculated flow fields. The most notable difference between models was within the separation region of the SBLI although due to the range of metrics used in the assessment it was difficult to determine a clearly superior method. LES simulations were found to produce similar error levels as RANS methods but the prediction of normal stresses was superior using LES (DeBonis et al., 2010). Investigations of SBLIs with separated flow fields using Goldberg's One-equation Rt model, SA, realizable k-l, Goldberg's realizable q-l, SST, Goldberg's k-ε-Rt and a seven equation second moment closure turbulence model in a thrust vector nozzle (Tian & Lu, 2013) showed that a k-ε model provided the most accurate results when calculating the separation point and shock wave position.

The Reynolds Stress Model (RSM) is a formulation of RANS model that solves the transport equations for Reynolds stresses directly, including an additional equation for dissipation rate, to close the RANS equations. This produces a seven-equation model that has greater computational demands yet is capable of providing higher accuracy for boundary layer flows when compared to the industry standard Eddy-Viscosity (EV) RANS models; SA, k-ε, k-ω and SST formulations. A study by Vallet (2007) using an RSM model with a 2D compression ramp reported improvements over k-ε models and was able to calculate mean-flow distributions for both separated and attached flows. The inability to correctly simulate experimentally observed low-frequency shock oscillations, also notably absent in computational simulations by Oliver et al. (2007), was again apparent in the foot of the shock-wave Reynolds stress profiles (Vallet, 2007). Further studies, such as those by Mendonca and Sharif (2010) found that surface roughness should not be ignored in SBLI flows due to the upstream influence and movement of the shock as surface roughness is increased. The RSM model was shown to provide the most accurate results although it was also shown that the k-ω model could provide useful results at a decreased computational cost. Comparison of RANS models for SBLIs generated with 2D oblique shocks on a flat plate (Bhagwandin & DeSpirito, 2011), showed that the RSM model is suitable for SBLI flows but that it requires a higher near-wall resolution grid than eddy-viscosity models such as SA and SST methods. This, coupled with the RSM model being a seven-equation model and hence requiring an increased amount of resources relative to the EV models, may be less desirable where rapid solutions are needed. Furthermore, numerical stability issues may arise which suggests that hybrid RANS/LES models may be useful to resolve some of the inherent unsteady effects within SBLIs. DeBonis et al. (2010) suggests that the capability of RSMs to determine the appropriate normal stress distributions warrants further investigation.

Recent case studies continue to prove useful in providing experimental data of complex SBLIs with computational comparisons. Holden et al. (2010) provided experimental results for flat plate and compression ramp SBLI datasets at Mach 4 to 11. Significant differences were reported between SA and SST models, with the former failing to predict separation for compression ramp flow and the latter over-predicting the size of the separated region. Subsequent "blind" code validation studies have provided direction for empirical modifications to RANS codes, such as those suggested by Wilcox (2006). SBLIs generated by crossing shocks from fins for Mach numbers 5 to 8 have been compared experimentally and computationally by Borovoy et al. (2013). The q-ω RANS turbulence model was found to agree well with the experimental results. It was seen that thermocouple sensors were more accurate than optical measurements and, thus, the CFD was in better agreement with the thermocouple results.

Unsteady RANS (URANS) methods have been investigated to address the low-frequency oscillations and, hence, to capture the flow unsteadiness which has been observed experimentally. An experimental oblique shock reflection study at Mach 2.0 (Sandham, 2010) was simulated computationally but failed to capture significant flow unsteadiness and the expected low-frequency oscillations. The study also demonstrated the superior ability of an LES model to capture these effects. Of further note for the RANS methods, is the effect that inlet turbulent intensity has on the size of the separation bubble. A reduction of inlet turbulence intensity from 3% to 0.1% resulted in a separation bubble with twice the length and in better agreement with experimental data (Sandham, 2010). In addition, there was better agreement with the measurements when modeling the system as 3D instead of 2D due to the inclusion of corner flows and the associated aerodynamic effects. This influence of the 3D and corner flows was also reported by Hirsch (2010b) with an oblique shock at Mach numbers 1.7, 2.0 and 2.25. URANS models were found to produce steady flow fields and did not capture high-frequency oscillations at the foot of the shock. Additionally, URANS methods were either unable to predict the natural shock motion or underestimated the effect. Distributions of velocity profiles in the region downstream of the interaction, as well as in the separated regions, were deemed to require improvement, although the upstream velocity profiles were calculated more accurately. It was reported that, to improve the accuracy of SBLIs, improvements need to be made to URANS turbulence models to address 3D separations and corner vortices (Hirsch, 2010b). Barakos et al. (2010) suggests that hybrid DES/LES models and data be used to propose URANS improvements through challenging the assumption of local equilibrium in URANS and the development of new anisotropic stress tensor methods.

Although LES methods are more advanced than RANS computations, studies on a 3D SBLI of an oblique shock impinging on a flat plate at Mach 2.75 concluded that there was little advantage of LES over RANS methods for this particular configuration (Eagle et al., 2012). It was reported that this was partially affected by the three-dimensional nature of the experimental data and the use of two-dimensional computational simulations. Two-dimensional studies of oblique shocks interacting with flat plates at Mach 2.75 (Eagle et al., 2012) note the more mature status and understanding of RANS methods as a driver for the similar accuracy and uncertainty in results when comparing RANS with LES. It was further observed that LES methods provide superior predictions of normal stresses when compared to RANS methods. Through experimental observation and computational simulation of a three-dimensional oblique shock interacting with a flat plate at Mach 2.28 with a shock incidence angle (β) of 32.4°, Hadjadj (2012) found excellent agreement between LES predictions and experimental results. The aforementioned low-frequency shock oscillations were observed and helped give credence to the hypothesis that the shock wave and separation bubble act as a coupled system.

Detached Eddy Simulation, originally proposed as a bridge between RANS and LES methods, has the ability to predict separated and complex turbulent flows. Shams and Comte (2010) studied SBLIs in nozzle flows and found very good agreement with experimental results. It was determined that the DES method calculated the unsteady low-frequency shock oscillations and is, thus, a potentially good method of predicting SBLIs. A SBLI accounting for the whole wind tunnel span at Mach 2.3 by Garnier (Garnier, 2009) using DES saw an improvement over 2D LES methods. Corner separations in 3D simulations were, again, noted to affect the results. Low-frequency oscillations, as with Shams and Comte (2010), were present and the results were in good agreement with the experimental data.

Direct Numerical Simulation has the potential to improve the detailed simulations of SBLIs. The computational and economical costs are, however, currently prohibitive for many engineering applications. Attempts have been made, such as by Adams (Adams, 2000), in which a compression ramp ($\beta = 18°$) at Mach 3 was directly simulated with a relatively low Reynolds number ($Re_\theta = 1685$), to enable calculations with the then current computational resources available. The simulation domain was found to be of insufficiently large enough scale to capture any large-scale shock motion and, therefore, the agreement between experimental and computational data was limited. An oblique SBLI at Mach 2.0 and $Re_{\delta * \theta} = 1000$ was predicted using DNS by Tokura and Maekwa (2011). Good agreement between previous studies was found, although comparisons were limited due to the low Reynolds number used in the study. The calculations provide access to flow structures such as lambda-vortices and broadband spectra seen experimentally but not present in most other CFD methods. In general, DNS still requires computational resources beyond current capabilities to be fully realized as an alternative to lower fidelity engineering models.

Overall, the range of CFD methods that have been applied to SBLI flows, indicate that the complexity of the flow field is a major challenge even for the most advanced tools. Although steady RANS calculations using eddy-viscosity-based turbulence models have well-known simplifying assumptions, in some cases they are able to calculate successfully some of the primary flow features. However, there is evidence that key aspects of the interaction such as separation size, reattachment locations and details of the flow topology are not fully resolved. In spite of this, some of the more recent studies indicate that there are only modest benefits of going to the much more demanding methods such as LES or DES. There are noted benefits of adopting a DNS approach, as seen in Tokura and Maekwa (2011), but these methods are particularly resource intensive. Within this context, and from the point of view of pragmatic engineering design, there is still a significant interest in understanding the capabilities of robust CFD tools that provide a measured balance between computational cost and solution fidelity.

1.2. Three-dimensional SBLI in multibody configurations

Although the previous experimental and computational work encompasses a wide range of SBLI configurations much of the work has focused on planar shocks impinging on flat plates, glancing and crossing shock interactions, as well as local normal and compression ramp configurations. Relatively little work has been done on the interaction of an impinging oblique shock with a nonplanar body. This is a configuration that is of particular interest within the context of multibody configurations, stores and submunition separation (Chaplin et al., 2011), sabot discard, and two-stage to orbit configurations (Pamadi et al., 2005).

For multibody configurations in close proximity under high-speed flow conditions, aerodynamic interference arises and this can significantly affect the force and moment characteristics (Chaplin, MacManus, & Birch, 2010; Hung, 1985; Wilcox, 1995). The complex flowfield is primarily dominated by the shock and expansion waves, which originate from one body and impinge upon the adjacent body. The interference aerodynamics is further complicated by multiple shock reflections, shock diffraction as well as shock interactions with the viscous body vortex and boundary-layer flows. The induced changes in static pressure and flow angularity across the impinging disturbances modify both the local and overall aerodynamics of a slender body in comparison with the isolated body case. Very limited information is available in the open literature on the effects of mutual interference between slender bodies at high speed. One previous investigation showed that a planar shock impinging on a cone-cylinder body at zero incidence induced changes in normal force and pitching moment coefficient of approximately 0.02 and 0.2 respectively (Wilcox, 1995). These changes were found to increase by up to an order of magnitude when the receiver body was placed at an incidence of $\sigma = 15°$. Interference effects of this order would modify the trajectory of the slender body, and would become significant if the slender body pitches (or translates) toward the generator resulting in a collision.

Previous experimental work showed that for multiple bodies in close proximity, the interaction of a conical oblique shock with an ogive-cylinder body at Mach 2.43 could result in a strong SBLI with local flow separations and notable changes in the pressure distributions around the body (Chaplin, 2010; Chaplin et al., 2010). The consequence of this was a significant change in the forces and moments on the body which leads to an adverse change in the body trajectory and ultimately in the two bodies colliding (Chaplin et al., 2010). Additional computational work by Chaplin also showed that steady RANS calculations using the k-w SST turbulence model showed very good agreement with the measured changes in forces and moments (Chaplin et al., 2011). Furthermore, comparisons between the calculated surface static pressure distributions and the measured data using pressure sensitive paints showed that these RANS calculations were able to capture the main aspects of this complex interaction.

In addition to the overall interference loads, it is important to understand the detailed underlying flow physics. Shock-body interactions have been studied previously for a number of pertinent configurations (Chaplin et al., 2011; Derunov, Zheltovodov, & Maksimov, 2008; Fedorov, Malmuth, & Soudakov, 2007; Malmuth & Shalaev, 2004; Volkov & Derunov, 2006). Of particular interest is the work of Brosh, Kussoy, and Hung (1985) and Hung (1985) who investigated a wedge-generated shock passing over a cylinder at $M_\infty = 2.85$. This particular configuration is the focus of the current research. Brosh et al. (1985) performed an experimental investigation using a configuration which comprised a prismatic wedge shock generator and a cylindrical body onto which the planar oblique shock impinged.

The measurements showed that the impinging shock footprint, in terms of local pressure rise, decreased as the shock diffracted around the body. In addition, the induced pressure rise on the windward reduces quickly due to the impact of expansion waves from the generator forebody. These expansion waves do not diffract to the same extent as the impinging shock and thus the leeward pressure rise associated with the diffracted shock is maintained along the body. Consequently, the difference between the strength and extent of the windward and leeward interactions significantly affects the local normal force distribution over the body. Finally, the windward pressure rise also resulted in a local boundary-layer separation and a double-reflected shock structure around the leeward separation bubble. Both studies found that due to the induced circumferential pressure gradient, a strong azimuthal crossflow occurred which resulted in a local separation on the farside of the receiver body. A similar effect was also noted by Morkovin, Migotsky, Bailey, and Phonney (1952).

This configuration of an oblique shock impinging onto a cylindrical body is of particular interest from the point of view of the specific applications as well as from the understanding of the detailed flow physics. Due to some of the specific flow physics aspects which are different from other SBLI configurations, it is also an interesting case from the point of view of understanding the capability of computational tools to calculate the pertinent flow features and characteristics. As discussed above, CFD has demonstrated mixed capabilities in this regard for other SBLI arrangements such as compression ramps, normal SBLIs, and glancing shocks. The work of Chaplin et al. (2010, 2011) showed that the overall effect of a multibody shock interaction on the forces and moments of an axisymmetric body could be determined using RANS calculations and a k-ω SST turbulence model.

The aim of this work is to examine the performance of different RANS turbulence models for the case of an impinging planar shock onto a cylindrical body. The effect of both eddy-viscosity and Reynolds Stress turbulence models is assessed and the CFD data is compared with the experimental dataset of static pressure distributions, total pressure traverses and oil flow visualizations. Following the assessment of the CFD methods, the impact of shock strength and angle as well as the flow Reynolds number is also examined. Complex 3D SBLIs are underrepresented in the literature in comparison to 2D studies, mainly due to difficulties in accurately representing many of the flow phenomena that occur beyond planar shocks impinging on flat surfaces. Additionally, current trends are in the direction of refining LES and DES methods for such aerodynamic flows which better account for some of the complex flowfield interactions. These, however, come at an increased computational cost in comparison to steady RANS methods. Consequently, it is of interest to evaluate steady RANS methods for these 3D interactions and to determine their appropriateness within the context of an engineering design application.

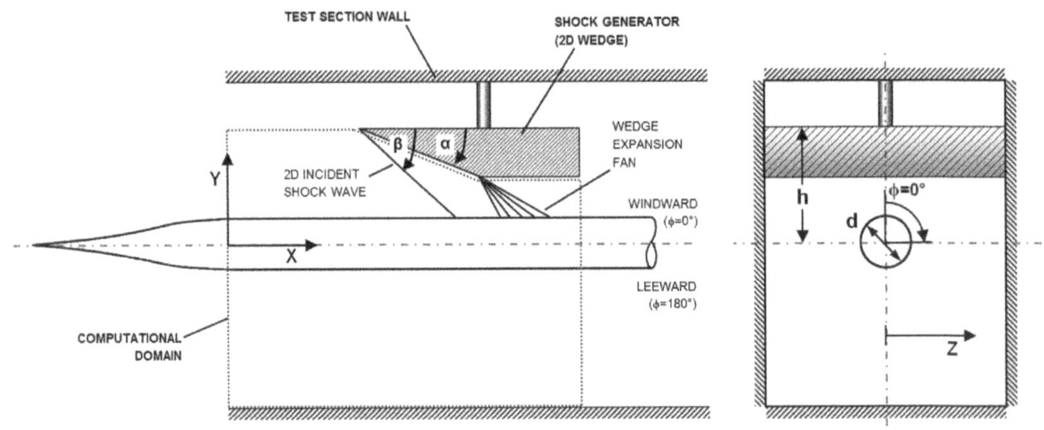

Figure 1. Schematic layout of experimental test setup with computational domain shown.

2. Methodology

2.1. Test case

The configuration of interest in this work was the subject of a previous experimental investigation as reported by Brosh and Kussoy (1983). The configurations comprised a circular cylinder (d = 50.8 mm) of 1.016 m length (L), mounted centrally in a 0.254 m wide and 0.381 m high test section of a wind tunnel. The shock generator was a 2D wedge which was positioned above the cylinder at a distance of h/d = 2.97 from the top surface (Figure 1).

The datum test flow conditions were M = 2.88, α = 16°, h = 150.9 mm, and a Reynolds number (Re_L) of 18.2 × 10^6 (Table 1) based on the free-stream velocity and length of the cylindrical body. This is the configuration for which the majority of the quantitative experimental data is available (Brosh & Kussoy, 1983). The effect of the flow Reynolds number was also examined across the range of Re_L = 7.28 × 10^6 to 58.3 × 10^6, although for these configurations the free-stream Mach number also changed by modest amounts across the range of M = 2.8 to 2.95, respectively (Table 1). The influence of the shock generator angle (α) was further experimentally examined for an increased alpha of 19°, at the same Mach number (2.88) and Reynolds number (Re_L = 18.2 × 10^6) as the datum configuration and for a lower Reynolds number (Re_L = 7.28 × 10^6) with Mach number of 2.80 and α of 13°. The total temperature was kept relatively constant across the configurations and ranged between 101.4 K and 108.2 K.

Surface pressure taps were positioned on the cylinder at 20 mm axial intervals (ΔX/d = 0.39), with additional azimuthal taps at ϕ = 90°, 180° and 270° at 0.2 m axial spacing (ΔX/d = 3.9), to verify flow symmetry. Time-averaged static and total pressure profiles were measured in the windward (ϕ = 0°) and leeward (ϕ = 180°) planes over a vertical distance from the surface of Y/d = 0.394 with a spatial resolution of Y/d = 9.8 × 10^{-4}. It was assumed that total temperature was constant throughout the boundary layer in accordance with Kussoy, Horstman, and Acharya (1978). The reported measurement uncertainties in the windward and leeward planes were ± 10% for static pressure, ± 1% for Pitot pressure, ± 6% for static temperature, 12% for density, ± 3% for velocity and ± 1.97 × 10^{-3}Y/d for the vertical probe movement in the Y-axis.

Table 1. Brosh and Kussoy (1983) geometric and nominal flow conditions of tested computational models.

Total Pressure $P_{T\infty}$ (psi / kPa)	T_∞ (K)	α (deg.)	h (mm)	Re_L (x10^6)	Mach No.
25 / 172.4	105.8	16	150.9	18.2	2.88
10 / 68.9	108.2	16	150.9	7.28	2.80
80 / 551.6	101.4	16	150.9	58.3	2.95
25 / 172.4	105.8	19	150.9	18.2	2.88

3. Computational model and method

The calculations were performed using Fluent V14.0 (SAS IP Inc., 2011a) using a steady, implicit density-based solver. The flow gradients were calculated using a least squares cell-based method with second order up-winding for the flow and turbulence properties. The convective fluxes are determined using the Roe flux difference splitting scheme. For the turbulence models four eddy-viscosity models were selected; a one equation Spalart-Allmaras model, two equation k-ε realizable and k-ω SST models and a four equation k-ω SST model with a γ-θ transition model (tSST). A seven equation RSM model using the linear-pressure strain sub-model was also evaluated. For the fluid, specific heat capacity (Cp) was set via a three-coefficient polynomial temperature equation, Hilsenralh et al. (1955), which results in γ varying with temperature. Thermal conductivity was calculated by the use of kinetic-theory using the molecular and material properties of the fluid to account for the user defined Cp value. The three

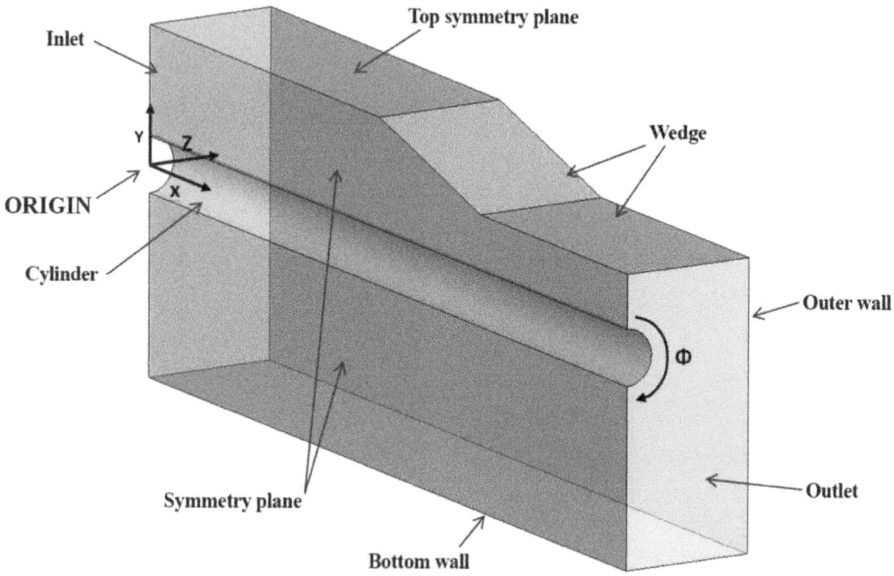

Figure 2. Model domain and boundary conditions.

coefficient method of Sutherland's law, based on a reference viscosity, temperature and effective temperature, was used to compute viscosity as a function of temperature. For all simulations, the double-precision solver was employed.

3.1. *CFD domain and boundary conditions*

The notation denotes 'X' as the axial axis, 'Y' as the vertical axis and 'ϕ' as the azimuthal direction around the cylinder and is shown in Figure 2. The computational domain is also highlighted in Figure 1. It comprises half the experimental domain, with the assumption of an X-Y symmetry plane intersecting the cylinder at $\phi = 0°$ and $\phi = 180°$ to reduce the computational requirements. No-slip wall boundary conditions are defined for the wind tunnel walls, the wedge and the cylinder surfaces. A symmetry plane was used for the 'top' surface of the computation domain which was limited to the same vertical position as the top of, and prior to, the wedge to prevent artificial boundary layer growth which could affect the shock wave characteristics. Experimental surface pressure data was limited axially, from X/d = 1.97 to X/d = 16.34. The computational domain originates as shown in Figure 2 and extends axially X/d = 16.34, with the origin X/d = 7.83 upstream of the wedge leading edge.

At the inlet plane the total temperature, total pressure and Mach number are as specified in Table 1. As no information was provided on the turbulence characteristics in the wind-tunnel experiments, the values for turbulence intensity, I, and length scale, l, were determined from approximate empirical correlations based on Re_{D_H} and the estimated boundary layer thickness ahead of the shock impingement point, δ_{99} (Equations 1, 2) (SAS IP Inc., 2011b). The turbulence intensity, I, was 1.98% and the length scale, l/d, was 0.096.

$$I = 0.16\left(Re_{D_H}\right)^{-1/8} \quad (1)$$

$$l = 0.4\delta_{99} \quad (2)$$

3.2. *CFD grid*

A structured multiblock approach was used with an O-grid blocking scheme around the cylinder with a radial expansion ratio of approximately 1.2 to grow the mesh and to provide between 40 and 90 cells across the boundary layer depending on the overall mesh resolution. The axial distribution of the mesh was clustered using an exponential contraction in cell size up to the axial position of the wedge, aft of which uniform cell sizes were used.

Three grids of increasing resolution were generated; coarse, medium and fine with a refinement ratio of 1.5 in all three dimensions in accordance with the methodology outlined by Roache (1998). This yielded meshes of approximately 9×10^5 (900 k), 3 x10^6 (3M) and 10×10^6 (10M) cells. To comply with turbulence models requirements, a y+ less than 1 was maintained for all grids and test conditions.

The effect of the spatial discretisation was assessed using the generalized Richardson Extrapolation method (Celik et al., 2008). Grid Convergence Indices (GCIs) were assessed using nodal points of free-stream Mach number and total pressure ratio, $P_T/P_{T\infty}$, at X/d = 9.84 on the windward surface, $\phi = 0°$, of the cylinder. A factor of safety of 2 was used with the data showing the GCI to be less than 1% for all cases and thus the grid was determined as mesh insensitive according to the generalized Richardson Extrapolation. The characteristics of the generated impinging shock was assessed using a position in between the wedge and the cylinder (Y/d = 1.56 and X/d= from 6.89 to 10.83). The static pressure and static temperature distributions in this region were evaluated relative to the theoretical values and the effect of the grid resolution on these parameters was assessed. It was determined that the static pressure and temperature ratios across the shock agreed to within ±1% of the theoretical value for the fine grid configuration.

3.3. Iterative convergence

Iterative convergence was assessed by considering the overall residual parameters throughout the flow domain in addition to integrated flow parameters. Overall, residuals were typically reduced to values below 10^{-4} for the continuity, momentum, energy and turbulence parameters. Due to the inherent unsteady nature of SBLIs and the assumption of 'steady' turbulence models, a simulation in which residual values below 10^{-4} were maintained for 300 consecutive iterations was considered to have converged.

The convergence of flow characteristics at key locations in the domain was monitored using a variety of metrics. The windward static pressure at locations ahead of the shock impingement position (P1) and in the centre of the interaction (P2), were assessed. Within the accepted convergence tolerance, the reported values of static pressure at these locations varied by typically 0.2% and −1.6% for P1 and P2 respectively between initial stabilization at approximately 7000 iterations and the final converged value.

Integrated drag and moment coefficients for the full cylinder, C_d and C_m respectively, were also considered as part of the convergence assessment, and the parameters typically converged to stable values after 5000 iterations where the variation in C_d and C_m was 0.0005 and 0.0, respectively. A similar characteristic was also observed for y+ when averaged over the cylinder the variation with further iterations was in the order of 3.8%. The total mass averaged vorticity magnitude at the outlet plane was also considered and it was found to stabilize to within 1.5%.

For the cases with different Reynolds numbers (Re_L, 7.28×10^{-6} and 58.3×10^{-6}), the same domain as the datum configuration, with the 3M cell mesh, was used. For these cases, the y+ values over the whole cylinder were assessed and also found to be less than 1. For the configuration where the shock generator wedge angle increased to $\alpha = 19°$, the geometry was slightly modified to accommodate this change. However, the main parameters of the 3 million cell mesh were preserved to ensure the same number of cells across the boundary layer, overall mesh resolution and axial spacing in the main regions of interest. Similarly the cylinder y+ was preserved at values less than 1.

4. Results

4.1. The datum configuration

The datum configuration was for a Mach number of 2.88, a Re_L of 18.2×10^6 with a 16 degree wedge shock generator. The various turbulence models provide a range of results for the cylinder boundary layer at the position ahead of the SBLI. The boundary layer thickness (δ_{99}) and compressible displacement thicknesses (δ^*) is assessed at X/d = 9.84 at $\phi = 0°$ where the boundary layer measurements showed a δ_{99} of 12.2 mm and a δ^* of 3.5 mm, with uncertainties of ±3% and ±12.4%. The SST and tSST models calculated a δ_{99} of 12.7 mm and 12.5 mm, respectively, with the latter being within the 3% experimental uncertainty range. The k-ε and RSM models calculated a slightly thicker boundary layer with δ_{99} of +16.9% and +7.1%, respectively, relative to the measurements. However, the calculated displacement thicknesses showed a different sensitivity where the k-ε and RSM models provided δ^* within the uncertainty level of -0.04 mm (-11.4%) and -0.04 mm (-12.3%), respectively. Conversely, the k-ω SST and γ-θ transition k-ω SST models under-predicted δ^* by approximately 30%.

The measurements of the surface static pressure ahead of the interaction indicate that there was a variation in the working section flow field conditions which is greater than the quoted uncertainty of the data acquisition system. Consequently, there are variations in the static pressure distributions on the cylinder ahead of the SBLI region, and therefore to enable a clear comparison of the pressure rise and SBLI extent for the CFD and experimental data sets, the cylinder surface static pressure has been nondimensionalized as shown in Equation (3):

$$\frac{P^*}{P_{T,\infty}} \equiv \frac{P_w - P_{ref}}{P_{T,\infty}} \tag{3}$$

where P_{ref} is the average surface static pressure in the region ahead of the shock across the area between X/d = 2 and 5.5.

The surface static pressure distributions on the windward ($\phi = 0°$) and leeward ($\phi = 180°$) show notable differences in the characteristics of EV and RSM models (Figure 3). On the windward side, the pressure ratio, $P^*/P_{T\infty}$, shows little difference between EV and RSM models and both are broadly in good agreement with the experimental results. Almost all of the data is within experimental uncertainty bounds although there are some key differences (Figure 3). All models calculated a peak pressure ratio which is larger than the nominal experimental data with the SA, SST, tSST and k-ε models over-predicting the windward peak $P^*/P_{T\infty}$ by approximately 8.7%, 11.7%, 9.7%, and 9.6%, respectively. All

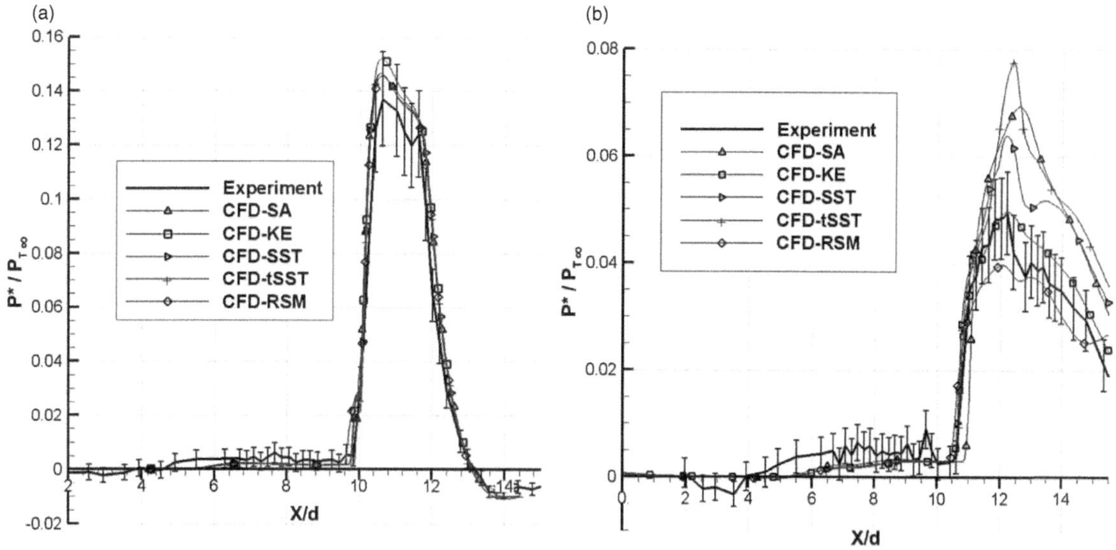

Figure 3. (a) Windward ($\phi = 0°$) and (b) leeward ($\phi = 180°$) surface pressure plots comparing experimental and numerical results of EV and RSM turbulence models at every $\sim 10^{th}$ data point. $M = 2.88$, $\alpha = 16°$, $Re_L = 18.2 \times 10^6$, $T_{T\infty} = 278$ K, $P_{T\infty} = 172.4$ kPa.

of the models determine the position of the windward $P^*/P_{T\infty}$ maximum to be at X/d = 10.61 which, within the spatial resolution, is in agreement with the experimental location of X/d = 10.63. The nominal inviscid point of impingement of the shock is at X/d = 12.21 and, as expected, the SBLI results in a compression region ahead of this point. The experimental measurements indicate that the footprint is approximately 2.41 X/d upstream of the nominal impingement point with the major initial pressure rise occurring at X/d = 9.84. Similarly, the k-ε, SST and tSST models all show evidence of this where the footprint is calculated to extend as far upstream as X/d = 9.80. There are small differences between the CFD models with the SA model exhibiting the rise further downstream at X/d = 9.89 and the RSM model further upstream at X/d = 9.61. In the expansion region downstream of the local maximum there are no notable characteristic differences in the CFD results where all of the $P^*/P_{T\infty}$ distributions are slightly higher than the measurements and generally at the upper limit of the measurement uncertainty. The experimental data indicates a small region of local re-compression at X/d = 11.5 which, although it is within the measurement uncertainty, is not reflected in any of the computations.

On the leeward side (ϕ = 180°), however, there are more notable differences between the CFD models as well as between the computational results and the measurements (Figure 3). Although all of the models broadly capture the locus of the initial pressure rise, there are large differences in the values of the local maximum in $P^*/P_{T\infty}$. In particular, peak values from the SA, SST and tSST models are 51%, 37% and 81% higher than the measurements, respectively. Most of the pressure ratio distribution from the k-ε calculations were within the measurement uncertainty and the peak, $P^*/P_{T\infty}$ was approximately 12% greater than the measurements. Finally, the best results were obtained using the RSM model where the pressure rise and peak value is within -8% of the measured value.

On the leeward side (Figure 3), the measurements also clearly exhibit a double peak in $P^*/P_{T\infty}$ due to a re-compression following the initial post-shock expansion with local maxima arising at X/d = 11.8 and 13.2. The SA, k-ε and tSST models do not capture this recompression. The SST and RSM simulations calculate this feature with the RSM getting the correct location, X/d = 12.0 and 13.1, and the SST indicating that the local maxima are slightly further aft at X/d = 12.1 and 13.6. The experiments indicate that the peak re-compression has a local relative magnitude of $P^*/P_{T\infty} = 0.05$. Although the SST and RSM models broadly capture the re-compression, the relative magnitude is around $P^*/P_{T\infty} =$ 0.065 and 0.04, respectively, or approximately +30% and -20% of the experimental values. Overall, based on the distributions of the surface $P^*/P_{T\infty}$ ratio, the k-ε and RSM models provide the best agreement with the measurements with perhaps slightly better characteristics for the RSM where the additional modeling fidelity provides some very minor advantages in comparison with the more standard EV models.

Figure 4 illustrates the SBLI system in the windward side symmetry plane for the k-ε calculation using the 3M cell grid. The expected main flow features are evident such as the upstream separation shock, a following expansion

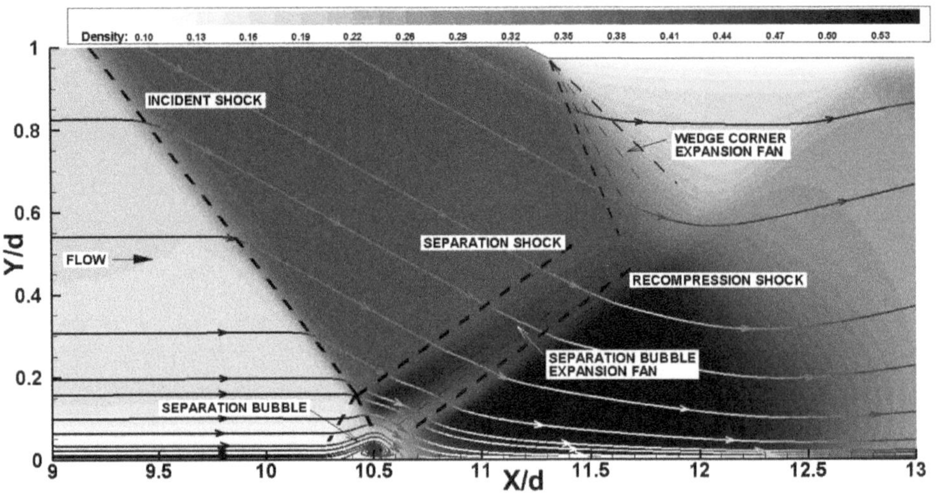

Figure 4. Calculated distribution of density on the windward side symmetry plane using the k-ε model 3M mesh. M = 2.88, α = 16°, $Re_L = 18.2 \times 10^6$, $T_{T\infty}$ = 278 K, $P_{T\infty}$ = 172.4 kPa.

region, the primary reflected shock and a local region of boundary layer separation. The other models also resolved these main flow features. For this configuration, there is a boundary layer separation which, from the experimental flow visualizations is estimated to extend from X/d = 10.1 to 10.7 along the cylinder symmetry plane. Both the EV and RSM models successfully calculate the occurrence of the separation region although with varying degrees of success. The SA model calculated a separated boundary layer region with an axial extent of 0.40 X/d which is approximately 31% shorter than that measured (0.58 X/d). For the other EV models, the k-ε, SST and tSST results showed separation lengths of X/d = 0.50, 0.55 and 0.60 which are approximately − 12%, − 4% and + 5% different from the measurements, respectively. The RSM model, however, provides the best calculation with a separation length of 0.58 X/d which is only -0.9% different from the measurements. This is a remarkably good level of agreement between the CFD and the experimental data given the complexity of the flowfield and is in disagreement with the over prediction of separation bubble size seen by Oliver et al. (2007). For the EV models, however, there is no clear relationship between the fidelity of the peak pressure ratio rise on the windward and leeward side and the accuracy of the calculation of the separation region length.

This SBLI is fundamentally three dimensional and the interaction is affected by the diffraction, and ultimate crossing, of the primary impinging shock around each side of the cylinder. These aspects result in a complex distribution of the surface static pressure on the cylinder which is characterized by a predominantly quasi 2D pressure rise on the cylinder windward side, the sweep of the diffracted impinging shock and the notable subsequent local pressure rises on the leeward side due to the crossing of the respective diffracted shocks. As also shown in the experimental results (Figure 5), the diffraction of the shock around the leeward side results in a region where the surface static pressure is affected by the impinging shock wave and its interaction with the curved cylinder surface. On the windward side the SBLI and pressure rise is affected by the local thickening of the boundary layer as well as the reflected shock. With increasing azimuthal angle the local pressure rise decreases as the shock reflection is reduced although this is a notable increase in the pressure rise again on the leeward side.

Where the shock interacts on the leeward side, it influences the upstream flow conditions (Figure 5), beginning experimentally at X/d = 11.0. By comparing this point of initial X/d Leeward Upstream Influence (LUI) in relation to the X/d position aft of the wedge, the CFD models can be compared. The X/d (LUI) position of the RSM model was effectively congruent with experimental data and within measurement uncertainty margins. The SA, k-ε, SST, and tSST EV models were + 12.8%, 1.3%, 3.1%, and 4.7% respectively. The effect of the transition model was modest. The SA model again illustrated its inaccuracy for such SBLIs, whilst the tSST model was slightly worse than the SST model.

The distribution of $p/P_{T\infty}$ over the cylinder highlights some of the aspects of the flow characteristics as the SBLI develops between the initial windward interaction (ϕ = 0°) and the eventual behavior on the leeward side. The measured distribution of $p/P_{T\infty}$ shows some of the main features although it is noteworthy that the resolution of the experimental data is ϕ = 10° steps between ϕ = − 10° to 190° (Figure 5, Figure 6). The spatial distribution along the cylinder surface (ϕ-X/d) highlights the main region of the strong SBLI pressure rise which is broadly concentrated

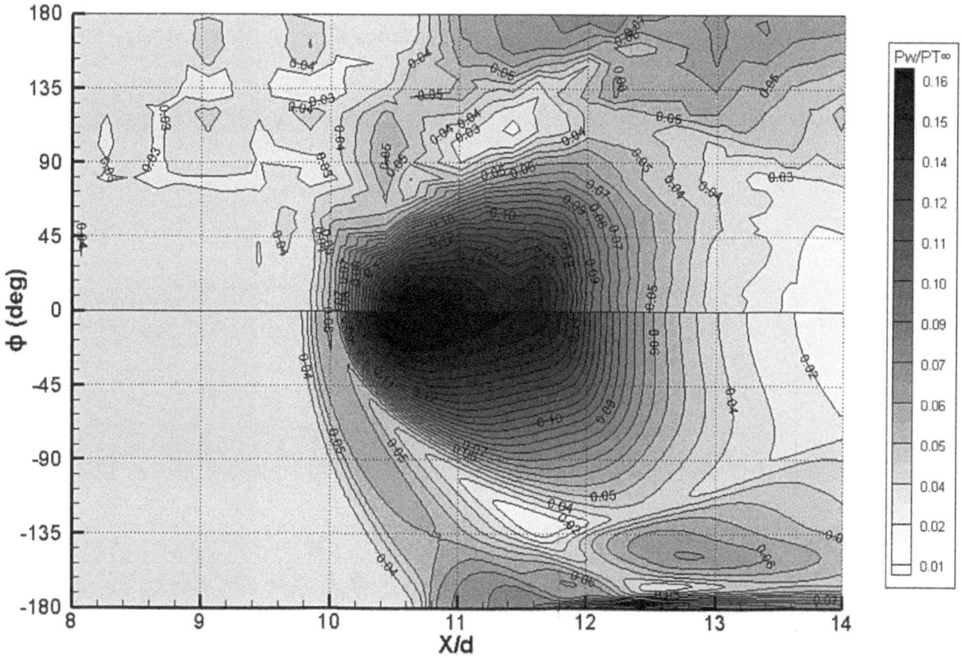

Figure 5. Comparison of experimental (top) and numerical (bottom) cylinder nondimensionalized surface pressure $P_w/P_{T\infty}$; SST model. 10M mesh. $M = 2.88$, $\alpha = 16°$, $Re_L = 18.2 \times 10^6$, $T_{T\infty} = 278$ K, $P_{T\infty} = 172.4$ kPa.

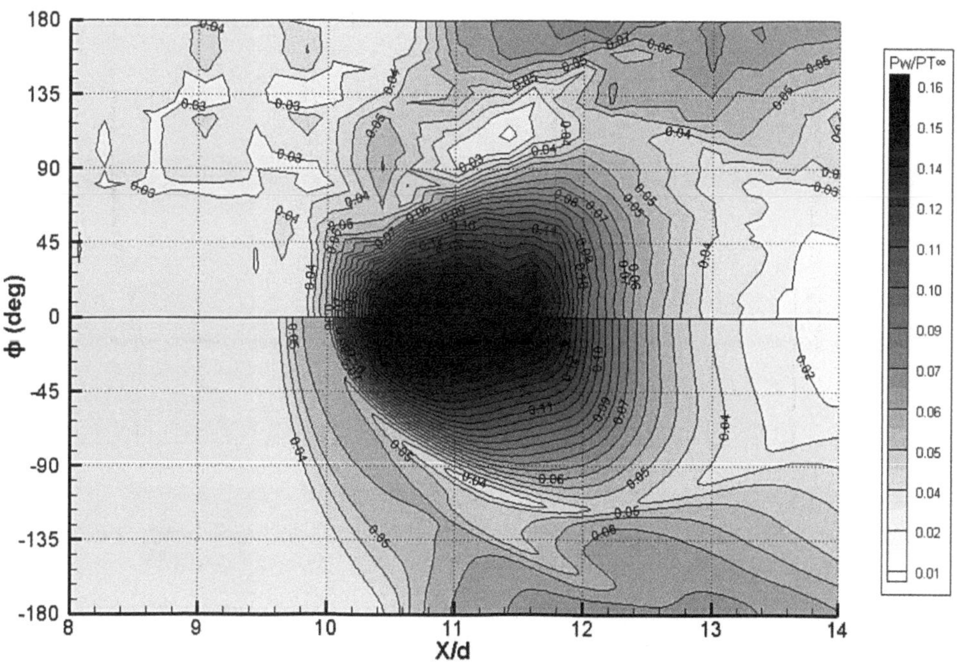

Figure 6. Comparison of experimental (top) and numerical (bottom) cylinder nondimensionalized surface pressure $P_w/P_{T\infty}$; RSM model. 10M mesh. $M = 2.88$, $\alpha = 16°$, $Re_L = 18.2 \times 10^6$, $T_{T\infty} = 278$ K, $P_{T\infty} = 172.4$ kPa.

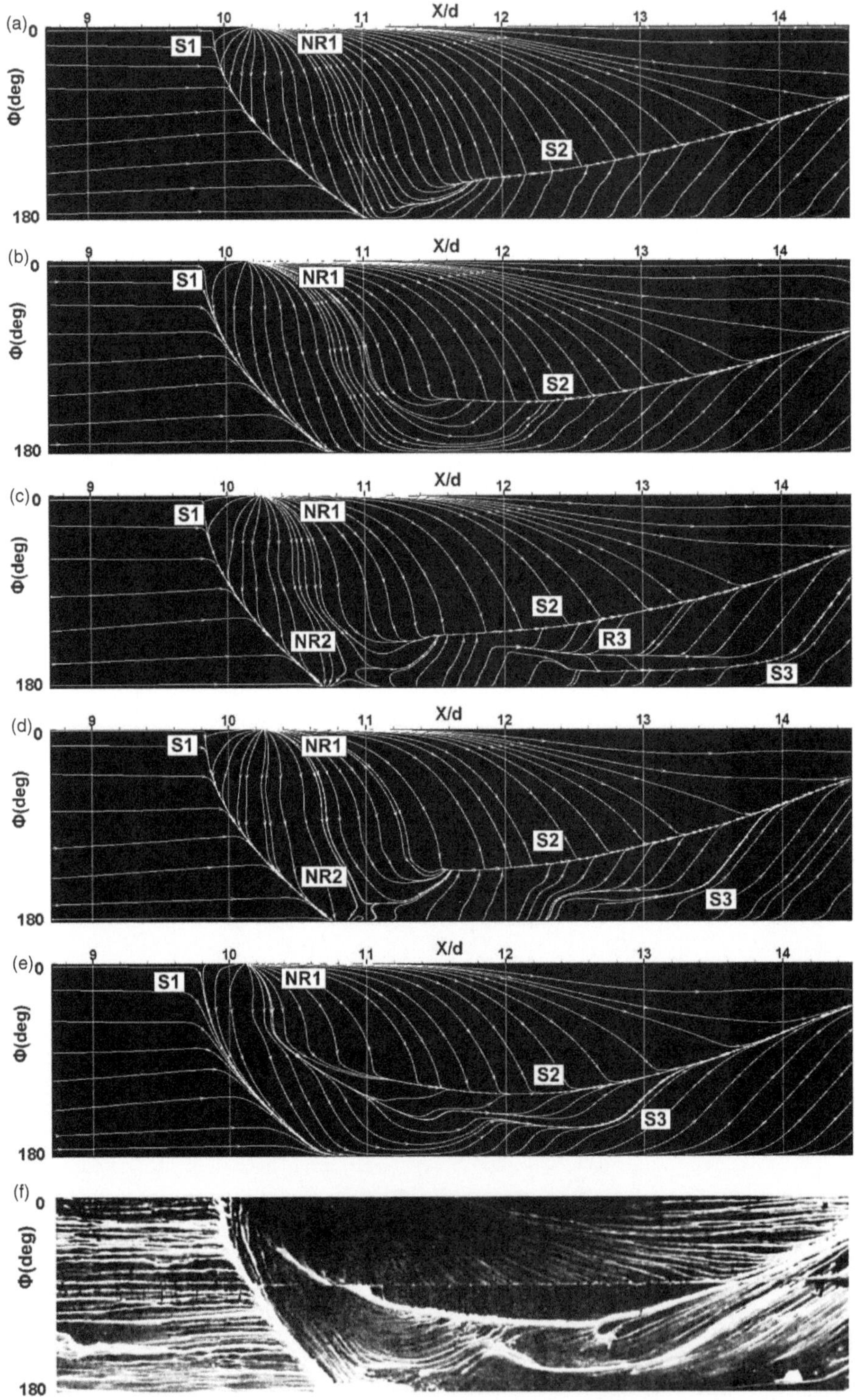

Figure 7. Comparison of computational streamlines, with experimental Mylar oil flow visualisation (a) SA, (b) k-ε, (c) SST, (d) tSST, (e) RSM. $M = 2.88$, $\alpha = 16°$, $Re_L = 18.2x10^6$, $T_{T\infty} = 278$ K, $P_{T\infty} = 172.4$ kPa. *Figure 7(f): This figure is taken from NASA-TM-84410, 'An Experimental Investigation of the Impingement of a Planer Shock Wave on an Axisymmetric Body at Mach 3', by A. Brosh, M. Kussoy, and used with permission of NASA.*

in the region of X/d = 10 to 12 and ϕ from 0° to ±90°. The distributions of $p/P_{T\infty}$ shows the reduction in $p/P_{T\infty}$ in the region aft of the main interaction as well as a region of relatively modest static pressure rise which borders the main shock interaction at around X/d = 11 to 12 and ϕ = 90° to 135°. On the leeward side and the aft region (X/d = 13 to 14), $p/P_{T\infty}$ increases again. This is where the diffracted shock has propagated around the cylinder body, crosses the centerline and augments the static pressure due to the crossing with the diffracted shock from the opposite side. All of these main features are calculated by both the SST and the RSM models (Figure 5, Figure 6) and overall there is good agreement in the topology of the surface static pressure distributions. In the leeward region there is a modest difference in the $p/P_{T\infty}$ distribution between the SST and RSM models where the SST shows regions with stronger rise and pressure gradients, particularly at ϕ = 140° and ϕ = 180° (Figure 5). The RSM model calculates more modest pressure gradients in this region and is in better agreement with the measurements (Figure 5, Figure 6).

As seen in the experiment, the computational results also show a strong lateral pressure gradient near the primary interaction at approximately X/d = 10 to 11 (Figure 5). This generates a strong crossflow and causes the flow to exhibit the wake type characteristics with separation (Brosh & Kussoy, 1983). The measured location point of minimum pressure occurs at ϕ = 110° and X/d = 11.4 (Figure 5). All of the eddy-viscosity models overpredict this azimuthal position of the minimum pressure point with the SA model having the largest difference ϕ = 140° (+27%) although the other models also had notable differences, with k-ε, SST and tSST positions of ϕ = 128° (+16%), ϕ = 132° (+20%), and ϕ = 135° (+23%), respectively. The RSM model provided the best results with the determined azimuthal position as ϕ = 105° (−5%). The prediction of the lateral cross-flows was also in better agreement on the subsequent leeward side pressure field, with these distributions and pressure gradients also affecting the local surface flow topology.

As part of the experimental studies, the surface flow topology was identified using a Mylar oil visualization method. This visualization highlighted the primary shock induced separation, which sweeps azimuthally around the cylinder, as well as the secondary separation line, notable crossflow regions and partial evidence of a tertiary separation (Figure 7).

The surface flow topology based on the skin friction distributions for the EV and RSM models are presented in Figure 7. In the experiment the flow visualization indicates a strong cross flow region associated with the initial interaction with the impinging shock which is also observed in all four eddy-viscosity models and best represented by the k-ε model. All EV simulations, the first separation line, S_1, and node of reattachment, NR_1, seen in the experiment are represented. The SA model, Figure 7(a), captures the fewest aspects of the surface flow features with a secondary separation line, S_2, as the only other main surface feature present. The k-ε model, Figure 7(b), calculates a surface flow topology which is very similar to that of the SA model, and also fails to simulate the tertiary separation line. The SST, Figure 7(c), and tSST, Figure 7(d), models show similar primary features as with the SA and k-ε models, but also include a tertiary separation line, S_3, and are more like the experimental data. This tertiary separation is less clear in the SST model although a secondary attachment node, NR_2, is present in both models and more clearly resolved in the SST model. The RSM model, Figure 7(e), shows the main features which are an improvement over the k-ε model. The SST model indicates a tertiary reattachment line (R_3) in the region between X/d = 12 to 13.5 although the transition model (Figure 7(d)), shows a slightly different structure without a clearly defined R_3. There is no clear evidence of such a feature in the flow visualization and overall the indications are that the RSM topology is more closely representative of the experimental data.

Overall it was found that for the RANS method, the turbulence model had a notable impact on the results. The turbulence model affected both the flow topology as well as the quantitative aspects of the SBLI. Of the EV models, the k-ω variants best represented the topology while the k-ε models gave the best quantitative agreement. The SA model performed the worst and demonstrated its unsuitability for this type of SBLI. This conclusion has also been reached in the literature by Oliver et al. (2007), Bhagwandin and DeSpirito (2011) and Tian and Lu (2013). The better agreement for the k-ε and k-ω formulations with experimental data, such as static surface pressures and separation position, have been seen by Oliver et al. (2007) and Bhagwandin and DeSpirito (2011) when comparing two and three-dimensional SBLIs on flat plates and compression ramps. Three-dimensional work by DeBonis et al. (2010) of a shock impinging on a flat plate found that k-ω models produced the lowest error when compared with SA, k-ε, k-ω and SST methods. The current work supports these findings within the context of shock interactions on three-dimensional geometries in addition to highlighting improvements that are possible using an RSM model when compared with the EV k-ε model. Vallet (2007) reported similar findings but for a simpler compression ramp configuration with varying incidence angles.

4.2. *Effect of Reynolds number*

The datum configuration was for a Reynolds number of 18.2×10^6 with a freestream Mach number of 2.88. In the experiment, the effect of Reynolds number was considered for values of 7.28×10^6 and 58.3×10^6 although in the experiment these also required a slight change in the freestream Mach number of 2.80 and 2.95, respectively

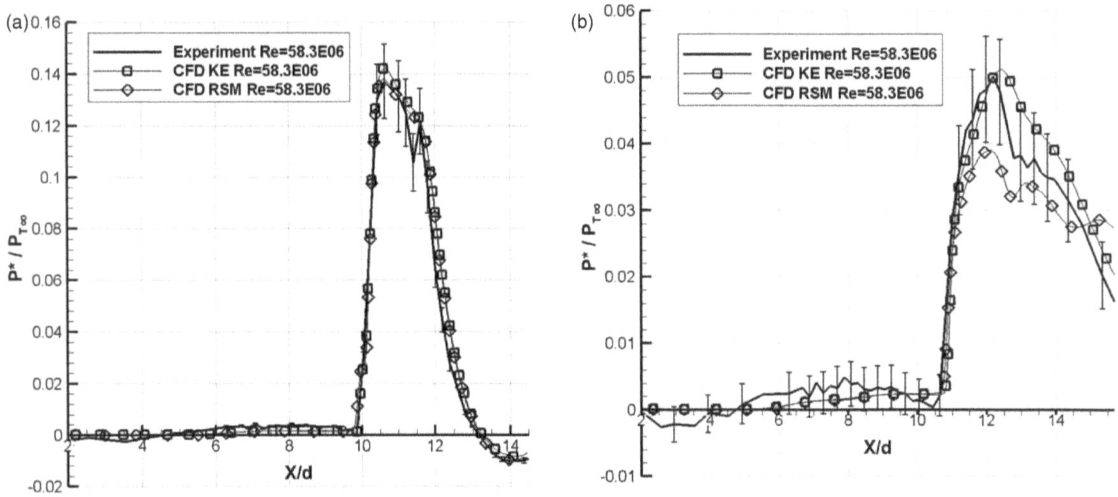

Figure 8. (a) Windward ($\phi = 0°$) and (b) leeward ($\phi = 180°$) surface pressure plots comparing k-ε and RSM turbulence models at $Re_L = 58.3 \times 10^6$ compared to experimental data. Computational results show approximately every 10th data point. 3M mesh. M = 2.95, $\alpha = 16°$, $T_{T\infty} = 278$ K.

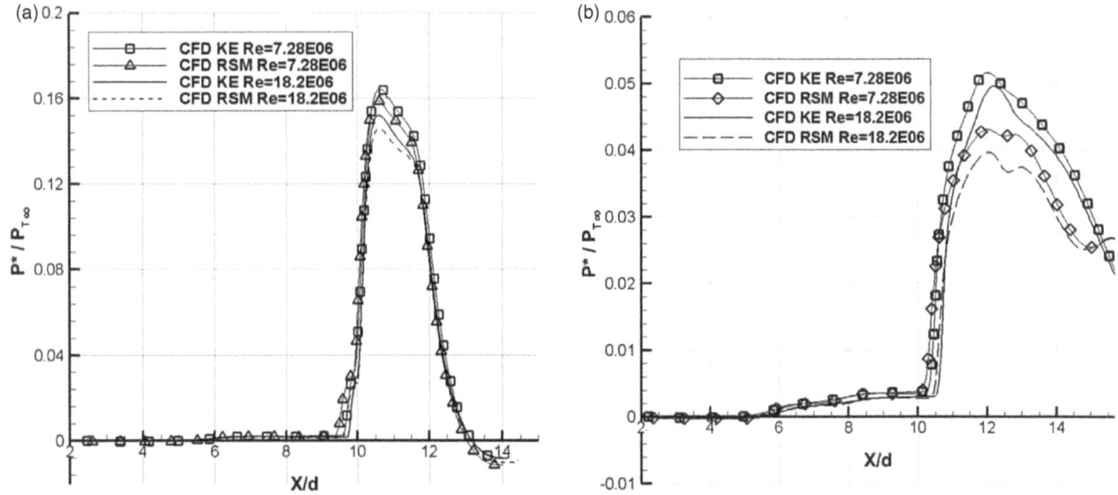

Figure 9. (a) Windward ($\phi = 0°$) and (b) leeward ($\phi = 180°$) surface pressure plots comparing k-ε and RSM turbulence models at $Re_L = 7.28 \times 10^6$ with datum CFD results (M = 2.88, $Re_L = 18.2 \times 10^6$). Computational results show approximately every 10th data point. 3M mesh. M = 2.80, $\alpha = 16°$, $T_{T\infty} = 278$ K.

(Table 1). Although this therefore intertwines the effects of Reynolds number, shock strength and shock angle it is still of interest to investigate the changes in the SBLI and the ability of the CFD models to calculate these changes. As these configurations were not the main focus of the previous experimental study, there is a reduced amount of experimental measurements available for these configurations with surface pressure data and flow visualizations for the higher Reynolds number (58.3×10^6) case.

For the configuration with the Reynolds number increased from the datum value of 18.2×10^6 to 58.3×10^6, the Mach number was also increased from 2.88 to 2.95 and therefore for a constant wedge angle of 16° there is a theoretical reduction in the shock angle from 34.2° to 33.6° and the static pressure ratio across the shock changes from 2.89 to 2.94. Overall these changes are relatively minor compared with the datum configuration (Figure 3) and both k-ε and RSM models produce results that are in close agreement with each other although neither captures the sudden drop in $P^*/P_{T\infty}$ at X/d = 11.5 (Figure 8). On the leeward side, although there is some variation, the k-ε shows better performance in terms of the peak pressure ratio rise (Figure 8(b)) but the RSM shows the local pressure maximum at about X/d = 13.5. The

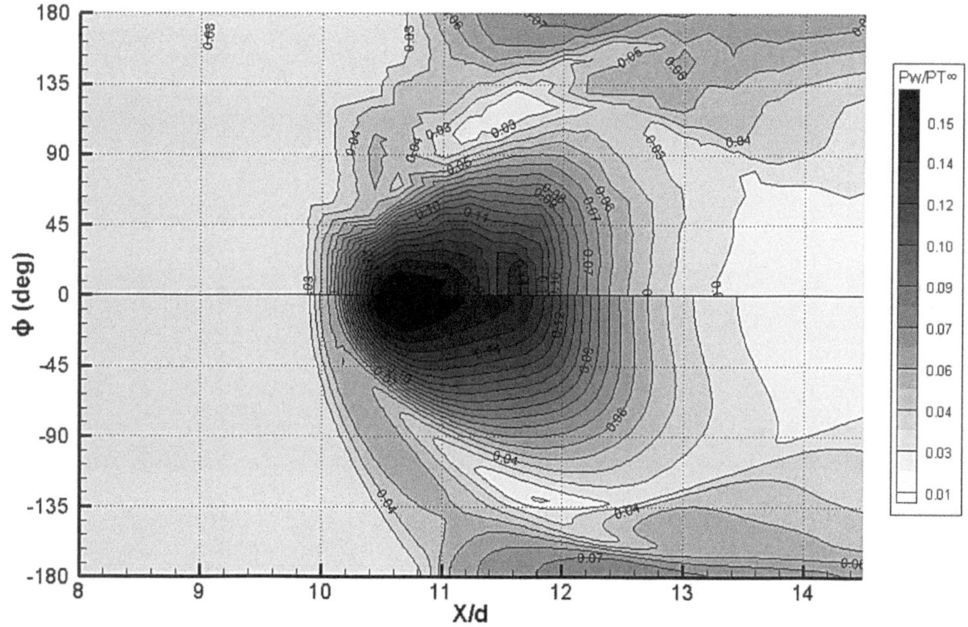

Figure 10. Comparison of experimental (top) and numerical (bottom) cylinder nondimensionalized surface pressure $P_w/P_{T\infty}$; k-ε model. 3M mesh. M = 2.95, $\alpha = 16°$, $Re_L = 58.3 \times 10^6$, $T_{T\infty} = 278$ K, $P_{T\infty} = 551.6$ kPa.

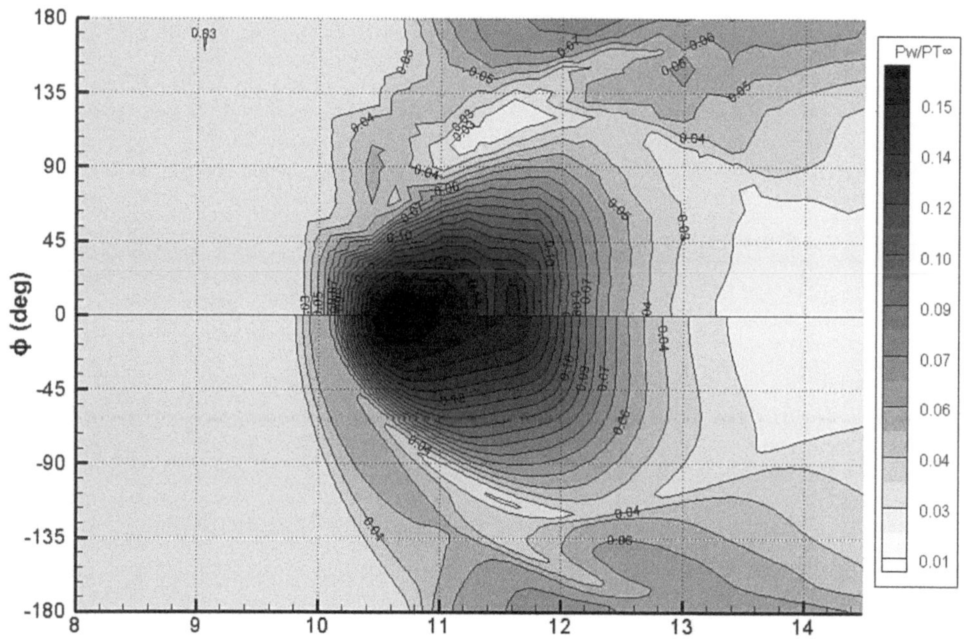

Figure 11. Comparison of experimental (top) and numerical (bottom) cylinder nondimensionalized surface pressure $P_w/P_{T\infty}$; RSM model. 3M mesh. M = 2.95, $\alpha = 16°$, $Re_L = 58.3 \times 10^6$, $T_{T\infty} = 278$ K, $P_{T\infty} = 551.6$ kPa.

effect of reducing the Reynolds number from 18.2×10^6 to 7.28×10^6 was partially investigated in the experimental work and computations of this lower Reynolds number case were also performed using the k-ε and RSM models (Figure 9). The calculations broadly show that there was a slight increase in the strength of the interaction with the peak pressure rise on the windward side from the RSM calculations increasing from around 0.14 to 0.16 when Re_L increased from 7.28×10^6 to 18.2×10^6. The point of the initial pressure rise is slightly further upstream at the lower

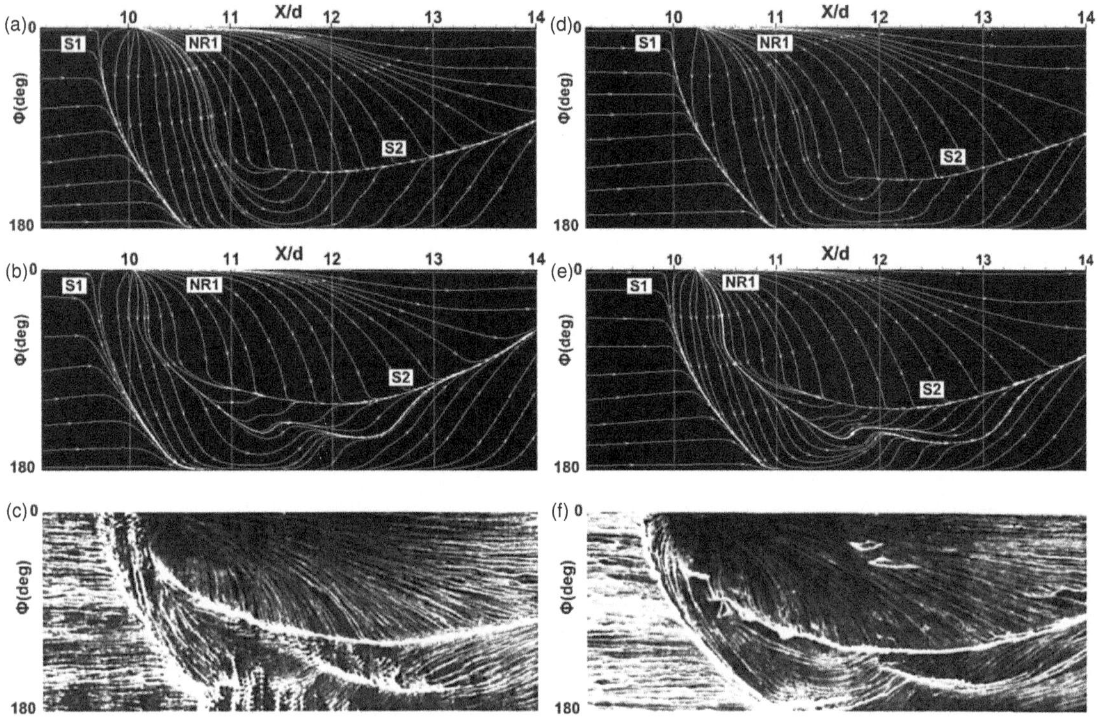

Figure 12. Comparison of computational streamlines, with experimental oil flow visualisation. Figures (a)–(c) M = 2.80, $\alpha = 16°$, $Re_L = 7.28 \times 10^6$, $T_{T\infty} = 278$ K, $P_{T\infty} = 68.9$ kPa. (a) k-ε, (b) RSM. Figures (d)–(f) M = 2.95, $\alpha = 16°$, $Re_L = 58.3 \times 106$, $T_{T\infty} = 278$ K, $P_{T\infty} = 551.6$ kPa. (d) k-ε, (e) RSM. *Figure 12(c)(f): These figures are taken from NASA-TM-84410, "An Experimental Investigation of the Impingement of a Planer Shock Wave on an Axisymmetric Body at Mach 3," by A. Brosh, M. Kussoy, and used with permission of NASA.*

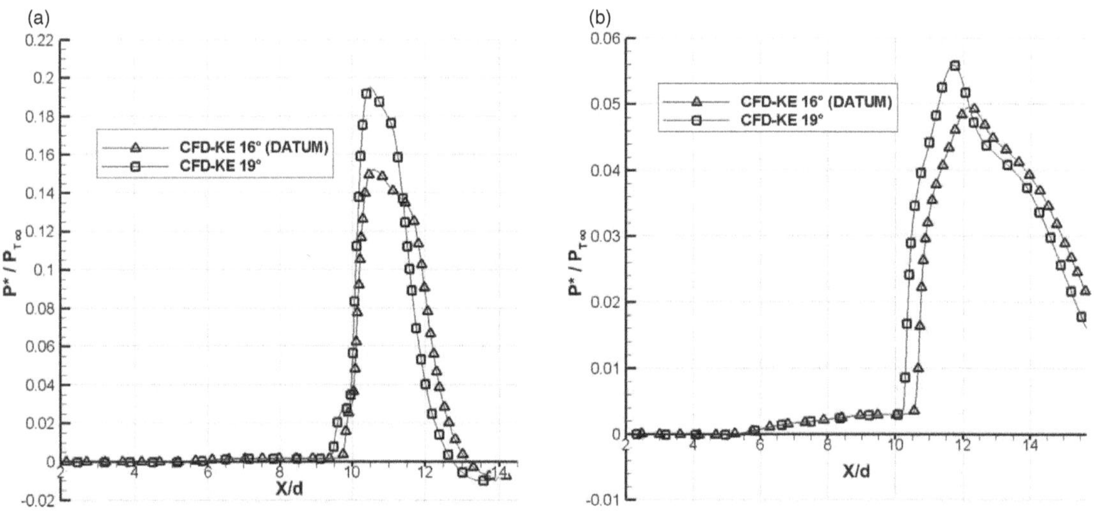

Figure 13. (a) Windward ($\phi = 0°$) and (b) leeward ($\phi = 180°$) surface pressure plot comparing k-ε model at $\alpha = 16°$ and $\alpha = 19°$ at every $\sim 10^{th}$ data point. 3M mesh. M = 2.88, $Re_L = 18.2 \times 10^6$, $T_{T\infty} = 278$ K, $P_{T\infty} = 172.4$ kPa.

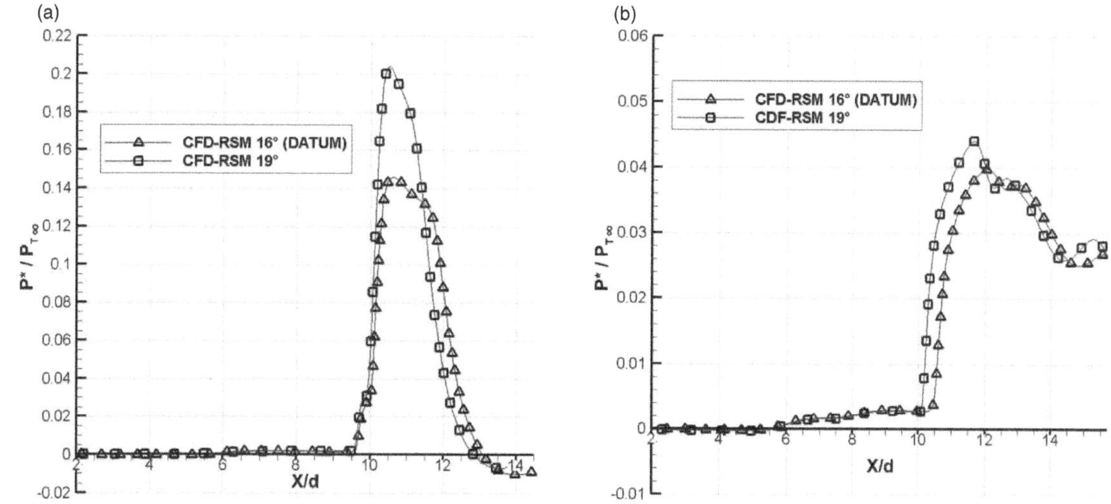

Figure 14. (a) Windward ($\phi = 0°$) and (b) leeward ($\phi = 180°$) surface pressure plot comparing RSM model at $\alpha = 16°$ and $\alpha = 19°$ at every $\sim 10^{th}$ data point. 3M mesh. $M = 2.88$, $Re_L = 18.2 \times 10^6$, $T_{T\infty} = 278$ K, $P_{T\infty} = 172.4$ kPa.

Re_L, but this is within the uncertainty of the arrangement and the slight changes in the flow conditions. On the windward side, the computations similarly show that the RSM model results in a lower peak pressure rise in comparison with the k-ε results. Concomitant with the slightly higher $P^*/P_{T\infty}$ on the windward side, both computations similarly show a slightly higher pressure rise on the leeward side.

The measured distributions of static pressure around the cylinder are shown in Figure 10 and Figure 11 for the high Re_L of 58.3×10^6 and compared with the results from the k-ε and the RMS models. The point of minimum pressure for the high Reynolds case occurs experimentally at $\phi = 120°$and X/d = 11.5. The EV and RSM models predict the minimum pressure position at X/d = 11.8, $\phi = 130°$ and X/d= 11.6, $\phi = 112°$, respectively (Figure 10 and Figure 11). In the low Re_L case ($Re_L = 7.28 \times 10^6$), the position of minimum pressure is found to be X/d=11.4 and 11.2, $\phi = 120°$ and 104° for EV and RSM models respectively. Whilst the axial positions are relatively close, the azimuthal positions vary more notably. In the datum simulations (Figure 5, Figure 6), the RSM model was found to calculate the axial position closer to the experimental results and this is borne out again for the high Reynolds case.

The impact of the change in Re_L across the range from 7.28×10^6 to 58.3×10^6 , for both the experimental flow visualizations as well as the k-ε and RSM computations is shown in Figure 12. The differences in the flow topologies are relatively minor and there is no notable change in either flow visualization of computational results. Nevertheless one of the notable features is in the separation bubble region at $\phi = 0°$ where the experiments indicate that the separation bubble is larger as Re_L is reduced. This characteristic is also observed in both the k-ε and RSM results (Figure 12(a,b,d,e)). For the RSM model there is a slight change in the topology at the highest Re_L (in Figure 12(e)), where there is evidence of a tertiary separation line starting at X/d = 13 and $\phi = 160°$. Overall the numerical methods show similar levels of agreement with the experimental data at the increased Reynolds number configuration and the RSM provides better agreement in the surface flow topology than the k-ε simulations.

4.3. Effect of shock strength

The results presented in section 5.1 above, indicate that for the datum configuration ($M = 2.88$, $\alpha = 16°$, $Re_L = 18.2 \times 10^6$) that the k-ε and RSM methods provided the best agreement with the surface static pressures along the windward ($\phi = 0°$) and leeward ($\phi = 180°$) centerlines. In addition, the RSM model shows generally better agreement with the wider surface static pressure distributions and the overall flow topology. Although the experiment considered the impact of the shock strength on the interaction, by performing flow visualizations for a configuration where the wedge angle, α, was increased to 19°, the centerline pressure measurements were not acquired.

Nevertheless, given the confidence in the application of the k-ε and RSM methods to this problem, it is of interest to assess the impact of the increase in the shock strength and change in incidence angle, on the simulated flow fields. The k-ε and RSM methods both show similar changes in the centerline $P^*/P_{T\infty}$ (Figure 13 and Figure 14) where the increase in shock strength from 0.14 to 0.20 for the RSM cases is broadly in agreement with the relative theoretical increase of 40% for an inviscid interaction. Of course, the overall levels of $P^*/ P_{T\infty}$ are lower than the inviscid value by approximately 30% based on the measurements for the datum configuration. Although the relative changes in $P^*/$

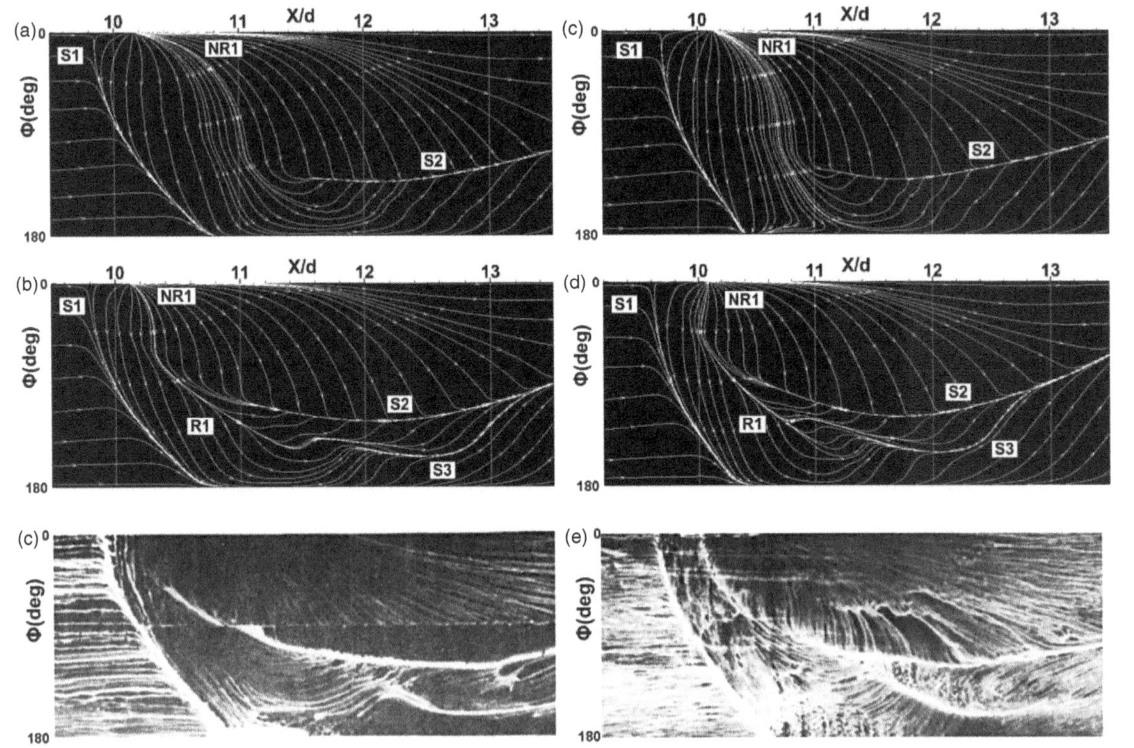

Figure 15. Comparison of computational streamlines, with experimental oil flow visualisation. Figures (a)-(c) M = 2.88, α = 16°, Re_L = 18.2 × 10^6, $T_{T\infty}$ = 278 K, $P_{T\infty}$ = 172.4 kPa. (a) k-ε, (b) RSM. Figures (d)-(f) M = 2.88, α = 19°, Re_L = 18.2 × 10^6, $T_{T\infty}$ = 278 K, $P_{T\infty}$ = 172.4 kPa. (d) k-ε, (e) RSM. *Figure 15 (c)(f): These figures are taken from NASA-TM-84410, "An Experimental Investigation of the Impingement of a Planer Shock Wave on an Axisymmetric Body at Mach 3," by A. Brosh, M. Kussoy, and used with permission of NASA.*

$P_{T\infty}$ on the windward side are broadly as expected, on the leeward side the impact of the increased shock strength is attenuated through the interaction so that the increase in this region is only approximately 13% for both the k-ε and RSM methods.

The increase in the impinging shock strength and change in shock angle has a small impact on the surface flow topology as illustrated in the flow visualizations (Figure 15) but was more difficult to resolve computationally. This has been noted by Oliver et al. (2007) and DeBonis et al. (2010) with Vallet (2007) observing significant failures in the k-ε model's SBLI prediction ability at higher ramp incidence angles in comparison to RSM methods. The flow topology effects are likely due to the increased size of the separation bubble: when α is increased from 16° to 19°, there is a slight increase in the size of the primary separation at ϕ = 0° and there are minor changes to the loci of the secondary and tertiary separation lines, S_2 and S_3, respectively. The k-ε model shows some of these characteristic changes, but the structure associated with the primary reattachment line (R_1) and the absence of the tertiary separation (S_3) indicates that the agreement with the experimental data is perhaps worse than the datum configuration. As with the datum case, the RSM model shows better agreement with the oil visualization also for this case of α = 19°.

5. Conclusion

A shock–boundary layer interaction test case based on the complex configuration of planar oblique wave impinging onto the curved surface of a prismatic cylinder was computationally investigated. The investigation is based around an established test case under a nominal freestream Mach number of 2.88 and the impact of Reynolds number and shock strength and angle were also considered. The study assessed the performance of a range of computational models and evaluated the results relative to quantitative experimental data as well as flow visualizations. The research considered a range of eddy viscosity RANS models including the Spalart-Allmaras, k-ε Realisable, k-ω Shear Stress Transport and the k-ω SST model with a γ-θ transition model. In addition a Reynolds Stress Model was also evaluated. A computational approach was developed to establish mesh independence as well as appropriate iterative convergence.

The main flow features of such a SBLI were observed; separation bubble, leeward upstream influence, lateral cross-flow, as well as primary and secondary flow separations. Of the EV models tested it was shown that the SA model is unsuitable for complex 3D SBLIs as it did not capture many important flow features. The surface flow topologies were better represented by k-ω variants (SST and the γ-θ SST model) whilst the magnitude was better represented by the k-ε two-equation model.

The RSM seven-equation model improved on the EV models in terms of the flow features such as the separation bubble size, and azimuthal pressure distribution. It replicated the leeward separation lines closer to the experimental results in comparison with the EV models but had difficulty, however, in predicting the leeward focal point. In terms of the magnitude of the local peak pressure values, both the k-ε EV and the RSM models provided the best results, with the RSM providing a better distribution of the flow features.

The sensitivity to changes in Reynolds number, shock strength and incidence were assessed and the RSM computations broadly agreed with the oil flow visualizations. The main aspects of the interaction were found to be relatively insensitive to Reynolds number with a slightly stronger interaction observed at the lowest Reynolds number where the static pressure rise was calculated to be approximately 10% greater. The effect of increasing shock strength showed a slightly larger primary separation region and at the primary impingement location, the static pressure rise increased as expected relative to the changes in the impinging shock. However, on the leeward side, the computations indicate that the interaction attenuates the influence of the impinging shock and the relative pressure rise is notably smaller than that on the windward side. Within the context of multibody shock interactions, this is a notable observation as it indicates that the impact on windward and leeward sides are not simply related and therefore it is expected that the overall resulting body forces and moments will similarly be sensitive to the shock strength.

All of the computational models were tested using a steady-state solver and, given that it is known that such SBLIs typically have low-frequency shock unsteadiness, the steady RANS method may have limitations in predicting this class of SBLIs. However, in spite of this, the computational methods investigated here show that steady RANS methods, specifically k-ε and RSM models, are capable of calculating the overall flow features and quantitative changes associated with an oblique shock impinging on an axisymmetric body. Clearly this conclusion comes with some caveats due to the known assumptions of the steady RANS method, however, given the relative computational cost in comparison with LES or DNS calculations, it indicates that within an engineering context it may be able to play a useful role. The S-A and k-ω variants were found to be unsatisfactory.

Beyond the aforementioned limitations of using steady RANS methods, improvements to the application of EV and RSM models to complex 3D SBLIs could be made by using unsteady-RANS to account for the low-frequency shock unsteadiness experimentally observed and inherent to this type of flow. Furthermore, existing literature suggests that LES and DES models will come to dominate the prediction of this class of flows, however, comparison of results between the different models could yield improvements to RANS methods and thus enable further improvements to this class of model which is typically computationally cheaper and continue their use in applied engineering contexts.

References

Adams, N. (2000). Direct simulation of the turbulent boundary layer along a compression ramp at $M = 3$, $Re\theta = 1685$. *Journal of Fluid Mechanics*, *420*, 47–83.

Allen, J. H., Heaslet, M. A., & Nitzberg, G. E. (1947). *The interaction of boundary layer and compression shock and its effect upon airfoil pressure distributions* (Report No. RM-A7A02). Washington: NACA.

Arnal, D., & Délery, J. (2004). *Laminar-turbulent transition and shock wave/boundary layer interaction* (NATO RTO AVT/VKI Lecture Series, RTO-EN-AVT-116, 4-1-4-46). Brussels, Belgium: NATO.

Babinsky, H. H., & Harvey, J. K. (2011). *Shock wave-boundary layer interactions*. New York: Cambridge University Press.

Barakos, G. N., Doerffer, P., Hirsch, C., Dussauge, J. P., & Babinsky, H. (2010). *WP-5 LES and Hybrid RANS/LES. Notes on numerical fluid mechanics and multidisciplinary design*. Berlin, Germany: Springer.

Bhagwandin, V. D., & DeSpirito, J. (2011). *Numerical prediction of supersonic shock-boundary layer interaction* (Report No. ARL-TR-5602. Aberdeen Proving Ground). MD: Army Research Laboratory.

Borovoy, V. E., Egorov, I., Maximenko, A., Mosharov, V., Radchenko, V., Skuratov, A., & Struminskaya, I. (2013). Three-dimensional shock-wave/boundary-layer interaction at the presence of entropy layer. *Progress in Flight Physics*, *5*, 327–348.

Brosh, A. K., & Kussoy, M. I. (1983). *An experimental investigation of the impingement of a planar shock wave on an axisymmetric body at Mach 3* (Technical Memorandum 84419 N84-18573). Moffett Field, CA: NASA.

Brosh, A. K., Kussoy, M. I., & Hung, C. M. (1985). Experimental and numerical investigation of a shock wave impingement on a cylinder. *AIAA Journal*, *26*(6), 840–846.

Celik, I. G., Ghia, U., Roache, P. J., Freitas, C. J., Coleman, H., & Raad, P. E. (2008). Procedure for estimation and reporting of uncertainty due to discretization in CFD applications. *Journal of Fluids Engineering-Transactions of the ASME*, *130*(7), 078001–1–078001–4.

Chaplin, R. M. (2010). *Aerodynamic interference between high-speed slender bodies* (PhD thesis). Cranfield, UK: University of Cranfield.

Chaplin, R. M., MacManus, D. G., & Birch, T. J. (2010). Aerodynamic interference between high-speed slender bodies. *Shock Waves*, *20*(2), 89–101.

Chaplin, R. M., MacManus, D., Leopold, F., Martinez, B., Gauthier, T., & Birch, T. (2011). Computational and experimental investigation into aerodynamic interference between slender bodies in supersonic flow. *Computers and Fluids*, *50*, 155–174.

DeBonis, J. O., Oberkampf, W. L., Wolf, R. T., Orkwis, P. D., Turner, M. G., & Babinsky, H. (2010). *Assessment of Computational Fluid Dynamics (CFD) models for shock boundary-layer interaction* (NASA-TM-2011-216832). Chicago, IL: NASA.

Délery, J. M., Marvin, J. G., & Reshotko, E. (1986). *Shock wave/boundary layer interactions* (AGARDograph No. 280). Brussels, Belgium: NATO.

Délery, J. M., & Coet, M. C. (1990). Experiments on shock wave/boundary layer interactions produced by two-dimensional ramps and three-dimensional obstacles. *Workshop on Hypersonic Flows for Re-entry Problems*. Berlin, Germany: Springer.

Délery, J. M. (1996). *Shock wave/boundary layer interactions in high Mach number flows* (AGARD/FDP WG 18 on Hypersonic Experimental and Computational Capability, Improvement and Validation, AGARD AR 319, Vol I). Brussels, Belgium: NATO.

Délery, J. (1999). Shock phenomena in high speed aerodynamics: still a source of major concern. *The Aeronautical Journal*, *103*(1019), 19–34.

Derunov, E. Z., Zheltovodov, A. A., & Maksimov, A. I. (2008). Development of three-dimensional turbulent separation in the neighborhood of incident crossing shock waves. *Thermophysics and Aeromechanics*, *15*(1), 29–54.

Dolling, D. S. (2001). Fifty years of shock-wave/boundary-layer interaction research: what next? *AIAA Journal*, *39*(8), 1517–1531.

Donaldson, C. (1944). *Effects of interaction between normal shock and boundary layer* (Report CB 4A27). Washington: NACA.

Eagle, W. D., Driscoll, J. F., & Benek, J. A. (2012, June 25–28). *3-D Inlet shock-boundary layer interactions - PIV database for the second SBLI workshop*. 30th AIAA Applied Aerodynamics Conference, New Orleans, LA, Paper No. AIAA 2012–3214.

Fedorov, A. M., Malmuth, N. D., & Soudakov, V. G. (2007). Supersonic scattering of a wing-induced incident shock by a slender body of revolution. *Journal of Fluid Mechanics*, *585*, 305–322.

Gai, S. T., & Teh, S. L. (2000). Interaction between a conical shock wave and a plane turbulent boundary layer. *AIAA Journal*, *38*(5), 804–811.

Garnier, E. (2009). Stimulated detached Eddy simulation of three-dimensional shock/boundary layer interaction. *Shock Waves*, *19*, 479–486.

Hadjadj, A. (2012). Large-Eddy simulation of shock/boundary-layer interaction. *AIAA Journal*, *50*(12), 2919–2927.

Hilsenralh, J., Beckett, C., Benedict, W., Fano, L., Hoge, H., Masi, J., Nuttoll, R., Touloukian, Y., & Woolley, H. (1955). *Tables of thermal properties of gases* (Circular 564). Washington, DC: National Bureau of Standards.

Hirsch, C. (2010a, June 28–July 1). *Lessons learned from the first AIAA-SWBLI workshop CFD simulations of two test cases*. 28th AIAA Applied Aerodynamics Conference, Chicago, IL, Paper No. AIAA 2010-4824.

Hirsch, C. (2010b). *WP-4 RANS/URANS simulations* (Notes on Numerical Fluid Mechanics and Multidisciplinary Design). Berlin, Germany: Springer.

Holden, M. M., Wadhams, T., MacLean, M., & Mundy, E. (2010, June 28–July 1). *Experimental studies of shock wave/turbulent boundary layer interaction in high reynolds number supersonic and hypersonic flows to evaluate the performance of CFD codes*. 40th Fluid Dynamics Conference and Exhibit, Chicago, IL, Paper No. AIAA 2010–4468.

Hung, C. (1985). *Computation of three-dimensional shock wave and boundary-layer interactions* (Techincal Memorandum 86780). Moffett Field, CA: NASA.

Kussoy, M. H., Horstman, C. C., & Acharya, M. (1978). *An experimental documentation of pressure gradient, and reynolds number effect, on compressible turbulent boundary layers* (Technical Memorandum 78488). Moffett Field, CA: NASA.

Malmuth, N., & Shalaev, V. (2004, January 5–8). *Theoretical modelling of interaction of multiple slender bodies in supersonic flows*. 42nd AIAA Aerospace Sciences Meeting and Exhibit, Reno, NV, Paper No. AIAA 2004–1127.

Mendonca, J., & Sharif, M. A. R. (2010). Effects of surface roughness on turbulent transonic flow over circular arc bumps in a chanel. *Engineering Applications of Computational Fluid Mechanics*, *4*(2), 164–180.

Morkovin, M. V., Migotsky, E., Bailey, H. E., & Phonney, R. E. (1952). Experiments on interaction of shock waves and cylindrical bodies at supersonic speeds. *Journal of Aerospace Science*, *19*(4), 237–248.

Oliver, A. B., Lillard, R. P., Schwing, A. M., Blaisdell, G. A., & Lyrintzis, A. S. (2007, January 8–11). *Assessment of turbulent shock-boundary layer interaction computations using the OVERFLOW code*. 45th AIAA Aerospace Sciences Meeting and Exhibit, NV, Paper No. AIAA 2007–104.

Pamadi, B. N., Neirynck, T. A., Hotchko, N. J., Scallion, W. I., Murphy, K. J., & Covell, P. F. (2005). *Simulation and analyses of stage separation of two-stage reusable launch vehicles*. AIAA/CIRA 13th International Space Planes and Hypersonics Systems and Technologies Conference, Paper No. AIAA 2005–3247.

Peake, D. J., & Tobak, M. (1982). *Three dimensional flows about simple components at angle of attack* (Technical Memorandum 84226). Moffett Field, CA: NASA.

Roache, P. J. (1998). *Verification and validation in computational science and engineering*. New Mexico: Hermosa.

Sandham, N. (2010). Oblique shock reflection at M = 2.0. *Unsteady Effects of Shock Wave Induced Separation, Notes on Numerical Fluid Mechanics and Multidisciplinary Design*, *114*, 263–285.

Saric, W. S., Dussauge, J. P., Smith, R. W., Smits, A. J., Fernholz, H., Finley, P. J., & Spina, E. F. (1996). *Turbulent boundary layers in subsonic and supersonic flow* (AGARD-AG-335). Brussels, Belgium: NATO.

SAS IP Inc. (2011a). *ANSYS FLUENT Theory Guide: Release 14.0*. Canonsburg, PA: ANSYS.

SAS IP Inc. (2011b). *ANSYS FLUENT User's Guide: Release 14.0*. Canonsburg, PA: ANSYS.

Sedney, R., & Kitchens, C. W. (1975). *The structure of three dimensional separated flows in obstacle, boundary layer interaction* (AGARD CP-168, Paper No: 37). Brussels, Belgium: NATO.

Shams, A., & Comte, P. (2010). *RDDES for strong shock-wave/boundary layer interaction* (Notes on Numberical Fluid Mechanics and Multidisciplinary Design, Volume 11 - Progress in Hybrid RANS-LES Modelling). Berlin, Germany: Springer.

Thivet, F. (2002, January 14–17). *Lessons learned from RANS simulations of shock-wave/boundary-layer interactions*. 40th AIAA Aerospace Sciences Meeting and Exhibit, Nevada, Paper No. AIAA 2002–0593.

Tian, C., & Lu, Y. (2013). Turbulence models of separated flow in shock wave thrust vector nozzle. *Engineering Applications of Computational Fluid Mechanics*, *7*(2), 182–192.

Tokura, Y., & Maekwa, H. (2011). Direct numerical simulation of impinging shock wave/transitional boundary layer interaction with separation flow. *Journal of Fluid Science and Technology*, *6*(5), 765–779.

Vallet, I. (2007). Reynolds stress-modelling of M = 2.25 shock-wave/turbulent boundary-layer interaction. *International Journal for Numerical Methods in Fluids*, *56*(5), 525–555.

Volkov, V. F., & Derunov, E. K. (2006). Interaction of a combination of bodies in supersonic flow interference and diffraction of shock waves in flow over two bodies of revolution. *Journal of Engingeering Physics and Thermophysics*, *79*(4), 712–721.

Wilcox, F. (1995). *Separation characteristics of generic stores from Lee side of an inclined Flat Plate at Mach 6* (Technical Memorandum 4652). Hampton, VA: NASA.

Wilcox, D. (2006). *Turbulence modeling for CFD* (3rd ed.). La Cañada, CA: DWC Industries.

8

Prediction of the bubble nucleation rate in a quasi-stable cavitating nozzle using 2D computational fluid dynamics and enhanced classical nucleation theory

Z.H. Ban, K.K. Lau* and A.M. Shariff

Chemical Engineering Department, Universiti Teknologi PETRONAS, Bandar Seri Iskandar, 31750 Tronoh, Perak, Malaysia

Most of the numerical studies of the cavitation process available were simulated based on the Rayleigh-Plesset equation, which has a presumed value for the number of bubble nuclei in the domain. However, in the computational domain for most cases, the bubble nuclei density varies, depending on the hydrodynamics and different bubble nucleation rates. Therefore, the modeling of the bubble nucleation rate based on the number of molecules required to form a critical cluster, which is affected by the hydrodynamics in the domain, is one of the challenges that must be overcome for bubble nuclei density prediction. In this work, an enhanced Classical Nucleation Theory (CNT) model was written as in-house code for integration in Computational Fluid Dynamics (CFD) to estimate the water vapor bubble nucleation rate across a quasi-stable cavitating flow nozzle. The enhanced CNT model included microscopic factors that specifically address water vapor bubble nucleation. Because these factors vary in the radial direction, this work was extended into the 2D domain to estimate the bubble nucleation rate in both the axial and the radial directions. The bubble nucleation rate was estimated through various numbers of molecules (depending on the hydrodynamics in the domain) required to form a critical cluster. The Population Balance Model was used to estimate the bubble nuclei density based on the bubble nucleation rate. The simulated results indicated that the bubble nucleation rate was highest at the throat of the nozzle, especially near the wall of the nozzle. Thus, a higher bubble nuclei density was found near the wall after the throat of the nozzle.

Keywords: bubble nucleation rate; number of bubble nuclei; CFD

1. Introduction

Nucleation of vapor bubbles occurs either by increasing the temperature of the liquid above its boiling point (superheated liquid) or reducing the external pressure below the saturated vapor pressure of the liquid (stretched liquid). This phenomenon, which is also known as cavitation, can be easily found in many places, such as marine propellers and fuel injectors.

Computational Fluid Dynamics (CFD) has been widely used in various industrial applications to predict the hydrodynamic flow in the relevant computational domain, such as in a nozzle (Cai, Fang, Xu, & Liu, 2007; Ding, Wang, & Chen, 2014; Peralta, Chew, & Wilson, 2011; Tian & Lu, 2013). In the numerical study of cavitation processes, the mathematical model based on the Rayleigh-Plesset equation is often used in the numerical simulation of cavitating flow due to its simplicity. Singhal, Athavale, Li, and Jiang (2002) has developed a full cavitation model by introducing vapor generation and condensation rates as source terms in the vapor transport equation. The effect of noncondensable gases was included in the model. The phase change rates in the model depend only on the pressure difference, vapor and gas mass fraction, density, turbulence kinetic energy, surface tension and empirical constant. This model was also integrated into CFD codes and successfully validated with experimental results. There are also several cavitation models proposed by other researchers, including Kunz, Boger, Stinebring, Chyczewski, Lindau, Gibeling, Venkateswaran, and Govindan (2000) and Shin, Yamamoto, and Yuan (2004). The factors considered in these cavitation models were similar to the model suggested by Singhal et al. (2002). In these models, the microscopic factors that influence the nucleation and growth of vapor bubbles, for example, cluster surface tension and the Zeldovich factor, were not included in the cavitation model. Therefore, the work required for forming a critical cluster was difficult to determine using the model. To date, there are limited numerical investigations on the cavitation of water in the bubbly flow considering the nucleation of vapor bubbles (Goncalves & Patella, 2009; Lindau, Kunz, Boger, Stinebring, & Gibeling, 2002; Preston, Colonius, & Brennen, 2002). Delale, Okita, and Matsumoto, (2005) successfully attempted to integrate the enhanced Classical Nucleation Theory (CNT) into the cavitating flow model in 1D steady state bubbly flow. However, one of the bubble nucleation determining factors, that is, the critical bubble radius, must be pre-set before the numerical simulation of the cavitating study. There are also works that attempted to include thermodynamic effects in the numerical study of cavitating

*Corresponding author. Email: laukokkeong@petronas.com.my

flow (Goncalvés, and Patella, 2010; Vortmann, Schnerr, & Seelecke, 2003).

CNT has been widely used in the studies of nucleation processes in various areas, such as in crystallizers, atomizers and bubble nucleation studies (Liška, Holubová, Černošková, Černošek, Chromčíková, & Plško, 2012; Lubetkin, 1995; Shukla, Mandal, & Ojha, 2001). In the early works of bubble nucleation studies, CNT was used for modeling a unary system (single component), either for dissolved gas bubbles or vapor bubbles (Lubetkin, 1995). This CNT model has been used by several researchers to calculate the cavitation pressure of water. From their calculations, large negative pressure (few hundreds of MPa) was required to initiate bubble nucleation in pure water without dissolved gas and impurities (Caupin & Herbert, 2006). Similar values were obtained using other methods of calculations for the cavitation pressure of water, such as the thin-wall approximation and density functional theory. However, the cavitation pressure values reported in the literature are widely scattered, with a maximum negative pressure of −120 MPa at 40°C reported (Zheng, Durben, Wolf, & Angell, 1991). The calculated negative pressure was larger than the values reported from experimental data due to impurities in water (Herbert, Balibar, & Caupin, 2006). In addition, the onset of nucleation for water with dissolved gas occurred at a much higher pressure (less negative) compared to the pressure for pure water, as discussed above. This happened due to the lowering of surface tension by the dissolved gas, which acted as a surfactant in this case and thus reduced the work of bubble formation (Lubetkin, 2003). Therefore, the effect of dissolved gas and impurities were suggested to be added into the CNT mathematical model as a correction factor. This correction factor was dependent on the quality of water, which can be determined by comparison with the experimental results (Delale, Hruby, & Marsik, 2003). The modification of the CNT model for bubble nucleation has enabled it to be used for predicting the nucleation process in water. Delale, Okita, and Matsumoto (2005) has also successfully tested and validated this enhanced version of the CNT model to predict the 1D steady-state cavitating nozzle flow. Later, the use of CNT was further extended to a multicomponent system (Vehkamäki, 2006). Another approach to predict the bubble nucleation rate in water for gas and vapor proposed by Kwak and Oh (2004) was based on a unified approach. This approach considered different factors of the bubble nucleation of gas and vapor.

To date, most of the cavitation processes have been described using the reduced Rayleigh-Plesset cavitation model, whereby the nuclei number density is predefined in the domain, regardless of the hydrodynamic effects. The famous CNT is one of the methods used in the effort to determine the bubble nucleation rate according to the hydrodynamic conditions. However, the coupling of CNT and the bubbly liquid flow model is complex and hinders the prediction of the bubble nuclei density, especially in the 2D flow domain. The prediction of bubble nuclei density is important for the modeling of bubble growth and mass transfer in the domain. In addition, modeling of the bubble nucleation rate based on the number of molecules required to form a critical cluster (which determines the critical cluster size) is also one of the challenges to obtain the bubble nucleation rate. Therefore, the main objective of this work is to model the bubble nucleation rate for water flowing across a 2D cavitating nozzle under quasi-stable condition in both the axial and the radial directions. Hence, the bubble nuclei density in the computational domain could be predicted. This study was accomplished with the aid of an enhanced CNT and CFD coupling approach. In addition, in-house code for bubble nucleation rate modeling was developed in the current work. The Population Balance Model (PBM) is applied to further estimate the distribution of the bubble nuclei. In addition, the bubble nucleation rate is estimated through the computed number of molecules required to form a critical cluster. Because the main focus of the current work are modeling the bubble nucleation rate using the coupled CFD and enhanced CNT model; bubble mass transfer and growth are not included in the current work.

2. Mathematical modeling

2.1. *Governing equation*

A multifluid model based on the Eulerian–Eulerian approach, as used by other similar works (Yan, Li, & Luo, 2012), was used in this work to describe the fluid flow across the nozzle. This model was assumed to be capable to describe the simple shear flow in the 2D domain. In addition, the 2D domain was sufficient to describe the flow in the nozzle because it is symmetric in the radial direction. Because the mass transfer between the two phases was not considered in the current work, the continuity equation for the system is expressed as

$$\frac{\partial \rho}{\partial t} + \nabla \cdot (\rho u) - 0 \tag{1}$$

where ρ is the density of the solution, u is the Reynolds-averaged velocity vector, and t is time.

The volume fraction transport equation for liquid phase is given as

$$\frac{\partial}{\partial t}(\alpha_l \rho_l) + \nabla \cdot (\alpha_l \rho_l u) = 0 \tag{2}$$

where α_l is the volume fraction of the liquid phase and ρ_l is the density of the liquid phase.

In the current work, the formation of bubble nuclei was assumed to have an insignificant effect on the hydrodynamics in the computational domain. Therefore, the source term that accounts for bubble nuclei was not included in Equations (1) and (2). The calculation for bubble nuclei formation and distribution was performed by using Population Balance Model based on Classical Nucleation Theory.

2.2. *Classical nucleation theory*

The nucleation rate, J_v, can be termed as the number of critical clusters formed per unit volume per unit time (nuclei/m^{3}s). CNT states that there is an energy barrier (or the work to form a critical cluster) in the nucleation process. The theoretical expression for the nucleation rate can be written as

$$J_v = J_0 exp\left[-\frac{W}{k_B T}\right] \quad (3)$$

where T is the liquid total temperature, k_B is the Boltzmann's constant, W is the work to form a critical cluster and J_0 is a kinetic pre-exponential factor. This expression suggests that the work to form a critical cluster plays an important role in determining the nucleation rate. Thus, the work to form a critical cluster must be perfectly evaluated so that the nucleation rate can be estimated accurately.

In general, for bubble nucleation, the energy barrier is determined by the surface tension of the liquid and also the pressure difference between the bubble and external liquid. The energy barrier is determined as

$$W = \frac{16\pi}{3}\frac{\gamma^3}{(P_v - P_l)^2} \quad (4)$$

where P_v and P_l are the pressure in the bubble and external liquid, respectively, and γ is the surface tension of the liquid-gas surface.

Regarding bubble nucleation, several microscopic parameters, for example, the critical bubble radius, was included in the improved bubble nucleation model based on the CNT (Delale, Hruby, & Marsik, 2003). In the current work, the mathematical model developed by Kwak and Oh (2004) for bubble nucleation was adopted (Equation (5)). They postulated that the physical mechanism behind gas and vapor bubble formation was completely different. The surface energy required to form a gas bubble was constrained by the dissolution due to solvent–solute interaction, while vapor bubble formation was linked with the potential energy between molecules. The expression for vapor bubble nucleation includes the effect from energy to separate a pair of molecules, the number of vapor molecules in the critical cluster, molecular volume, and so on. The equations involved in the calculations of vapor bubble formation was discussed in detail by Kwak and Oh (2004).

$$J_v = Z_{fv} D_v N_v exp\left(-\frac{1}{2} n_{cv}^{2/3}\right) \quad (5)$$

with

$$Z_{fv} D_v = \frac{N_v}{\sqrt{6\pi}}\left(\frac{k_B T}{2\pi m_v}\right)^{1/2} 4\pi \left(\frac{3v'_m}{4\pi}\right)^{2/3} \times \beta'_v exp\left[-\frac{\Delta H_{vap}}{RT} - \frac{\Delta H_f}{RT_f}\right]$$

where

J_v = Vapor bubble nucleation rate, nuclei/m^3
n_{cv} = Number of molecules required to form a vapor critical cluster
N_v = Number density of the liquid, 3.3428×10^{28} molecules/m^3
m_v = Mass of the liquid molecules, 2.99151×10^{-26} kg
v'_m = Transformed volume of vapor molecules, m^3
β'_v = Accommodation coefficient of the vapor molecules to the cluster
ΔH_{vap} = Enthalpy for the evaporation of the solution
ΔH_f = Enthalpy for the freezing of the solution
T_f = Melting temperature for the solution, K
R = Gas constant, 8.3144621 J/ K·mol

Water under normal conditions will most likely contain another dissolved gas, such as air, which affects the bubble nucleation rate. Hence, the number of vapor molecules required to form a critical cluster is affected. To extend the application of the nucleation model to account for the dissolved gas in water, the partial pressure of the dissolved gas was added in the model, as shown in Equation (6); this modification is analogous with the modification implemented by Delale et al. (2005). The effect of dissolved air was considered in the current work.

$$-(P_l - P_v - P_{air}) n_{cv}^{'\frac{1}{3}} = \frac{k_B T}{v'_m} \quad (6)$$

where P_{air} is the partial pressure of air at atmospheric pressure.

Other than the effect of dissolved air on the number of vapor molecules required to form a critical cluster, it can also affect the surface tension, as it may act as a surfactant. In addition, some other impurities or contaminants may exist as well in the water, which would influence the nucleation rate of the water. Consequently, a correction factor, C, was introduced into the work of bubble formation, as also introduced by Delale et al. (2005). This correction factor can be obtained by comparing the nucleation rate with the experimental results. Here, the correction factor was determined to be 2.396×10^8, which was obtained through the nucleation rate results from the work of Delale et al. (2005). The modified bubble nucleation rate is then expressed as

$$J_v = Z_{fv} D_v N_v exp\left(-\frac{1}{2}\frac{n_{cv}^{2/3}}{C}\right) \quad (7)$$

2.3. *Population balance model (PBM)*

The PBM was used in the current work to predict the bubble nuclei density in the computational domain through the bubble nucleation rate and the number of bubble nuclei that exist at the particular position. Because the main focus of this work was to study the bubble nucleation phenomenon across the nozzle, only the nucleation rate was taken into consideration in the current work in the population balance equation (PBE), where the expression in terms of the number density function $n(V,t)$ is reduced to

$$\frac{\partial n(V,t)}{\partial t} + \nabla \cdot [un(V,t)] = -\frac{\partial}{\partial V}[G(V)n(V,t)] \quad (8)$$

where $n(V,t)$ is the bubble number density distribution per unit volume (V) for a given time (t). On the right-hand side term, G (V) is the growth rate of bubble number density, which acted as a source term, in this case, for the bubble nucleation.

The quadrature method of moments (QMOM) was widely used to study the particulate system because it can solve the PBE accurately but requires a lower number of equations. Therefore, QMOM was used in this work to solve the PBE. Based on QMOM, the moments of the particle size distribution (PSD) are approximately obtained as

$$m_k = \int_0^{\infty} n(V,t)V^k dV \approx \sum_{i=1}^{Q} w_i L_i^k \quad (9)$$

where m_k is the kth moment of the number density function, abscissas (Li) and weights (wi) are determined through the product-difference (PD) algorithm, Q is the order of quadrature approximation, and k is the specified number of moments (ANSYS, 2011).

After applying the moment transformation, the PBE is expressed as

$$\frac{dm_k}{dt} = \int_0^{\infty} G(V)n(V,t) \cdot V^k dV \quad (10)$$

Equation (10) shows that the rate of change of moment is directly proportional to the bubble nucleation rate in this case.

3. Numerical method

For reasons of comparison with the quasi-stable flow in the cavitating nozzle studied by Wang and Brennen (1998), the geometry and boundary conditions for the current work were changed based on the reference. In addition, the axisymmetric model was used for the nozzle in solver. The normalized area of the nozzle, A(x), is given by

$$A(x) = \left\{1 + 0.5\left[1 - \cos\left(\frac{2\pi x}{L}\right)\right]\right\}^{-1/2} \quad \text{for } 0 \leq x \leq L \quad (11)$$

with x as the dimensionless normalized length of the nozzle. The normalized length of the nozzle was set as 5 in this study. Thus, the throat of the nozzle is located at $x = 2.5$, which is the middle of the nozzle, as shown in Figure 1 (a).

Figure 1 (b) shows the mesh of the computational domain. A total of 13100 cells were generated in the computational domain. In addition, a finer mesh was generated near the wall region. Figure 1 (c) shows the predicted bubble nucleation rate along the axis in the computational domain by using three different mesh densities (1200, 13,100 and 28,500 cells). The results obtained by Mesh 2 and Mesh 3 are both close to each other. Because Mesh 3 requires more computational cost compared to Mesh 2, the computations in current work was performed based on Mesh 2.

Commercial available CFD software, ANSYS FLUENT 14.5, was used in the current work to perform the investigation (Payri, Ruiz, Gimeno, & Martí-Aldaraví, 2015). Table 1 shows the boundary conditions and the settings used in the current work. The governing equations were solved by a coupled pressure-based algorithm and the finite volume approach. The QUICK scheme (Li & Li, 2011) was chosen for the discretization of the governing equations because it can provide third-order accuracy. The 2D space and steady state formulation was used in the simulation. The k-ε turbulence model with a standard wall treatment (Vaidyanathan et al., 2003) was used in the current work. Because the flow in the domain was a simple shear flow without a high swirling component or a large pressure gradient, the standard k-ε turbulence model was assumed to be capable to provide a reasonably accurate prediction in the simulation. In addition, the standard wall function can also perform well in the simple shear flow condition.

Water, at 20°C, together with dissolved air, was considered to be flowing through the nozzle from the inlet at velocity of 10 m/s. Several properties of water at this temperature were assumed to be constant, such as the density of water, at 1000 kg/m^3; the viscosity of water, at 0.001 Ns/m^2; and the vapor pressure of water, at 2340 Pa. The volume fraction of dissolved air was set as 2.5×10^{-6} in current work because the cavitation in the nozzle would be quasi stable flow as reported by Wang and Brennen (1998). The dissolved air was assumed to have a negligible effect on the hydrodynamics in the nozzle. The effect of the dissolved air was included in the bubble nucleation rates model as discussed.

To use the equations (Equations (6) & (7)) for the calculation of bubble nucleation rate, a user-defined function (UDF) was adopted. An in-house code (written in UDF) to for the calculation of bubble nucleation rate was developed. This code was incorporated with the commercial CFD codes to be coupled along with the hydrodynamics equations. Thus, the UDF was integrated in the PBM in the solver. Consequently, the bubble nucleation rate based on enhanced CNT was coupled with CFD-PBM. The bubble nucleation rate can be calculated directly from the hydrodynamics simulated from the CFD codes. One of the unique aspects of the use of this method to calculate the bubble nucleation rate was that the varied number of molecules that was required to form a critical cluster in

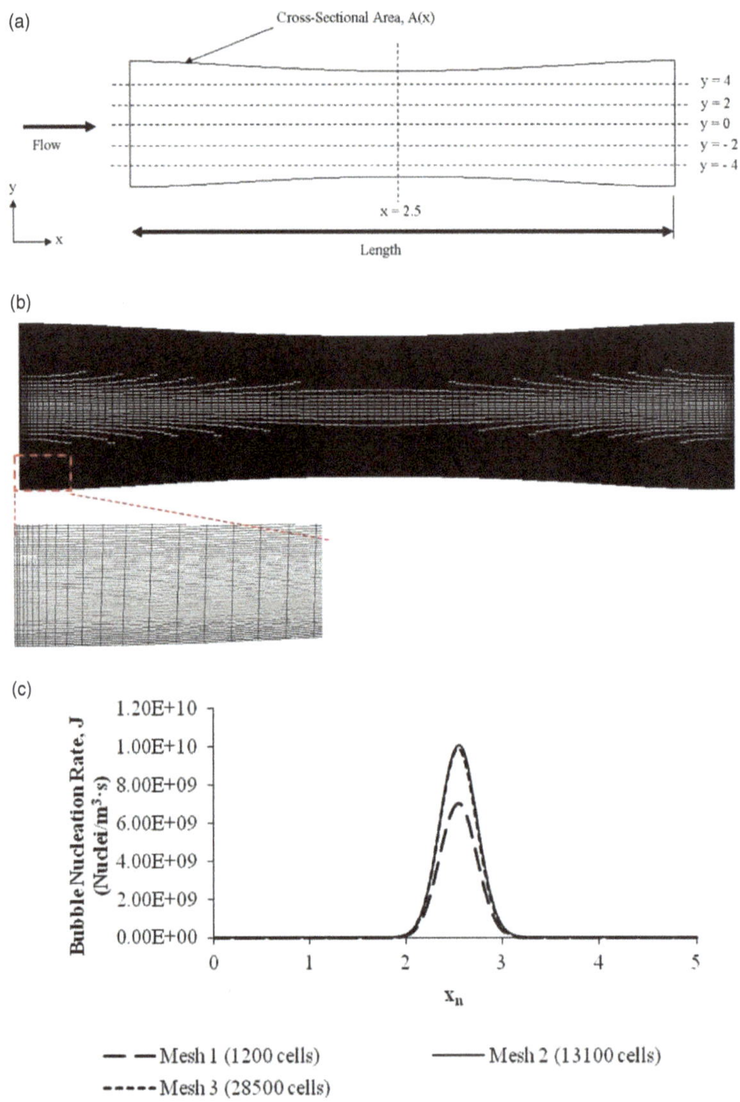

Figure 1. (a) Geometry, (b) mesh of the computational domain, and (c) mesh independence study.

the domain could be computed. The number of molecules is a very important parameter that greatly affects the work of formation of the critical cluster and, consequently, the bubble nucleation rate. The equation for the calculation of the number of molecules required to form a critical cluster was found to depend on the pressure at a specific position. Therefore, the hydrodynamics profile in the nozzle is important to determine the bubble nucleation rate.

For comparison, two additional parameters were introduced in the current work. First, a dimensionless parameter, namely, pressure coefficient, C_p, was introduced. This parameter has been widely used in cavitating flow analysis to determine the pressure profile across the nozzle with regards to the inlet flow velocity. The pressure coefficient is defined as

$$C_P = \frac{(P_l - P_i)}{0.5\rho_l U_i^2} \tag{12}$$

where P_i and U_i are the liquid pressure and velocity at the inlet of the nozzle, respectively. This expression is analogous to the definition of the cavitation number, which is used to characterize the potential of the flow to cavitate.

Meanwhile, the second parameter is the number of vapor molecules required to form a critical cluster. This parameter acted as the main factor to determine the bubble nucleation rate because it will directly affect the work of formation of the critical cluster. The parameter can be calculated by using Equation (6), as suggested by Kwak and Oh (2004).

Table 1. Boundary conditions and settings for the numerical simulation.

Property	Boundary Conditions and Settings
Materials	Water, density = 1000 kg/m^3
Dimensionality	2D
Solution method	Coupled pressure-based solver
Solution formulation	Implicit
Time dependence	Steady state
Turbulence model	k-ε model, standard wall function
Velocity inlet	Velocity = 10 m/s, Temperature = 20°C
Volume fraction of gas phase	2.5×10^{-6}
Pressure outlet	Pressure = 0, Temperature = 20°C
Population balance model	Quadrature method of moments (QMOM)

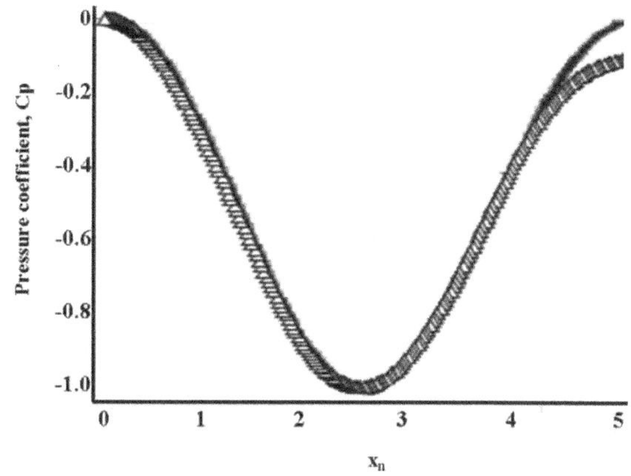

Figure 3. Distributions of the pressure coefficient C_p along the axis of the nozzle in the axial direction in the current work (Dotted line) and the results from Wang and Brennen (1998) (Solid line).

4. Results and discussion

4.1. *Model validation*

The CFD simulation coupled with the improved CNT was used to simulate the case of cavitating nozzle flow based on the conditions discussed by Wang and Brennen (1998). The hydrodynamic profiles of the simulated results in the present investigation were compared with the results obtained from Wang and Brennen (1998) for the case without dissolved air. The comparisons in flow velocity and pressure coefficient along the axis of the nozzle are plotted in Figure 2 and Figure 3, respectively. Based on the figures, the results from the current work exhibit good agreement with the results obtained from Wang and Brennen (1998), with the highest fluid flow velocity at the throat of the nozzle recorded as 14 m/s, while the lowest pressure coefficient was found to be −1.0 m/s. There is a slight difference in the fluid flow velocity and the pressure coefficient near the nozzle exit. Under 2D simulation, uniform inlet velocity profile would develop in both radial and axial direction. Therefore, the velocity profile at axis in 2D simulation is higher as compared velocity profile in 1D simulation since 1D simulation neglects the velocity profile development in radial direction. Because the simulation results in the axial direction were validated, the results in the radial direction were assumed to be valid because the same governing equations were used for the simulation. In addition, this approach has been applied in several similar works available (Cai, Fang, Xu, & Liu, 2007; Ding, Wang, & Chen, 2014; Tian & Lu, 2013).

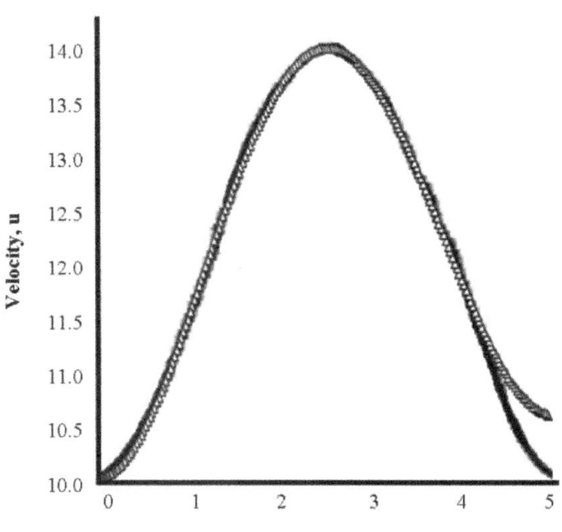

Figure 2. Distribution of the flow speed along the axis of the nozzle in the axial direction in the current work (dotted line) and the results from Wang and Brennen (1998) (solid line).

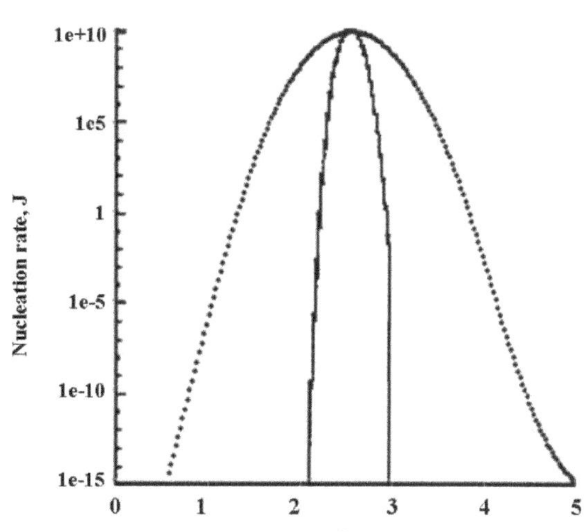

Figure 4. Distributions of the nucleation rate J along the axis of the nozzle in the axial direction in the current work (diamond dots) and the results from Delale et al. (2005) (solid line).

Figure 4 shows the bubble nucleation rate profile of water vapor along the axis of the nozzle. Based on the figure, the result for the current work was compared with the simulation result from Delale et al. (2005), who also attempted to integrate CNT into the cavitation simulation with the same geometry but in the 1D domain. Both works exhibited the same maximum bubble nucleation rate along the axis of the nozzle, which was recorded to be 10^{10} nuclei/m^3 s at the throat of the nozzle. In their study, bubble nucleation only occurred between positions $2 < x < 3$. According to Equation (3), the onset of cavitation is actually determined by the work of formation for a critical cluster. In the current study, this work of formation for a critical cluster is represented by the number of molecules required to form a critical cluster. However, Delale et al. (2005) used the critical cluster size to indicate the work of formation for a critical cluster. They set a fixed value for the critical cluster size in their study. In other words, the onset of cavitation was dependent on the preset minimum pressure difference. Therefore, the bubble nucleation was ignored for the case when the pressure difference has not achieved the minimum value. Their approach caused the bubble nucleation rate to increase and decrease sharply at the onset and end of cavitation because only the pressure difference in this region was sufficient to create a critical cluster.

The predicted bubble nucleation process in the current work occurred almost throughout the entire nozzle, with a

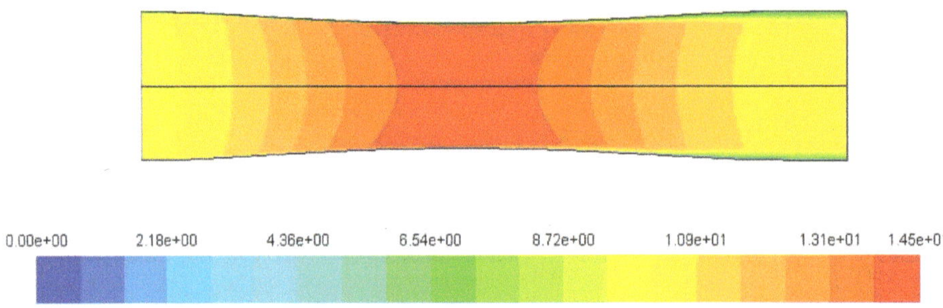

Figure 5. Contour of the water flow velocity distribution across the nozzle (m/s).

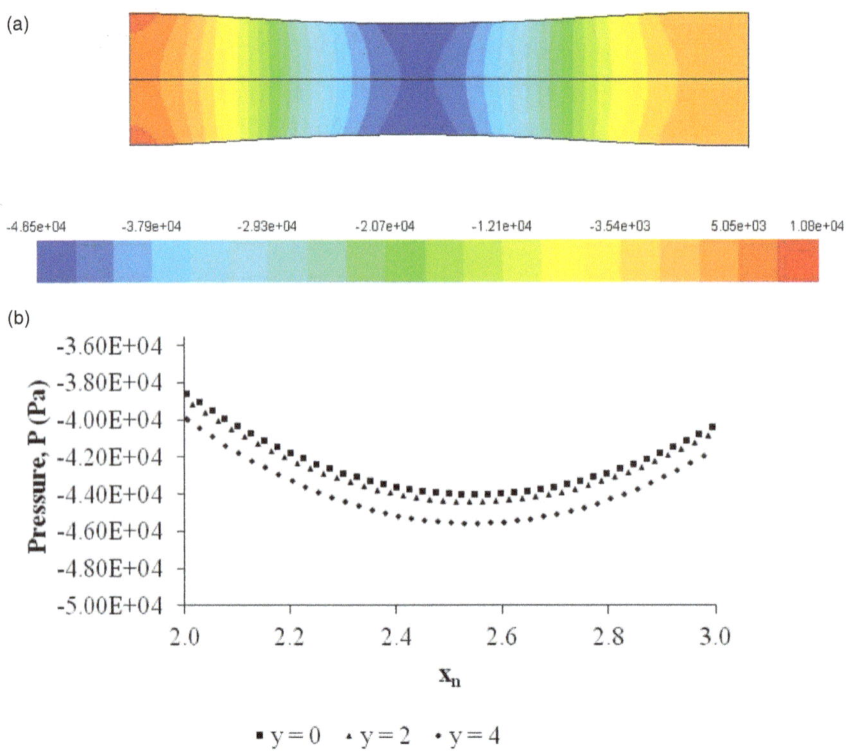

Figure 6. Liquid pressure: (a) contour plot across the nozzle and (b) along the axial direction at the throat of the nozzle ($2 < x < 3$), at the radial positions of $y = 0$, 2 and 4.

significant rate at the throat of the nozzle. In the current work, the mathematical model of the number of molecules required to form a critical cluster, which was specifically designated for vapor bubble nucleation, as suggested by Kwak and Oh (2004), was used. The computed number of molecules required is used to determine the bubble nucleation rate in the domain. Therefore, the onset of bubble nucleation does not depend on the predefined minimum pressure difference.

4.2. CFD analysis

Using the approach of coupling CFD and bubble nucleation theory in the 2D computational domain, the distribution of various parameters, including the bubble nucleation rate, pressure, and number of bubble nuclei, can be estimated both in the axial direction and the radial direction. The bubble nucleation rate is strongly dependent on the pressure drop across the nozzle, whereby the hydrodynamics information can be obtained through the CFD approach. In this section, the distribution of pressure and velocity, the number of molecules required to form a critical cluster, and the bubble nucleation rate are presented.

4.2.1. Velocity and pressure

The contour plots of the velocity distribution for water flowing across the nozzle are shown in Figure 5. Based on Figure 5, the water flows in at a velocity of 10 m/s and the speed increases at the throat of the nozzle to nearly 14 m/s due to the constriction. Subsequently, the speed of the water flow gradually reduces again to approximately 10 m/s. The increase and decrease of velocity of the water was predicted based on the laws of mass and momentum conservation, with the assumption of an incompressible fluid flow.

Figure 6 (a) shows the contour plot of the pressure distribution in the nozzle. Based on the law of momentum conservation because there was no external force applied to the fluid flow and because of the difference in height, the

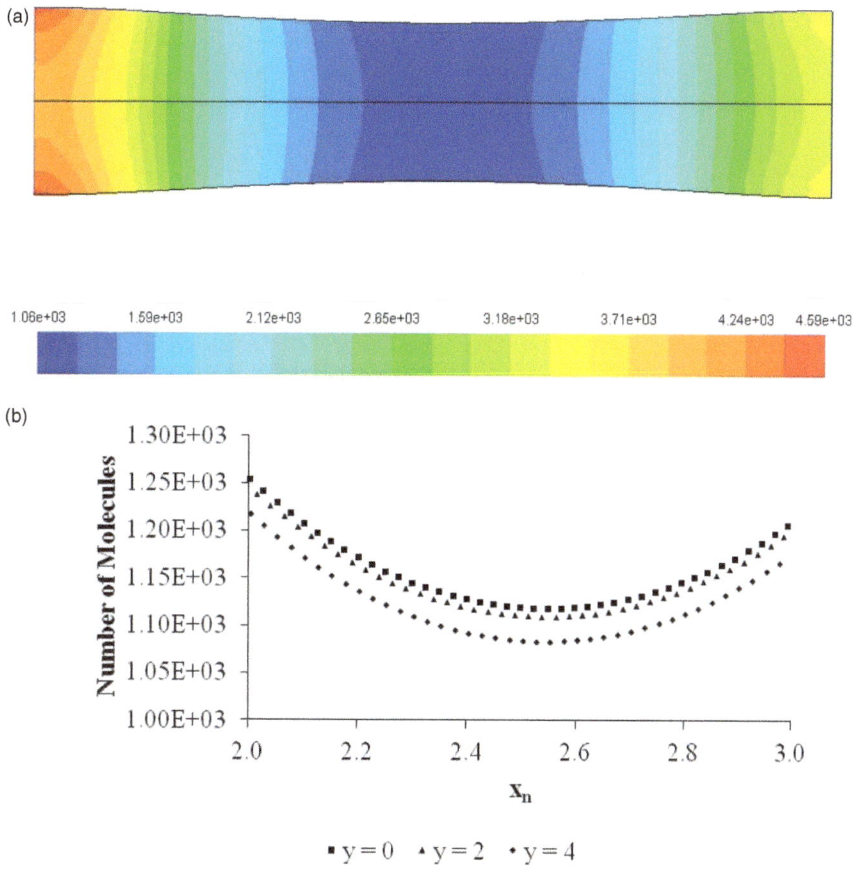

Figure 7. Number of molecules required to form a critical cluster: (a) contour plot across the nozzle and (b) along the axial direction at the throat of the nozzle ($2 < x < 3$) and at radial positions of $y = 0$, 2 and 4.

increase of velocity in fluid flow must be compensated by a pressure drop. As a consequence of the increase of the velocity of the water flowed across the nozzle, the pressure of the water at the throat of the nozzle has dropped to nearly − 46 kPa in the current work, that is, as if the nozzle was being stretched.

Furthermore, Figure 6 (b) shows the liquid pressure in the axial direction at the radial positions of y = 0, 2 and 4. Based on the figure, the lowest liquid pressure is decreasing from the middle of the throat (y = 0), at approximately − 44.0 kPa, towards the wall of the nozzle, recorded at approximately − 45.5 kPa at y = 4. The difference in the vapor and liquid pressures, ΔP, is larger when the liquid pressure is lower. According to Equation (6), the difference in the vapor and liquid pressures is the main factor that determines the number of molecules required to form a critical cluster. Therefore, the pressure difference was the key to initiate the bubble nucleation process. In addition, the pressure difference also acted as driving force for bubble nucleation, whereby the larger pressure difference would generate a higher bubble nucleation rate.

4.2.2. *Number of molecules required to form a critical cluster*

Figure 7 (a) shows the contour plot of the number of molecules required to form a critical cluster. The number of molecules required decreases towards the throat of the nozzle and increases after the region. Figure 7 (b) shows the number of molecules required to form a critical cluster at the throat of the nozzle for the radial position of y = 0, 2 and 4 in the axial direction. The graph shows that the number of molecules required is not constant in the radial direction and that the number of molecules decreases when approaching the wall. The lowest number of molecules required decreases from approximately 1150 at the axis (y = 0) to approximately 1140 at y = 2 and further decreases to approximately 1100 at y =4. The same trend is observed in the axial direction for the above positions. The profile of the numbers of molecules required to form a critical cluster was dependent on the pressure difference in the vapor and liquid pressures, ΔP. A higher pressure difference was determined to require a lower number of molecules to form a critical cluster.

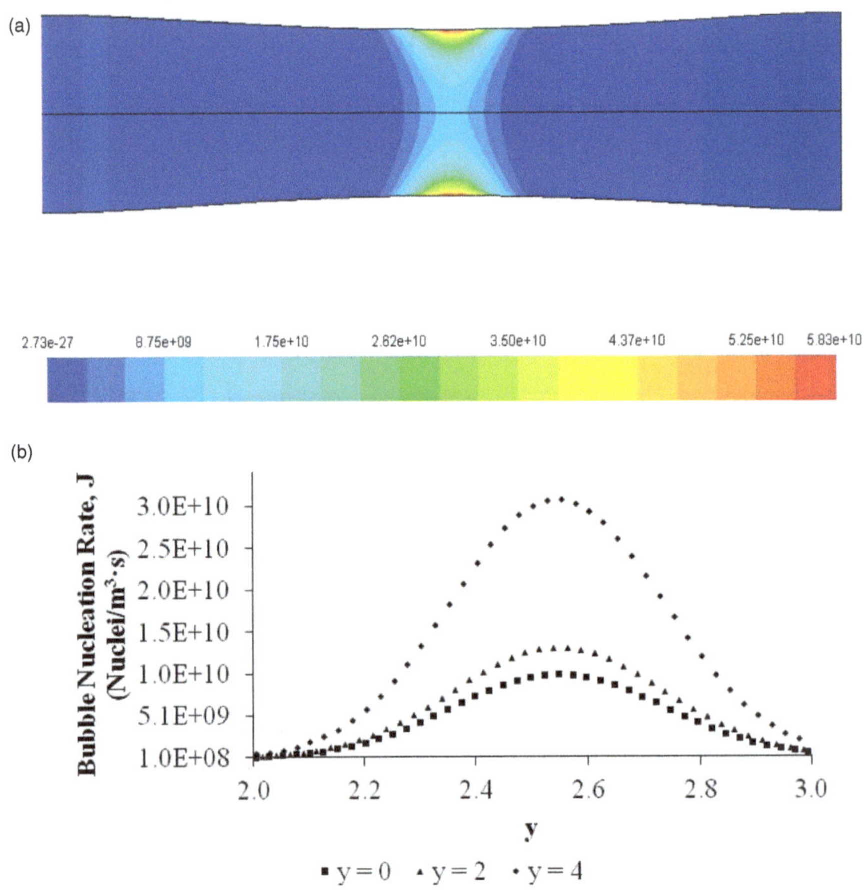

Figure 8. Bubble nucleation rate: (a) contour plot across the nozzle and (b) along the axial direction at the throat of the nozzle (2 < x < 3) and at the radial positions of y = 0, 2 and 4.

4.2.3. *Bubble nucleation rate*

The contour of the bubble nucleation rate across the nozzle, which was the main parameter under study in the current work, is presented in Figure 8(a). Based on the figure, the bubble nucleation rate was predicted to be the highest at the throat of the nozzle. Figure 8(b) shows the bubble nucleation rate at the throat of the nozzle in the axial direction at radial positions of y = 0, 2 and 4. The maximum bubble nucleation rate increases from the middle of the throat (y = 0) at approximately 1×10^{10} nuclei/m$^{3\cdot}$s to approximately 1.3×10^{10} nuclei/m$^{3\cdot}$s at y = 2 and further increases to 3×10^{10} nuclei/m$^{3\cdot}$s at y = 4. The trend of the bubble nucleation rate exhibited a similar trend to the number of molecules required to form a critical cluster. Hence, the real maximum bubble nucleation rate was determined not only to be 10^{10} nuclei/m^3 s, as predicted by Delale et al. (2005); in addition, the rate was found to be 5.8×10^{10} nuclei/m$^{3\cdot}$s at the wall of the throat of the nozzle, as shown in Figure 8 (a).

The higher pressure difference between the vapor and liquid pressures reduces the number of molecules required to form a critical cluster, thereby increasing the nucleation rate. Figure 8 (a) shows that the liquid pressure is lowest at the throat of the nozzle in the axial direction, while in Figure 8 (b), the pressure difference, ΔP, increases in the radial direction. Therefore, the bubble nucleation rate was found to be highest at the throat of the nozzle and increased in the radial direction.

4.2.4. *Bubble nuclei distribution*

The distribution of bubble nuclei formed across the nozzle is presented in Figure 9(a). From the figure, the bubble nuclei developed significantly at the region after the throat of the nozzle. This behavior was due to the high bubble nucleation rate at the throat of the nozzle. Consequently, a significant number of bubble nuclei were formed and dispersed across the nozzle. Therefore, the bubble nuclei distribution was determined by both the local bubble nucleation rate and also the hydrodynamics in the nozzle. In accordance with the bubble nucleation rate profile (high bubble nucleation rate near the wall of the throat of the nozzle), a high density of bubble nuclei was found to be concentrated near to the wall of the nozzle.

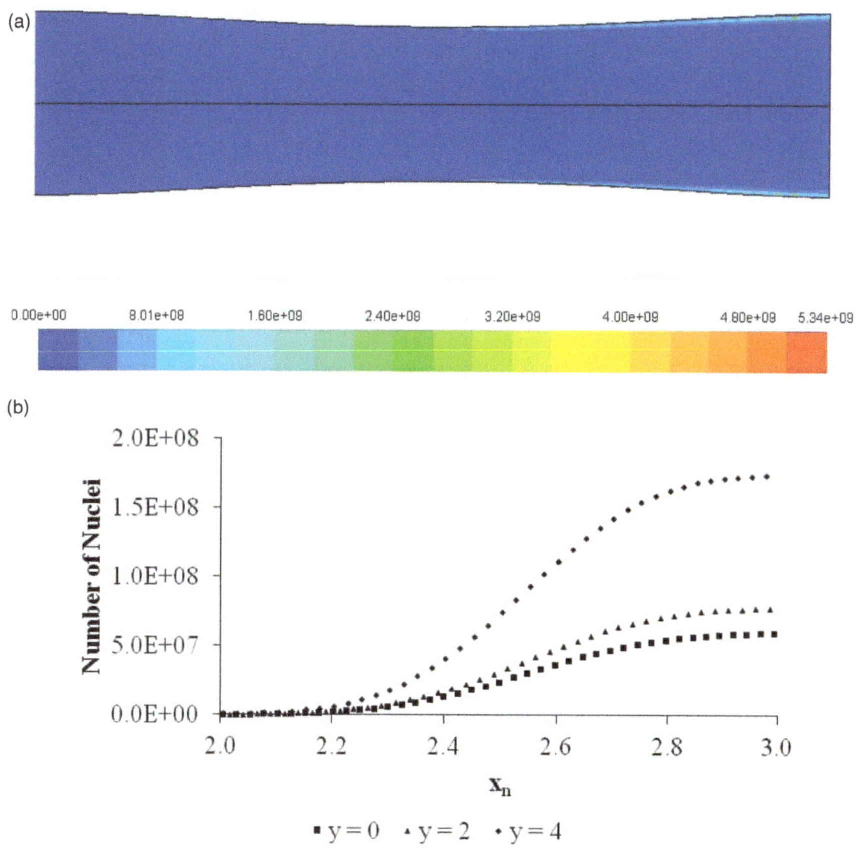

Figure 9. Number of bubble nuclei: (a) contour plot across the nozzle and (b) along the axial direction at the throat of the nozzle (2 < x < 3) and at the radial positions of y = 0, 2 and 4.

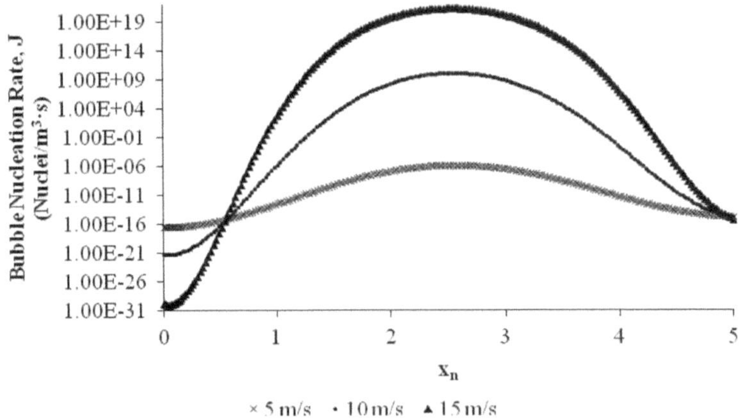

Figure 10. Bubble nucleation rate along the axis of the nozzle at the inlet velocities of 5 m/s, 10 m/s and 15 m/s.

The number of bubble nuclei estimated at the throat of the nozzle in the axial direction at the radial positions of y = 0, 2 and 4 is presented in Figure 9 (b). Generally, the number of bubble nuclei formed increases sharply at the throat of the nozzle and remains almost constant after the throat. The number of bubble nuclei formed after the throat of the nozzle increases from approximately 6×10^7 at y = 0 to approximately 8×10^7 at y = 2 and further increases to approximately 1.7×10^8 near the wall of the nozzle at y = 4. The number of bubble nuclei formed in the radial direction was found to be different due to the differences of the bubble nucleation rate in the radial direction.

The bubble nuclei distribution estimated in the current work was predicted using the Population Balance Model (PBM). The PBM was capable to consider the local bubble nucleation rate, and the nuclei existed at particular positions and times within the computational domain. This approach could predict the bubble nuclei distribution, which is an important element to estimate the mass transfer from liquid to gas phase when a bubble is formed.

4.3. *Bubble nucleation rate at different inlet velocities*

Figure 10 shows the bubble nucleation rate along the axis of the nozzle at different inlet velocities. A similar approach was used to predict the bubble nucleation rate along the axis of the nozzle at different inlet velocities of 5 m/s, 10 m/s and 15 m/s. From the figure, the bubble nucleation rate is generally higher for cases with a higher inlet velocity. The highest bubble nucleation rate is recorded as 1×10^{-6}, 1×10^{10} and 1×10^{21} with inlet velocities of 5 m/s, 10 m/s and 15 m/s, respectively, at the throat of the nozzle (x = 2.5). The bubble nucleation rate increases with the inlet velocity because a higher inlet velocity will generate a higher pressure drop at the throat of the nozzle. Hence, the higher bubble nucleation rate is developed due to the high pressure drop.

5. Conclusion

The bubble nucleation rate in a quasi-stable cavitating nozzle flow in the 2D computational domain was successfully estimated via the coupling of CFD and the enhanced CNT. In addition, the bubble nucleation rate was estimated using the computed number of molecules required to form a critical cluster. Because the pressure was found to be lowest at the throat of the nozzle, the bubble nucleation rate was predicted to be the highest at the throat of the nozzle (up to 6×10^{10} nuclei/m^3 s). Furthermore, PBM was used to predict the bubble nuclei density across the nozzle. In addition, the bubble nucleation rate was determined to increase with inlet velocity through the same approach. Because the use of a 1D simulation cannot represent the bubble nucleation phenomenon in the cavitating nozzle, the prediction of the bubble nucleation rate and the nuclei distribution in the 2D domain is important because these properties increase in the radial direction. Because bubble growth modeling was not included in the current work, it will be implemented in future work.

Acknowledgements

This work was supported by the UTP Research Centre for CO_2 Carbon Capture (RCCO_2C).

Nomenclateure

A(x)	Normalized area of the nozzle
C	Correction factor
D	Rate of molecules striking on a surface area of the cluster, molecules/ m^2·s
G (V)	Growth rate of the bubble number density, number/m^3·s
ΔH_{vap}	Enthalpy for the evaporation of the solution
ΔH_f	Enthalpy for the freezing of the solution
J	Bubble nucleation rate, nuclei/m^3·s
J_0	Kinetic pre-exponential factor
k_B	Boltzmann's constant, J/K
k	Specified number of moments
Li	Abscissas
m_k	kth moment of the number density function
m_v	Mass of vapor molecule, kg

n	Number of molecules
N	Number density of the liquid, molecules/m^3
$n(V,t)$	Bubble number density distribution per unit volume (V) for a given time (t)
P	Pressure, Pa
Q	Order of the quadrature approximation
r	Radius, m
R	Gas constant, J/ K·mol
t	Time, s
T	Liquid total temperature, K
T_f	Melting temperature of the solution, K
U	Velocity, m/s
u	Reynolds-averaged velocity vector, m
v'_m	Transformed volume for the vapor molecules, m^3
W	Work to form a critical cluster, J
wi	Weights
x	Dimensionless length of the nozzle
Z_f	Zeldovich nonequilibrium factor

Greek letters

α	Volume fraction
β'_v	Accommodation coefficient of the vapor molecules to the cluster
ρ	Density, kg/m^3
γ	Surface tension of the liquid-gas surface, Pa

Subscript

air	Air
c	Critical cluster
i	Inlet
l	Liquid
v	Vapor

References

ANSYS FLUENT 14.0. (2011). *ANSYS FLUENT population balance module manual.* ANSYS Inc.

Cai, G., Fang, J., Xu, X., & Liu, M. (2007). Performance prediction and optimization for liquid rocket engine nozzle. *Aerospace Science and Technology, 11*(2–3), 155–162.

Caupin, F., & Herbert, E. (2006). Cavitation in water: A review. *Comptes Rendus Physique, 7*(9–10), 1000–1017.

Delale, C. F., Hruby, J., & Marsik, F. (2003). Homogeneous bubble nucleation in liquids: The classical theory revisited. *The Journal of Chemical Physics, 118*(2), 792–806.

Delale, C. F., Okita, K., & Matsumoto, Y. (2005). Steady-state cavitating nozzle flows with nucleation. *Journal of Fluids Engineering, 127*(4), 770–777.

Ding, H., Wang, C., & Chen, C. (2014). Non-equilibrium condensation process of water vapor in moist air expansion through a sonic nozzle. *Flow Measurement and Instrumentation, 40*, 238–246.

Goncalves, E., & Patella, R. F. (2009). Numerical simulation of cavitating flows with homogeneous models. *Computers & Fluids, 38*(9), 1682–1696.

Goncalvés, E., & Patella, R. F. (2010). Numerical study of cavitating flows with thermodynamic effect. *Computers & Fluids, 39*(1), 99–113.

Herbert, E., Balibar, S., & Caupin, F. (2006). Cavitation pressure in water. *Physical Review. E, Statistical, Nonlinear, and Soft Matter Physics, 74*(4 Pt 1), 041603.

Kunz, R. F., Boger, D. A., Stinebring, D. R., Chyczewski, T. S., Lindau, J. W., Gibeling, H. J., Venkateswaran, S., & Govindan, T. R. (2000). A preconditioned Navier–Stokes method for two-phase flows with application to cavitation prediction. *Computers & Fluids, 29*(8), 849–875.

Kwak, H.-Y., & Oh, S.-D. (2004). Gas-vapor bubble nucleation-a unified approach. *Journal of Colloid and Interface Science, 278*(2), 436–446.

Li, C., & Li, Y. Z. (2011). Investigation of entrainment behavior and characteristics of gas–liquid ejectors based on CFD simulation. *Chemical Engineering Science, 66*(3), 405–416.

Lindau, J. W., Kunz, R. F., Boger, D. A., Stinebring, D. R., & Gibeling, H. J. (2002). High Reynolds number, unsteady, multiphase CFD modeling of cavitating flows. *Journal of Fluids Engineering, 124*(3), 607–616.

Liška, M., Holubová, J., Černošková, E., Černošek, Z., Chromčíková, M., & Plško, A. (2012). Nucleation and crystallization of an As2Se3 undercooled melt. *Physics and Chemistry of Glasses - European Journal of Glass Science and Technology Part B, 53*(6), 289–293.

Lubetkin, S. D. (1995). The fundamentals of bubble evolution. *Chemical Society Reviews, 24*(4), 243–250.

Lubetkin, S. D. (2003). Why is it much easier to nucleate gas bubbles than theory predicts? *Langmuir, 19*(7), 2575–2587.

Payri, R., Ruiz, S., Gimeno, J., & Martí-Aldaraví, P. (2015). Verification of a new CFD compressible segregated and multi-phase solver with different flux updates-equations sequences. *Applied Mathematical Modelling, 39*(2), 851–861.

Peralta, J. M., Chew, Y. M. J., & Wilson, D. I. (2011). Effect of nozzle external geometry on the pressure and shear stress exerted on the surface being gauged in fluid dynamic gauging. *Chemical Engineering Research and Design, 89*(12), 2540–2551.

Preston, A. T., Colonius, T., & Brennen, C. E. (2002). A numerical investigation of unsteady bubbly cavitating nozzle flows. *Physics of Fluids, 14*(1), 300–311.

Shin, B. R., Yamamoto, S., & Yuan, X. (2004). Application of preconditioning method to gas-liquid two-phase flow computations. *Journal of Fluids Engineering, 126*(4), 605–612.

Shukla, P., Mandal, R. K., & Ojha, S. N. (2001). Non-equilibrium solidification of undercooled droplets during atomization process. *Bulletin of Materials Science, 24*(5), 547–554.

Singhal, A. K., Athavale, M. M., Li, H., & Jiang, Y. (2002). Mathematical basis and validation of the full cavitation model. *Journal of Fluids Engineering, 124*(3), 617–624.

Tian, C., & Lu, Y. (2013). Turbulence models of separated flow in shock wave thrust vector nozzle. *Engineering Applications of Computational Fluid Mechanics, 7*(2), 182–192.

Vaidyanathan, R., Senocak, I., Wu, J., & Shyy, W. (2003). Sensitivity evaluation of a transport-based turbulent cavitation model. *Journal of Fluids Engineering, 125*(3), 447–458.

Vehkamäki, H. (2006). *Classical nucleation theory in multicomponent systems.* New York, NY: Springer.

Vortmann, C., Schnerr, G. H., & Seelecke, S. (2003). Thermodynamic modeling and simulation of cavitating nozzle flow. *International Journal of Heat and Fluid Flow, 24*(5), 774–783.

Wang, Y. C., & Brennen, C. E. (1998). One-dimensional bubbly cavitating flows through a converging-diverging nozzle. *Journal of Fluids Engineering, 120*(1), 166–170.

Yan, W. C., Li, J., & Luo, Z. H. (2012). A CFD-PBM coupled model with polymerization kinetics for multizone circulating polymerization reactors. *Powder Technology, 231*, 77–87.

Zheng, Q., Durben, D. J., Wolf, G. H., & Angell, C. A. (1991). Liquids at large negative pressures: Water at the homogeneous nucleation limit. *Science, 254*(5033), 829–832.

9

Experimental and numerical study of an offset jet with different velocity and offset ratios

Ali Assoudi[a], Sabra Habli[b], Nejla Mahjoub Saïd[b]*, Hervé Bournot[c] and Georges Le Palec[c]

[a]*LGM, National Engineering School of Monastir, Mechanical Engineering Department, University of Monastir, Monastir, Tunisia;* [b]*LGM, Preparatory Institute for Engineering Studies, Physics Department, University of Monastir, Monastir, Tunisia;* [c]*IUSTI, UMR 7343, Faculty of Science, Mechanical Engineering Department, University of Aix-Marseille, Marseilles, France*

The present study examines the configuration of an offset jet issuing into either a quiescent medium or a moving stream (co-flowing). The mean velocity and turbulence characteristics of the turbulent offset jet have been investigated using a particle image velocimetry technique at three velocity ratios and for two offset ratios. A numerical simulation of a three-dimensional offset jet has also been carried out by comparing the corresponding results with previous experimental data and our measurements. The numerical investigation was performed by means of the finite volume method together with a second-order turbulent closure model – the Reynolds stress model (RSM) – to examine the behavior of the flow for different velocity ratio and offset ratios. Results give a satisfactory agreement between the experimental data and the calculations. Data from the early flow region clearly show a significant influence of the velocity ratio and the offset ratio on the mean flow and turbulence characteristics.

Keywords: offset jet; co-flow stream; particle image velocimetry; numerical simulation; velocity ratio; offset ratio

1. Introduction

Jet flows have been extensively explored in the literature because of their presence in more than one area. These types of flows are either free or bounded, depending upon the distance of the jet boundaries containment. When borders (parallel to the jet axis) are sufficiently far from the output of the jet, it is defined as a free jet. However, a bounded jet occurs when the flow interacts with a parallel wall. Bounded jets can be classified into three families based on the orientation of the containment wall: (a) an impinging jet aimed toward the boundary, (b) a wall jet where the fluid is discharged tangentially to the boundary, and (c) an offset jet, which is formed when a jet discharges into a medium above a horizontal solid wall parallel to the axis of the jet exit but which is offset by a certain distance.

The flow field of an offset jet is complex and has gained extensive research interest, primarily due to their diverse and highly useful practical and engineering applications, such as vertical takeoff and landing (VTOL) aircraft engines, heating, ventilation, and air conditioning (HVAC) systems, and fuel injection systems. The behavior of an offset jet is affected by the presence of a co-flow stream. Indeed, the offset jet is characterized by three main regions (Figure 1). In the first zone, the entrainment of the fluid between the jet, the offset wall and the horizontal wall creates a low pressure zone, forcing the jet to deflect towards the wall and eventually attaching to it at the impingement point. This is called the 'Coanda effect'. Within a very short distance from the nozzle exit, the jet flow is dominated by momentum and has the properties of a free jet. In the impingement region, as the jet deflects towards the wall, the flow has features qualitatively closed to other reattaching shear layers flows, like the flow over a backward facing step (BFS). Far away, downstream from the attachment point, the offset jet has the characteristics of a wall jet.

Two-dimensional offset jets have been experimentally investigated by several authors (Bourque & Newman, 1960; Levin & Manion, 1962; Rajaratnam & Subramanya, 1968; Sawyer, 1960, 1963). These authors used various measurement techniques such as hotwires, Pitot tubes and laser Doppler anemometry (LDA) to develop a more general uniform equation that predicts the impingement of the jet to the wall as a function of the offset ratio (the vertical offset distance normalized by the jet nozzle diameter). They also included the measurements of the mean velocity and static pressure distributions.

Most studies on a turbulent offset flow involved jets with large offset ratios ($h/d > 5$). Hoch and Jiji (1981b) experimentally examined an offset jet in the presence of a free-stream motion using the hotwire technique. They reported measurements of the maximum velocity decay which are in good agreement with results reported by Rajaratnam and Subramanya (1968). For the same previous configuration, Hoch and Jiji presented the analytical solution of the thermal characteristics. They measured

*Corresponding author. Email: nejla.mahjoub@fsm.rnu.tn

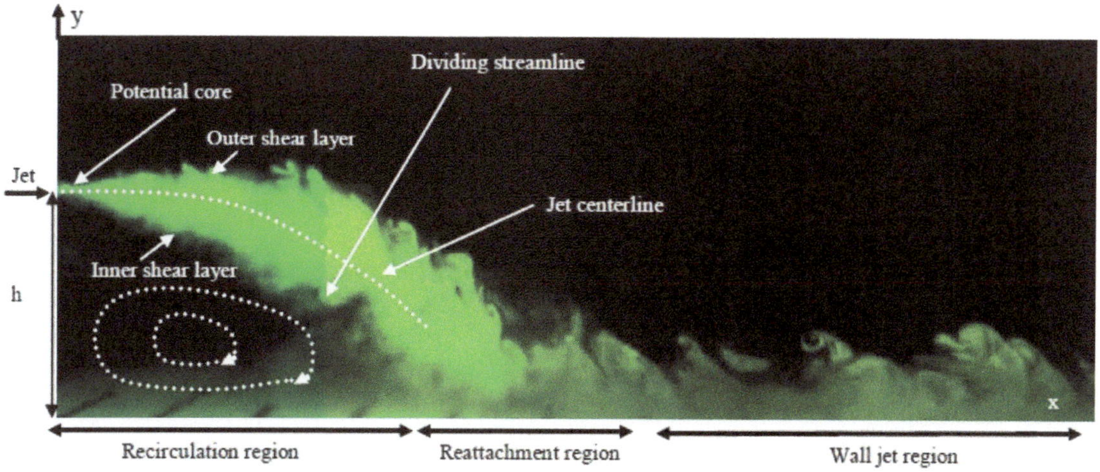

Figure 1. Visualization of an offset jet.

the temperature distributions of this flow relative to an adiabatic surface, and they found that the temperature within the recirculation zone is uniform and decreases with the increase of the injection height.

Pelfrey and Liburdy (1986a, 1986b) studied the mean and turbulent flow of an offset jet. They employed the LDA for the measurement of the mean velocity profiles at different locations, ratio of curvature to shear strain rate and the entrainment parameter in the recirculation and attachment regions. In this study, the parallel wall was adiabatic and the offset ratio and Reynolds number considered are 7 and 15,000 respectively. They have mentioned how entrainment and local pressure are affected by the jet curvature in the recirculation and impingement regions.

Holland and Liburdy (1990) studied the thermal characteristics of a heated offset jet over an adiabatic impingement surface for different offset ratios and jet exit Grashof numbers. They examined the temperature profile at different downstream locations, the surface temperature distribution in three regions (the recirculation, impingement and developing wall jet regions) and the decay of the maximum temperature. The experimental results showed that the temperature within the recirculation region is close to the impingement location temperature. However, there is evidence that the thermal distribution is dependent on the offset ratio. They also showed that downstream of the impingement region temperature, profiles become locally similar.

There have been fewer investigations of jets with small offset ratios ($h/d < 5$) (Gao & Ewing, 2007; Lund, 1986; Nasr & Lai, 1998). Lund (1986) dealt with the reattachment length with a wide range of offset ratios, and these results showed that the ratio of the reattachment to the offset ratio changed with the offset distance. Nasr and Lai (1998), in their work, determined the mean velocities and turbulence characteristics of a turbulent plane offset jet with a small offset ratio of 2.125 and a Reynolds number of 11,000 using LDA. A numerical simulation of the problem was also carried out. Three different turbulence models were considered (the standard k-ε model, the k-ε model RNG and the second-order turbulence model RSM). The comparison of the corresponding results to the experimental data showed that the standard k-ε model is more appropriate for the prediction of the turbulent offset jet with small offset ratios. In the study of Gao and Ewing (2007), an experimental investigation was provided to characterize the development of planar jets discharging from a nozzle adjacent to a parallel wall with small offset ratios, of up to 1, and a Reynolds number of 44,000. Their results mentioned that the initial development of the mean flow field could be divided into five regions (three regions associated with the jet attaching to the wall and two associated with the resulting planar wall jet region).

Many investigations are devoted to the study of the heat transfer characteristics of offset jet flows in a turbulent regime. The work of Vishnuvardhanarao and Das (2008a, 2008b, 2009) and Song et al. (2000) examined the characteristics of conjugate heat transfer of a two-dimensional, steady and incompressible offset jet flow. They show that the local Nusselt number is maximum at the reattachment point and reduces with the offset ratio. However, the conjugate heat transfer study involving a laminar flow has received little attention. Kanna and Das (2005, 2008) carried out a steady state heat transfer study for a two-dimensional, laminar and incompressible offset jet flow. The detailed heat transfer results are presented in terms of isotherm contour and Nusselt number for three offset ratios (0.5, 1 and 1.5), with the Reynolds number varying from 200 to 600 and for three Prandtl numbers (Pr = 0.01, 1 and 100). The results revealed that the maximum Nusselt number increases as the Reynolds number and Prandtl number increase. Further, in the recirculation region, a significant effect on the thermal boundary layer was observed for Pr = 0.01 and 100, whereas for Pr = 0.1 it was absent. Kumar Raja et al. (2009) numerically investigated the mixed convection flow and heat transfer characteristics in a two-dimensional laminar offset jet issuing parallel to an isothermal flat plate. The goal of their study was to explore the associated heat transfer process in the mixed convection regime in the range of a Reynolds number of 300 to 600 a and Grashof number of 10^3 to 10^7. Their results show that the reattachment length is strongly dependent on both the Reynolds number and the Grashof number for the range considered.

Y. Yang and Yeh (1994) have numerically studied the flow field of an offset jet with a Reynolds number of 15,000 for offset ratios of 3, 7 and 11. Their results reported that a zero normal gradient boundary condition could be used at the entrainment and exit boundary only when the flow is fully developed near the exit and when the boundaries are sufficiently far away from the region of interest.

Further, a comparative study of the flow field and turbulence characteristics of a turbulent offset jet has been carried out by Rathore and Das (2013). The numerical results obtained from the standard k-ε model were compared with those proposed by Launder and Sharma (1974) and Z. Yang and Shih (1993). They observed a similar profile in the wall jet region due to the resemblance of an offset jet with that of a wall jet flow in the wall jet region.

Li, Huai, and Han (2011) reported a numerical investigation for the interaction between a plane wall jet and a parallel offset jet using large eddy simulation (LES). They used the finite-volume method to discretize the governing equations for a Reynolds number of 10^4. The profiles of the mean streamwise velocity at some locations show a similarity after a certain distance where the two jets are fully merged. The half-width of the jet decreases near the exit due to the presence of the low-pressure zone and then increases as the downstream distance increases. The authors also noticed that the turbulent intensity increases rapidly to the maximum and then decreases along the downstream direction, which indicates that the interaction is strong near the jet exit.

A global observation shows that most studies are two-dimensional. However, the three-dimensional offset jet has received little attention in the literature. Davis and Winarto (1980) and Agelin-Chaab and Tachie (2011) used a circular jet exit, while Nozaki, Hatta, Nakashima, and Matsumura (1979, 1981) utilized a rectangular nozzle jet to study the three-dimensional configuration. The three-dimensional experimental investigation of Agelin-Chaab and Tachie examined the effects of the offset ratio and Reynolds number on the structure of the turbulent offset jet. The use of particle image velocimetry (PIV) measurements has revealed that, in the early region of the flow development, the Reynolds number and the offset ratio have significant effects on the decay of the maximum mean velocity and growth of the shear layer. However, at large downstream distances, the decay and spread rates were found to be nearly independent of the offset ratio. These authors also reported that in the self-similar region, the location of the maximum mean velocity increased nearly linearly with the streamwise distance. For the 3D offset jet from a square nozzle, Nozaki (1983) reported values for the reattachment length (x_r). He obtained $x_r/d = 3.8, 6.7$, and 9.9 for $h/d = 1.5, 2.5$, and 3.5 respectively.

In the following, we propose to study a turbulent three-dimensional offset jet discharged into a co-flow stream with different velocity ratios and offset ratios. To achieve this purpose, an experimental study is fulfilled using PIV to describe the dynamic and turbulent characteristics of the flow. Then, a numerical simulation is carried out with the Reynolds stress second-order turbulent model (RSM). After validation of the results by the comparison of numerical and experimental data, the characterization of the resulting flow field is examined in order to investigate both the mean velocity and turbulence characteristics of the jet discharging for different velocity ratios and offset ratios.

2. Experimental setup

The experimental measurements were performed in a wind tunnel (Figure 2). The test section was of rectangular dimensions of 3000 × 200 × 300 mm in the longitudinal, vertical and lateral directions, respectively. The top wall of the tunnel was open and the ground was made of wood and covered with plastic to ensure optimal visibility. One of the side walls was made of wood and painted black to prevent possible light reflections. The second side was made of Plexiglas to allow better visibility, and therefore better flow visualizations and data measurements. The turbulence intensity level of the external flow was less than 0.2%. The motor allowed, through a controller, the introduction of a velocity range of about 0 to 16 m/s. The jet was produced by a smooth pipe with an outer diameter of 12 mm and an internal diameter of $d = 10$ mm, through which air was injected. The pipe was placed at an adjustable height (h) of the horizontal wall. The jet was subjected to an external flow at rest or in uniform co-flow of velocity u_∞. The velocity profiles of the co-flow were established by a hot wire anemometry (Ben Kalifa, Habli, Mahjoub Saïd, Bournot, & Le Palec, 2014). The thickness of the boundary layer of the main flow was about 10 mm. Indeed, the thickness of the boundary layer was defined as the point where the local velocity was 99% of the velocity of the main flow.

To make a quantitative study of the velocity field of a moving fluid, the experimentally tracked values were depicted by means of PIV (Mahjoub Saïd, Mhiri, Le Palec, & Bournot, 2005). This technique provides instantaneous velocity fields over global domains. The PIV technique records the position over time of small tracer particles introduced into the flow to extract the local fluid velocity (Radhouane, Mahjoub Saïd, Mhiri, Le Palec, & Bournot, 2009). In this study, the fluid ejected by the nozzle was

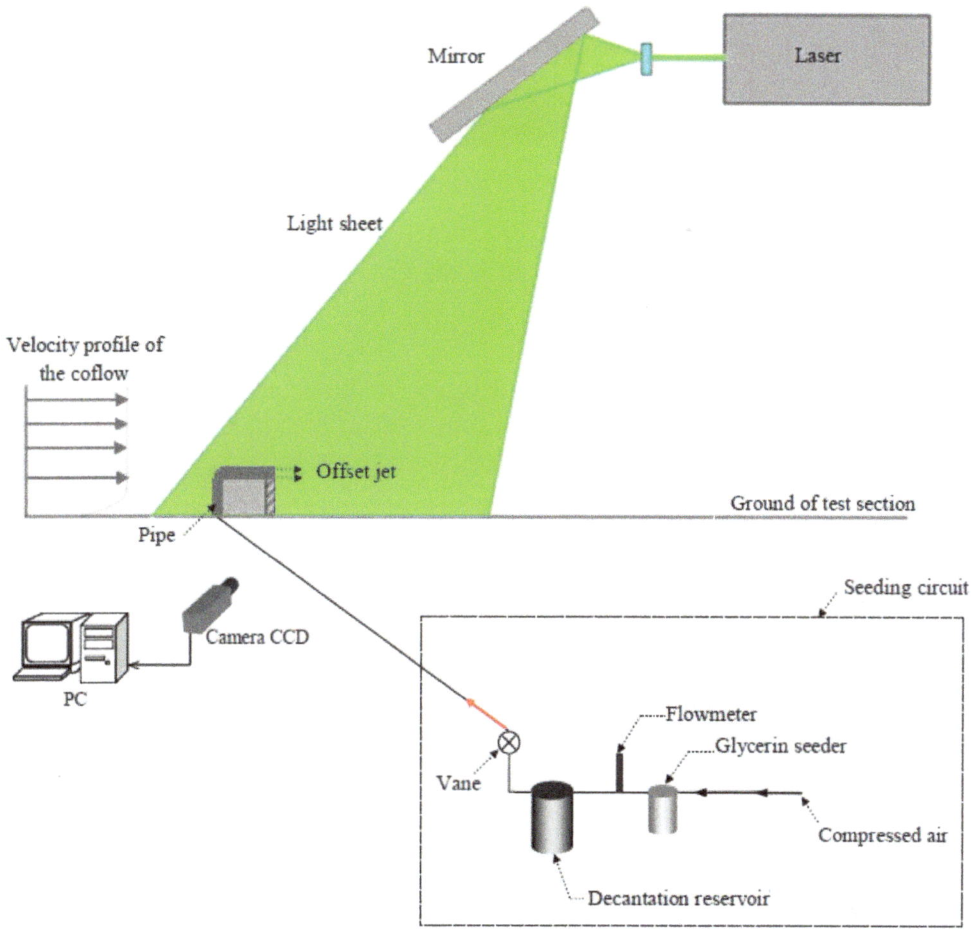

Figure 2. Experimental setup.

seeded with approximately 1 μm diameter glycerin particles (seeding density of approximately 30 particles per ml of pure jet fluid; see Mahjoub Saïd, Habli, Bournot, & Le Palec, 2012). The main air co-flow was not seeded. The seeding process is fully described in the circuit sketch (see Figure 2). The flowmeter helped to regulate the flow rate of the seeding particles while the decantation reservoir was used to retain eventual contaminants present within the discharged jet. A vane was used to control the flow rate of the discharged offset jet.

The experimental setup of the PIV system was based upon a TSI Incorporated power view system, including a 50 mJ dual YAG laser which produced two flat pulses, the duration of one ranging from 5×10^{-9} to 10^{-8} s, a Power View 4M high-resolution cross-correlation camera (2×2 k resolution, 12 bits), a synchronizer and *Insight* Windows-based software for acquisition, processing and post-processing. This software allows the synchronization of pulsations according to the observed phenomena, and the adjustment of the time step between two images. The time step in the experiment was 70 μs. In order to avoid errors, the velocity vectors were calibrated at 130 μm/pixel and limited to the representation of the velocity field in the regions where the luminance was strong enough.

Laser pulses from an ND-YAG 'Neodymium-Doped Yttrium Aluminum Garnet' laser were expanded to form a thin light sheet using a cylindrical lens. The light sheet was brought from the top of the wind tunnel using a 45° mirror. The light scattered from the seeded particles was captured and the resulting scattering signal was collected by an interline transfer Charge-Coupled Device (CCD) camera. The camera was oriented perpendicular to the plane of the sheet. The collected images were transferred to the control computer. A target image was used to provide accurate scaling of the flow. However, in this process, after the vector field is initially processed, filters are applied to the data to eliminate spurious vectors, and the remaining vectors are linearly interpolated. Then the vector field is reprocessed. This permits the use of a cross-correlation PIV algorithm (Hasselbrink, 1999) which eliminates directional ambiguity and yields improved resolution over single-image autocorrelation techniques. In this way, the in-plane loss of correlation is minimized and therefore the chance of valid vector detection is maximized. Once an ensemble of instantaneous vector fields has been obtained from a set of images, several post-processing calculations are made. Raw data are cast from pixel units into physical coordinates, and turbulence statistics and derived quantities are calculated, using custom software. The final fields were averages performed with 100 successive acquisitions (Knapp, Bertrand, & Mouret, 2003; Mahjoub Saïd et al., 2005; Mahjoub Saïd, Mhiri, Bournot, & Le Palec, 2008).

For each point, experimental uncertainties were estimated as follows: $(v_{\max} - v_{\min}/v_{av})$, where $v_{\max}$, $v_{\min}$ and v_{av} are, respectively, the maximum velocity, the minimum velocity and the average velocity measured over the whole process. This uncertainty rate may be accounted for as follows:

- the total number of flow fields used to compute time averages and root means square (RMS);
- the development of reverse flows and the deformation of the flow within target areas which may deteriorate the algorithm correlations; and
- excessive velocity gradients and out-of-boundary particle motion.

In addition to these factors, problems associated with particle image overlap, non-uniform illumination of the sample volume, reflections of surfaces, particle coalescence, non-uniform flow seeding and image discretization all contribute to the number of correlation anomalies. These problems, although not as serious as velocity gradient and out-of-boundary effects, contribute to sub-pixel bias errors and are often the cause of spurious vectors (Hart, 1998). The uncertainties of the experiments in the present study did not exceed 5%.

The velocity profile of the co-flow was measured with a hot wire anemometer together with a high-precision electronic pressure transducer. With the help of an on-line micropressure calibration system, the uncertainty in the co-flow velocity was estimated to be as large as 3% of a reading.

3. Computational setup

3.1. Governing equations

In this work, the incompressible flow is assumed to be steady, three-dimensional and turbulent. Under these assumptions, the governing dimensional equations can be written as:

$$\frac{\partial \bar{u}_i}{\partial x_i} = 0$$

$$\bar{u}_j \frac{\partial \bar{u}_i}{\partial x_j} = -\frac{1}{\rho}\frac{\partial \bar{p}}{\partial x_i} + \frac{\partial}{\partial x_j}\left(\nu \frac{\partial \bar{u}_i}{\partial x_j} - \overline{u_i' u_j'}\right)$$

The resolution of these equations requires a turbulent closure model capable of properly characterizing the fluctuating functions. Based on our previous work (Mahjoub Saïd, Mhiri, El Golli, Le Palec, & Bournot, 2003), we chose the second-order closure model RSM. In this model, the following equation is solved (Lien & Leschziner, 1994; Roland, 1993):

$$\rho \bar{u}_k \frac{\partial \overline{u_i' u_j'}}{\partial x_k} = \frac{\partial}{\partial x_k}\left(\mu \frac{\partial}{\partial x_k}(\overline{u_i' u_j'})\right) - \rho\left[\overline{u_i' u_k'}\frac{\partial \bar{u}_j}{\partial x_k} + \overline{u_j' u_k'}\frac{\partial \bar{u}_i}{\partial x_k}\right] + D_{ij}^T + \varphi_{ij} + \varepsilon_{ij}$$

where:

$$C_{ij} = \rho \bar{u}_k \frac{\partial \overline{u_i' u_j'}}{\partial x_k}\text{: consecutive term,}$$

$$P_{ij} = -\rho\left[\overline{u_i' u_k'}\frac{\partial \bar{u}_j}{\partial x_k} + \overline{u_j' u_k'}\frac{\partial \bar{u}_i}{\partial x_k}\right]\text{: stress production,}$$

$$D_{ij}^T = -\frac{\partial}{\partial x_k}\left[\frac{\mu_t}{\sigma_k}\frac{\partial \overline{u_i' u_j'}}{\partial x_k}\right]\text{: turbulent diffusion,}$$

$$\varphi_{ij} = \overline{p'\left[\frac{\partial u_i'}{\partial x_j} + \frac{\partial u_j'}{\partial x_i}\right]}\text{: pressure strain,}$$

$$\varepsilon_{ij} = 2\mu \overline{\frac{\partial u_i'}{\partial x_k}\frac{\partial u_j'}{\partial x_k}}\text{: dissipation rate.}$$

The equations describing the turbulent kinetic energy (k) and the dissipation rate of the kinetic energy (ε) associated with the second-order model are defined as follows:

$$\bar{u}_j \frac{\partial k}{\partial x_j} = \frac{\partial}{\partial x_j}\left[\left(\upsilon + \frac{\upsilon_t}{\sigma_k}\right)\frac{\partial k}{\partial x_j}\right] + \frac{1}{2}P_{ii} - \varepsilon$$

$$\bar{u}_j \frac{\partial \varepsilon}{\partial x_j} = \frac{\partial}{\partial x_j}\left[\left(\frac{\upsilon_t}{\sigma_\varepsilon} + \upsilon\right)\frac{\partial \varepsilon}{\partial x_j}\right] + C_{\varepsilon 1}\frac{1}{2}\frac{\varepsilon}{k}P_{ii} - C_{\varepsilon 2}\frac{\varepsilon^2}{k}$$

where $P_{ii} = 2\upsilon_t(\partial \bar{u}_i/\partial x_i)^2$ and $\upsilon_t = C_\mu(k^2/\varepsilon)$.

The empirical constants appearing in the above equations are given by the following values:

$$C_{\varepsilon 1} = 1.44,\ C_{\varepsilon 2} = 1.92,\ \sigma_\kappa = 0.82 \text{ and } \sigma_\varepsilon = 1.0.$$

3.2. Boundary conditions

An air jet from a circular nozzle of a diameter $d = 10$ mm with an ejection velocity u_0 is subjected to a co-flow stream at rest or in uniform motion with a velocity u_∞. The jet is discharged from a vertical wall offset by a height h above a horizontal wall. To simulate the different flows and scale the different features, a Cartesian coordinate system and its set are used so that the origin is at the intersection of the offset wall and the boundary wall. The x-axis is measured along the horizontal wall and the jet symmetry plane

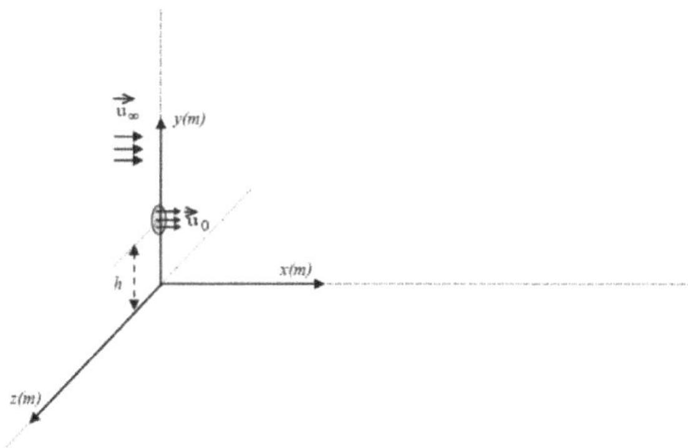

Figure 3. Geometric configuration of the offset jet among the whole domain.

is at $z = 0$ (Figure 3). The dimensions of the computational domain are chosen so that they do not affect the numerical solutions. After several tests, the dimensions chosen in this study are respectively: $L_x = 50\,d$, $L_y = 20\,d$, and $L_z = 20\,d$ in the streamwise (flow) direction (x), normal to the ground wall (y) and spanwise direction (z). The height of the co-flow is noted as H_T.

In order to close the problem, the incorporation of boundary conditions associated with the above-described system of differential equations is summarized in Table 1:

3.3. *Numerical simulation*

The numerical code uses the finite volume method based on the algorithm SIMPLER proposed by Patankar and Spalding (1972). The differential equations governing the problem are then integrated over each control volume and discretized using an implicit scheme to obtain a system of algebraic equations. The convection terms are interpolated from the values at the center of every face of the control volume, with a second-order upwind scheme, and the discretization of the diffusion terms is made with a second-order finite differences method. The elimination method of Gauss Seidel associated with an under-relaxation technique is used to solve the resulting tridiagonal matrix.

The iterative process is stopped when the convergence criterion is defined as $|f^{m+1} - f^{m}/f^{m+1}| \leq 10^{-5}$ is satisfied. In the previous equation f stands for velocity components, turbulent kinetic energy and its dissipation rate, while m is the iterative number.

The resolution of the system of equations needs a fine mesh in zones where interesting interactions take place. In order to describe well any velocity distribution and the turbulent parameters, particularly near and immediately downstream of the jet, we adopted a non-uniform grid system, strongly tightened near the jet nozzle and near walls.

4. Results and discussion

4.1. *Validation and comparison of the experimental and numerical results*

A comparison between the experimental data of Agelin-Chaab and Tachie (2011) and the computational results obtained by a numerical simulation is shown in Figure 4, for an offset ratio $h/d = 4$ and a Reynolds number $\mathrm{Re} = 10{,}000$. The experimental results were measured into a stream at rest. This figure presents the variation of the longitudinal velocity component normalized by the local maximum velocity $\bar{u}/u_m$ as a function of the dimensionless vertical position $Y^{**} = (y - y_m)/(y_{0.5} - y_m)$ for different locations $x/d = 2$, $x/d = 6$, $x/d = 20$ and $x/d = 40$ in the symmetry plane $z = 0$. Respectively, y_m and $y_{0.5}$ denote the location where the streamwise velocity is maximum and where it is half the local maximum velocity. The agreement with both the experimental data and the numerical calculation is acceptable. However, for the longitudinal location $x/d = 20$, in the region $(-2 < Y^{**} < -1)$, we note a slight difference, probably due to the turbulent character of this zone.

The development of the mean flow was examined by studying the decay of u_m. Thus, the maximum velocity decay normalized by the ejection velocity u_m/u_0 along the dimensionless longitudinal coordinate x/d in the symmetry plane $z = 0$ for an offset ratio $h/d = 4$ and a Reynolds number $\mathrm{Re} = 10{,}000$ is presented in Figure 5. The behavior of the maximum velocity decay describes the different regions of the flow. We note the existence of the potential core region followed by the zone of established flow region where the velocity u_m/u_0 decreases while progressing downstream. We note a satisfying agreement between our numerical results and the experimental data

Table 1. Boundary conditions.

Boundaries	Velocity	Turbulent kinetic energy (k)	Dissipation rate of the turbulent kinetic energy (ε)
Nozzle section	$u = u_0,\ v = 0,\ w = 0$	$k = k_0 = 10^{-3}u_0^2$	$\varepsilon = (k_0{}^{3/2}/0.5d)$ (Demuren & Rodi, 1987)
Co-flow stream	$u = u_\infty,\ v = 0,\ w = 0$	$k = k_\infty = 5.10^{-3}u_\infty^2$	$\varepsilon = (k_\infty{}^{3/2}/0.2H_T)$ (Demuren & Rodi, 1987)
Vertical wall	$u = 0,\ v = 0,\ w = 0$	$k = 0$	$(\partial\varepsilon/\partial x) = 0$
Horizontal wall	$u = 0,\ v = 0,\ w = 0$	$k = 0$	$(\partial\varepsilon/\partial y) = 0$
Other boundaries of the domain	$(\partial u/\partial n) = 0,\ (\partial v/\partial n) = 0,\ (\partial w/\partial n) = 0$	$k = 0$	$(\partial\varepsilon/\partial n) = 0$

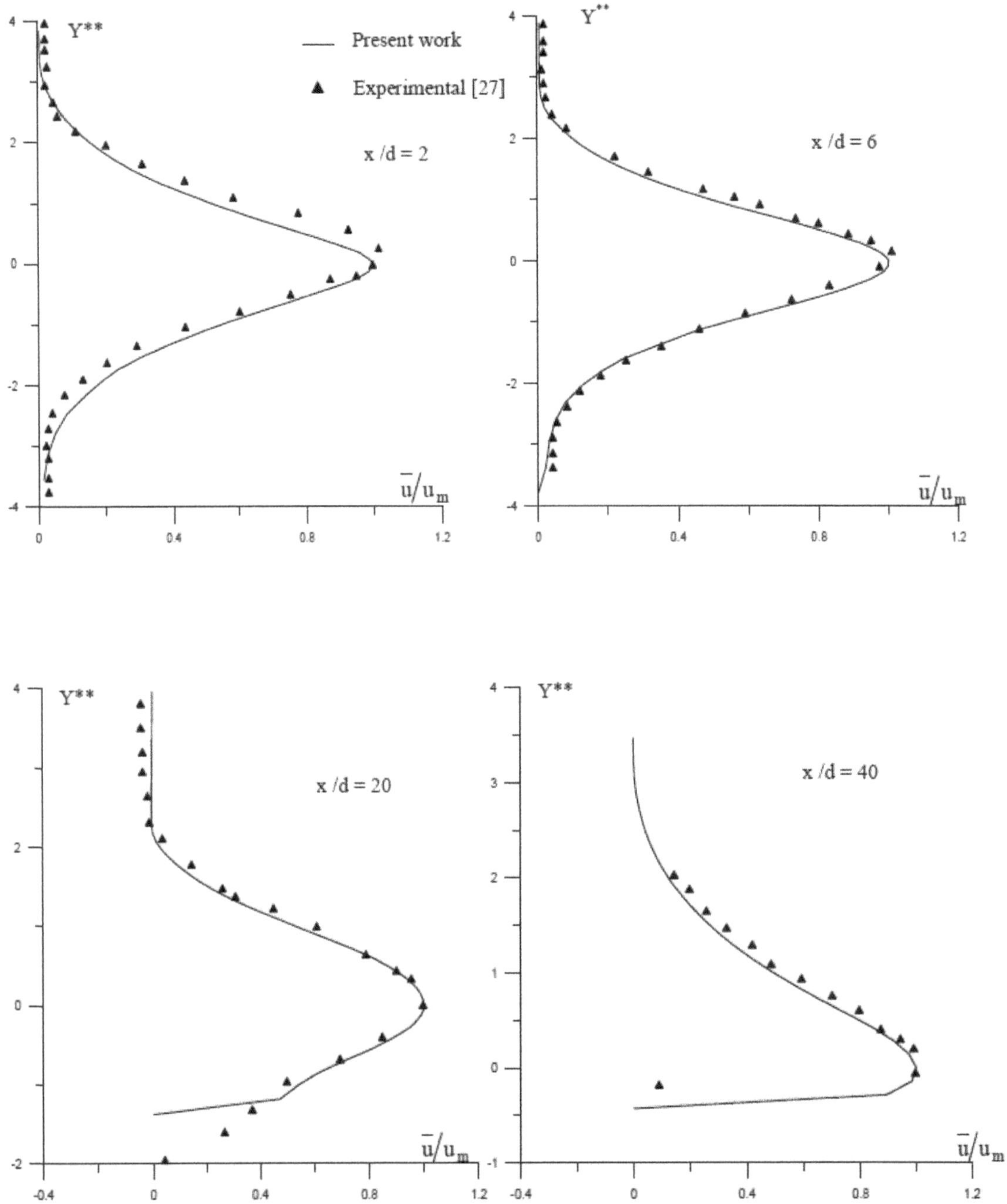

Figure 4. Longitudinal velocity profiles at different longitudinal locations in the symmetry plane: $u_0 = 1.43\,\text{m.s}^{-1}$, $h/d = 4$.

(Agelin-Chaab & Tachie, 2011). Hence, the chosen turbulence model predicts the measured data well and is accurate enough for a parametric study.

Before detailing any of the obtained results, we need to compare our PIV measurements with the numerical results obtained by reproducing the same dimensions as the experimental ones.

Figures 6 and 7 show the evolution of the longitudinal velocity component normalized by the maximum velocity ($\bar{u}/u_m$) for two velocity ratios: $R = 1$ ($u_0 = 3\,\text{m.s}^{-1}$, $u_\infty = 3\,\text{m.s}^{-1}$) and $R > 1$ ($u_0 = 4\,\text{m.s}^{-1}$, $u_\infty = 3\,\text{m.s}^{-1}$) at an offset ratio $h/d = 5.3$. The profiles were presented at $x/d = 4$ and $x/d = 5$ along the vertical distance y/d. The confrontation of the calculated and measured results gives

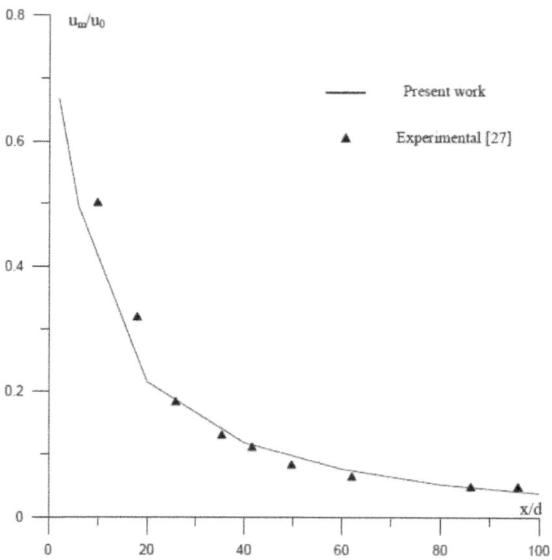

Figure 5. Streamwise distribution of the local maximum mean velocity in the symmetry: $u_0 = 1.43$ m.s^{-1}, $h/d = 4$.

a satisfying matching along all its variation processed with the RSM. These results show the minima just behind the vertical wall, in the cavity area, and rising to near constant velocity further away vertically. This figure shows a typical curve of a zero velocity at the horizontal wall, peak in the recirculation and wake area. This peak is more important at the longitudinal coordinate $x/d = 5$ for the two velocities' ratios.

According to the above-discussed comparisons, and in accordance with the better description of the experiments with the RSM, we can conclude that our numerical modeling is acceptable and models the experimental configuration of the offset jet issuing within a co-current flow well.

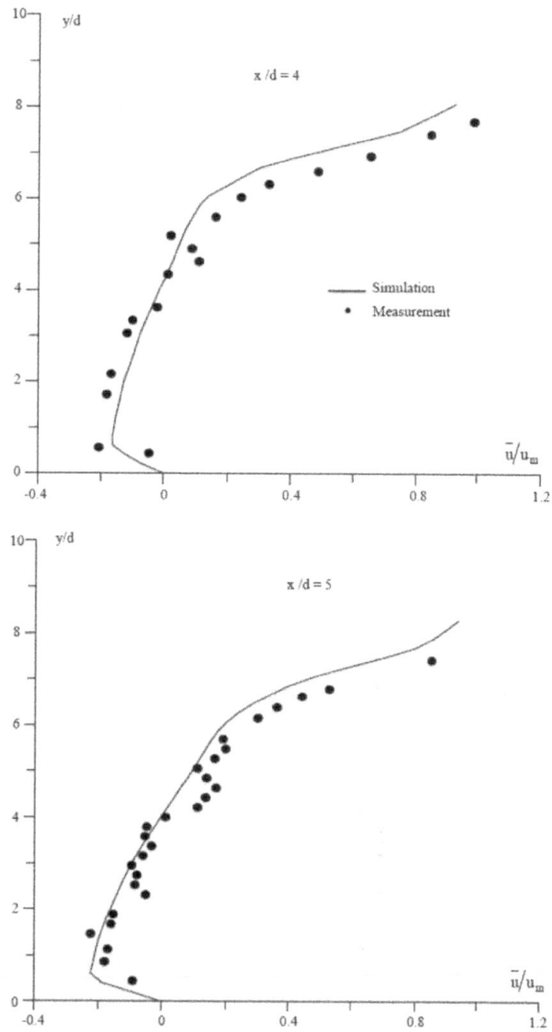

Figure 6. Mean longitudinal velocity profile in the symmetry plane: $R = 1$ ($u_0 = 3$ m.s^{-1}, $u_\infty = 3$ m.s^{-1}), $h/d = 5.3$.

4.2. *Experimental results*

4.2.1. *Results for different velocity ratios*

We propose formulating a detailed description of the flow field of a turbulent offset jet issuing into a moving stream (co-flowing) for different velocity ratios $R = u_0/u_\infty$ ($R = \infty$, $R = 1$ and $R > 1$).

4.2.1.1. Flow structure for $R = \infty$ ($u_0 = 3$ m.s^{-1} and $u_\infty = 0$ m.s^{-1}). We analyze, in Figure 8, the structure of the flow issued from an offset discharged into a quiescent medium. Figure 8(a) illustrates the flow visualization of the jet for an offset ratio $h/d = 5.3$.This image shows that, for a jet discharging in a stream at rest, the flow at the outlet of the nozzle is predominant and the jet elapses horizontally parallel to the wall for very large distances downstream and behaves as a free jet. In this case, and due to the difference between the jet core and the ambient fluid, a shearing layer is created from the nozzle exit. Downstream of the exit nozzle, we note the formation of vortex structures in the upper and lower boundaries of the jet frontier. Furthermore, we noticed the absence of wake zone vortices located downstream of the nozzle, delimited by the jet flow, the offset wall and the horizontal wall.

In Figure 8(b), streamlines of the flow are presented in the symmetry plane ($z = 0$). This figure shows that the horizontal penetration of the jet is maintained within the domain for large distances downstream of the jet exit. In this region, we can easily see that downstream of the nozzle exit, the velocity (u_0) is the local maximum streamwise velocity for a given location x/d, just like a single free jet. This figure showed that the region with a high velocity decreases in width downstream of the nozzle exit. The

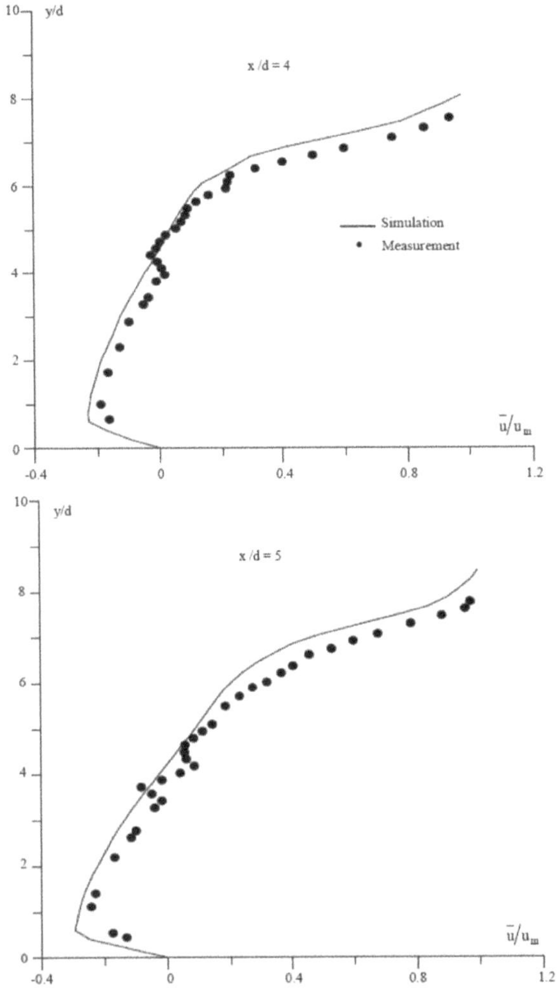

Figure 7. Mean longitudinal velocity profile in the symmetry plane: $R > 1$ ($u_0 = 4$ m.s^{-1}, $u_\infty = 3$ m.s^{-1}), $h/d = 5.3$.

streamlines close to the wall are slowed and stagnate at surface impact with the attachment point. One of the characteristics of the shear layer is the reattachment length, which was estimated as the distance from the nozzle exit to the location where the jet reattached to the wall. It is noticed that the reattachment length is increased for a jet flow issuing into an ambient flow at rest. The size of the recirculation eddy is increased considerably in this case. In the wall region, the streamlines become parallel to the horizontal wall.

Figure 9 shows the evolution of the normalized mean velocity ($\bar{u}/u_m$) at different longitudinal positions ($x/d = 4$, $x/d = 5$, $x/d = 7$ and $x/d = 9$) in the symmetry plane ($z = 0$). The profiles of the longitudinal velocity adopt a Gaussian shape, nearly symmetrical relative to the axis of the nozzle exit ($y/d = 5.3$). The symmetry of the profiles is more pronounced close to the nozzle exit ($x/d \leq 4$), indicating that the injected flow behaves as a free jet in this region. In addition, this figure reveals that, for the different longitudinal positions, the values of the longitudinal velocity are positive, indicating the absence of the recirculation zone in this region. The advancement in the streamwise direction similarly promotes an overall decrease in the velocities range.

4.2.1.2. Flow structure for $R = 1$ ($u_0 = 3$ m.s^{-1} and $u_\infty = 3$ m.s^{-1}). The flow visualization and experimental streamlines of the flow in the symmetry plane ($z = 0$ m) for an injection ratio equal to unity and for an offset ratio $h/d = 5.3$ are shown in Figure 10.

Figure 10(a) illustrates the behavior of the flow downstream of the nozzle exit by an instantaneous visualization captured by the CCD camera. When the injection ratio R is equal to the unity, the jet and the outer flow have comparable strengths, giving rise to equivalent interactions, so we can presume the dominance of a jet flow relative to the exterior flow. The jet in this case takes less time to deflect towards the horizontal wall, which leads to a reduction of the location of the impact point. Therefore, the reattachment length becomes smaller compared to the previous case ($u_\infty = 0$ m.s^{-1}). This figure also allows a quantitative analysis of the resulting flow field, such as the formation of Kelvin Helmholtz instabilities. This phenomenon occurs when the flow is subjected to shearing between two fluids which slip over one other at different velocities. These instabilities lead to the ripple and the winding within the interface zone until the formation of the vortex to generate the mixing-layer between the two flows. Depending on wind velocity, these patterns turn clockwise or counterclockwise. In fact, the rotational direction depends on the R factor.

We analyze in Figure 10(b) the experimental streamlines for $R = 1$. This figure reveals that, close to the jet exit, many singular streamlines are horizontal and keep following this direction. Thereafter, these streamlines deflect to the horizontal wall in a region near the exit nozzle as a result of the surrounding flow. By examining streamlines in this figure, there are three main zones. In the first zone (the recirculation zone), we noticed the deflection of the jet by the wall and the decrease of the velocity as the jet approaches the horizontal wall. After the jet impacts on the wall, a part of the fluid is deflected upstream from the attachment point into the recirculation zone to form the wake vortices. Herein the velocity is very weak, justifying the weak intensity of the developed vortices. In the attachment region (zone 2), the flow stretches from the impact point to form another flow type. Here, the flow begins to transform into a wall jet flow. In the third and last region (zone 3), downstream of the impact point, the streamlines develop parallel to the horizontal wall to resemble a wall jet flow.

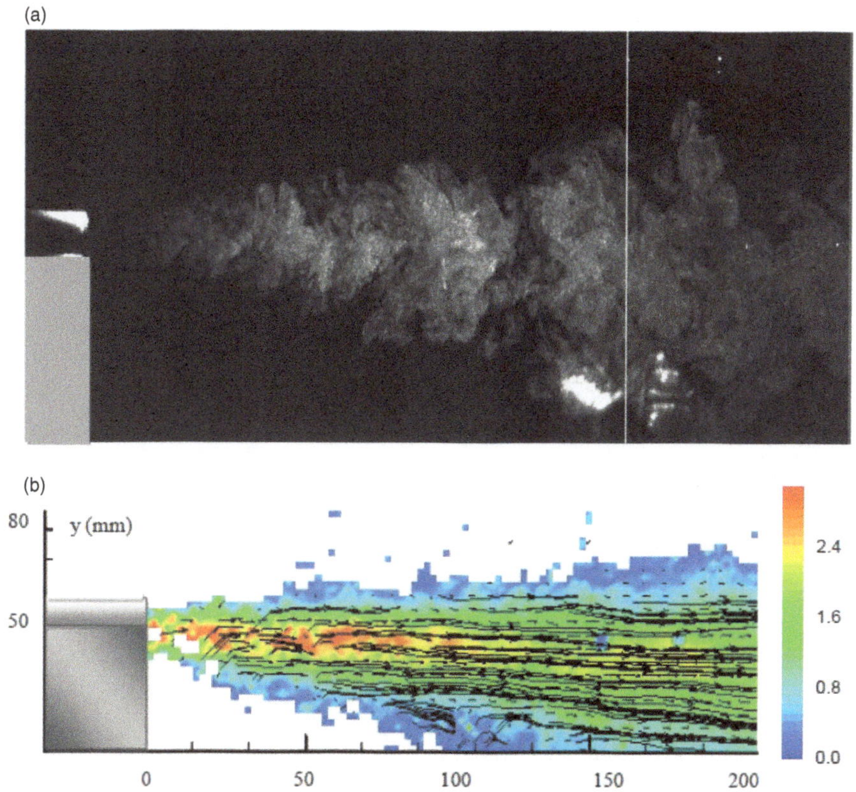

Figure 8. (a) Flow visualization and (b) streamlines in the symmetry plane: R = ∞ (because $u_\infty = 0$) $u_0 = 3$ m.s^{-1}, $u_\infty = 0$ m.s^{-1}, $h/d = 5.3$.

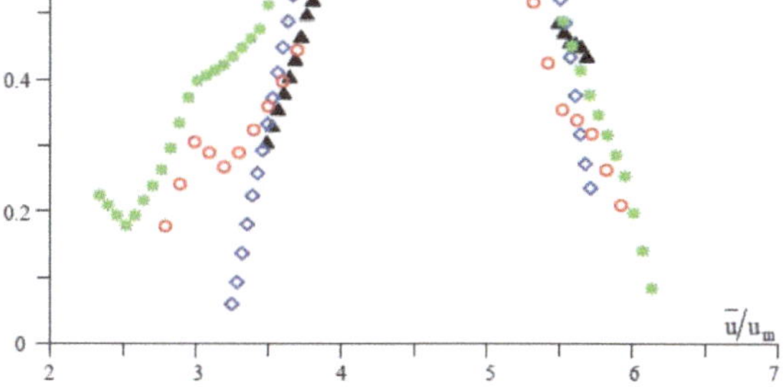

Figure 9. Profiles of the mean longitudinal velocity for different longitudinal locations in the symmetry plane: R = ∞ (because $u_\infty = 0$) $u_0 = 3$ m.s^{-1}, $u_\infty = 0$ m.s^{-1}, $h/d = 5.3$.

4.2.1.3. Flow structure for $R > 1$ ($u_0 = 4$ m.s^{-1} and $u_\infty = 3$ m.s^{-1}). In Figure 11(a), the flow visualization in the symmetry plane ($z = 0$) for an offset ratio ($h/d = 5.3$) and for an injection ratio greater than the unity $R > 1$ is plotted. This figure reveals that the penetration and mixing of the jet with the outer flow has a significant effect. The entrainment of the ambient fluid in this case generates a shear jet flow leading to the appearance of turbulent structures of Kelvin-Helmholtz instabilities. These structures lead to the formation of a 'swirling sheet' on the border between the two fluids. The rotation of this vortex structure is in the anticlockwise direction. The orientation of the vortices is more apparent on the downstream jet leading edge due to its direct confrontation with the outer flow in this region. The jet deflected towards the horizontal wall and attached to it at a position, downstream from the nozzle exit, greater than that found in the previous case ($R = 1$).

Figure 11(b) illustrates the sectional view of flow streamlines for the velocity ratio $R > 1$ in the symmetry plane. We noticed that, immediately downstream of the nozzle exit, the jet flow is weakly influenced by the wind flow, which allows it to rise a little more in the outer flow. Streamlines indicate the motion of the exterior flow fluid. This figure shows three distinct regions. In the area close to the jet exit, the streamlines start horizontally in the ground wall. In this region, the velocity is maximum along the centerline of the nozzle exit until the deflection of the jet. Then, because of the 'Coanda effect', these streamlines are

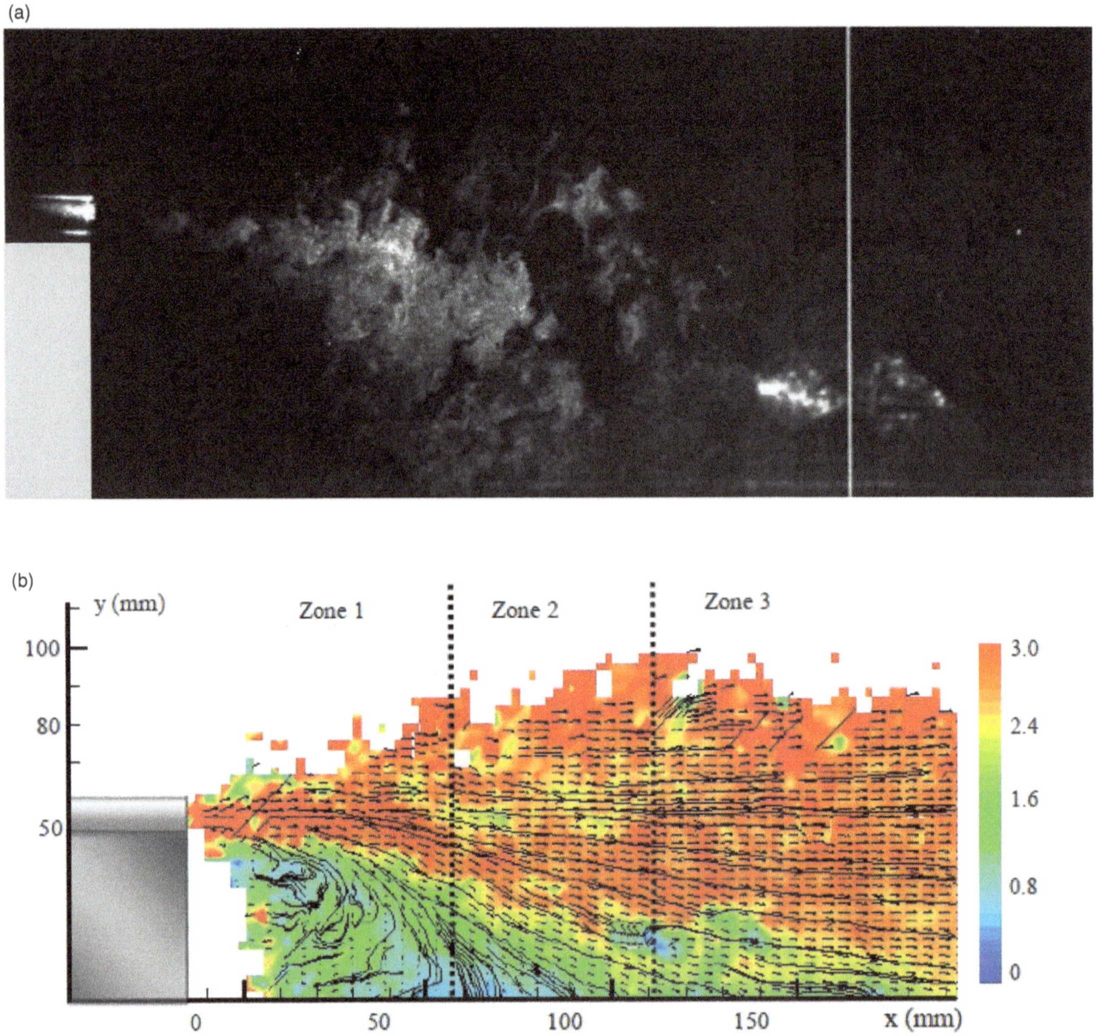

Figure 10. (a) Flow visualization and (b) streamlines in the symmetry plane: $R = 1$ ($u_0 = 3$ m.s^{-1} and $u_\infty = 3$ m.s^{-1}), $h/d = 5.3$.

curved towards the horizontal wall and finally to bind with it at the attachment point. Herein, the velocity decreases to a minimum value where the mean streamlines reattach to the wall. At this point, some of streamlines change direction upstream towards the recirculation region (the blue zone) while others continue to develop to resemble a wall jet flow.

4.2.2. *Results for different offset ratios*

Flow visualization and mean velocity vectors in the symmetry plane ($z = 0$) are shown in Figure 12 for different offset ratios ($h/d = 5.3$ and $h/d = 7.8$) and for a velocity ratio $R > 1$. The flow visualization reveals the existence of vortex structures of Kelvin-Helmholtz instabilities. The rotation direction depends on the value of the velocity ratio R. For $R > 1$ these instabilities turn in the counterclockwise direction.

For a higher offset ratio ($h/d = 7.8$), the jet flow is released from the vertical wall attachment, allowing the jet to spread more horizontally. Indeed, the small offset ratio leads to a gradual decrease of the jet wake and a reduction in the size of the recirculation zone. Consequently the standing vortex was found to be very small compared with a higher offset height. In addition, it is observed that the reattachment length of the jet is longer with the increase of the offset ratio. Consequently, the attachment point is located at a further larger position for $h/d = 7.8$ than for $h/d = 5.3$. The mean velocity vectors for two offset ratios indicate that the jet converges towards the horizontal wall as it proceeds downstream from the nozzle ejection, forming a recirculation flow region upstream

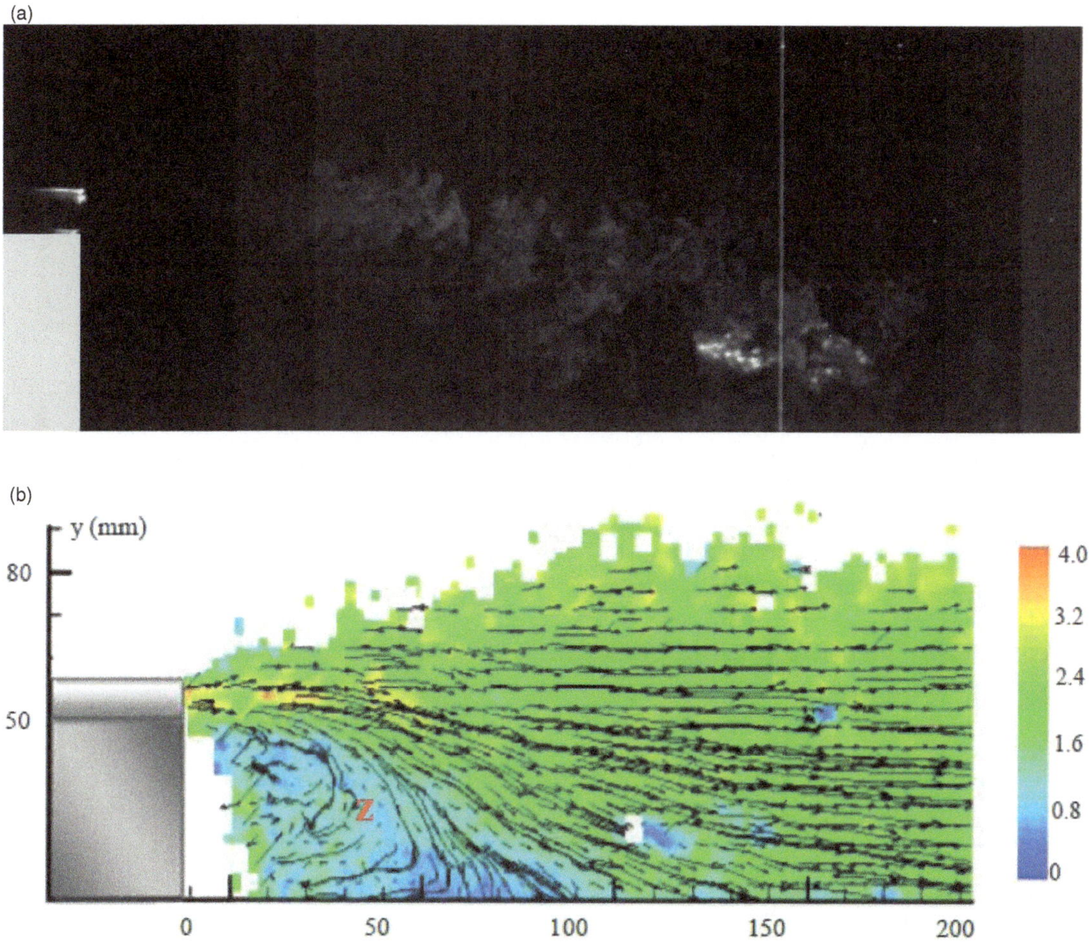

Figure 11. (a) Flow visualization and (b) streamlines in the symmetry plane: $R > 1$ ($u_0 = 4$ m.s^{-1} and $u_\infty = 3$ m.s^{-1}), $h/d = 5.3$.

of the reattachment point. Downstream from the recirculation zone, the jet continues its development to resemble a wall jet flow. For an offset ratio $h/d = 7.8$, the jet develops and progresses horizontally across the flow for greater distances than for a jet with a small offset ratio $h/d = 5.3$ before impingement with the horizontal wall. Thereafter, the recirculation zone induced by the greater offset ratio ($h/d = 7.8$) is larger than that given by the small offset ratio ($h/d = 5.3$), which causes the remoteness of the impact point of the jet with the horizontal wall. The reduced pressure in the recirculation region is created by the big vortex and the air entrainment into the jet flow. We also noticed that, before the attachment region, the mean velocity for an offset height $h/d = 5.3$ decreases more rapidly compared to an offset ratio $h/d = 7.8$. Indeed, the jet is released from the effect of the horizontal wall when it is discharged from a nozzle that is far away from the horizontal wall. This figure highlights that the global mean flow configuration is governed mainly by the offset distance, the extension of which defines the dimensions of the recirculating zone. The outer shear layer spreads faster than the inner shear layer, which is limited by the presence of the attachment wall.

4.3. *Numerical results*

4.3.1. *Results for different velocity ratios*

The vertical distributions of the longitudinal and vertical velocity components in the symmetry plane ($z = 0$) for different longitudinal locations ($x/d = 4$, $x/d = 7$ and $x/d = 10$) and for different velocity ratios ($R = 1$, $R < 1$, $R > 1$) are shown in Figure 13. This figure shows that, downstream of the exit nozzle, the profiles of the longitudinal velocity component indicate negative values in the vicinity of the horizontal wall for different velocity ratios, indicating the presence of a recirculation zone (this is a region characterized by the presence of vortices). Then, the longitudinal velocity increases as a function of the vertical coordinate (y/d) to achieve the uniform velocity of the surrounding flow prevailing outside the interaction zone.

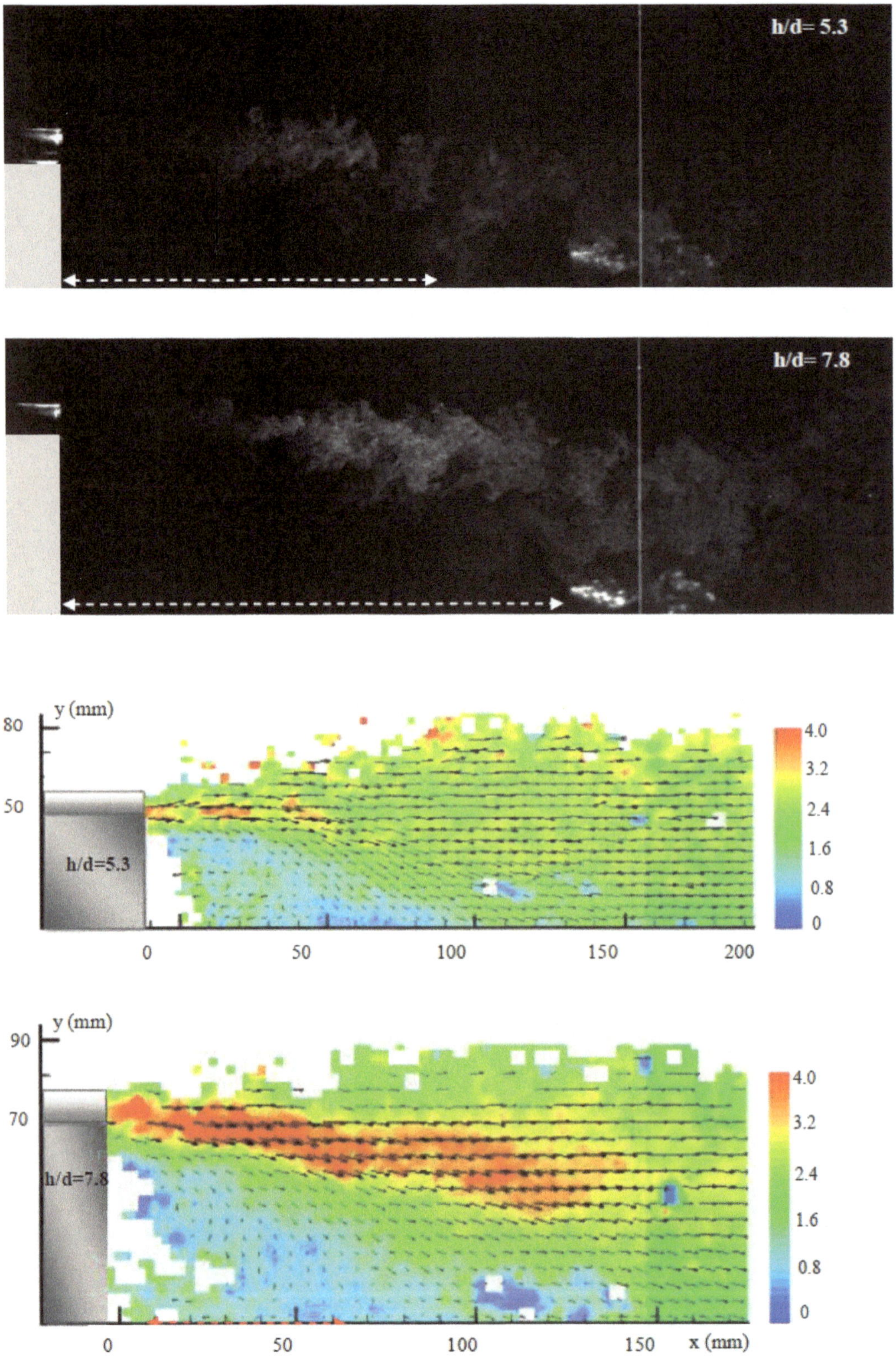

Figure 12. (a, b) Flow visualization and (c, d) velocity vectors in the symmetry plane: $R > 1(u_0 = 4\ \mathrm{m\,s^{-1}}$ and $u_\infty = 3\ \mathrm{ms^{-1}})$.

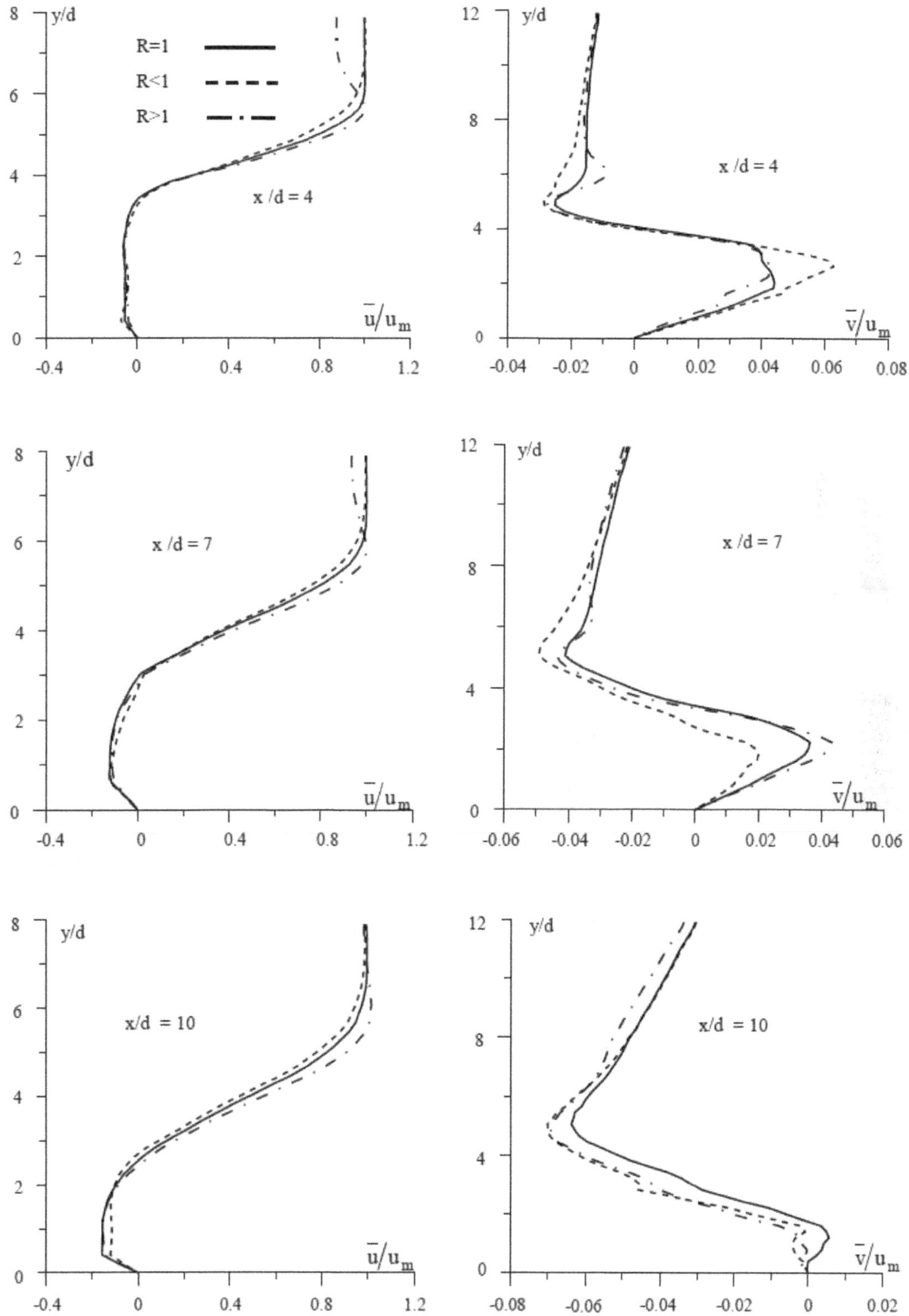

Figure 13. Profiles of mean longitudinal and normal velocities in the symmetry plane for different velocity ratios and for $h/d = 5.3$.

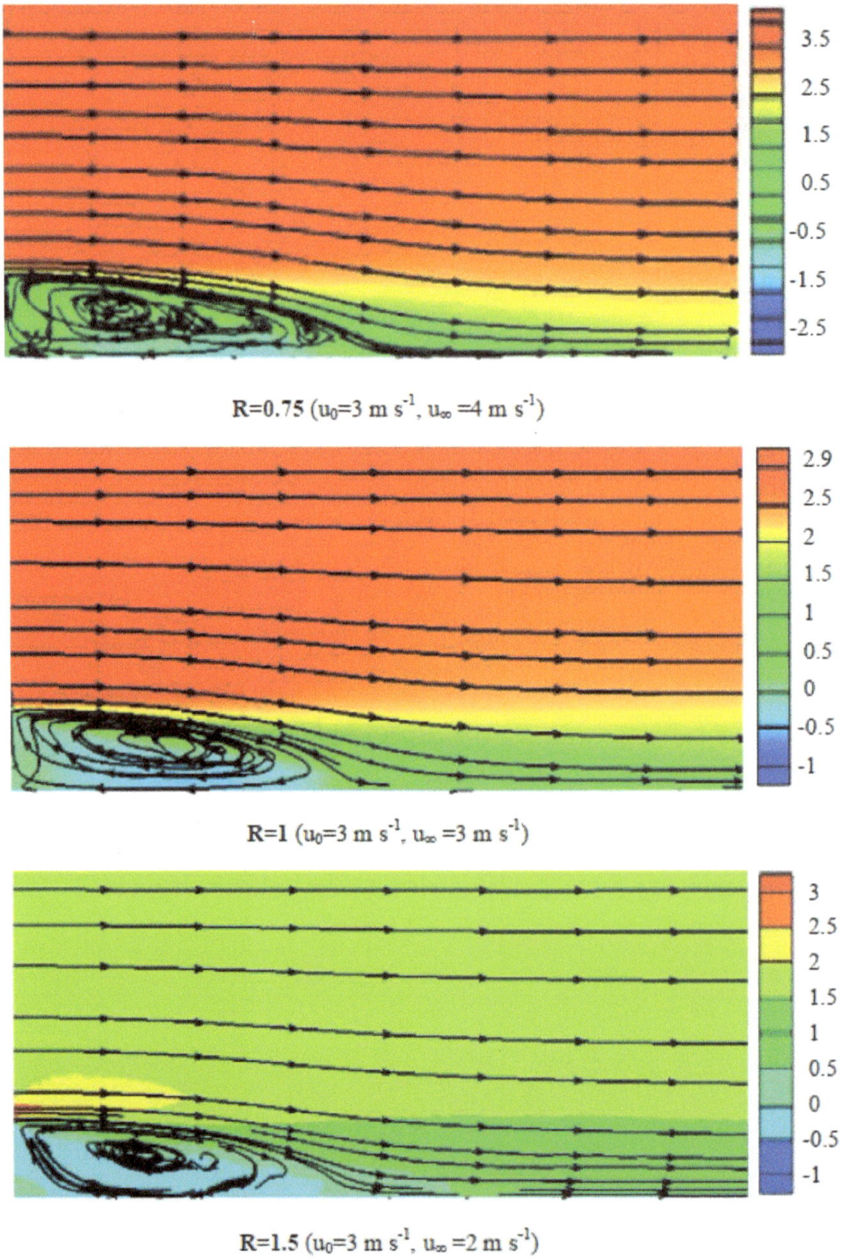

Figure 14. Flow streamlines in the symmetry plane for different velocity ratios and for $h/d = 5.3$.

In this figure, it can also be seen that, for different velocity ratios, the longitudinal component increases to compensate for the supplying of fluid through the jet, and it increases until a maximum value ($\bar{u}/u_m = 1$) downstream from the jet exit. The examination of the vertical velocity component shows the existence of two peaks – a peak of positive value, located in the recirculation zone ($y/d = 2$) due to the deflection of the jet towards the horizontal wall, and a second peak of negative value, located close to the nozzle axis ($y/d = 5.3$), indicating that the fluid is entrained from the ambient medium towards the horizontal wall.

Figure 14 introduces numerical streamlines in the symmetry plane for different velocity ratios. The streamline contours illustrate the main, recirculation and entrainment flows of the jet. The flow is characterized by a longitudinal variation of the streamline curvature and skewed

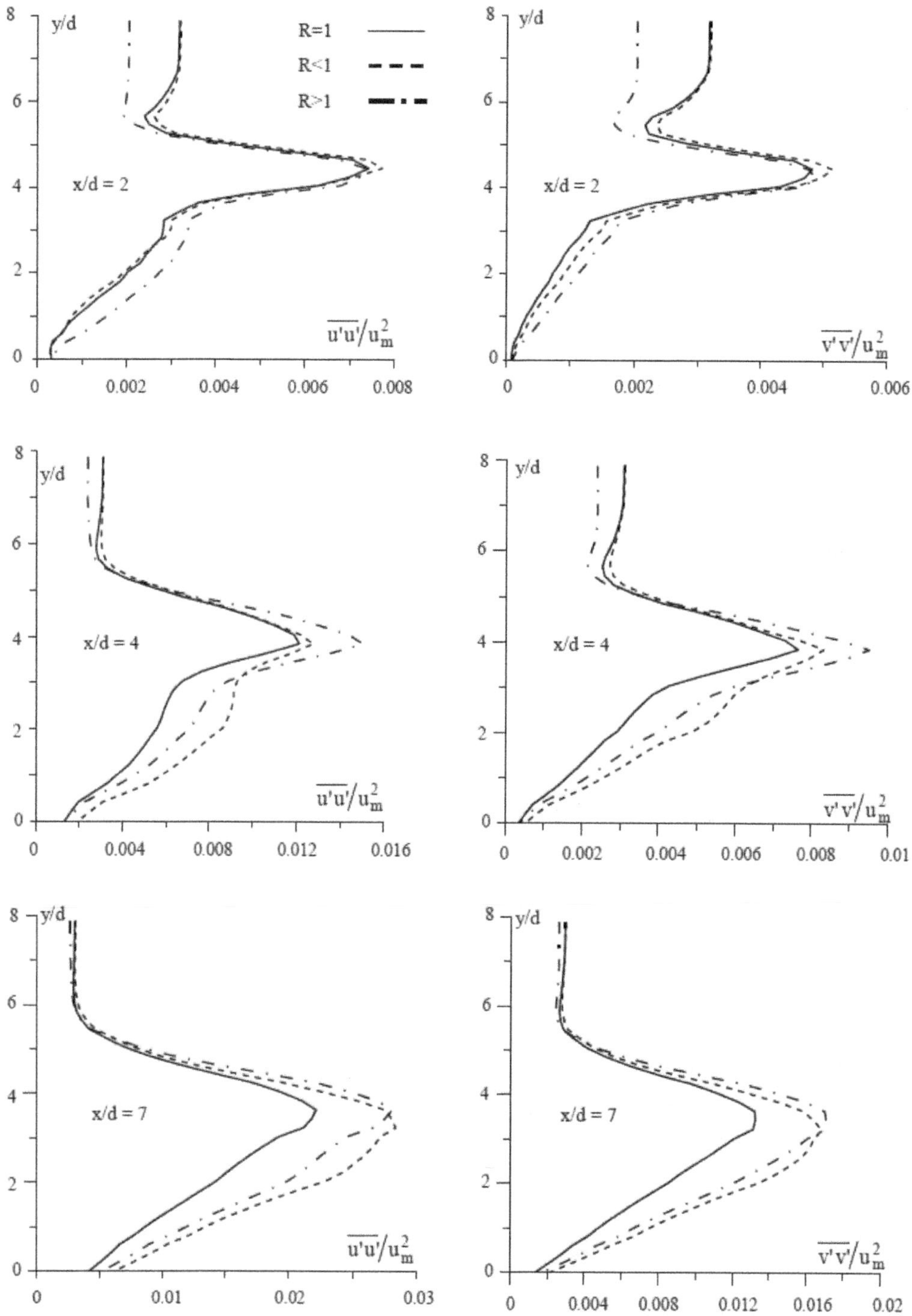

Figure 15. Shear stress components $\overline{u'u'}/u_m^2$ and $\overline{v'v'}/u_m^2$ in the symmetry plane for different velocity ratios and for $h/d = 5.3$.

impingement onto the horizontal wall. Streamlines indicate the motion of the surrounding medium. Streamlines entering the domain from the exterior flow boundary, above the jet, appear to bend in the direction of the jet fluid and almost merge with the leading edge boundary of the jet. These streamlines show the deflection of the jet towards the boundary wall and reattach to the surface in the impingement zone, enclosing a region known as the recirculation region. It is noticed that the increase in the velocity ratio reduces the reattachment length and the position of the attachment point. Downstream from the attachment point, these streamlines continue to develop parallel to the horizontal surface to resemble a wall jet flow. This figure also shows that the qualitative behavior of the creation and propagation of the reverse movement (recirculation region) is similar in the experiment and in the calculation.

In Figure 15, we represent the effect of the velocity ratio on the vertical distribution of the longitudinal ($\overline{u'u'}/u_m^2$) and the normal ($\overline{v'v'}/u_m^2$) shear stress components in the symmetry plane ($z = 0$) as a function of the vertical coordinate y/d, for several positions of x/d (2, 4 and 7) for an offset ratio ($h/d = 5.3$). This figure depicts that the profiles of the turbulent fluctuations have a similar behavior for different velocity ratios. This figure also highlights the presence

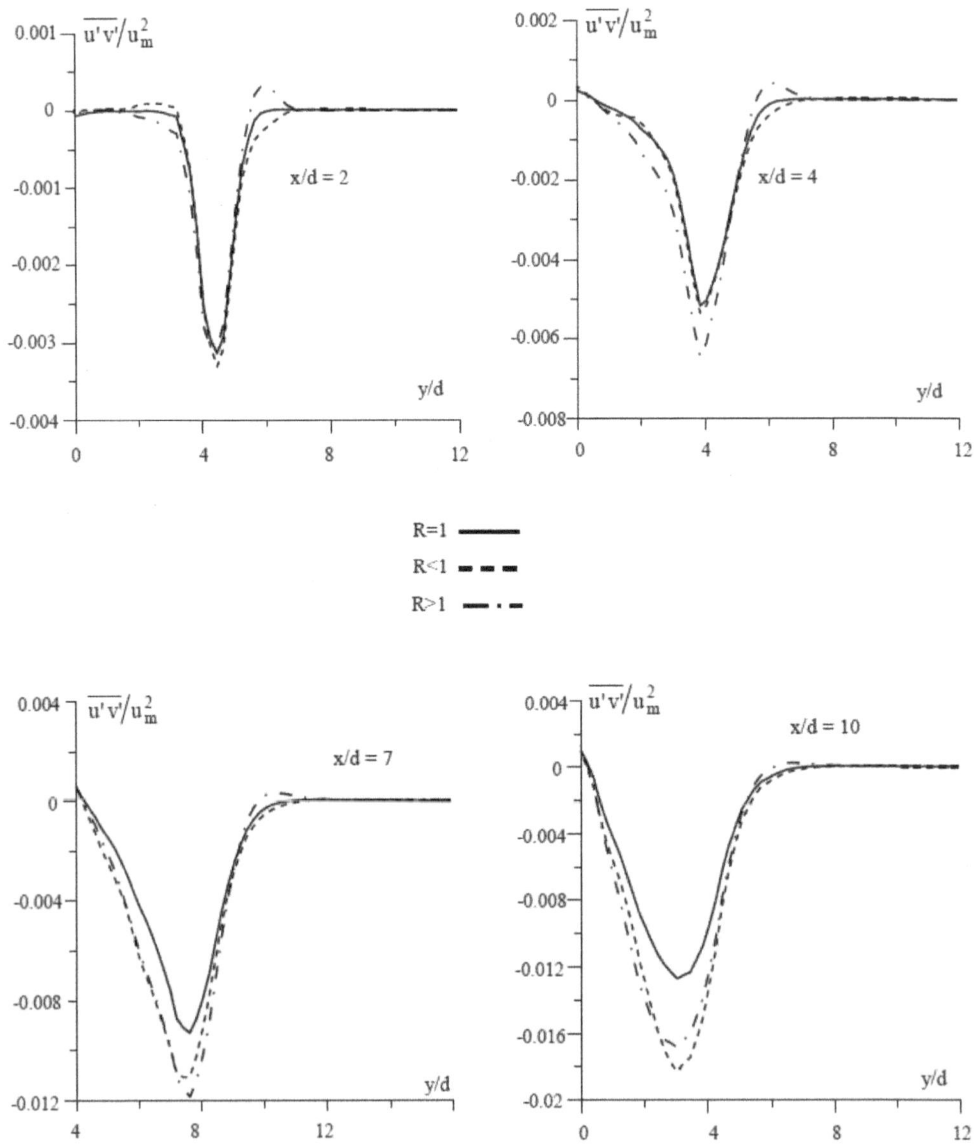

Figure 16. Normal shear stress component $\overline{u'v'}/u_m^2$ in the symmetry plane for different velocity ratios and for $h/d = 5.3$.

of peaks at the same position but with different fluctuating intensities. These peaks are located near the outlet of the nozzle within the area of interaction of the jet and the co-flow. It is also noticed that the velocity fluctuations for a low velocity ratio ($R < 1$) are stronger compared to those for $R = 1$ and $R > 1$. This is due to the effect of the down-wash jet for a lower velocity ejection to the wind flow. We also see that the magnitude of the turbulent fluctuations in the inner shear layer increases as the flow moves downstream, until the jet attaches to the horizontal wall (in the recirculation zone). Thereafter, these turbulent fluctuations decrease near the impact point due to the interaction of the jet flow and the horizontal wall. This result is in good agreement with that found by Gao and Ewing (2007), who reported that the magnitude of turbulent fluctuations in the inner shear layer increases downstream from the attachment point and then decreases due to interaction with the wall.

Figure 16 presents the vertical distribution of the normal shear stress component $\overline{u'v'}/u_m^2$ in the symmetry plane ($z = 0$) as a function of the vertical coordinate y/d, for several longitudinal positions ($x/d = 2$, $x/d = 4$, $x/d = 7$ and $x/d = 10$). Three velocity ratios ($R = 1, R < 1, R > 1$) are illustrated in this figure for an offset ratio $h/d = 5.3$. This figure shows that the most significant variations occur in the vicinity of the jet outlet. Indeed, the jet flow penetrates

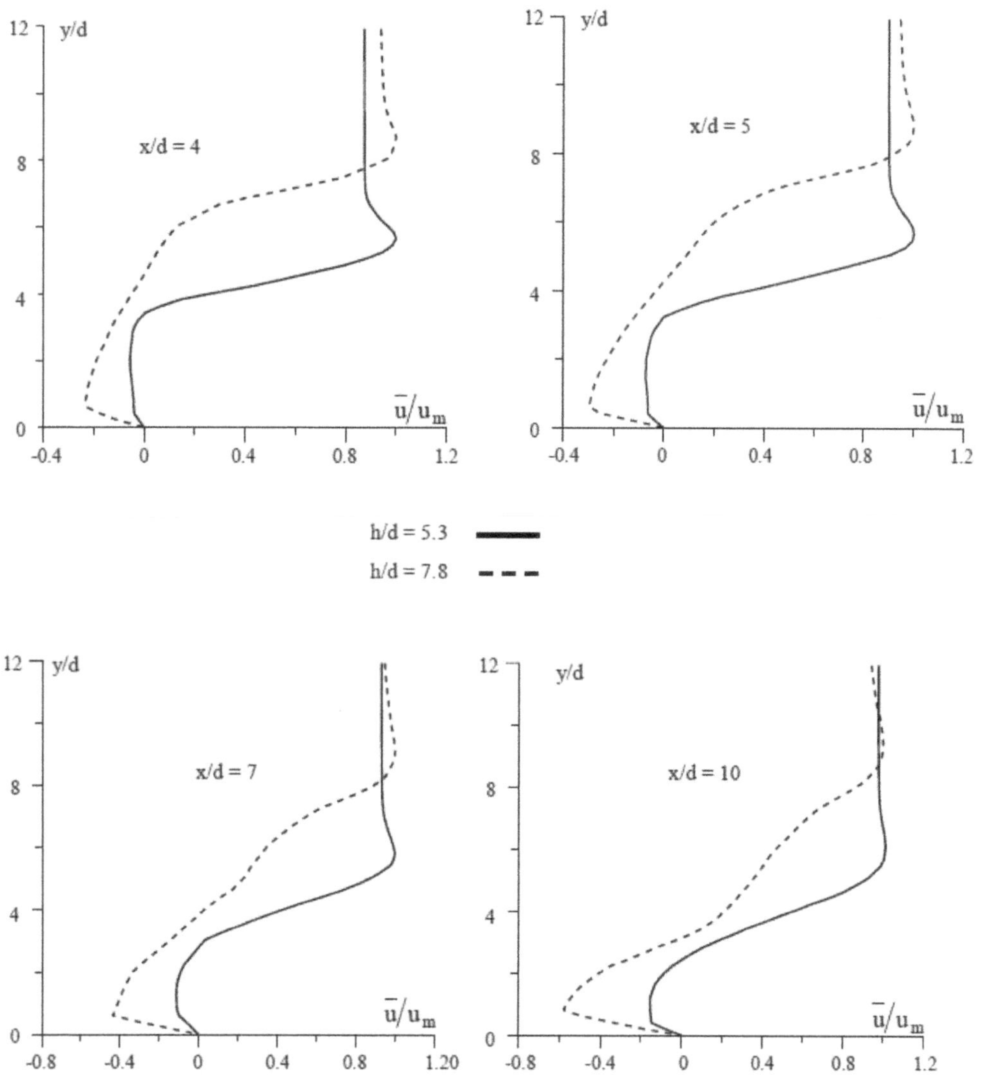

Figure 17. Profiles of the longitudinal mean velocity in the symmetry plane for different offset ratios and for $R > 1$ ($u_0 = 4$ m.s^{-1} and $u_\infty = 3$ m.s^{-1}).

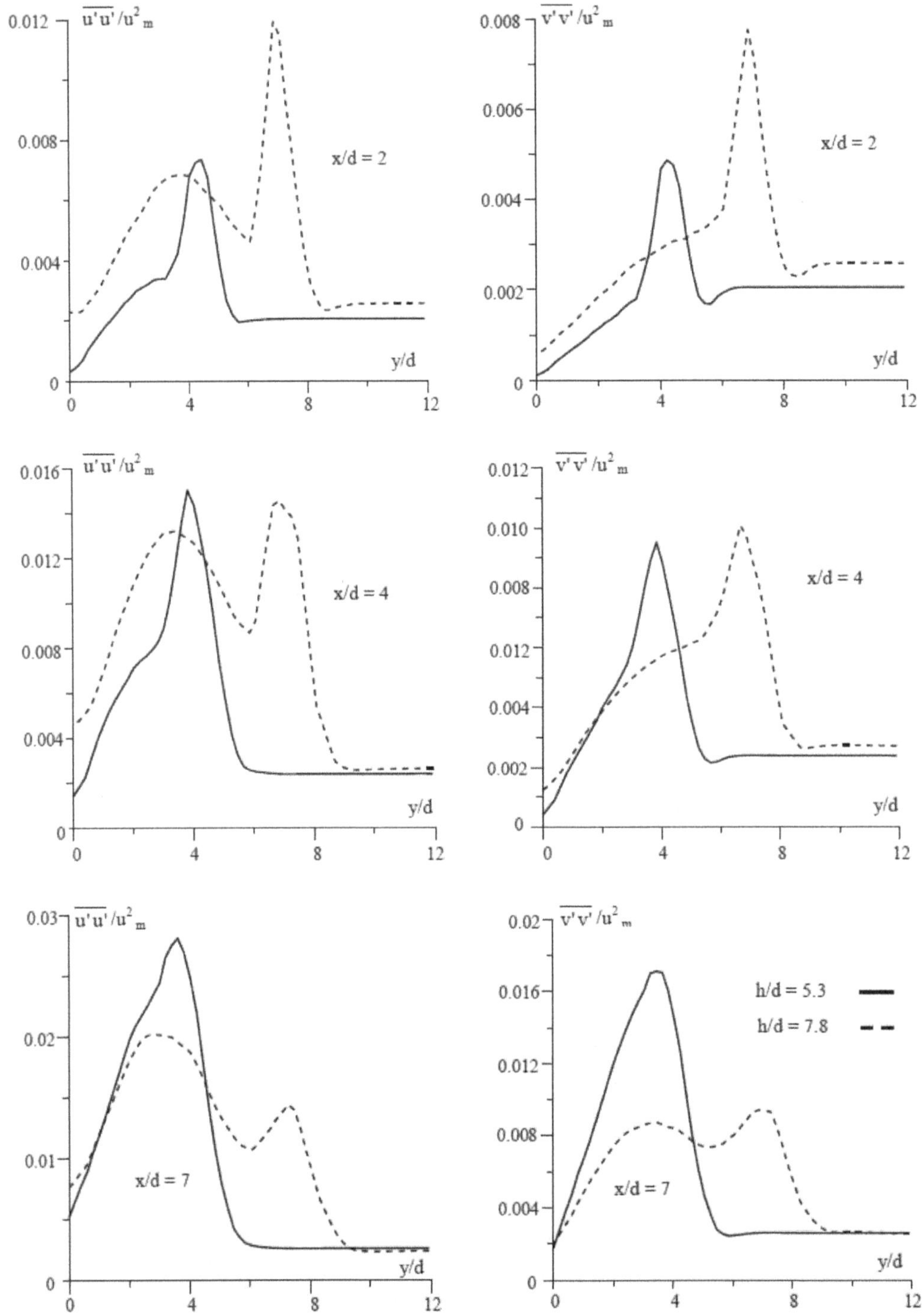

Figure 18. Shear stress components $\overline{u'u'}/u_m^2$ and $\overline{v'v'}/u_m^2$ in the symmetry plane for different offset ratios and for $R > 1$ ($u_0 = 4$ m.s^{-1} and $u_\infty = 3$ m.s^{-1}).

within the surrounding flow in this region. These profiles indicate peaks with different values for different velocity ratios, which are located in the recirculation region. Close to the nozzle exit (for $x/d = 2$), there are similar values of these peaks for different velocity ratios. Far downstream from the nozzle ($x/d = 10$), the recorded values are

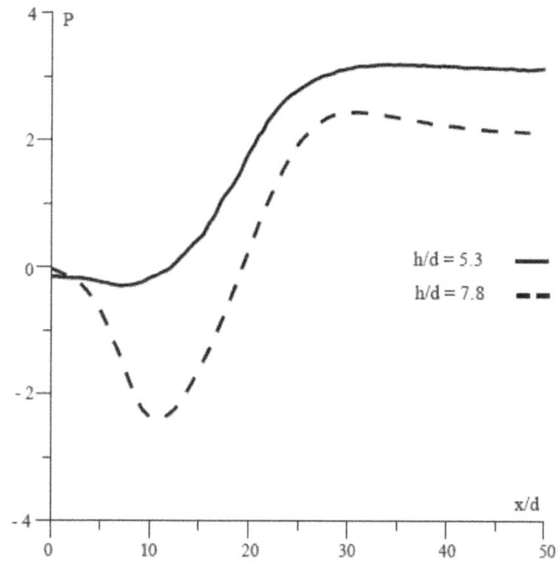

Figure 19. Wall pressure distribution for different offset ratios in the symmetry plane: $R > 1$ ($u_0 = 4$ m.s^{-1} and $u_\infty = 3$ m.s^{-1}).

larger for a velocity ratio less than unity ($R < 1$). Progression through the flow direction (near the attachment point) favors an overall decrease in the shear stress component for different velocity ratios.

4.3.2. *Results for different offset ratios*

Profiles of the mean longitudinal velocity (u/u_m), are shown in Figure 17, for different offset ratios ($h/d = 5.3$ and $h/d = 7.8$) and for a velocity ratio $R > 1$ in the symmetry plane ($z = 0$). This figure indicates negative values of the longitudinal velocity related to the presence of a recirculation zone. In this region, values of the longitudinal velocity are higher for $h/d = 7.8$, showing that the size of the recirculation zone is larger for a higher velocity ratio. This indicates that the maximum streamwise velocity of the reverse flow is very large for a higher offset height. The relatively lower maximum reverse flow velocity for the small offset jet is due to the retarding effect of the wall on the recirculating flow. Hence, the reattachment length is higher for a jet emitted with an offset ratio $h/d = 7.8$. Therefore, the position of the impact point of the jet on the horizontal wall is located at a position farther from the nozzle exit.

Figure 18 represents the vertical distribution of the normal $\overline{u'u'}/u_m^2$ and $\overline{v'v'}/u_m^2$ shear stress components in the symmetry plane ($z = 0$) for different offset ratios ($h/d = 5.3$ and $h/d = 7.8$) and for a velocity ratio $R > 1$. These distributions are plotted in different longitudinal positions ($x/d = 2$, $x/d = 4$ and $x/d = 7$). Profiles of the shear stress show that the distance from the offset of the jet to the impact wall causes significant slowing and suppression effects in the recirculation and attachment regions. It is also noted that these fluctuations have maximum values in the inner shear layer and occur upstream of the impact point. This indicates a strong interaction between the flow in the recirculation region and the inner shear layer near the impact point. It is also noticed that the Reynolds stresses in the internal shear layer close to the attachment point increase with the offset ratio. This figure shows a similarity of the profiles of the longitudinal and vertical shear stress components for the two offset ratios.

Figure 19 presents the wall pressure distribution for two offset ratios ($h/d = 5.3$ and $h/d = 7.8$) tested in the symmetry plane ($z = 0$). For both of the offset ratios, it is noticed that the static pressure decreases in the direction of the flow to a minimum value located close to the center of the recirculation region ($x/d = 10$), indicating the end of the first region when the jet is deflected towards the horizontal wall. The strength of the recirculation region decreases with the offset ratio, as can be seen from the decrease in the magnitude of the minimum pressure. Then, the static pressure increases rapidly through the second region (the impingement region) to a maximum value at a downstream location in the vicinity of the attachment point. In this region, the pressure is generated by the deflection in the streamline after attaching to the horizontal wall, at which point it changes its direction to be parallel to the wall. Figure 19 also highlights that the position of the negative-pressure core and the attachment point move downstream with an increase in the offset ratio. This is probably due to the difference in the trajectory of the jet towards the horizontal wall.

Figure 20 shows the contours of the static pressure for two offset ratios ($h/d = 5.3$ and $h/d = 7.8$) in the symmetry plane ($z = 0$). At the jet exit, and because of the

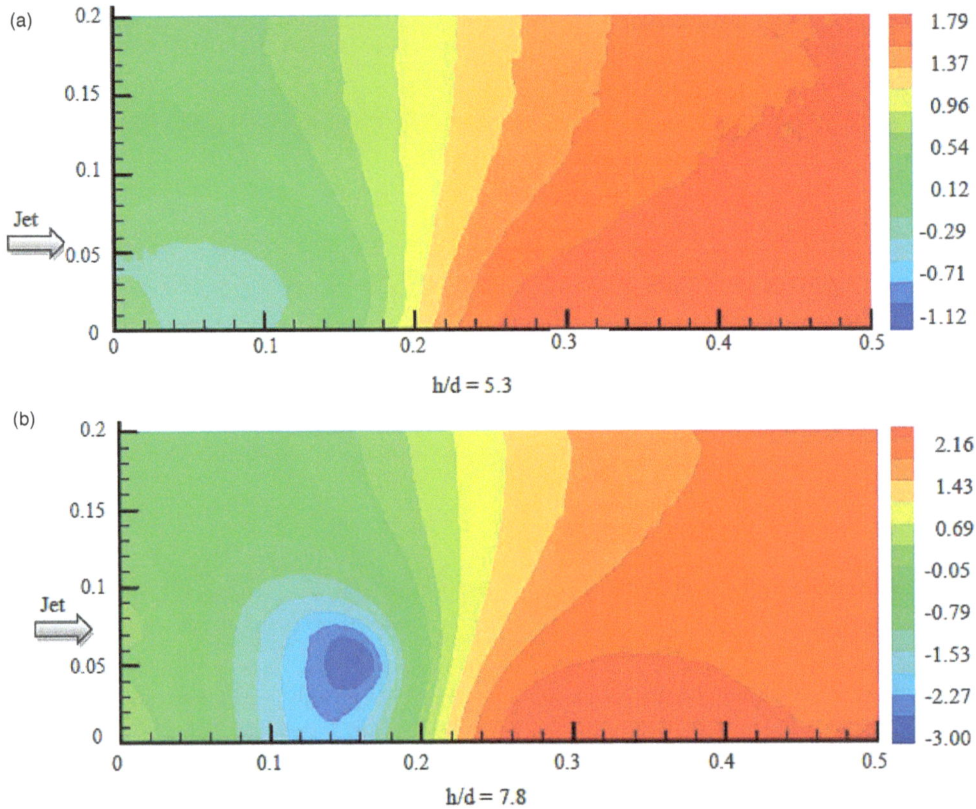

Figure 20. Contours of the static pressure in the symmetry plane for different offset ratios.

entrainment of the air limited by the jet and the offset wall, we note the creation of a sub-atmospheric pressure region below the jet, forcing the jet to curve towards the boundary wall and impact it at the attachment point, which leads to the formation of three regions (recirculation, impingement, and wall jet region). This phenomenon is known as the Coanda effect. The existence of a negative static pressure zone upstream of the attachment point can be seen in the recirculation region, owing to the existence of the vortex structures. In the impingement region, the pressure is higher than the ambient pressure and the static pressure increases rapidly to a maximum value close to the attachment point.

5. Conclusion

The structure of the flow resulting from the interaction of the offset jet and the exterior flow is very complex. This complexity is generated by the manifestation of various diversified vortices: the vortices established under the jet, the eddies caused by the recirculation zone and the vortices of the shear layer formed in the upper interface jet with the external flow. We examined in this paper the dynamic and turbulent characteristics of a flow issuing from a circular offset jet discharged in a co-flow stream or in a quiescent medium. The flow field was studied for different velocity ratios and offset ratios. This study was conducted both experimentally by means of the PIV technique and numerically by means of the finite volume method, together with a turbulent closure model. A close agreement was achieved between the experimental and numerical results, which show that, in the presence of a co-flow stream, the vertical expansion of the offset jet is limited. It was also observed that the value of the velocity ratio affects the size of the recirculation zone, which begins downstream from the nozzle exit, between the vertical and horizontal walls and the lower border of the jet.

Similarly, we have studied the influence of the offset ratio on the flow. The results indicate that increasing the offset ratio gives a better distribution of the jet within the flow field, giving rise to a better dynamic mixture. It was also found that the reattachment length increases linearly with the offset ratio and the impact of the jet with the horizontal wall is located at a position farther from the offset of the jet. However, for a lower offset ratio, the attachment point is at a closer distance from the nozzle, which implies a spatial reduction of the recirculation zone.

Disclosure statement

No potential conflict of interest was reported by the authors.

References

Agelin-Chaab, M., & Tachie, M. F. (2011). Characteristics and structure of turbulent three-dimensional offset jets and wall jets. *International Journal of Heat and Fluid Flow*, *32*(3), 608–620.

Ben Kalifa, R., Habli, S., Mahjoub Saïd, N., Bournot, H., & Le Palec, G. (2014). Numerical and experimental study of a jet in a crossflow for different velocity ratio. *Journal of the Brazilian Society of Mechanical Sciences and Engineering*, *36*, 743–762.

Bourque, C., & Newman, B. G. (1960). Reattachment of a two-dimensional incompressible jet to an adjacent flat plate. *Aero.Quart.*, *11*, 201–232.

Davis, M. R., & Winarto, H. (1980). Jet diffusion from a circular nozzle above a solid plane. *Journal of Fluid Mechanics*, *101*, 201–221.

Demuren, A. O., & Rodi, W. (1987). Three-dimensional numerical calculations of flow and plume spreading past cooling towers. *Journal of Heat Transfer*, *109*, 113–119.

Gao, N., & Ewing, D. (2007). Experimental investigation of planar offset attaching jets with small offset distances. *Experiments in Fluids*, *42*, 941–954.

Hart, D. H. (1998). The elimination of correlation errors in PIV processing. Proc. of the 9th intl symposium on applications of laser techniques to fluid mechanics. Lisbon, Portugal, 8p.

Hasselbrink, E. F. (1999). *Transverse jets and jet flames: structure, scaling and effects of heat release* (Ph. D. Thesis). Stanford University.

Hoch, J., & Jiji, L. M. (1981a). Theoretical and experimental temperature distribution in two dimensional turbulent jet-boundary interaction. *Journal of Heat Transfer*, *103*, 331–335.

Hoch, J., & Jiji, L. M. (1981b). Two-dimensional turbulent offset jet-boundary interaction. *Journal of Fluids Engineering*, *103*, 154–161.

Holland, J. T., & Liburdy, J. A. (1990). Measurements of the thermal characteristics of heated offset jet. *International Journal of Heat and Mass Transfer*, *33*, 69–78.

Kanna, P. R., & Das, M. K. (2005). Conjugate heat transfer study of two dimensional laminar incompressible offset jet flows. *Numerical Heat Transfer, Part A: Applications*, *48*(7), 671–691.

Kanna, P. R., & Das, M. K. (2008). Heat transfer study of two-dimensional laminar incompressible offset jet flows. *International Journal of Thermal Sciences*, *47*, 1620–1629.

Knapp, Y., Bertrand, E., & Mouret, F. (2003). *2D-PIV measurements of the pulsatile flow in a left heart simulator*. Proceedings of PSFVIP-4, June 3–5, Chamonix, France.

Kumar Raja, K., Manab, K. D., & Rajesh, K. P. (2009). Numerical study of mixed convection in a two dimensional laminar incompressible offset jet flow. *International Journal of Heat and Mass Transfer*, *52*, 1023–1035.

Launder, B. E., & Sharma, B. I. (1974). Application of the energy dissipation model of turbulence to the calculation of flow near a spinning disc. *International Communications in Heat and Mass Transfer*, *1*, 131–137.

Levin, S. G., & Manion, F. M. (1962). Jet attachment distance as a function of adjacent wall offset and angle. Technical report, Harry Diamond Laboratories, Army Materiel Command, Washington DC.

Lien, F. S., & Leschziner, M. A. (1994). Assessment of turbulent transport models including non-linear RNG eddy viscosity formulation and second-moment closure. *Computers and Fluids*, *23*, 983–1004.

Li, Z.-W., Huai, W.-X., & Han, J. (2011). Large eddy simulation of the interaction between wall jet and offset jet. *Journal of Hydrodynamics*, *23*(5), 544–553.

Lund, T. S. (1986). Augmented thrust and mass flow associated with two-dimensional jet reattachment. *AIAA Journal*, *24*, 1964–1970.

Mahjoub Saïd, N., Habli, S., Bournot, H., & Le Palec, G. (2012). The near field in the mixing of a three-dimensional inclined pollutant jet with a cross flow. *Journal of Enhanced Heat Transfer*, *19*(2), 161–178.

Mahjoub Saïd, N., Mhiri, H., Bournot, H., & Le Palec, G. (2008). Experimental and numerical modelling of the three-dimensional incompressible flow behaviour in the near wake of circular cylinders. *Journal of Wind Engineering & Industrial Aerodynamics*, *96*(5), 471–502.

Mahjoub Saïd, N., Mhiri, H., El Golli, S., Le Palec, G., & Bournot, P. (2003). Three-dimensional numerical calculations of a jet in an external crossflow: Application of dispersion of pollutants. *Journal of Heat Transfer*, *125*, 510–522.

Mahjoub Saïd, N., Mhiri, H., Le Palec, G., & Bournot, P. (2005). Experimental and numerical analysis of pollutant dispersion from a chimney. *Atmospheric Environment*, *39*, 1727–1738.

Nasr, A., & Lai, J. (1998). A turbulent plane offset jet with small offset ratio. *Experiments in Fluids*, *24*, 47–57.

Nozaki, T. (1983). Reattachment flow issuing from a finite width nozzle (Report 4. Effects of Aspect Ratio of the Nozzle). *Bull JSME*, *26*(221), 1884–1890.

Nozaki, T., Hatta, K., Nakashima, M., & Matsumura, H. (1979). Reattachment flow issuing from a finite width nozzle. *Bull JSME*, *22*(165), 340–347.

Nozaki, T., Hatta, K., Nakashima, M., & Matsumura, H. (1981). Reattachment flow issuing from a finite width nozzle, (Report 2. Effects of Initial Turbulence Intensity). *Bull JSME*, *24*(188), 363–369.

Patankar, S. V., & Spalding, D. B. (1972). A calculation procedure for heat, mass and momentum transfer in three dimensional parabolic flows. *International Journal of Heat and Mass Transfer*, *15*, 1787–1806.

Pelfrey, J. R. R., & Liburdy, J. A. (1986a). Effect of curvature on the turbulence of a two-dimensional jet. *Exp Fluids*, *4*, 143–149.

Pelfrey, J. R. R., & Liburdy, J. A. (1986b). Mean flow characteristics of a turbulent offset jet. *Journal of Fluids Engineering*, *108*, 82–88.

Radhouane, A., Mahjoub Saïd, N., Mhiri, H., Le Palec, G., & Bournot, P. (2009). Impact of the initial streamwiseinclination of a double jet emitted within a cool crossflow on its temperature field and pollutants dispersion. *Heat and Mass Transfer*, *45*(6), 805–823.

Rajaratnam, N., & Subramanya, K. (1968). Plane turbulent reattached wall jets. *J. Hydr. Div.*, *94*, 95–112.

Rathore, S. K., & Das, M. K. (2013). Comparison of two low-Reynolds number turbulence models for fluid flow study of wall bounded jets. *International Journal of Heat and Mass Transfer*, *61*, 365–380.

Roland, S. (1993). *Modélisation Et Simulation Des Ecoulements Turbulents*. Paris: Hermes.

Sawyer, R. A. (1960). The flow due to a two dimensional jet issuing parallel to a flat plate. *Journal of Fluid Mechanics*, *9*, 543–561.

Sawyer, R. A. (1963). Two dimensional reattaching jet flows including the effect of curvature on entrainment. *Journal of Fluid Mechanics*, *17*, 481–498.

Song, H. B., Soon, H. Y., & Dae, H. L. (2000). Flow and heat transfer characteristics of a two-dimensional oblique wall attaching offset jet. *International Journal of Heat and Mass Transfer*, *43*, 2395–2404.

Vishnuvardhanarao, E., & Das, M. K. (2008a). Computation of mean flow and thermal characteristics of incompressible turbulent offset jet flows. *Numer Heat Transf Part A*, *53*(8), 843–869.

Vishnuvardhanarao, E., & Das, M. K. (2008b). Study of conjugate heat transfer from a flat plate by turbulent offset jet flow. *Numerical Heat Transfer, Part A: Applications*, *53*(5), 524–542.

Vishnuvardhanarao, E., & Das, M. K. (2009). Study of the heat transfer characteristics in turbulent combined wall and offset jet flows. *International Journal of Thermal Sciences*, *48*, 1949–1959.

Yang, Y. T., & Yeh, Y. Y. (1994). Numerical study of turbulent offset jets with entrainment boundary. *Journal of Thermophysics and Heat Transfer*, *8*(1), 182–184.

Yang, Z., & Shih, T. H. (1993). New time scale based k-ϵ model for near-wall turbulence. *AIAA Journal*, *31*(7), 1191–1198.

10

Effects of residual riblets of impeller's hub surface on aerodynamic performance of centrifugal compressors

Dong Lei* and Cao Lixin

School of Mechanical Engineering, Dalian University of Technology, Dalian 116023, China

In this article, the effects of residual riblets of impeller's hub surface on the aerodynamic performance of a compressor are studied in detail. Several accurate geometrical models of impellers after their hub surface are machined, with various tool passes are built. The number of tool passes for the hub surface finish of a single passage of the impeller is denoted by N. Meanwhile, a dimensionless parameter B_{max} is introduced to characterize the size of the ribs of the arbitrary impeller, which is the percentage ratio rate of the maximum cusp height of the hub surface to the exit blade width. The geometrical models are numerically studied by solving the Reynolds averaged Navier–Stokes equations using the SST turbulent model. The predictions show that a group of secondary cross-stream vortices are generated by the interaction of streamwise vortices above the hub surface with riblets' tips. For the impeller researched in this article, with the decline of B_{max}, the average viscous shear stress on the hub surface decreases gradually at first and then starts increasing and leveling off; accordingly, the polytropic efficiency of the compressor, on the contrary, increases gradually and then starts leveling off later. When B_{max} comes to 0.58, the average viscous shear stress of the fluids on the hub surface reaches its minimum, and it is found that the polytropic efficiency of the compressor reaches a peak and stable value. The simulation results are validated against experimental data at the design Mu, and the maximum discrepancies are less than 3.8%.

Keywords: residual riblets; numerical study; average viscous shear stress; polytropic efficiency; experiment

1. Introduction

The centrifugal compressor has been widely used in several areas, such as metallurgy, chemical and petroleum, natural-gas transporting, and refrigeration. It has features of simple structure, high reliability, efficiency, pressure ratio of a single stage, and a wide range of working conditions. The impeller is one of the most important components of the centrifugal compressor. The machining precision of the impeller has a great impact on the compressor's aerodynamic performance; thus, the designing and manufacturing of impellers plays a key role in centrifugal compressors.

The aerodynamic performance of the compressor mainly relies on the flowing status of transported fluids. In general, the flow in the centrifugal compressor is turbulent and more complicated near solid walls. Several factors such as fluid viscosity, appearance, and surface roughness of solid walls affect the fluids' flowing status. The turbulent viscosity is the most important factor. Previous research had been proposed by Crowe, Elger, and Roberson (2005), Shereena, Vengadesan, Idichandy, and Bhattacharyya (2005). The turbulent viscosity generated not only the skin friction drag but also the pressure drag. Since the boundary layer separation is unlikely to occur at a design point for the centrifugal impeller, skin friction drag is more pronounced than pressure drag. Therefore, it is more important to reduce skin friction drag in the turbulent boundary layer in the research of wall drag reduction for centrifugal impellers.

During the last decades, fine surface finish has been proved not to always get more drag reduction effectiveness, and riblets had shown potential for the reduction of skin friction drag. From the research presented by Oehlert and Seume (2006), tiny riblets aligned in the streamwise direction had been proved to be able to reduce skin friction and wall shear stress in turbulent flows compared with smooth surfaces. Chu and Karniadkis (1993) numerically simulated a channel flow over riblets using a spectral element Fourier method and obtained approximately 6% drag reduction for the riblet-mounted wall compared with the smooth wall. For the principle of riblets' surface drag reduction, two important mechanisms were proposed in the literature. Bechert and Bartenwerfer (1989) introduced a so-called "protrusion height" mechanism for different types of riblets and related it to the ability of riblets to impede the cross-flow. Bacher and Smith (1985) put forward the generating mechanism of secondary cross-stream vortex, which suggested that the effect of the riblets is to modify and effectively reduce the momentum exchange properties of the streamwise vortices due to a consequent reduction in surface shear stress. The mechanism

*Corresponding author. Email: donglei2003599@126.com

of secondary cross-stream vortex had been validated by Lee (2001) and Klumpp, Meinke, and Schröoder (2010). Through an extended analysis of the two mechanisms, it has been found that the appearance and size of riblets is the key to determining the drag reduction effectiveness from various studies, such as Walsh and Lindmann (1984), Walsh (1982), Walsh (1983), Lazos and Wilkinson (1988), El-Samni, Yoon, and Chun (2005), Choi, Moin, and Kim (1993), Qamar and Sanghi (2012), and Boese and Fottner (2002). Drag reduction effectiveness will be improved if the riblet structure is immersed into the lower part of the viscous sublayer and transition layer; otherwise, it will be weakened if the sharp corner of the riblet is completely beyond the turbulent boundary layer. However, most research on riblets' surface drag reduction is focused on plates. If the riblets' surface drag reduction is applied on complex surfaces, it will be very difficult to solve the problem.

For semi-open three-dimensional impellers, solid walls are composed of hub surfaces and blade surfaces. The blade surfaces of most three-dimensional impellers are ruled surfaces and can be machined by the flank milling method using a cylindrical cutter or a conical cutter. This method can obtain high machining precision and low surface roughness from Gong, Cao, and Liu (2005). As mentioned earlier, the fine surface finish would not always produce higher drag reduction effectiveness. Klumpp, Meinke, Schröoder, Feldhaus, and Klocke (2009), Büttner and Schulz (2010) developed the rolling technique and electrodeposition technique, respectively, to manufacture riblets on turbine blades; total drag reduction of 4.7% and 4.9% were obtained, respectively. Furthermore, Lietmeyer, Oehlert, and Seume (2013) had clarified the best choices where riblets should be applied on the blade surface. The results showed a higher potential of the profile loss reduction by riblets on the suction side. Another conclusion of this research is that more drag reduction effectiveness would be got to adopt segmented riblets compared with a constant riblet geometry along the compressor blade surfaces.

The hub surfaces of the impeller are usually machined by ball-end mill or filleted end-mill and the surface roughness is high, whereas some tiny riblets are left on the hub surfaces as shown in Figure 1. Currently, there are no related works on this kind of tiny riblets, which would affect the boundary flow and the aerodynamic performance of compressors.

As shown in Figure 2, Kang and Hirsch (2001) conducted an in-depth study on a number of centrifugal impellers and observed different vortices in the passage of the centrifugal impeller. Among them, PVH and CV are streamwise vortices near the hub surface. These streamwise vortices are able to concentrate low-speed fluid in the spanwise sense and trigger an outward eruption of low-momentum fluid that has unquestionably adverse effects on the flow in the centrifugal impeller. However, for impellers

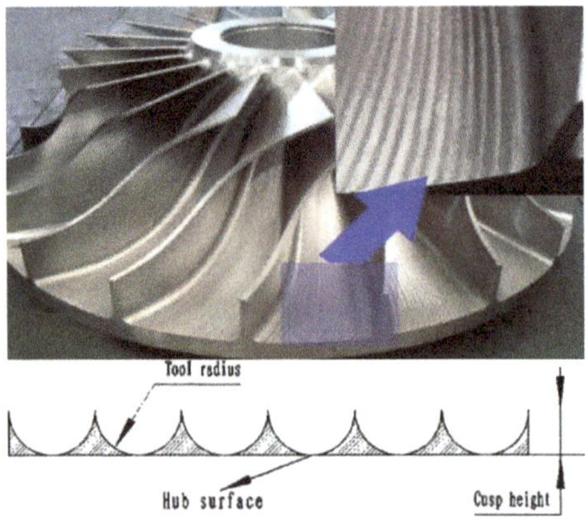

Figure 1. Residual riblets of impeller hub surface after finish.

with riblets on their hub surface, the results may be different. According to the mechanism that the secondary cross-stream vortex is generated by the interaction of the riblet tip and the flow above, it can be speculated that a group of secondary cross-stream vortices will be generated at the peaks of the riblets, which can inhibit the spanwise concentration of low-speed fluid into streak formations, weaken PVH and CV to a certain degree, reduce the wall shear stress of turbulent boundary layer flow, as well as restrain the development of the boundary layer so as to realize the drag reduction effectiveness. One of the key points that is used to determine the drag reduction effectiveness is the sizes of riblets. Drag reduction effectiveness will be weakened with a large cusp height left after hub surface finish; moreover, the aerodynamic performance of the compressor will deteriorate due to a larger friction area, so B_{max} should not be too large. On the other hand, drag reduction effectiveness will also be weakened even if the machining precision and surface roughness were improved; moreover, the processing costs will also increase correspondingly. In order to obtain the optimal drag reduction effectiveness of the residual riblets on the hub surface and avoid the increase in processing costs from the unnecessary decrease in the B_{max} value, drag reduction effectiveness of different cusp height and its effect on the aerodynamic performance of a compressor are investigated in this article, which aims at obtaining the optimal B_{max} value.

There are several difficulties in our research. First, due to the complexity of the impeller's hub surface, it is difficult to distribute the geometric modeling and mesh generation for riblets. To accurately predict the flow condition of fluids near the residual riblets, a finer grid scheme in the boundary layer should be used. Furthermore, since the size and density of the riblets of different positions are variable, these bring big troubles to flow field analysis. Meshes are divided by manual operation in this article. Every rib element is set as an independent block to guarantee the density and quality of the grid. The flow field

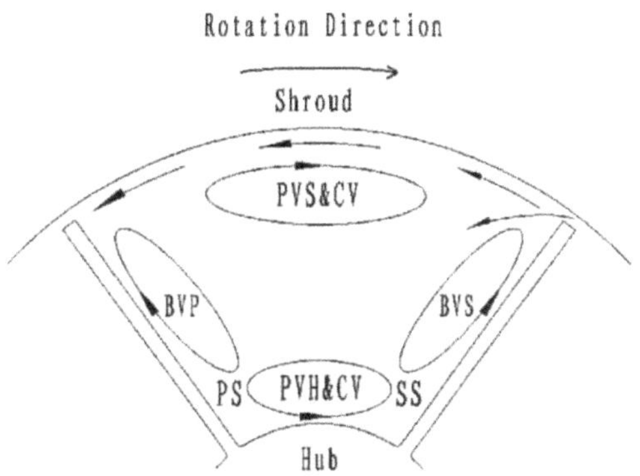

Figure 2. Sketch of secondary vortices in centrifugal impeller (Extracted from Kang and Hirsch (2001)).

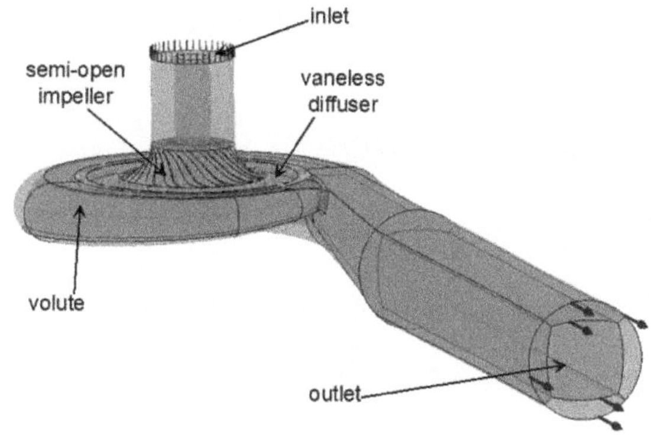

Figure 3. Configuration sketch of centrifugal compressor.

Table 1. Flow conditions.

Mu	φ	n (r/min)	T_{inlet} (K)	P_{inlet} (atm)	Q_m (Kg/s)
0.85	0.09	23,408	298	1	1.4200
0.95	0.09	26,176	298	1	1.5649
1.05	0.09	28,930	298	1	1.7301
1.15	0.09	31,687	298	1	1.8948

analysis primarily focuses on the analysis of wall shear stress; in-depth analyses are conducted on the wall shear stress of the rib surface from the whole to the part. The content of this article is managed as follows:

In this article, the 3D steady viscous flow field of a small centrifugal impeller is numerically studied. Impellers with different values of B_{max} after the hub surface finish are modeled and meshed in §2.2 and §2.3, followed in §3.1 by a grid independence test. After the single-flow passage numerical simulation presented in §3.2, the generation of secondary cross-stream vortices is confirmed by an analysis of velocity vector distribution over the cross-section at 90% of the flow passage from inlet to outlet in §3.3.1; meanwhile, the hub surface wall shear stress is analyzed from the whole to the part in §3.3.2. By comparing the polytropic efficiency of compressors that apply different values of B_{max} after hub surface finish, drag reduction characteristics of residual riblets on the hub surface are also reported in §3.3.2. A performance experiment is carried out and described in §4; the result is well matched with simulation results, followed by a brief conclusion in section 5.

2. Numerical simulations

2.1. *Research object*

As shown in Figure 3, a small-sized centrifugal compressor is selected as the research object, which consists of a semi-open impeller, a vaneless diffuser, and a volute as its major components. The primary structure dimensions and flow conditions are summed up as follows.

Geometric parameters of the impeller: the diameter of the impeller is 240 mm, and the number of blades is 16. The exit blade width is 15.96 mm, and average blade tip clearance is 0.5 mm. The flow conditions of the design flow coefficient points for four Mu are listed in Table 1.

2.2. *Generation of computational domain*

Generally, the hub surface of the impeller is machined by ball-end milling cutter and the circular arc riblets will be left on the hub surface, the distribution of which will be different based on different methods of tool-path planning. In this article, the method of iso-parametric tool-path planning is applied to the hub surface finish. Finally, the residual riblets on the hub surface are the enveloping surface that is formed by the motion of the ball-end milling cutter along the iso-parametric tool path. In this study, a ball-end milling cutter whose radius is 3 mm and half cone angle is 3 degrees is applied to the hub surface finish. Five computational domains where N is 7,13,17,23 and infinite are modeled, respectively, and the maximum cusp height of the hub surface is 0.75 mm, 0.167 mm, 0.093 mm, 0.0489 mm, and 0 mm, respectively. Obviously, the optimization in terms of the number of riblets (N) cannot be generalized, as it depends on the impeller size. Thus, a non-dimensional parameter B_{max} is introduced, which is the percentage ratio rate of the maximum cusp height of the hub surface to the exit blade width. For the impeller researched in this article, when N is 7,13,17,23 and infinite, B_{max} is equal to 4.70, 1.06, 0.58, 0.31, and 0, respectively. The computational domain where B_{max} is equal to 1.06 is shown in Figure 4.

2.3. *Numerical research method*

The steady numerical simulations mentioned earlier are performed for the computational models of the centrifugal impeller using the commercial software CFX. Its computational method is based on the finite volume method. The Shear Stress Transport (SST) turbulence model (1994) is employed, which is a two-equation eddy-viscosity model. The use of a k-ω formulation in the inner parts of the boundary layer makes the model directly usable all the

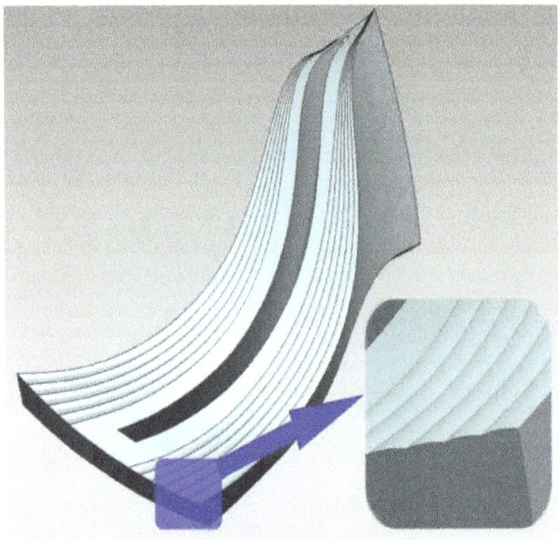

Figure 4. Computational domain where B_{max} is equal to 1.06.

way down to the wall through the viscous sub-layer. The SST formulation also switches to a k-ε behavior in the free stream, thereby avoiding the common k-ω problem that the model is too sensitive to the inlet free stream turbulence properties. So, it can accurately predict the flow condition of fluids near the residual riblets on the hub surface and obtain precise numerical solutions of separated flows by solving the transport problems of turbulent shear stress. The transport equations for the SST model are listed next:

$$\frac{\partial}{\partial t}(\rho k) + \frac{\partial}{\partial x_i}(\rho k u_i) = \frac{\partial}{\partial x_j}\left(\Gamma_k \frac{\partial k}{\partial x_j}\right) + \tilde{G}_k - Y_k + S_k \quad (1)$$

$$\frac{\partial}{\partial t}(\rho \omega) + \frac{\partial}{\partial x_i}(\rho \omega u_i) = \frac{\partial}{\partial x_j}\left(\Gamma_\omega \frac{\partial \omega}{\partial x_j}\right) + G_\omega - Y_\omega + D_\omega + S_\omega, \quad (2)$$

where $\tilde{G}_k$ represents the generation of turbulence kinetic energy due to mean velocity gradients; G_ω represents the generation of ω; Γ_k and Γ_ω represent the effective diffusivity of k and ω, respectively; Y_k and Y_ω represent the dissipation of k and ω due to turbulence; $D_\omega$represents the cross-diffusion term; and S_k S_ω are user-defined source terms. Furthermore, the high resolution scheme put forward by Barth and Jesperson (1989) is applied to discretize the turbulence equation in order to obtain the flow information of a structured grid within the boundary layer more accurately. The impeller is solved on the rotating coordinate system, and all stators are solved on the absolute reference frame. The type of interfaces between stators and rotors is set to be frozen rotor.

The all-hexahedral meshes for the computational domain of the impeller, vaneless diffuser, and volute are created using commercial software ICEM. Meshes near all solid walls are refined geometrically in the normal direction of the solid walls and y+ of all the solid walls are maintained below 2. A meshed computational domain of single-flow passage of the impeller is shown in Figure 5, where B_{max} is equal to 1.06.

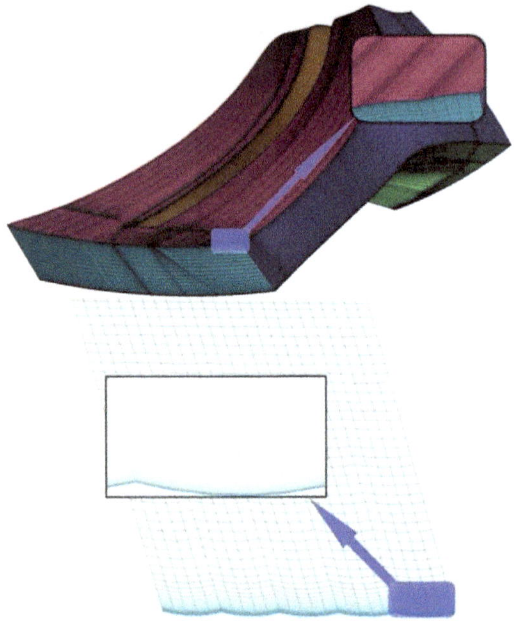

Figure 5. A meshed computational domain where B_{max} is equal to 1.06.

Boundary conditions are treated as follows: Total temperature and total pressure are specified at the inlet, and mass flow rate is specified at the outlet; the adiabatic and non-slip condition is applied at all the solid walls; the blades and hub surfaces are set as rotating surfaces; and the other solid walls are set as stationary surfaces.

3. Results and analysis

3.1. Grid independent test

In order to obtain the grid independent results, a grid sensitivity study should be conducted for all models. Mesh1, 2, and 3 are, respectively, divided by manual operation for the five computational domains, which are mentioned earlier. For mesh1 to mesh3, the grid is refined by scaling up the number of nodes on each edge, so the grid quantity of mesh1, 2, and 3 increased successively, which is shown in Figure 6. Numerical simulations of a single-flow passage are performed for all computational models at the design flow coefficient point of Mu = 0.95; the volute is not included in the numerical simulations, and the outlet boundary of the computational domain is on the exit plane of the vaneless diffuser. The predictions are shown in Figure 7. Three points in each line are results of mesh1,

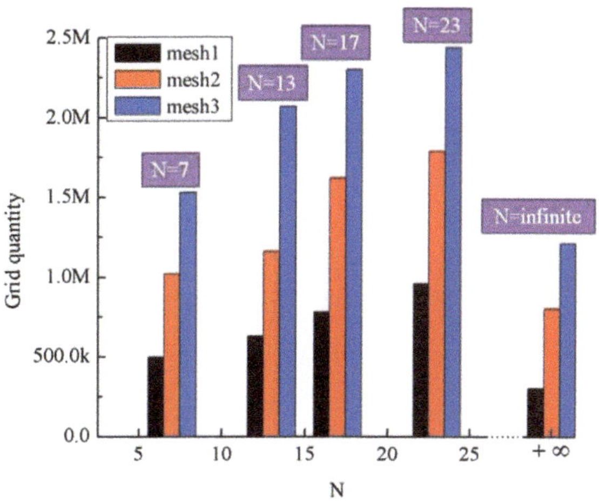

Figure 6. Grid quantity of all models in grid independence test.

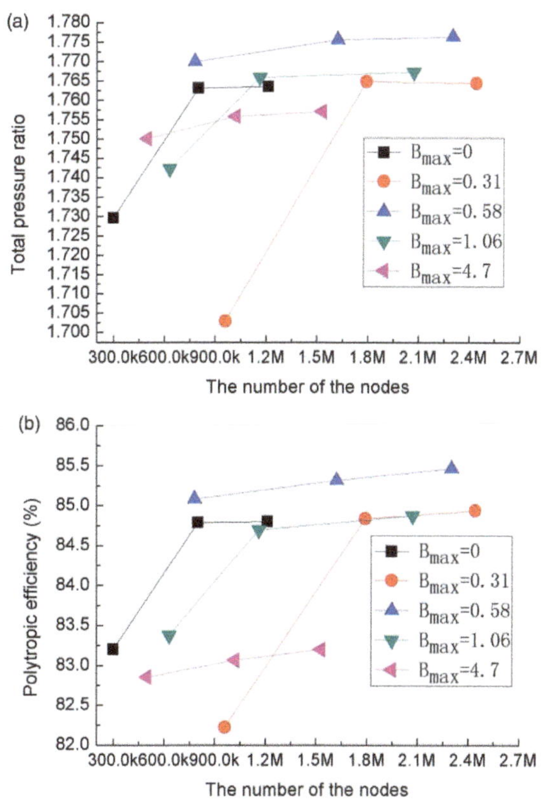

Figure 7. Results of grid independence test.

mesh2, and mesh3, respectively. It is obvious that the predicted results tend to be stable along with the gradually refined mesh; error for inlet-to-outlet total pressure ratio and compressor polytropic efficiency between mesh2 and mesh3 are less than 0.21%. In order to confirm whether mesh2 of all models can accurately predict the flow conditions of fluids near the residual riblets on the hub surface, the quantity of grid nodes in the residual riblets of all models' mesh2 is further scaled up, and numerical simulations are performed again using methods mentioned earlier for all computational models. Error for inlet-to-outlet total pressure ratio and compressor polytropic efficiency are less than 0.06%, based on which the credibility of all models' mesh2 is confirmed. The results imply that more accurate results can be obtained to perform the numerical simulation by using mesh2 from each model, and the quantity of grid nodes is 1.02 million, 1.16 million, 1.62 million, 1.79 million, and 0.8 million, respectively, when N is 7,13,17,23 and 0. Furthermore, as shown in Figure 6, the grid quantity of mesh2 from all models increases with an increase in N value. Thus, it can be seen that it will need a larger grid quantity to obtain more accurate results with the increase of the density of riblets aligned on the hub surface except that N is infinite.

3.2. *Simulation results of a single-flow passage*

Comparing the aerodynamic performance of impellers with different sizes of riblets aligned on the hub surface, the numerical simulations of a single-flow passage for five computational domains are performed at the design flow coefficient points of four Mu using mesh2 from each model, the predicted results are depicted in Figures 8–10, which show that the inlet-to-outlet total pressure ratio and polytropic efficiency of the compressor are the lowest when B_{max} is equal to 4.7. With the decline of the B_{max} value, the two parameters gradually increase and reach a maximum when B_{max} is equal to 0.31. However, the discrepancy of the two parameters is very small when B_{max} is equal to 0.31, 0.58, and 0, based on which it is considered that the two parameters reach a peak and stable value when B_{max} is reduced to 0.58. It can be seen from Figure 10 that the average shear stress on the hub surface reaches a maximum when B_{max} is equal to 4.7, then gradually decreases with

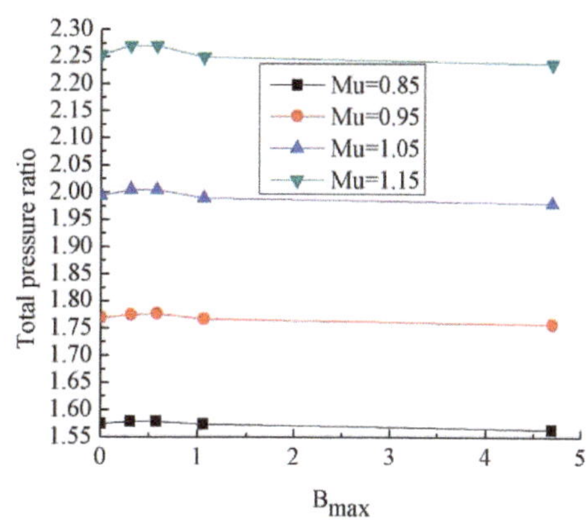

Figure 8. Comparative results of total pressure ratio.

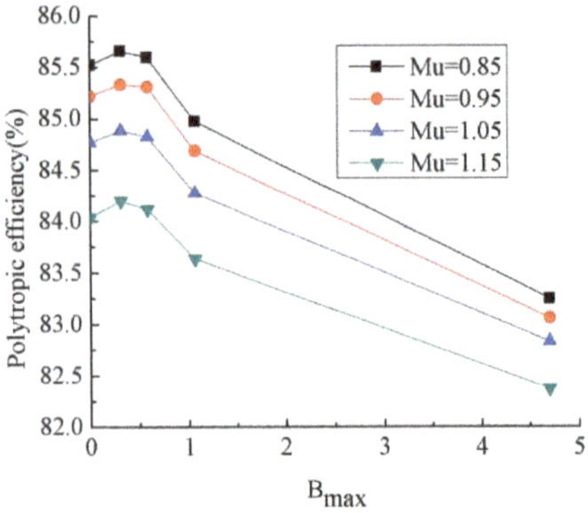

Figure 9. Comparative results of polytropic efficiency.

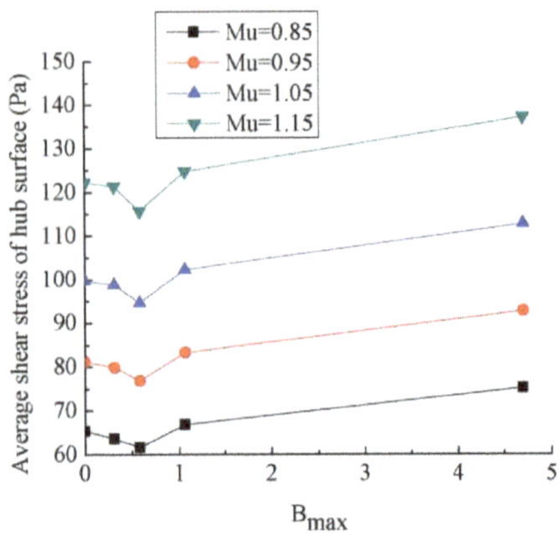

Figure 10. Comparative results of average shear stress on hub surface.

the decline of the B_{max} value, and reaches a minimum when B_{max} is equal to 0.58, which means that the flow resistance is the smallest at this moment. After this minimum, it starts increasing again with the decline of the B_{max} value. When B_{max} comes to 0, the average shear stress and the aerodynamic performance of the compressor are very close to the ones when B_{max} is equal to 0.31. So, it can be considered that the aerodynamic performance of the compressor reaches a peak and stable value when B_{max} is reduced to 0.58, and it is the optimum B_{max} value. The varying tendencies of the three parameters are the same for the four Mu mentioned earlier, which implies that the effect of residual riblets of the impeller's hub surface on the aerodynamic performance of the compressor is closely related to the size of the riblets; constantly, there should be an optimum value for the cusp height, which corresponds to an optimum B_{max} value and N value. Under this number of tool passes, the average shear stress on the hub surface reaches a minimum and the drag reduction effectiveness is the best; the total pressure ratio and polytropic efficiency of the compressor also reach a peak and stable value.

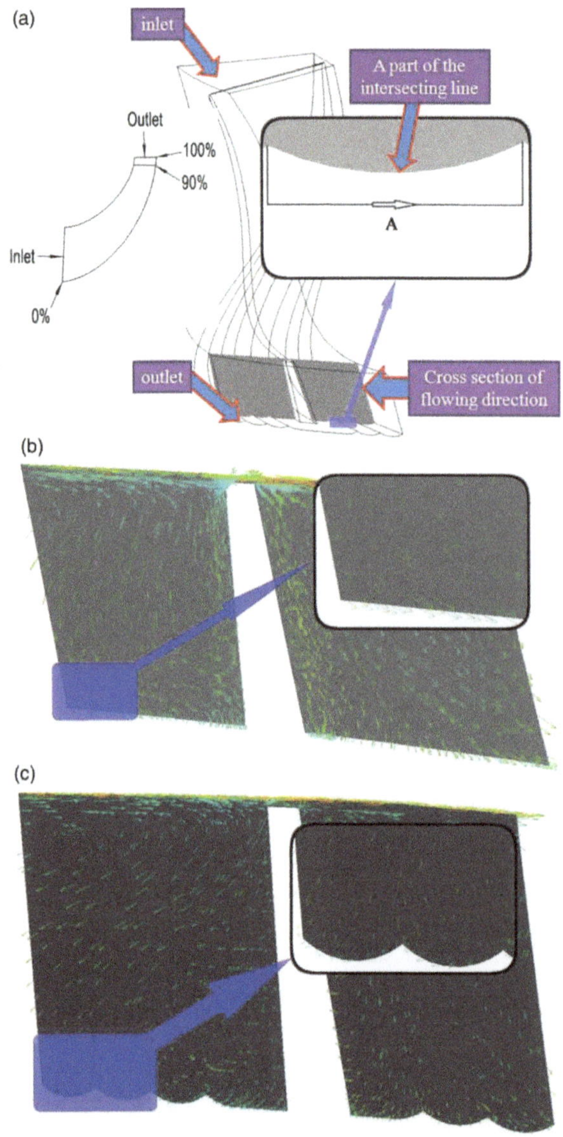

Figure 11. Comparison diagram of velocity vector.

3.3. *Flow field analysis*

3.3.1. *Flow structure analysis*

To compare and analyze the near-wall flow structures of the turbulent boundary layer over the residual riblets' surface, a cross-section of the flow passage at 90% of the flow passage from the inlet to the outlet is taken as shown in Figure 11(a), and velocity vector distribution over the cross-section of the flow passage of smooth hub surface and riblets' hub surface is indicated in Figure 11(b) and Figure 11(c), respectively. Based on Figure 11(c), an interaction between streamwise vortices and the riblets' structure enable flow direction of the fluid to deflect at

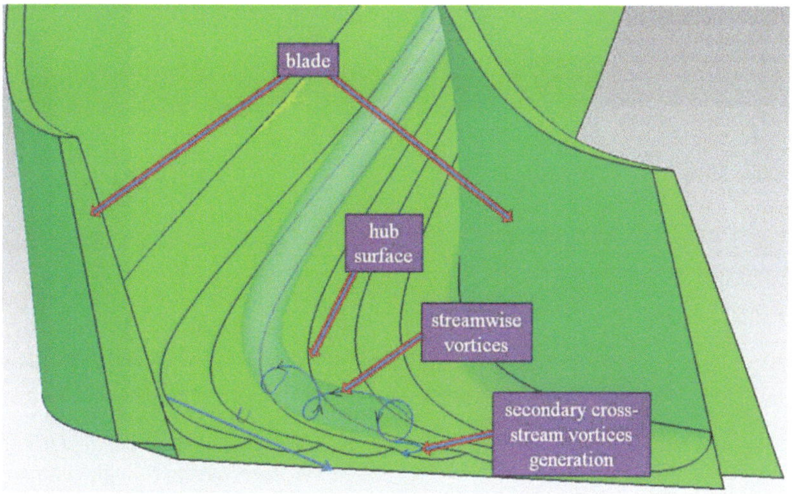

Figure 12. Perspective schematic showing the interaction of streamwise vortices with riblets.

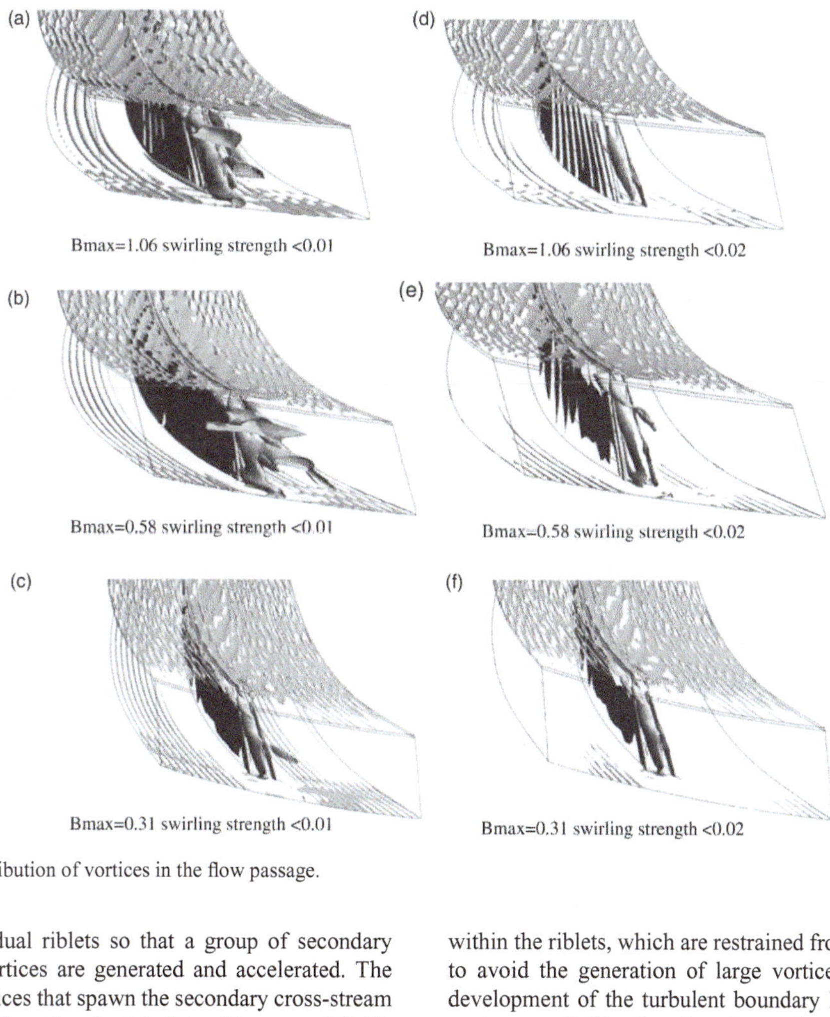

Figure 13. Distribution of vortices in the flow passage.

the tips of residual riblets so that a group of secondary cross-stream vortices are generated and accelerated. The streamwise vortices that spawn the secondary cross-stream vortices are weakened and retain lots of low-speed fluids within the riblets, which are restrained from rising in order to avoid the generation of large vortices and retard the development of the turbulent boundary layer. This is the root cause of riblets' surface drag reduction. A schematic

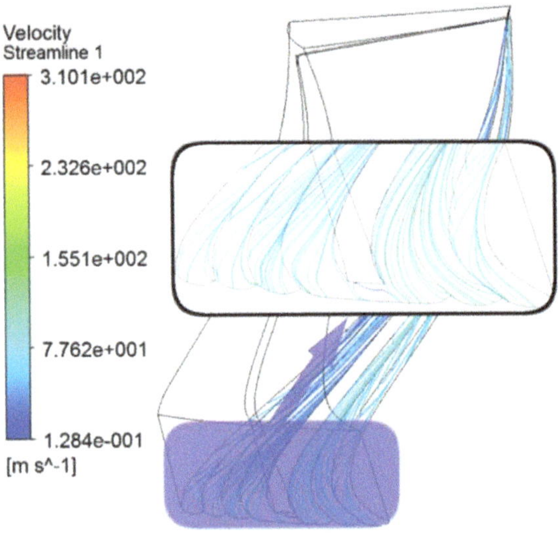

Figure 14. Limiting streamline diagram of hub surface.

drawing illustrating the interaction of riblets' structure with streamwise vortices is shown in Figure 12; where U is the flow direction of the fluid.

In order to compare the influence of different sizes of riblets on the strength and distribution of vortices above the hub surface, the distribution of vortices whose swirling strength is below 0.01 and 0.02 in the flow passage when B_{max} is equal to 1.06, 0.58, and 0.31 is shown in Figure 13(a)–Figure 13(f). The value of swirling strength represents the strength of swirling flow. The distribution of vortices whose swirling strength is below 0.01 when B_{max} is equal to 1.06, 0.58, and 0.31 is shown in Figures 13(a), (b), and (c), respectively, and the distribution of vortices whose swirling strength is below 0.02 when B_{max} is equal to 1.06, 0.58, and 0.31 is shown in Figure 13(d), (e), and (f), respectively, based on which it is found that a group of secondary cross-stream vortices are generated at the tips of the riblets, and with an increase in the depth of the cusp height, more secondary cross-stream vortices are distributed on the tips of the riblets. From the limiting streamline, which is shown in Figure 14, it is obvious that velocity of the fluids will be deflected due to an interaction between the fluids and riblets' surface, and secondary cross-stream vortices will be generated.

Furthermore, the influence of the riblets on the jet-wake structure of the impeller is analyzed. The "jet-wake" structure which is mainly caused by the secondary vortices is an accumulation of comparatively low-energy fluids at the corner of the casing and blade suction surface. The low-energy fluids are transported to the big vortices zone by the secondary vortices, then the "jet-wake" structure is formed. In this article, the low-energy fluids near the hub surface will be transported to the main flow region by the secondary vortices (PVH and CV) which is shown in Figure 2, then the "jet-wake" structure will be intensified. However, for the hub surface with residual riblets, PVH and CV above residual riblets on the hub surface interact with the tips of the riblets, which makes the velocity deflexion of fluids happen in the rotating direction of the impeller. The secondary cross-stream vortex forms above the sharp corner of the riblet. All secondary cross-stream vortices above the riblets constitute the structure of the secondary cross-stream vortices group, which cause lots of low-energy fluids to be retained inside the riblets. With the weakening of PVH and CV and the inhibition of the secondary cross-stream vortices group to the low-energy fluids, rising of low-energy fluids to intensify the "jet-wake" structure is weakened to a certain extent. Therefore theoretically speaking, the "jet-wake" structure could be improved to a certain extent by the residual riblets on the hub surface. It will be certain that the "jet-wake" structure have the characteristic of high-entropy. Cross section a, b and c are taken at the same position when Bmax = 0 and Bmax = 0.58, respectively, as shown in Figure 15(a). The contour of static entropy of the cross section a, b and c when Bmax = 0 is shown in Figure 15(b), Figure 15(d), Figure 15(f), respectively, and the contour of static entropy of the cross section a, b and c when Bmax = 0.58 is shown in Figure 15(c), Figure 15(e), Figure 15(g), respectively, from which it could be found that the high-entropy region was obviously weakened when Bmax = 0.58 compared with that when Bmax = 0. As can be seen, the "jet-wake" structure could be improved to a certain extent by the residual riblets on the hub surface.

3.3.2. *Analysis of wall viscous shear stress*

Figures 16 and 17 show the shear stress contour on the impellers' hub surface, where B_{max} is equal to 0 and 0.58, respectively.

The internal flow field of the centrifugal impeller near the hub surface is more complicated, compared with plate flow. Shear stress on the smooth hub surface and hub surface aligned with residual riblets is not uniformly distributed as shown in Figure 16 and Figure 17. It seems that they are similar as a whole, shear stress on the hub surface gradually increases from the inlet to the outlet of the impeller. Shear stress on the smooth hub surface is more evenly distributed, analyzed from the local region; however, when it comes to shear stress on the hub surface with residual riblets, some large differences between the tips and valleys of riblets will emerge. The inlet of flow passage is narrower than the outlet, which results in the gradual increase of cusp height from the inlet to the outlet. The height of riblets is lower at the upper and middle part of the hub surface, where the wall shear stress is smaller as well compared with that at the middle and lower part, so that fluids are less likely affected by the residual riblets and shear stress is more evenly distributed in this region. However, the distribution of wall shear stress is obviously affected by the residual riblets and differs a lot between the tips and valleys of riblets at the middle and lower part of the hub surface. In order to further analyze the drag reduction effectiveness of the residual riblets, local wall shear stress distribution in the same place of the hub surface when B_{max} is equal to 4.7, 0.58, 0.31, and 0 is comparatively analyzed. A part of the intersecting line of the cross-section with the hub surface between cartesian coordinates of (− 0.007, 0.12, $Z1_{Bmax}$) and (− 0.013, 0.12, $Z2_{Bmax}$) is taken, which is shown in Figure 11(a), where $Z1_{Bmax}$ and $Z2_{Bmax}$ are the coordinates at the axis direction of the impeller. Due to different sizes of residual riblets, $Z1_{4.7}$, $Z1_{0.58}$, $Z1_{0.31}$, and $Z1_0$ are different, and $Z2_{4.7}$, $Z2_{0.58}$, $Z2_{0.31}$, and $Z2_0$ are also different. Then, points are taken uniformly from the part of the intersecting line. Horizontal ordinate 'A' represents the position of a point on the intersecting line that is also shown in Figure 11(a). The shear stress of every point is taken to analyze the distribution rule of the shear stress on the hub surface with different B_{max} values, and the results are shown in Figure 18.

For the smooth hub surface, the shear stress distribution is more uniform, but the nonuniform distribution for shear

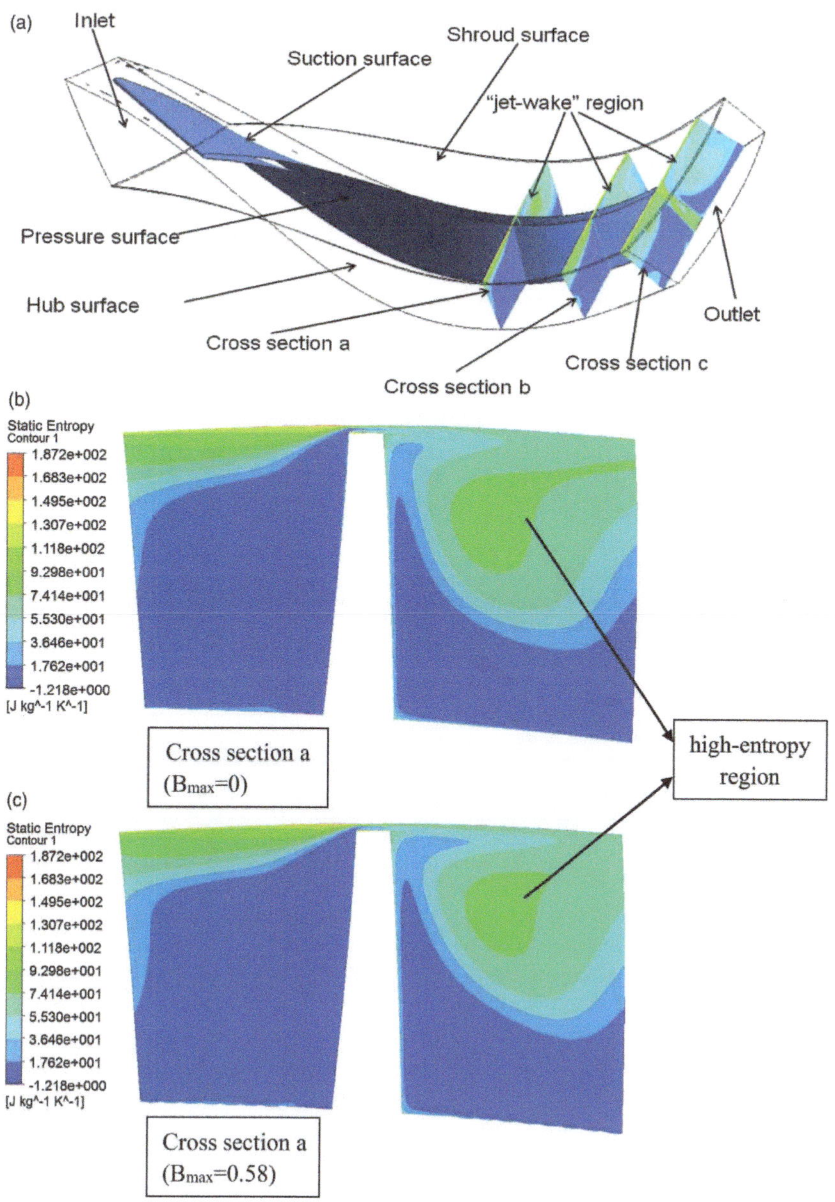

Figure 15. Jet-wake structure analysis.

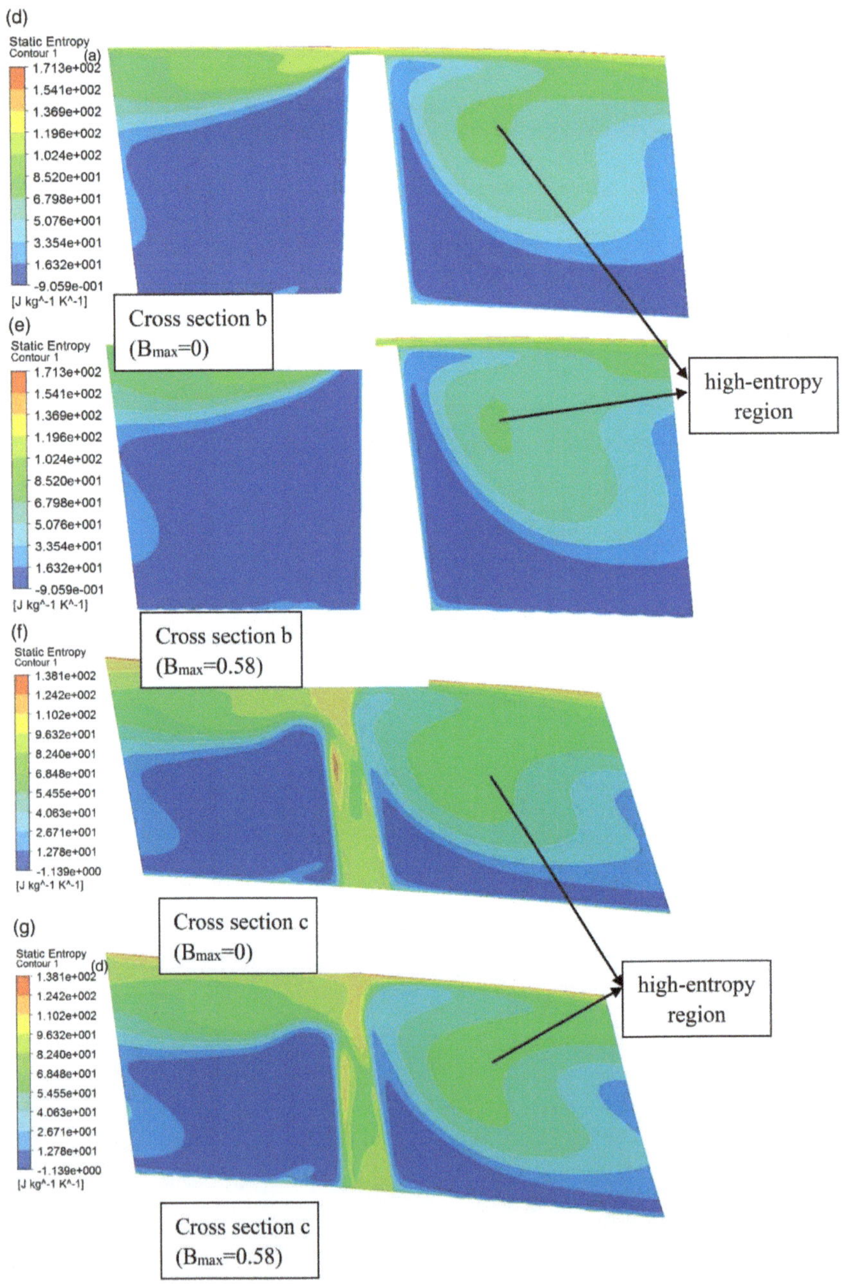

Figure 15. Continued.

stress on the riblets surface is obvious. Regardless of what the B_{max} value is, the minimum points of the shear stress are at the bottom of the riblets and significant upward shifts are obtained at both sides of them, which reach a maximum at the tip points of the riblets. Through the comparison of different residual riblets on the hub surface where B_{max} is equal to 4.7, 0.58, 0.31, and 0, it is seen that shear stress is the smallest around the riblets' valley but is the largest around the riblets' tips when B_{max} is equal to 4.7, on comparing with the ones when B_{max} is equal to 0.58 and 0.31. With the decline of B_{max}, the height and width of residual riblets decrease accordingly. When B_{max} is equal to 0.58, shear stress around riblets' valley is larger than that when B_{max} is equal to 4.7, but the shear stress around the

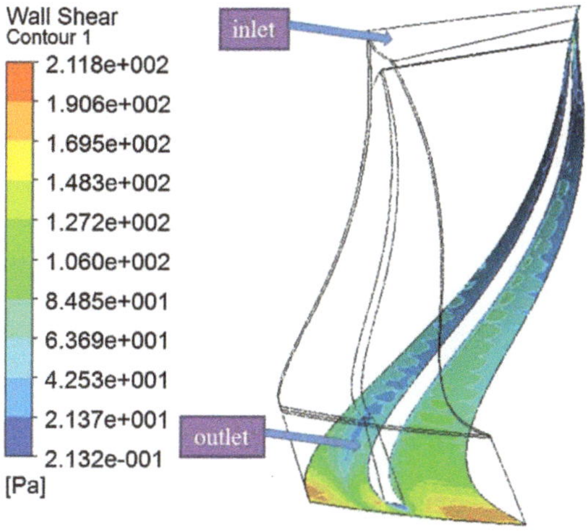

Figure 16. Shear stress contour of smooth hub surface.

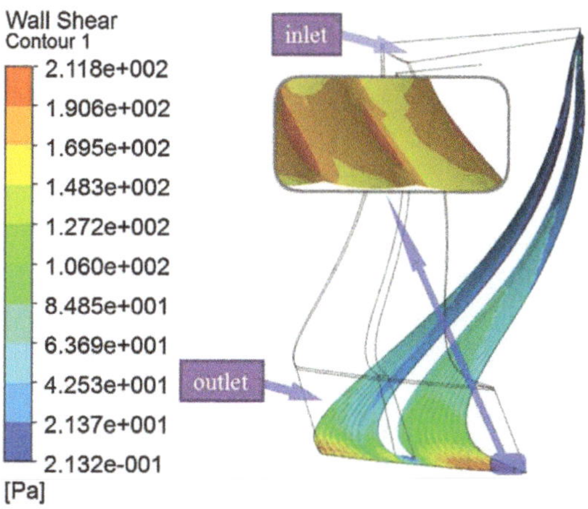

Figure 17. Shear stress contour of hub surface with residual riblets.

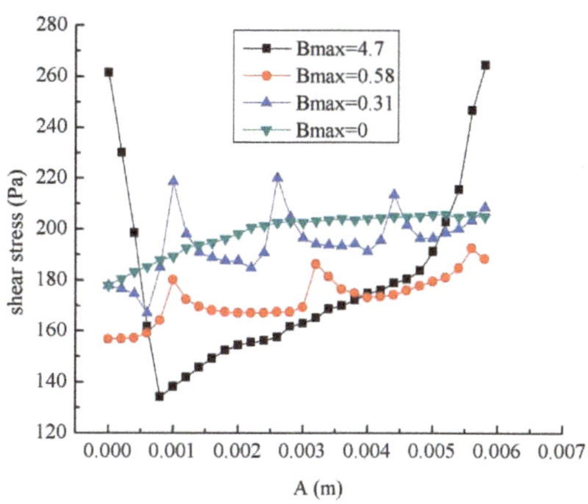

Figure 18. Partial shear stress of hub surface.

riblets' tips is much smaller than that when B_{max} is equal to 4.7. Shear stress around the riblets' valley continues to increase with B_{max}'s decline, and shear stress around the riblets' tips also increases. So, when B_{max} reaches 0.31, the drag reduction effectiveness becomes very small. By integration, results are explained as follows: Total shear stress on the intersecting line when B_{max} is equal to 0.58 is much smaller than that when B_{max} is equal to 4.7, 0.31, and 0, and it is consistent with the results of average shear stress of the whole hub surface shown in Figure 10.

4. Numerical simulation of full flow passages and experimental verification

4.1. *Experimental rig*

Three impellers are machined in order to verify the accuracy of the earlier results, which apply 13(B_{max} is equal to 1.06), 17(B_{max} is equal to 0.58), and 23(B_{max} is equal to 0.31) as the number of tool passes in a single-flow passage in hub surface finish, respectively. The ball-end milling cutter whose radius is 3 mm and half cone angle is 3 degrees is applied. The residual riblets left on the hub surface are shown in Figure 19. It is well known that the tip clearance has a great impact on the performance of the compressor. The tip clearance contains two parts, the radial clearance and the axial clearance, as shown in Figure 20(a). The radial clearance is secured by machining precision. The machining precision of processing equipment in our factory is so high that the machining error could be constrained below 0.01 mm. So, the differences of the radial clearances among the tested impellers could be neglected. The axial clearance is secured during the assembling process, which could have a greater impact than the radial clearance. In order to reduce the influence of the axial clearance, some coating was added to the internal surface of the rings before the performance experiments as shown in Figure 20(b). The impeller was pushed along the axial direction during the assembling process to cling to the internal surface of the rings. During the experiment, the rotating impeller will shave off some of the coating. In fact, the axial clearance is almost 0 for all the tested impellers. So, the differences of the axial clearances of different impellers are almost negligible. In conclusion, the efficiency change due to different tip clearances could be neglected.

Measurements are performed in a single-stage centrifugal compressor on a high-speed test bed of Shenyang Blower Works Group Co, LTD, China. Figure 21 is a photograph of experimental arrangements.

4.2. *Experimental method*

The overall performance of the experimental rig is measured for the first three Mu mentioned earlier, and three measurements near the design flow coefficient points are

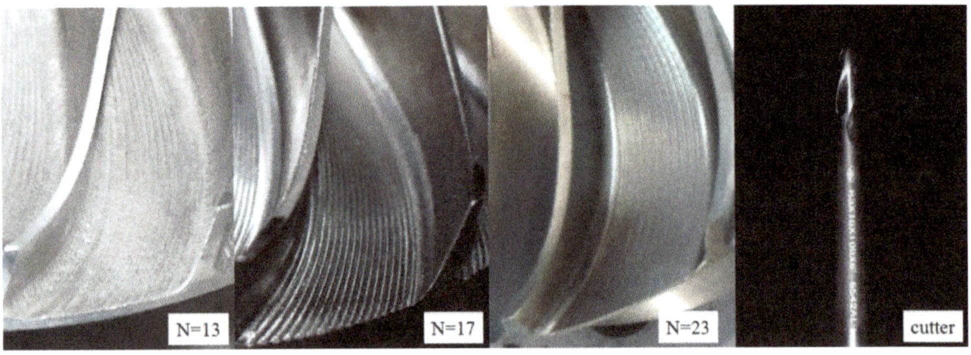

Figure 19. Residual riblets left on hub surface.

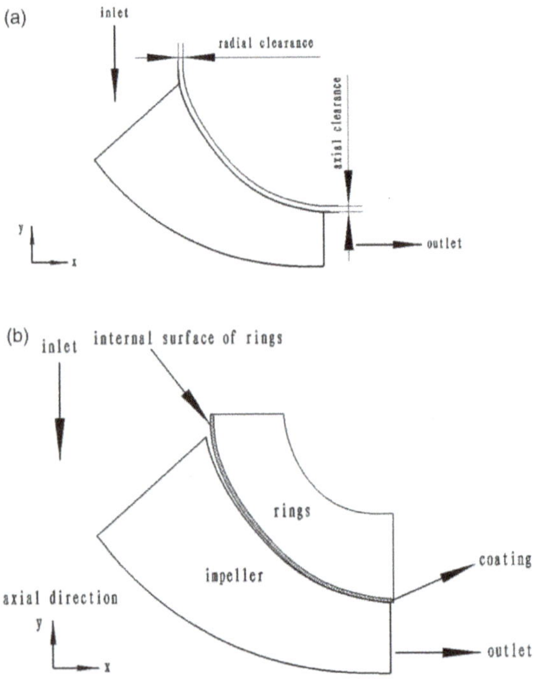

Figure 20. Tip clearance.

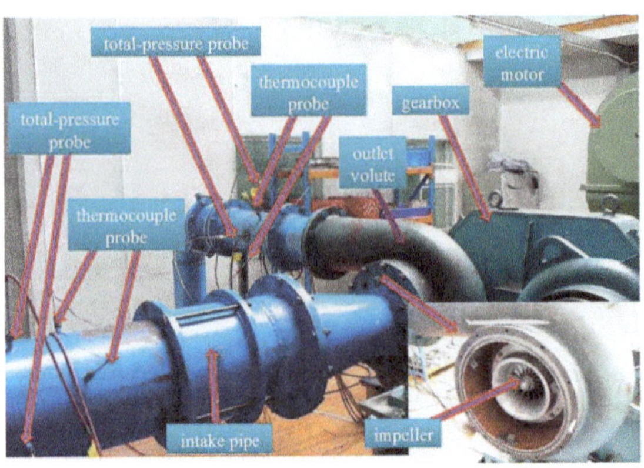

Figure 21. Photograph of experimental arrangements.

taken for each Mu. Four thermocouple probes and four total-pressure probes are arrayed along the circumference of the intake pipe to measure the total temperature and total pressure of the inlet of the rig; another four thermocouple probes and four total-pressure probes are arrayed along the circumference of the outtake pipe to measure the total temperature and total pressure of the outlet of the rig. The overall performance is calculated from the arithmetic mean values of the total pressure and total temperature of the inlet and outlet measured earlier.

4.3. *Numerical simulation of full flow passages*

The volute is not included in the numerical simulations of a single-flow passage; in order to be consistent with the experimental conditions, a numerical simulation of full flow passages is also performed, because numerical simulation of full flow passages is time consuming. Therefore, it is conducted only for Mu = 0.95, and the calculation domain is shown in Figure 3. Mesh 2 of the three computational domains when B_{max} is equal to 1.06, 0.58, and 0.31 and mesh of volutes are applied. The boundary conditions (T_{inlet}, P_{inlet}, and Q_m) are set to ones measured in the experiment, and boundary conditions are maintained the same as the ones in the numerical simulations of a single-flow passage.

4.4. *Results and discussion*

In Figure 22(b), Figure 22(e), a comparison of polytropic efficiency and total pressure ratio is made between performance experiments and numerical simulations of full flow passages when Mu = 0.95 for three impellers when B_{max} is equal to 1.06, 0.58, and 0.31, which indicate that the experimental trend is in reasonable accord with the numerical trend of the full flow passage and the discrepancies between them at each test point are less than 4.2%. In Figure 22(a), Figure 22(c), Figure 22(d), and Figure 22(f), experiment results of Mu = 0.85, 1.05 are shown, respectively, which indicate that the experimental trend is also

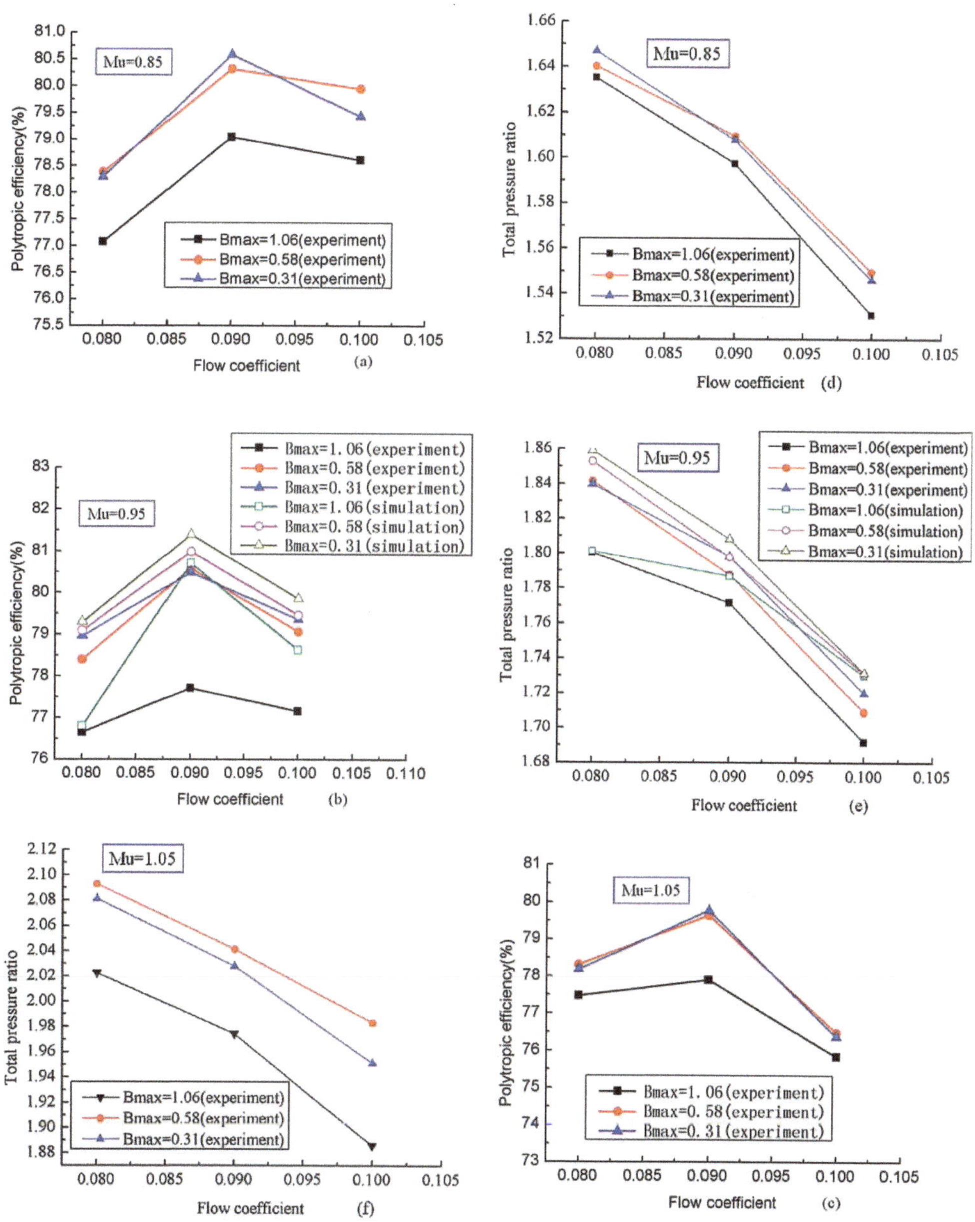

Figure 22. Results of performance experiments and numerical simulations of full flow passages of three Mu for three impellers.

in reasonable accord with the numerical trend of a single-flow passage. From the comparison, it can be found that the influence of the size of the residual riblets on the performance of impellers in the experiments is more evident than that in the numerical simulations, especially when B_{max} is equal to 1.06. According to the lab record, there is a problem that the radial clearance is about 0.4 mm when B_{max} is equal to 0.58 and 0.31, but it is 0.5 mm when B_{max} is equal to 1.06; it is the operator's mistake in causing it, which is the most important reason that the performance curves when B_{max} is equal to 1.06 are lower in the experiments. While taking into account the presence of value differences between results of performance experiments and numerical simulations, analyses are made as follows. Four thermocouple probes and four total-pressure probes are arrayed along the circumference of the intake pipe and out-take pipe, respectively, in the performance experiments. First, measurement errors of probes are inevitable; furthermore, value differences coming from different data processing methods should be taken into consideration, because the arithmetic mean values of measured values are taken as the results to calculate the overall performance of the compressor. Due to limited number of the probes, it is difficult to obtain accurate values of total temperature and total pressure, and the results in numerical simulation are calculated by the mass flow average method, so there are value differences between results of experiments and simulations, but there is no influence on the judgment of the overall trend of the performance curves.

Comprehensive analysis results of experiments and simulations are given as follows. Take the results when Mu = 0.85, the average wall shear stress on the hub surface is, respectively, 75.21 Pa, 66.87 Pa, 61.73 Pa, 63.65 Pa, and 65.35 Pa when B_{max} is equal to 4.7, 1.06, 0.58, 0.31, and 0 in the single-flow passage numerical simulation, which indicate that the drag reduction effectiveness of riblets improves when the B_{max} value declines from 4.7 to 0.58, and reaches almost the best when B_{max} is around 0.58, and the polytropic efficiency of the compressor also comes to a peak value at this moment. Furthermore, the drag reduction effectiveness of riblets declines slightly, and the polytropic efficiency of the compressor changes little when the B_{max} value declines from 0.58 to 0.31 in the simulations. The polytropic efficiency of the compressor when B_{max} is equal to 0.58 and 0.31 is almost the same obtained in the performance experiments. However, the polytropic efficiency of the compressor when B_{max} is equal to 0.58 and 0.31 is significantly better than that when B_{max} is equal to 1.06 in both simulations and experiments. In conclusion, in the hub surface finish of the impeller researched in this article, the optimum B_{max} value is around 0.58 in all the working conditions; thus, it is unnecessary to apply a smaller B_{max}.

5. Conclusion

(1) During the working process of the three-dimensional impeller, streamwise vortices above residual riblets on the hub surface interact with the tips of the riblets, which makes the velocity deflexion of fluids happen in the rotating direction of the impeller. The secondary cross-stream vortex forms above the sharp corner of the riblet. All secondary cross-stream vortices above the riblets constitute the structure of the secondary cross-stream vortices group, which cause lots of low-speed fluids to be retained inside the riblets. Since the low-speed fluids are restrained from rising, the generation of large vortices is avoided, so as to produce the drag reduction effectiveness.

(2) The reasonable size and structure of residual riblets can reduce the average shear stress on the hub surface and the viscous drag of the boundary layer on the hub surface, so as to increase the efficiency of the whole compressor.

(3) For the impeller studied in this article, drag reduction effectiveness and aerodynamic performance of the compressor will reach the optimum value when B_{max} is around 0.58 and the maximum of the cusp height is about 0.093 mm. However, B_{max} is only one factor to affect the drag reduction effectiveness of the residual riblets. So, even for the impellers with similar size, there are many other factors to influence the drag reduction effectiveness of the residual riblets, such as specific speed, meridional channel shape and so on. But the research methods in this paper offer an idea on confirming the optimal B_{max} of impellers. A lot more work would need to be done through further research and experiments in the future.

(4) Designers should consider the influence of hub surface residual riblets after finish on the aerodynamic performance of compressors during the arrangement of processing accuracy. If the B_{max} value of the residual riblets on the hub surface is carefully chosen, it can result in drag reduction effectiveness and improve the compressor efficiency by reducing the flow loss.

Acknowledgments

This research was supported by National Natural Science Foundation of China (51175065); and Shenyang Blower Works Group Co, LTD, China, under the research project NO: KYG2012-14.

Nomenclature

N	number of tool passes in a single-flow passage in hub surface finish
B_{max}	percentage ratio rate of the maximum cusp height of hub surface after finish to the exit blade width
n	rotating speed of impeller
D_2	diameter of impeller
u_2	linear velocity of impeller outlet
c	local speed of sound
k	adiabatic exponent of air, 1.4
ρ	air density at room temperature and atmosphere pressure, 1.1691Kg/m^3
T_{inlet}	total temperature of compressor inlet
T_{outlet}	total temperature of compressor outlet
P_{inlet}	total pressure of compressor inlet
P_{outlet}	total pressure of compressor outlet
Q_v	volumetric flow rate of impeller inlet
Q_m	mass flow rate of the whole machine, $Q_v \cdot \rho$
φ	flow coefficient, $Q_v/(\pi \cdot D_2^2 \cdot u_2/4)$
Mu	machine Mach number, u_2/c
η_{pol}	polytropic efficiency of compressor, $(k-1)\lg(P_{outlet}/P_{inlet})/k \cdot \lg(T_{outlet}/T_{inlet})$

References

Bacher, E. V., & Smith, C. R. (1985). A combined visualization-anemometry study of the turbulent drag reducing mechanisms of triangular micro-groove surface modifications. AIAA-85-0548.

Barth, T. J., & Jesperson, D. C. (1989). *The design and application of upwind schemes on unstructured meshes*. AIAA-89-0366.

Bechert, D. W., & Bartenwerfer, M. (1989). The viscous flow on surfaces with longitudinal ribs. *Journal of Fluid Mechanics*, *206*, 105–129.

Boese, M., & Fottner, L. (2002, June 3–6). *Effect of riblets on the loss behavior of a highly loaded compressor cascade*. Proceedings of ASME Turbo Expo, Amsterdam, The Netherlands. ASME-GT-2002-30438.

Büttner, C. C., & Schulz, U. (2010, September 28–October 1). *Fabrication, oxidation and characterization of shark skin inspired*. Proceedings of the ASME 2010 Conference on Smart Materials (SMASIS2010). Philadelphia, PA, USA. ASME-SMASIS2010-3626.

Choi, H. C., Moin, P., & Kim, J. (1993). Direct numerical simulation of turbulent flow over riblets. *Journal of Fluid Mechanics*, *255*, 503–539.

Chu, D. C., & Karniadakis, G. E. (1993). A direct numerical simulation of laminar and turbulent flow over riblet-mounted surfaces. *Journal of Fluid Mechanics*, *250*, 1–42.

Crowe, C. T., Elger, D. F., & Roberson, J. A. (2005). *Engineering fluid mechanics* (8th ed.). New York: John Wiley & Sons.

El-Samni, O. A., Yoon, H. S., & Chun, H. H. (2005). Turbulent flow over thin rectangular riblets. *Journal of KSME International*, *19*(9), 1801–1810.

Gong, H., Cao, L. X., & Liu, J. (2005). Improved positioning of cylindrical cutter for flank milling ruled surfaces. *Journal of Computer-Aided Design*, *37*(12), 1205–1213.

Kang, S., & Hirsch, C. (2001). Numerical simulation and theoretical analysis of the 3D viscous flow in centrifugal impellers. *Journal of TASK Quarterly*, *5*(4), 433–458.

Klumpp, S., Meinke, M., & Schröoder, W. (2010). Numerical simulation of riblet controlled spatial transition. *Journal of High Performance Computing in Science and Engineering*, *9*, 311–324.

Klumpp, S., Meinke, M., Schröder, W., Feldhaus, B., & Klocke, F. (2009, June 8–12). *Riblets in turbulent flow regimes of 2-D compressor blades*. Proceedings of ASME Turbo Expo, Orlando, FL, USA. ASME-GT2009-59352.

Lazos, B., & Wilkinson, S. P. (1988). Turbulent viscous drag reduction with thin-element riblets. *AIAA Journal*, *26*(4), 496–498.

Lee, S. J., & Lee, S. H. (2001). Flow field analysis of a turbulent boundary layer over a riblet surface. *Journal of Experiments in Fluids*, *30*, 153–166.

Lietmeyer, C., Oehlert, K., & Seume, J. R. (2013). Optimal application of riblets on compressor blades and their contamination behavior. *ASME Journal of Turbomachinery*, *135*, 011036.

Oehlert, K., & Seume, J. R. (2006, July 17–20). *Exploratory experiments on machined riblets on compressor blades*. Proceeding of Fluids Engineering Division Summer Meeting (FEDSM2006). Miami, FL. ASME-FEDSM2006-98093.

Qamar, A., & Sanghi, S. (2012). Aerodynamics characteristic of axissymmetric surface protuberance supersonic regime. *Journal of Engineering Applications of Computational Fluid Mechanics*, *6*(3), 316–335.

Shereena, S. G., Vengadesan, S., Idichandy, V. G., & Bhattacharyya, S. K. (2005). CFD study of drag reduction in axisymmetric underwater vehicles using air jets. *Engineering Applications of Computational Fluid Mechanics*, *7*(2), 193–209.

Walsh, M. J. (1982, January 11–14). Turbulent boundary layer drag reduction using riblets. AIAA 20th Aerospace Sciences Meeting, Orlando, FL. AIAA-82-0169.

Walsh, M. J. (1983). Riblets as a viscous drag reduction technique. *AIAA Journal*, *21*(4), 485–486.

Walsh, M. J., & Lindmann, A. M. (1984). *Optimization and application of riblet for turbulent drag reduction*. AIAA-84-0347.

11

A three dimensional in *SILICO* model for the simulation of inspiratory and expiratory airflow in humans

A.F. Tena[a], J.F. Francos[b]*, E. Álvarez[b] and P. Casan[a]

[a]*Área de Medicina, Departamento de Medicina, C/Julián Clavería, 6, 3ª planta, Campus del Cristo, 33006 Oviedo, Asturias, España;* [b]*Área de Ingeniería Hidráulica, Departamento de Energía, Universidad de Oviedo, Escuela Politécnica de Mieres. Universidad de Oviedo, C/ Gonzalo Gutiérrez Quirós s/n. 33600-Mieres, Asturias, España*

A three-dimensional, computational fluid dynamics model of the human respiratory system is put forward for consideration. This *in silico* model is based on existing state-of-the-art geometric approximations. As a complete simulation of the geometry would have unaffordable high computational costs, the model is composed of a single path of air from the mouth to the 16th bronchial generation, truncating the symmetrical branches that were not included. The innovation housed in this model is the development of a function that is applied to the truncated branches, which contemplates the behavior of the airflow within those branches, by copying cell by cell the velocity vectors existing in the symmetrical face of the developed branch. This model can reproduce the complete airflow process: inspiration and expiration cycles, through simulation in unsteady flow conditions. In normal breathing the air velocity varies within a narrow range, with steady changes. Therefore the model has been validated simulating a spirometry maneuver, in which an inspiration and expiration are performed in the most demanding conditions in order to obtain a wide speed range with faster variations.

Keywords: computational fluid dynamics; human respiratory system; inhalation; exhalation; spirometry

1. Introduction

For decades, mathematical models that reveal the morphology and workings of the human respiratory system have been developed. The first approaches were mainly focused on a mathematical description of the airways that make up the system.

The most commonly used anatomical model of the lower airways was developed by Weibel (1963). In this unidimensional model, the directions of bifurcation are indicated, designating the trachea as the first airway (generation 0) and assumed that each airway leads to two symmetrical branches (regular dichotomy). Weibel described a minimum of 23 bronchial generations up to the alveolar sacs, specifying lengths and diameters. Alternative approaches based on Weibel's morphology, which enhance the description of the lungs, have been proposed. The most notable are the work by Horsfield and Cumming (1968) who introduced some asymmetries in the model, modifying the diameter of the airways and the branching angles; and the morphology developed by Hammersley and Olson (1992) which is suitable for the smaller airways (from generations 6 to 12).

Other research work carried out focus on detailed mathematically descriptions of the airways: Sauret, Goatman, Fleming, and Bailey (1999) and Hegedüs, Balásházy, and Farkas (2004) and Schmidt, Zidowitz, Kriete, Denhard, Krass, and Pietgen (2004) proposed high-resolution models of morphologically lifelike airways up to the fifth bronchial generation; Kitaoka, Takaki, and Suki (1999) and Tawhai and Burrowes (2003) proposed algorithmic techniques to describe the geometry of the lower airways.

More recently, the use of computational fluid dynamics (CFD) modeling techniques and the increased calculation capacity of computers have facilitated the modeling and simulation of the airflow throughout the airways. However, few studies to date have attempted to simulate the flow in the whole airway or even through the conducting airways: it can be deduced from the basic morphology of Weibel (1963) that the number of paths for the airflow in the conducting airways is equal to 2 to the power of 23, i.e., more than 8 million individual flow segments. A computational resolution of the complete flow at all scales with pinpoint accuracy would require a mesh size of the order of several billions of elements, making it impossible to simulate withexisting computer capabilities.

A first attempt to solve this problem was made by Nowak, Kadake, and Annapragada (2003). Nowak used a series of increasingly smaller models of the airways. An entry boundary condition obtained from a simulation of the section immediately above was applied to each section of

*Corresponding author. Email: jffrancos@uniovi.es

the airways. One of the difficulties that arises in this type of sequential procedure is that the output boundary condition at the end of each section must be prescribed before running calculations. The authors found that an equal mass division in the outlets was more appropriate as a boundary condition than a constant pressure value. A similar methodology was applied by Zhang, Kleinstreuer, and Kim (2008) to simulate a model based on Weibel's model. Although both studies represented a significant advance in the simulation of the airflow in the lungs, this sequential procedure does not allow a full-flow coupling at all levels.

Later, researchers opted for a partial resolution of the airway geometry to perform a continuous 3D Computational Fluid Dynamics (CFD) simulation. In these cases only a reduced number of the flow paths were taken into account, providing identical conditions at the areas that were not modeled.

The first significant approach to apply a reduction in the number of resolved flow paths was the work of Gemci, Ponyavin, Chen, Chen, and Collins (2008), which uses the seventeen-generation lung model of Schimdt et al. (2004). The authors studied the inhalation process in a reduced number of resolved flow paths from trachea (generation 0) up to the terminal bronchioles (generation17), through quasi steady-state simulations. An equal and constant pressure condition at all flow outlets (corresponding to areas not modeled)has been used.

The current state of the art for CFD simulations is the use of partially resolved models of the lung airways, in which some branches are removed (truncated) from the airway tree at different levels of the scale. The truncated branches are substituted by boundary conditions. The challenge is to provide realistic conditions for these geometries.

In keeping with such criteria, Walters and Luke (2010) uses Weibel's morphology and the anatomical data of Hammersley and Olson, truncating 50% of the airway paths in each generation from the second generation (bronchi)to the twelfth generation (bronchioles) to create steady state simulations of the inhalation process. The truncations were executed in the middle of the branch segment and a calculated value of static pressure was used as a boundary condition. This pressure value is obtained using a probabilistic approach using pressure data from resolved flow paths. In another paper (Walters, Burgreen, Lavallee, Thompson, & Hester, 2011), the authors continue the investigations of the airflow in steady state conditions using the same static pressure calculation approach for truncated branches in a CT based model. Later on, Walters, Burgreen, Hester, Thompson, Lavallee, Pruett, and Ford-Green (2012), using the same model, presents the results obtained in a complete breathing cycle using unsteady-state simulations, where boundary conditions at truncated branches are prescribed based on an presupposed mass flow split at each bifurcation (it assumes uniform velocity profile).

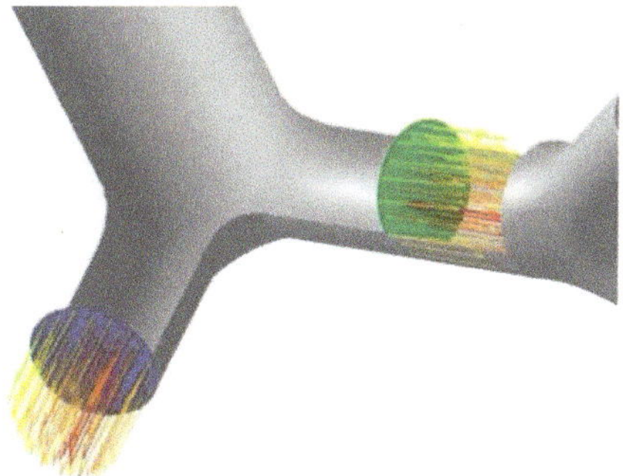

Figure 1. Velocity vectors in the exit of the truncated branch (marked in blue colour) and its symmetrical face in the opened airway (marked in green colour).

Another approach to simplify the complete lung model is to define a single-path model. This was indicated by Spencer, Schroeter, and Martonen (2001) as a way to optimize computational costs in simulations with realistic airway surface features. Longest also demonstrated that the approach including a stochastic individual path (SIP) CFD model from the secondary bronchi to the terminal bronchioles (Tian, Longest, Su, Walenga, & Hindle, 2011) permits accurate studies of the transport and deposition of pharmaceutical aerosols from inhalers (Tian, Hindle, & Longest, 2014). The authors specify mass flow rate boundary conditions at truncated outlets, based on a presupposed even split of the air flow at each bifurcation.

Following the single-path approach, this paper endeavors to provide an investigation that includes a model combining existing CT- based geometries of the upper airways (nose, mouth and throat) with a designed single-path model from the trachea up to the terminal bronchioles. The branches aresymmetric, based on the geometries proposed by Weibel and Kitaoka. This symmetrical state allows the use of a boundary condition for each simulation step that imposesthe same velocity vector field in the truncated-branches as in the developed branches through which the air-flow passes (Figure 1). This boundary condition is considered much more realistic than predefined flow-rate values or pre-calculated pressure conditions. The model is validated with a complete spirometry test which represents the most demanding conditions for unsteady simulations of both the inhalation and exhalation cycles.

2. DEVELOPED MODEL

2.1. The basis

For the elaboration of the three-dimensional airway model, the model proposed by Kitaoka et al. (1999) is combined with Weibels model (1963).

Weibel's classic model indicates the bifurcation modes, designing the trachea as the first airway (order 0) and assuming that each airway generates two new branches (regular dichotomy). Thus, the left and right main bronchi

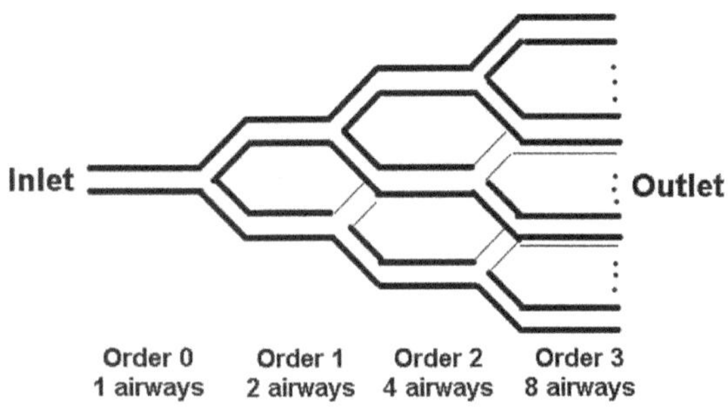

Figure 2. Airway scheme.

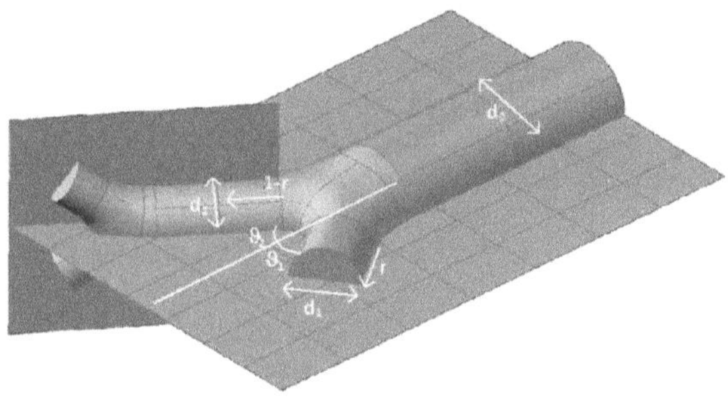

Figure 3. Three dimensional model of the lung proposed by Kitaoka.

are order 1 and so on. There are at least 23 airway generations until reaching the alveolar sacs. In accordance with this model, there is one order 0 generation ($2^0 = 1$), 2 airways of order 1 ($2^1 = 1$), 4 airways of order 2 ($2^2 = 4$), etc. Therefore there are 2^{23} airways of order 23; the number of ramifications increases exponentially to a total of 8,388,608 (2^{23}). Figure 2 shows a scheme of these airways.

The diameter of each ramification diminishes according to the formula $d_n = d_0 \cdot 2^{-1/n}$, n represents the generation order and d the diameter. This relationship is maintained up to generation 16 at the end of the conductive zone. From generation 17 to 23 there are only small changes in the dimensions of the airways, because they are more or less alveolized. In Weibeĺs model the spatial disposition of the branches was not taken into account, therefore it only allows one-dimensional modeling of the lung.

Kitaoka et al. (1999) brought about the design of a realistic three dimensional model of the lung, with nine basic rules:

- Rule 1: Branching is dichotomous.
- Rule 2: The parent branch and its two daughter branches lie on the same plane, called the branching plane.
- Rule 3: The volumetric flow rate through the parent branch is conserved after branching; that is, the sum of the flows in the daughter branches is equal to the flow in the parent branch.
- Rule 4: The region supplied by a parent branch is divided into two daughter regions by a plane called the "space-dividing plane." The space-dividing plane is perpendicular to the branching plane and extends out to the border of the parent region. There is a supplementary rule (rule 4a) for correcting the space division scheme whenever the space in that region requires it.
- Rule 5: The flow-dividing ratio "r" is set to be equal to the volume-dividing ratio, defined as the ratio of the smaller daughter region to that of its parent. Figure 3 illustrates these rules.
- Rule 6: Diameters and branching angles of the two daughter branches are determined by substituting r from rule 5 into the following equations.

With no symmetry:

$$d_1 = d_0 r^{1/n}, d_2 = d_0(1-r)^{1/n}$$

$$cos\theta_1 = \frac{1 + r^{4/n} - (1-r)^{4/n}}{2r^{2/n}};$$

$$cos\theta_2 = \frac{1 - r^{4/n} + (1-r)^{4/n}}{2(1-r)^{2/n}}$$

With symmetry:

$$\theta_1 = \theta_2; r = 1/2;\ d_1 = d_2 = d_0 2^{-1/n};\ cos\theta = 2^{\left(\frac{2}{n}-1\right)}$$

Where:

- d is the airway diameter
- subindex 0 for mother branch
- subindexes 1 and 2 for daughter branches
- n is a constant called diametrical exponent (for laminar flow its value is 3 and for turbulent flow its value is 2.6)

There is a supplementary rule (rule 6a) to correct the branching angle according to the shape of the daughter region.

- Rule 7: The length of each daughter branch has a value three times its diameter. There is a supplementary rule (7a) to correct the length according to the shape of the region.
- Rule 8: If branching continues in a given direction, the daughter branch becomes the new parent branch, and the associated branching plane is set perpendicular to the branching plane of the old parent. There is a supplementary rule (8a) to correct the rotation angle according to the shape of the region.

- Rule 9: The branching process in a given direction stops whenever the flow rate becomes less than a specified threshold or the branch extends beyond its own region.

2.2. *Geometric model*

The geometric model used is composed of a conductive airway model designed following Weibel's (1963) and Kitaoka's et al. (1999) symmetrical cases, as well as existing upper airways models. It was developed up to the 16th bronchial generation (end of the conductive airways) because, according to Weibel's model, from the 17th generation to the end, the airways are alveolized. The development of the geometry of the alveolar units is much more complex and is beyond the scope of this work.

a) Conductive airways model

The numerical model of the conductive airways was designed to be as realistic as possible while avoiding prohibitive computational costs when it is simulated. Therefore the model has been simplified in two ways:

1. Considering a model of a single active branch: when the trachea divides into the two main bronchi, the model will continue through the right main bronchus while the pathway through the left main bronchus will be "truncated." When the right main bronchus bifurcates into the right upper bronchus and the intermediate bronchus, the model will continue through the intermediate bronchus and the upper right bronchus will be "truncated," and so on. This kind of model was researched and developed by Walters and Luke (2010) and Walters et al. (2012) and has proven to be effective to simulate the entire lung, saving time and computational costs.
2. Overlooking the changes in the diameter of the airway in accordance with the respiratory-cycle phase.

The diameter and the length of the airways were deduced from the relations proposed by Weibel (1963) and Kitaoka et al. (1999):

$$d = 0.018^{-0.388n};\, l = 0.012^{-0.92n} \text{ if } n \leq 3$$

$$d = 0.013^{-(0.2929-0.00624n)\,n};\, l = 0.025^{-0.17\,n} \text{ if } n > 3$$

Being, d the airway diameter, n the generation order and l the airway length.

A summary of the main geometric dimensions in the branches is presented in Table I.A (see Annex).

With this data and other restrictions from Weibel's (1963) and Kitaoka's et al. (1999) model, the geometry was constructed using the commercial software Ansys Gambit 2.4.6. The numerical model of the lung, composed ofan open branch (or single path), is shown in Figure 4. The complete morphology of the lung can be generated from

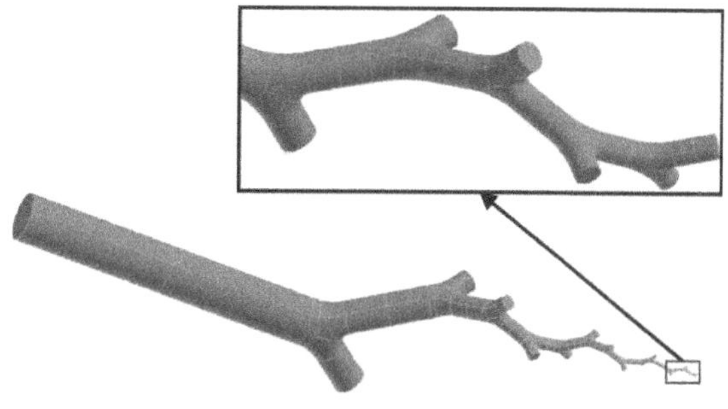

Figure 4. Geometric model of the lung.

the numerical model. This can be done by establishing symmetry at each of the truncated branches.

b) Upper airways model

Once the model has been designed up to the 16th bronchial generation, the next step is to complete the conductive airway by adding the upper airway components: nose, mouth and throat (larynx and pharynx).

The nose is constructed using the simplified geometrical model used in Castro-Ruiz (2003) obtained from CT images of a thirty-year-old woman. The volume of each duct of the nose is 19,568 mm^3. Figure 5 shows the model used (right image) and seven coronal cuts (left image), with the following areas:1: 482 mm^2, 2: 205 mm^2, 3: 164 mm^2, 4: 198 mm^2, 5: 154 mm^2, 6: 33 $mm^2$and 7: 13 mm^2. The perimeter of the cuts allows the construction of the longitudinal surfaces using the following distances between sections:1 – 2: 18 mm, 2 – 3: 10 mm, 3 – 4: 36 mm, 4 – 5: 16 mm, 5 – 6: 17 mm and 6 – 7: 7 mm.

The mouth and throat are a reproduction of the Stapleton model (Stapelton, Guentsch, Hoskinson, & Finlay, 2000) constructed by using simple geometric shapes. The advantage of this approach is that extremely high mesh resolutions are not needed near the walls to resolve all the small airway irregularities.

The geometry of the mouth used is shown in Figure 6. In the left image seven coronal cuts in the mouth model (right image) are shown. The areas are: 1: 503 mm^2, 2: 471 mm^2, 3: 574 mm^2, 4: 638 mm^2, 5: 653 mm^2, 6 and 7: 257 mm^2. The distances between sections are: 2–3: 25 mm, 3–4: 25 mm, 4–5: 30 mm, 5–6: 4 mm and 6–7: 11 mm. The walls of the mouth are smoothly joined with the walls of the pharynx. The shape of the tongue changes while speaking or swallowing. This model has a flat tongue in the foremost part of the mouth.

The throat geometry is more complicated because its three points of entry (two nasal passages, the choanae; and the mouth passage, the fauces) become a single conduct (the pharynx) whose output is the trachea. Inside this conduct is the epiglottis, a cartilage separating the trachea from the esophagus. The epiglottis is smoothly attached to the trachea. The epiglottal folds form a tube with a diameter of approximately 18 mm.

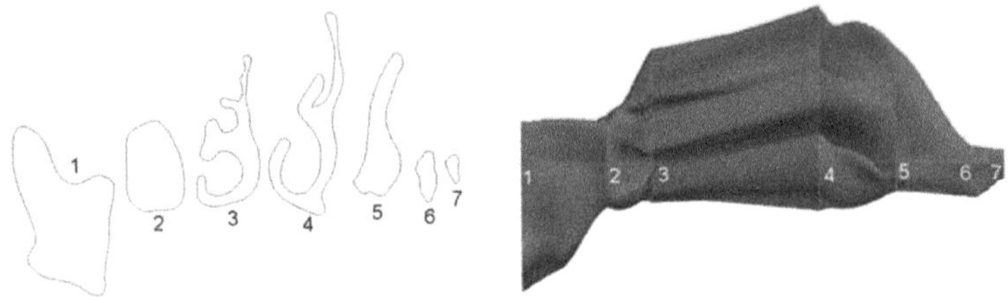

Figure 5. Geometry of the nose.

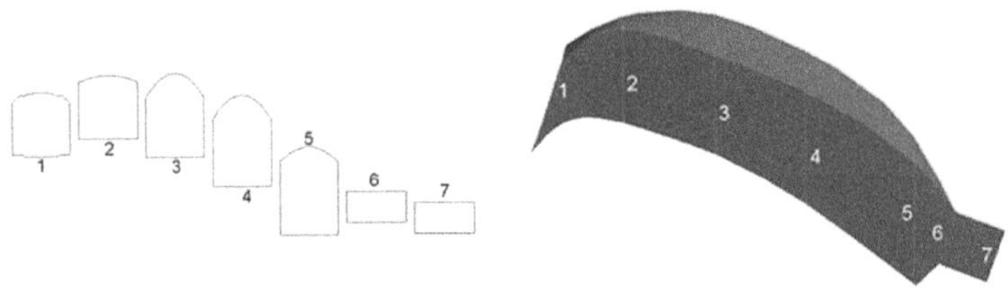

Figure 6. Geometry of the mouth.

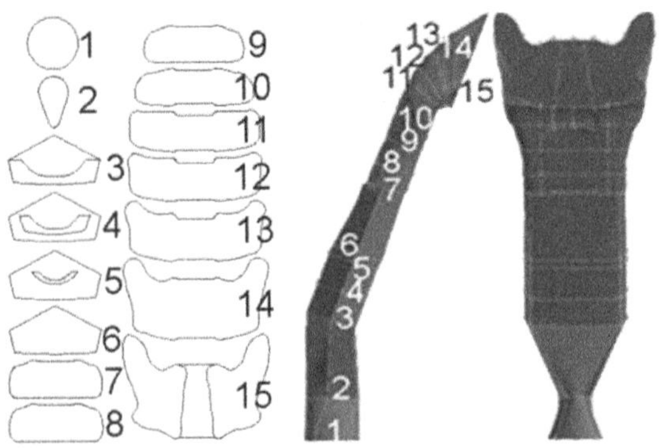

Figure 7. Geometry of the throat.

Figure 7 represents the throat model (right image) divided intosections (left image). The pharynx includes cross sections from 7 to 15, with the corresponding areas:7: 393 mm^2, 8: 391 mm^2, 9: 392 mm^2, 10: 458 mm^2, 11: 572 mm^2, 12: 658 mm^2, 13: 752 mm^2, 14: 1020 mm^2 and 15: 964 mm^2.The larynx includes cross sections from 1 to 6 with the corresponding areas:1: 254 mm^2, 2: 126 mm^2, 3: 242 mm^2, 4: 315 mm^2, 5: 370 mm^2 and 6: 395 mm^2.

The entire geometric airways model is presented in Figure 8.

2.3. *Numerical model*

Once the geometry has been done, the next step is to create the numerical model. This is determined by dividing, or meshing, the geometry into different cells in which the flow equations must be solved. Therefore a mesh will be applied, choosing the appropriate cell type.

a) Meshing

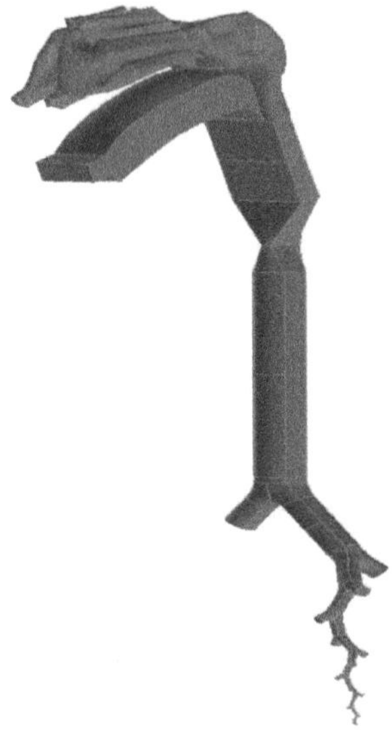

Figure 8. Geometric airways model.

The model of the lung was meshed with tetrahedral cells because of their favorable adaptation to complex geometries. The size of the tetrahedrons diminishes while descending from high-order to low-order generations. Before meshing the volumes a boundary layer mesh was built using structured meshing in order to obtain a better description of the boundary layer in the numerical calculations. The size of the tetrahedrons was consistent with the size of the boundary layer cells.

The total size of the grid is about 10^6 cells. A quality analysis of the mesh yielded very satisfactory results, indicating a magnitude of the equitize skew below 0.6 for 98% of the cells in the mesh. This parameter shows the shape of the cells formed. Values near 0 indicate more regular cells, which have a higher probability of obtaining satisfactory results.

Figure 9 shows an example of the surface mesh at some branches with a detail of the boundary layer mesh that includes numerical information needed for its definition. The boundary mesh has equal number of cells for each branch as the number of rows (6rows) and other dimensionless parameters are evenly set throughout the model.

b) Flow equations

The equations that describe a fluid (air in this case) in movement can be deduced from the mass and momentum conservation laws. Applying these conservation laws to an incompressible fluid element, the Navier-Stokes equations are obtained:

Continuity:

$$\nabla \cdot \vec{v} = 0$$

Momentum:

$$\rho \frac{d\vec{v}}{dt} = -\nabla p + \rho \vec{g} + \mu \nabla^2 \vec{v}$$

where:

- $\vec{v}$is the air velocity.
- p is the air pressure.
- μ is the dynamic viscosity of the air.
- $\rho$is the air density
- $\vec{g}$ is the constant gravity

Resolving these equations by using a turbulence model, the values of pressure and velocity of the fluid in each cell may be obtained.

c) Boundary conditions

To solve the flow equations in the inhalation and exhalation cycles, a set of boundary conditions have been selected.

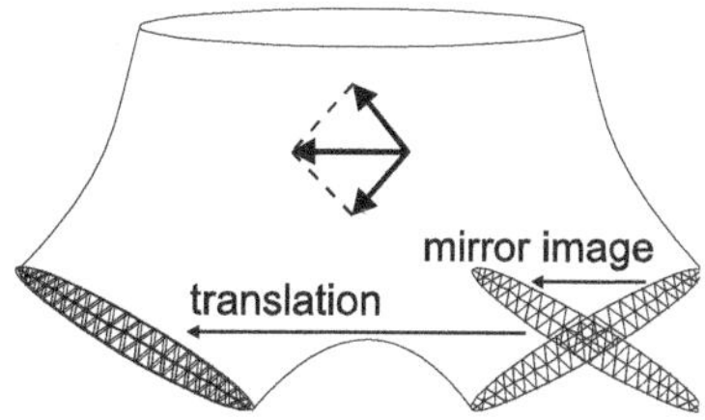

Figure 10. UDF for truncated branch.

- Mouth (model inlet) and 16th bronchial generation (model outlet): values of static pressure, total pressure (sum of static and dynamic pressures) or flow rate conditions have been used depending on the test carried out.
- Truncated branches conditions: these need a special boundary condition because the same conditions which were imposed at the inlet and outlet of the model cannot be applied. This special condition must unwaveringly represent the effect of the branches removed. A user-defined function (UDF) is used to impose a symmetric operation of two branches. This UDF obtains the velocity vector field at each open branch (either in inhalation or exhalation) from the calculations, and prescribes the same vector field in the corresponding truncated branch (Figure 10). This methodology is repeated repeatedly until reaching convergence in the flow field.

d) Solution parameters

The numerical calculations were resolved with the commercial code Ansys Fluent 6.3.26. This code was used to solve the unsteady Reynolds-averaged Navier-Stokes

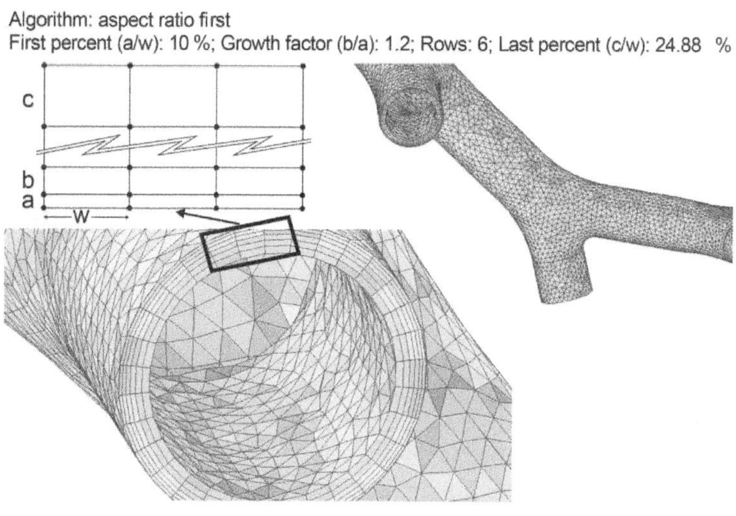

Figure 9. Surface meshes at some branches: a detail of the boundary layer mesh.

equations (URANS) by the finite volume method, converting them from differential equations into their numerical analogues (Eulerian method).

The solutions were obtained, assuming air with a constant density 1.225 kg/m^3 and dynamic viscosity 1.9×10^{-5} kg/m s.

The SIMPLE (Semi-Implicit Method for Pressure Linked Equations) algorithm was used to resolve the coupling between pressure and velocity fields, the same solver used by Walters in his research (Walters & Luke, 2010 and Walters et al., 2012). Thediscretization of the spatial and temporal derivatives for the velocities in the equations was carried out by means of second-order schemes. The PRESTO! discretization scheme was used for the pressure.

Since the flow in human airways is both laminar and turbulent, with transition zones, the turbulence model chosen for this work is the shear stress transport k- ω (SST k-ω), which is the best suited to these conditions. While the standard k-ω turbulence model performs better than k-ε in the near wall region, it is worse in the far field; this is why the shear stress transport formulation has been selected: it combines the best of both (Versteeg & Malalasekera, 2007). The SST k-ω model is defined by the following two equations:

$$\frac{\partial}{\partial t}(\rho k) + \frac{\partial}{\partial x_i}(\rho k u_i) = \frac{\partial}{\partial x_j}\left(\Gamma_k \frac{\partial_k}{\partial x_j}\right) + \tilde{G}_k - Y_k + S_k$$

$$\frac{\partial}{\partial t}(\rho \omega) + \frac{\partial}{\partial x_i}(\rho \omega u_i) = \frac{\partial}{\partial x_j}\left(\Gamma_\omega \frac{\partial_\omega}{\partial x_j}\right) + G_\omega - Y_\omega + D_\omega + S_\omega$$

Where k is the turbulence kinetic energy; ω is the specific dissipation rate; ρ is the air density; $\tilde{G}_k$ represents the generation of turbulence kinetic energy due to mean velocity gradients; G_ω represents the generation of ω; Γ_k and Γ_ω represent the effective diffusivity of k and ω respectively; Y_k and Y_ω represent the dissipation of k and ω due to turbulence; D_ω represents the cross-diffusion term and S_k and S_ω are user-defined source terms.

3. SIMULATION METHODOLOGY

The designed model has been validated by two tests: a first test in which a real spirometry is introduced in the model and a second test used to check the results obtained in the first one. A forced spirometry test has been used for the evaluation of the model as it presents the maximum variations of flow rate possible. Therefore the model is tested in the most challenging conditions.

Different forced spirometry tests of individuals without obstructive pulmonary diseases were performed to obtain data under realistic conditions. The spirometry that best matched the ERS (European Respiratory Society) quality criteria was chosen (Miller, Hankinson, Brusasco, Burgos, Casaburi, Coates, Crapo, Enright, van der Grinten, Gustafsson, Jensen, Johnson, MacIntyre, McKay, Navajas, Pedersen, Pellegrino, Viegi, & Wanger, 2005). The spirometry data was obtained using a spirometer model CPFS/D USB, with a MedGraphicspreVentTMpneumotachometer and BREEZESUITETM diagnostic software (Medical Graphics Corporation 2004, 350 OakGrove Parkway, St. Paul, Minnesota 55127–8599).

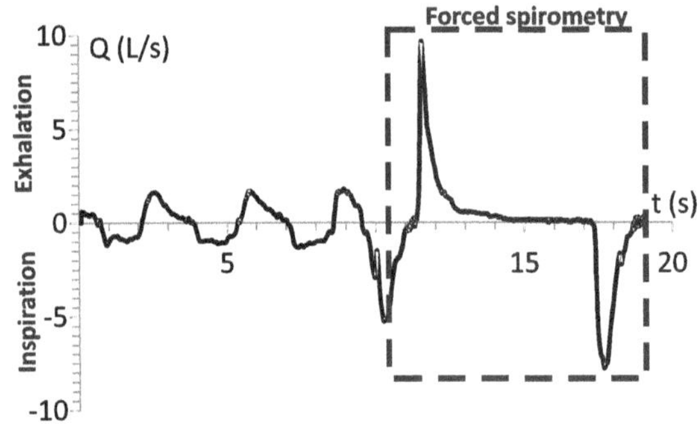

Figure 11. Relationship between flow rate and time from the chosen spirometry.

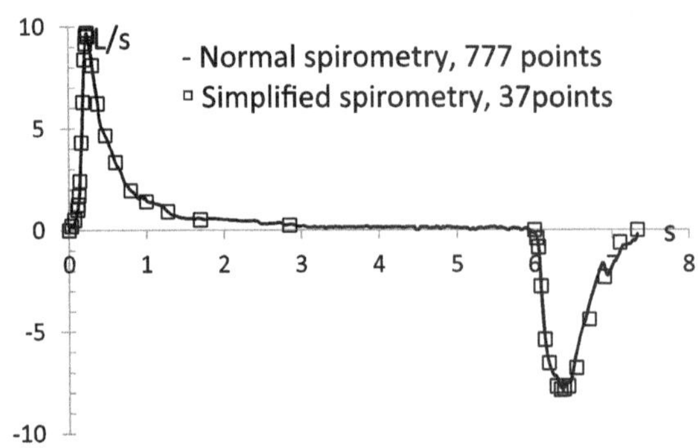

Figure 12. Normal and simplified spirometry.

In a spirometry test the spirometer measures pressure differences in a venturi tube during the test period. Then a software tool calculates the time evolution of the mean velocity and the flow rate. The relationship between the flow rate and time isrepresented in Figure 11. From this data, the portion corresponding to the forced spirometry corresponds to the data obtained between seconds 11.32 and 19.09.

For calculation purposes the original spirometry has been simplified, reducing the number of points from 777 to 37 and adjusting the new curve (Figure 12). The number of points has been reduced to include the important changes in the inlet flow so that the simulation is not overburdened with irrelevant tasks.

To reproduce the same conditions as in a real spirometry, the boundary conditions values used/adopted imposed were:

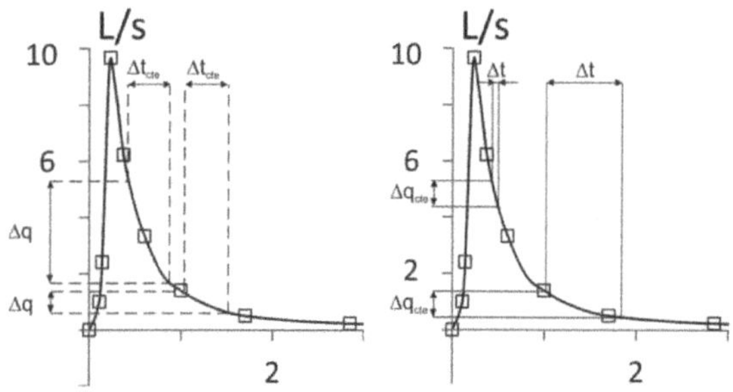

Figure 13. The graph on the left shows the option with a constant time step (variable flow rate step), the graph on the right shows the option with a constant flow rate step (variable time step).

- Mouth (inlet): flow rate varying over time using the detailed function, and atmospheric pressure. During the respiratory cycle (inhalation and exhalation) the inlet of the model, the source of the trachea, is set at atmospheric pressure. In fact, it is slightly lower as it has to pass through the mouth. However, this does not affect the calculations.In order to apply the values of the flow rate obtained in the spirometryto the model, it was necessary to design a user-defined function (UDF) as Ansys Fluent has no established functions to simulate time-dependent variables. Due to the non-uniformity of the breathing cycle and in order to include all the representative points, this UDF must include a subroutine to adapt the magnitude of the time step to a calculated percentage of change in the flow rate (adaptive time step). This value is also calculated by the UDF using the data of the number of steps defined and the spirometry function (Figure 13). The UDF automatically changes the conditions from inhalation to exhalation.
- 16th bronchial generation (outlet): a static pressure in the inhalation cycle and a total pressure (sum of static and dynamic pressures) in the exhalation cycle (Figure 14).

Finally, to check the previous test and the results found, the unsteady pressure function obtained was set as the inlet boundary condition at the 16th bronchial generation. Figure 15 shows the new boundary conditions employed:

- Mouth (inlet): static pressure on the exhalation cycle and a total pressure on the inhalation cycle.
- Sixteenth bronchial generation (outlet): a total pressure (sum of static and dynamic pressures) varying over time using the same function (UDF) that was used to introduce the variable flow rate in the previous test.

The convergence criterion established was to reduce the scaled residuals for all the variables below 10^{-5}.One

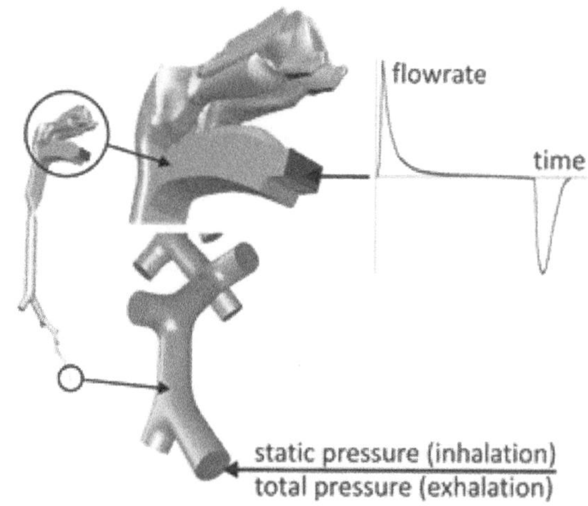

Figure 14. Boundary conditions imposed in the first test.

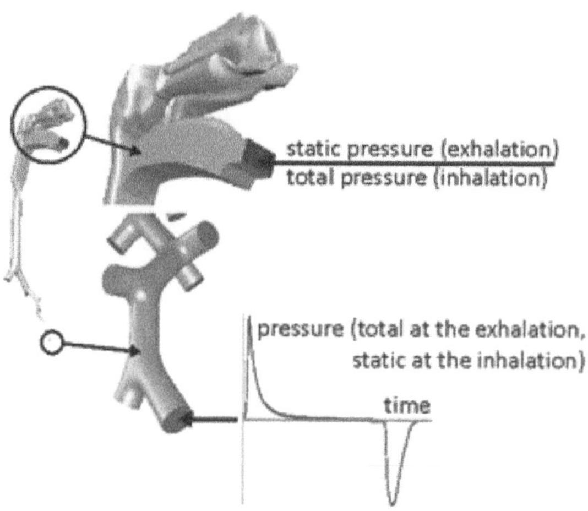

Figure 15. Boundary conditions established at the checking test.

thousand steps were used. A maximum of 1,400 iterations were required in order to ensure the convergence of the results at each time step. The spirometry maneuver in forced conditions was covered in 1000 flow steps with the corresponding time adaptations.

4. RESULTS AND DISCUSSION

With these values the time required for each simulation was 5 days working in parallel on a computer with an I-7 processor of 8 cores. The results of the tests provide data of the lung that cannot be obtained with a real forced spirometry. For example, Figure 16 shows static pressure obtained at the 16th bronchial generation of the lung as result of the first test. It is compared with the boundary condition used (unsteady flow rate obtained from the chosen forced spirometry test) finding that the static pressure curve at the 16th bronchial generation obtained has a similar tendency to the one of the flow rate curve. This indicates a suitable performance of the numerical model.

Upon further examination and checks carried out, it was noticeable how different variables (pressures and velocities) have similar values in both cases, meaning that the solutions are unique and representative for the inhalation and exhalation cycles. For example, Figure 17 shows that the maximum velocity occurs at time 0.23 s, being 38.02 m/s the value obtained experimentally and 37.04 m/s the simulated value. Therefore, the maximum error is less than 2.6%.

The contours of the instantaneous static pressure and velocity vector field during inhalation are shown in Figure 18 and Figure 19. It can be observed that the magnitude of the pressure drops from the mouth (zero pressure) to the narrow area close to the glottis where a minimum of about -3.6E-3 Pais obtained. This corresponds to an increase in velocity from 0 m/s (in the mouth) to a maximum value of around 45 m/s in the mentioned area. After this section, the air velocity descends in the trachea, and from the trachea the velocity increases again up to a maximum value around 35 m/s at the third level and therefore a reduction in the static pressure is produced. For the following generations the velocity decreases as the lung

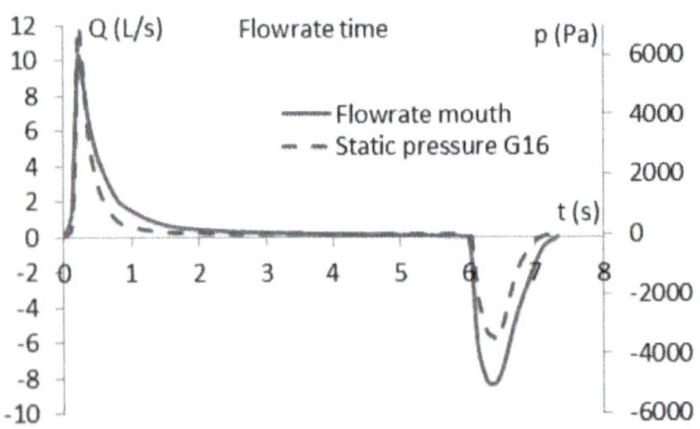

Figure 16. Unsteady flow rate established at the mouth in the simulation whilst static pressure is obtained in the 16th bronchial generation.

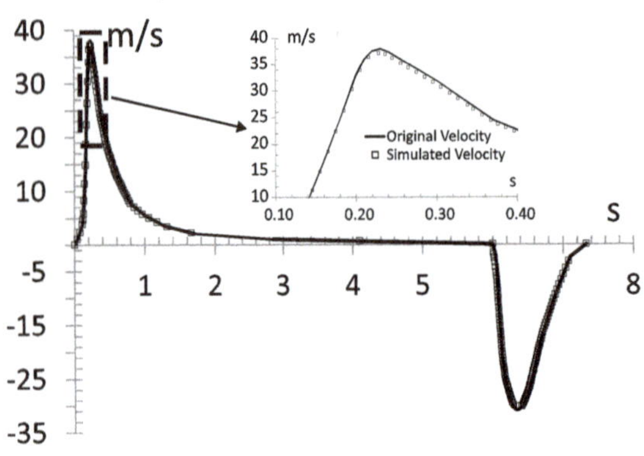

Figure 17. Real and simulated velocities at the mouth.

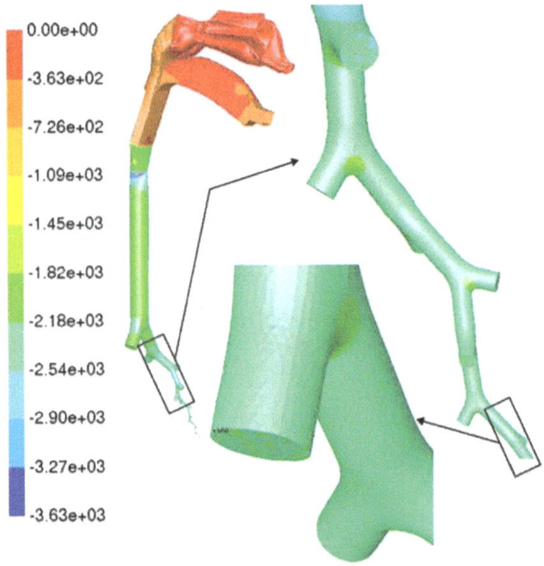

Figure 18. Pressure contours at maximum inhalation flow rate (Pa).

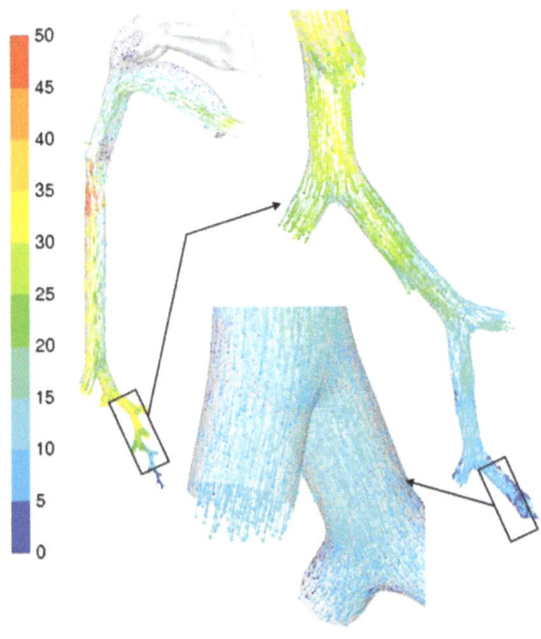

Figure 19. Velocity vector field at maximum inhalation flow rate (m/s).

surface increases, though there is no appreciable variation in terms of pressure.

The contours of the instantaneous static pressure during exhalation are shown in Figure 20. As one may observe, the magnitude of the pressure diminishes while ascending from the lowest generation to the trachea, where a magnitude of zero pressure is reached (i.e., atmospheric

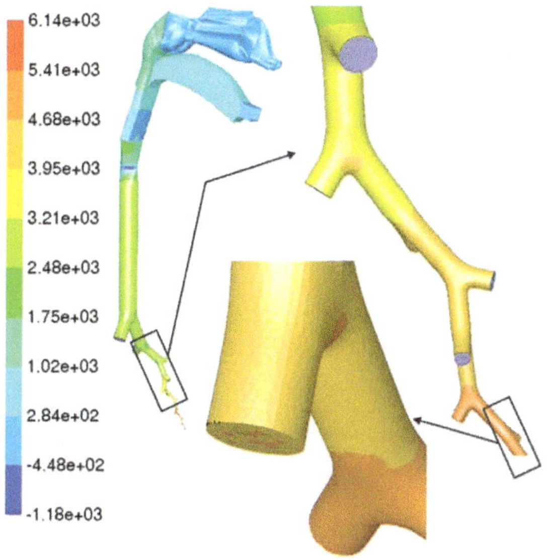

Figure 20. Pressure contours at maximum exhalation flow rate (Pa).

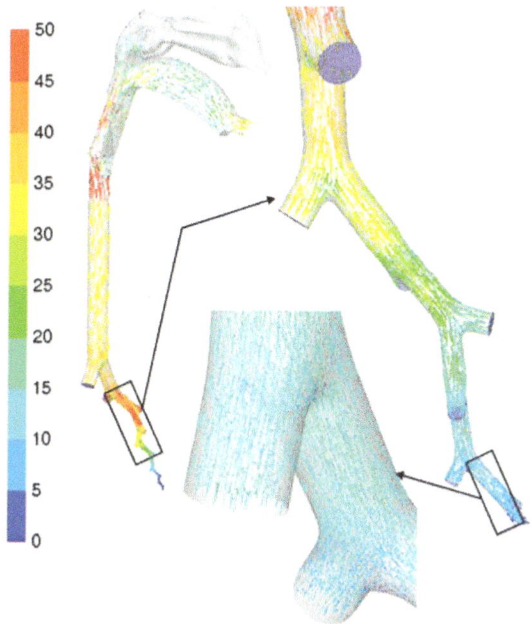

Figure 21. Velocity vector field at maximum exhalation flow rate (m/s).

conditions). The symmetry of the velocity field at the bifurcations may also be noted. As expected, the velocity has maximum values in the area close to the glottis and the 3rd level (around 50 m/s). The velocity decreases from the third level to the trachea as the surface of the lung increases in the flow direction (Figure 21) while from other levels to third the velocity gradually increases since there is an overall decrease in the lung's surface. This explains the reduced pressure in the different zones (Figure 20, Figure 21). These results comply with those already quoted Sbirlea-Apiou, Katz, Caillibotte, Martonen and Yang (2007) and Gemci, Ponyavin, Chen, Chen & Collins (2008).

Using post processing data, two typical parameters that describe turbulence modeling: y^+ and a Reynolds number (Re), have been obtained. With regard to the near-wall resolution, the average value of y^+ obtained was 0.91. Reynolds numbers in the range of 0–8700 for normal breathing and 0–47,000 for spirometry tests were found. The maximum values correspond to the conditions in the trachea (diameter of 0.018 m) with velocity ranges of between 0–7 m/s (normal breathing) and 0–38 m/s (spirometry test) respectively.

5. CONCLUSION

This paper reveals a three-dimensional *in silico* CFD model of the human airways designed following Weibel's and Kitaoka's symmetrical models as well as existing upper airways models. It consists of a single air pathfrom the mouth to the 16th bronchial generation in addition to a user-defined-function (UDF) that represents the effect of the branches that were not contemplated. The UDF copies cell by cell the velocity vectors existing in the cross section of a developed branch to the symmetrical truncated one while consolidating simulation. This approach permits a continuous unsteady simulation of the respiratory system both in the inspiration and the exhalation cycles, using limited computational resources, making it affordable.

In the investigations, in order to achieve a fully reliable CFD simulation, the airway model was meshed using 10^6 tetrahedral cells, which offer satisfactory results with a suitable adaptation to the geometry. The model has been validated using flow data of a real forced spirometry. This involved introducing the flow rate values through the model inlet (mouth) with a user-defined function designed for this purpose. The results were very lifelike values of pressure and velocity at all airways levels. These results were checked by applying the reverse conditions to the model (total pressure in the outlet of the model at 16th bronchial generation) thereby obtaining similar results to those of the spirometry test.

Further research work could be carried out with the model obtained in order to investigate specific diseases, such as chronic bronchitis and lung emphysema, as well as the study of the deposition of pollutants or drugs in the airways.

References

Castro-Ruiz, P. (2003). *Análisis computarizado del flujo aéreo en cavidad nasal* [Thesis doctoral]. Universidad Autónoma de Madrid.

Gemci, T., Ponyavin, V., Chen, Y., Chen, H., & Collins, R. (2008). Computational model of airflow in upper 17 generations of human respiratory tract. *Journal of Biomechanics*, *41*, 2047–2054.

Hammersley, J., & Olson, D. (1992). Physical models of the smaller pulmonary airways. *Journal of Applied Physiology*, *72*(6), 2402–2414.

Hegedüs, C. J., Balásházy, I., & Farkas, Á. (2004). Detailed mathematical description of the geometry of airway bifurcations. *Respiratory Physiology & Neurobiology*, *141*(1), 99–114.

Horsfield, K., & Cumming, G. (1968). Morphology of the bronchial tree in man. *Journal of Applied Physiology*, *24*(3), 373–383.

Kitaoka, H., Takaki, R., & Suki, B. (1999). A three-dimensional model of the human tree. *Journal of Applied Physiology*, *87*, 2207–2217.

Miller, M. R., Hankinson, J., Brusasco, V., Burgos, F., Casaburi, R., Coates, A., ... Wanger, J. (2005). Standardisation of spirometry. *European Respiratory Journal*, *26*, 319–338.

Nowak, N., Kadake, P. P., & Annapragada, A. V. (2003). Computational fluid dynamics simulation of airflow and aerosol deposition in human lungs. *Annals Biomedical Engineering*, *31*(4), 374–390.

Sauret, V., Goatman, K. A., Fleming, J. S., & Bailey, A. G. (1999). Semi-automated tabulation of the 3D topology and morphology of branching networks using CT: application to the airway tree. *Physics in Medicine and Biology*, *44*(7), 1625–1638.

Sbirlea-Apiou, G., Katz, I., Caillibotte, G., Martonen, T., & Yang, Y. (2007). Deposition mechanics of pharmaceutical particles in human airways (2007). *Lung Biology in Health and Disease*, *221*, 1–30.

Schmidt, A., Zidowitz, S., Kriete, A., Denhard, T., Krass, S., & Pietgen, H. O. (2004). A digital reference model of the human bronchial tree. *Computerized Medical Imaging and Graphics*, *28*(4), 203–211.

Spencer, R. M., Schroeter, J. D., & Martonen, T. B. (2001). Computer simulations of lung airway structures using data-driven surface modeling techniques. *Computers in Biology and Medicine*, *31*(6), 499–511.

Stapelton, K. W., Guentsch, E., Hoskinson, M. K., & Finlay, W. H. (2000). On the suitability of the κ-ε turbulence modeling for aerosol deposition in the mouth and throat: a comparison with experiment. *Journal of Aerosol Science*, *31*, 739–749.

Tian, G., Longest, P., Su, G., Walenga, R. L., & Hindle, M. (2011). Development of a stochastic individual path (SIP) model for predicting the tracheobronchial deposition of pharmaceutical aerosols: effects of transient inhalation and sampling the airways. *Journal of Aerosol Science*, *42*(11), 781–799.

Tian, G., Hindle, M., & Longest, P. W. (2014). Targeted lung delivery of nasally administered aerosols. *Aerosol Science and Technology*, *48*(4), 434–449.

Tawhai, M. H., & Burrowes, K. S. (2003). Developing integrative computational models of pulmonary structure. *Anatomical Record*, *275B*(1), 207–218.

Versteeg, H. K., & Malalasekera, W. (2007). An introduction to computational fluid dynamics. Harlow, England: Pearson.

Walters, D. K., & Luke, W. H. (2010). A method for three-dimensional Navier-Stokes simulations of large-scale regions of the human lung airway. *Journal of Fluids Engineering*, *132*(5), 051101-1–051101-8.

Walters, D. K., Burgreen, G. W., Lavallee, D. M., Thompson, D. S., & Hester, R. L. (2011). Efficient, physiologically realistic lung airflow simulations. *Biomedical Engineering, IEEE Transactions*, *58*(10), 3016–3019.

Walters, D. K., Burgreen, G. W., Hester, R. L., Thompson, D. S., Lavallee, D. M., Pruett, W. A., & Ford-Green, J. Simulations of Cyclic Breathing in the Conducting Zone of the Human Lung. *ASME Paper 2012*, No. FEDSM2012-72474. 2012.

Weibel, E. R. (1963). *Morphometry of the human lung*. Berlin, Germany: Springer-Verlag.

Zhang, Z., Kleinstreuer, C., & Kim, C. S. (2008). Airflow and nanoparticle deposition in a 16-generation tracheobronchial airway model. *Annals of Biomedical Engineering*, *36*, 2095–2110.

Annex

Table I.A. Geometric dimensions of the lung branches.

Gen	Diameter (m)	Length (m)	Branches (m2)	Total (m2)
0	1.800000E-02	1.200000E-01	2.5447E-04	2.5447E-04
1	1.221142E-02	4.782228E-02	1.1712E-04	2.3424E-04
2	8.284380E-03	2.485314E-02	5.3903E-05	2.1561E-04
3	5.620225E-03	1.686068E-02	2.4808E-05	1.9847E-04
4	4.451270E-03	1.266542E-02	1.5562E-05	2.4899E-04
5	3.512926E-03	1.068537E-02	9.6923E-06	3.1015E-04
6	2.807205E-03	9.014874E-03	6.1893E-06	3.9611E-04
7	2.271429E-03	7.605532E-03	4.0522E-06	5.1868E-04
8	1.860991E-03	6.416519E-03	2.7201E-06	6.9634E-04
9	1.543866E-03	5.413392E-03	1.8720E-06	9.5847E-04
10	1.296864E-03	4.567088E-03	1.3209E-06	1.3526E-03
11	1.103061E-03	3.853092E-03	9.5563E-07	1.9571E-03
12	9.500025E-04	3.250718E-03	7.0883E-07	2.9033E-03
13	8.284568E-04	2.742516E-03	5.3905E-07	4.4159E-03
14	7.315347E-04	2.313764E-03	4.2030E-07	6.8862E-03
15	6.540636E-04	1.952042E-03	3.3599E-07	1.1010E-02
16	5.921409E-04	1.646869E-03	2.7538E-07	1.8048E-02

12

Calibration of Horizontal Acoustic Doppler Current profilers by three dimensional CFD simulations

Kari Bråtveit* and Nils Reidar B. Olsen

Department of Hydraulic and Environmental Engineering, The Norwegian University of Science and Technology, S.P. Andersens veg 5, N-7491 Trondheim, Norway

This study presents a calibration method of horizontal acoustic Doppler current profilers (H-ADCP), based on 3D Computational Fluid Dynamics (CFD) simulations. Rigidly mounted H-ACCPs are currently used to continuously monitor water discharges. In this study three instruments were employed at separate sites in an unlined rock-blasted hydro power tunnel. The total discharge in the tunnel was predicted by fitting numerical and measured velocity profiles. The accuracy of the method was evaluated by comparing discharge predictions to results obtained by thermodynamic efficiency measurements. This study shows that 3D CFD simulations not only can be used to accurately calibrate H-ADCP, but also to assess the flow conditions at the locations of installation. An average error of 2% was obtained for the tunnel stretch that possessed the most uniform flow. However, the precision of the discharge estimates were confined by the CFD code's capability to correctly resolve the flow pattern.

Keywords: Hydro power; waterways; discharge; calibration of H-ADCP; numerical models

1. Introduction

Hydropower is an important renewable energy source worldwide, and 98% of the electric energy in Norway is hydroelectricity (Cheng, Shen, Wu, & Chau, 2012). A large portion of the waterways feeding the hydro power plants in Norway are unlined rock-blasted tunnels, making up a total distance of 3500 km. Many power plants are fed by multiple reservoirs and streams, and the system of tunnels can therefore be vast and complex. Reliable discharge data are crucial for water resource planning, since continuous series of water discharges are essential inputs both for hydrological models and operation simulations (McMillan, Freer, Pappenberger, Krueger, & Clark, 2010). Hence, an accurate monitoring system of the water discharges in the different branches of a tunnel system is valuable.

The current article presents an approach where H-ADCP instruments are applied in a hydro power tunnel. The transducers on an H-ADCP transmit acoustic signals in horizontal lines. These signals are reflected by suspended particles in the flow, and the Doppler effects can be utilized to calculate the water velocities. A two beam H-ADCP measures the flow velocities in two horizontal lines. The beam velocities are transposed into one horizontal velocity profile. Since the H-ADCP does not measure the velocity distribution over the whole cross section, it needs to be calibrated in order to determine the discharge.

In this study a calibration procedure founded on 3D CFD simulations is suggested. A 3D CFD model can be used to compute the velocities and the corresponding discharges. Thus, the total discharge can be predicted based on the best fit between measured and computed velocity profiles. The main advantage of utilizing a 3D numerical approach is that it enables a more detailed evaluation of how a complex geometry affects the flow field, because a 3D CFD simulation provides an estimate of the spatial variation of the velocities. In order to compare the accuracy of resulting CFD discharges, a secondary field measurement was conducted. The secondary discharge measurement was part of a conventional thermodynamic efficiency measurement.

Recent studies described how H-ADCP can be calibrated and employed to continuously monitor river discharge (Nihei & Kimizu, 2008; Sassi, Hoitink, Vermeulen, & Hidayat, 2011). Often the discharge is found by applying the index-velocity method. The index velocity method normally computes the discharge by two separate ratings, the index velocity rating and the stage-area rating. The mean channel velocity and the cross-sectional area are then multiplied together to compute the discharge. The mean channel velocity is a function of the streamwise velocity, stage, cross-stream velocity and velocity head. In a pressurized flow system the water filled area will be constant, which eliminates the need of establishing a stage-area rating curve. The index-rating is established by regression, where the relationship between the mean cross-sectional velocity for the standard section and the

*Corresponding author. Email: kari.bratveit@ntnu.no

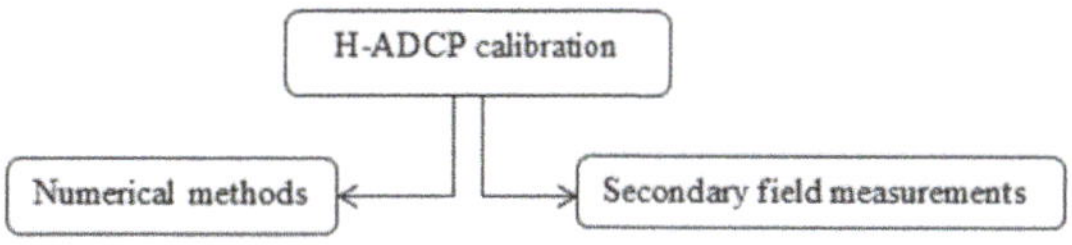

Figure 1. Calibration methods of ADCPs.

measured index velocity is defined (Gandhi, Verma, & Patnaik, 2008; Levesque & Osberg, 2012). The index velocity rating must be developed through a calibration process. This calibration process can be divided into two main categories; numerical methods or secondary field measurements, as illustrated in Figure 1. In most operated flows several discharge levels can be determined by secondary field measurements. In contrast, numerical methods become useful for natural flows and flows in structures that are dangerous to access for field measurements. However, a reliable relationship between the velocities and discharge is sometimes hard to establish, regardless of the method chosen, for example, for flood events.

This issue has been resolved in different ways. An automatic system that changes the vertical position of the H-ADCPs, and in this way collects the velocities for the entire cross section, has been developed. However, such a system is expensive and spacious. Another method applied is to interpolate and extrapolate the horizontal velocity profile or point velocity measurements over the whole cross section on the basis of the law of the wall (Gandhi, Verma, & Patnaik, 2008; Nihei & Kimizu, 2008). However, the law of the wall is only valid for a steady, uniform and fully developed turbulent flow, and does not capture the dynamic principle for fluid motion. The accuracy of this approach therefore decreases distinctly (Nihei & Kimizu, 2006).

The most recently developed calibration method is based on assimilating a river flow simulation to H-ADCP field data. Assimilations of river flow data are infrequently performed (Sulzer, Rutschmann, & Kinzelbach, 2002). However, Nihei and Kimizu (2006) have developed a river flow model that incorporates the measured velocities across a horizontal line at a fixed height. In these numerical simulations the measured velocity profiles are interpolated and extrapolated over the cross section. This method is referred to as the dynamic interpolation and extrapolation method (DIEX). The fundamental equation in a DIEX simulation calculates the velocity field in the cross-sectional domain. The equation is a simplified version of the two-dimensional momentum equation. In order to solve the equation a zero-equation-turbulence model is applied (Nihei & Kimizu, 2008).

In this study, a new calibration method has been tested. This method is based on determining the flow field by 3D CFD simulations. This has earlier been done by Olsen and Kjellesvig (1999), who computed flow in a desilting basin of the Svartisen Hydropower plant in Norway. Recently a number of numerical studies have been conducted where CFD is used to predict the flow fields for different hydraulic cases, as documented by Baghalian, Bonakdari, Nazari, and Fazil (2012), Haun, Olsen, and Feurich (2011), Chau and Jiang (2004), Andersson, Andreasson, and Lundström (2013), Erduran, Seckin, Kocaman, and Atabay (2012), Chau and Jiang (2001), and Baranya, Olsen, Stoesser, and Sturm (2012).

A detailed 3D numerical model incorporates the fluid-structure interaction, and enables an evaluation of the cross-stream velocities. In the presented study, the computations were performed by the commercial 3D CFD package, Star-CCM+. This numerical application can solve the Reynolds-averaged Navier-Stokes (CD-adapco, 2009), equation with a standard two equation k-ε turbulence model (Haun, Olsen, & Feurich, 2011). A k-ε model computes the turbulent eddy-viscosity by solving two partial differential equations. A k-ε turbulence model requires no calibrations contrary to a zero-equation model.

2. Measurements

2.1. Study site

This study was conducted in a fully pressurized rock-blasted unlined hydro power tunnel, which is part of the supply system of Tonstad Hydro Power Plant; see Figure 2. Three 1200 kHz Broadband Channel master H-ADCPs from RDI were installed and deployed. During the measurements the section was pressurized with an approximately 60 meter water head. The water in this system was clear, the sediment concentration was low and few floating obstacles, such as fish or pieces of wood or plastic, were observed.

Each of the ADCPs was rigidly mounted to the tunnel wall at different locations. An overview of the geometry of the tunnel and the location of each instrument can be seen in Figure 3.

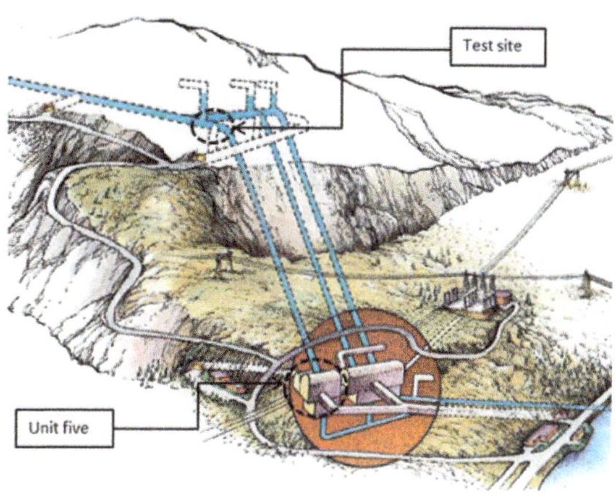

Figure 2. The lower part of Tonstad Hydro Power Plant (Helge Steinnes, 1993).

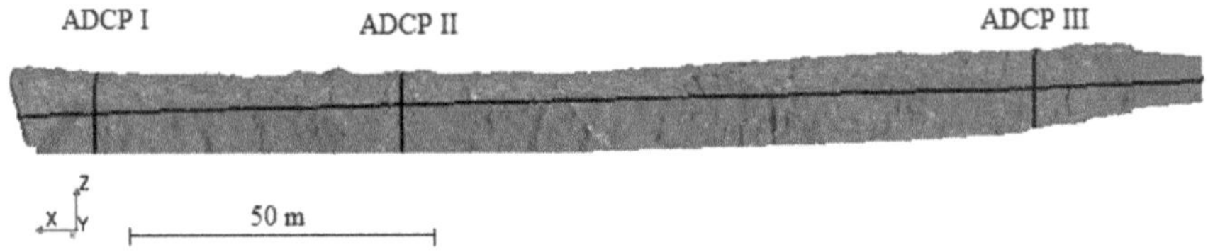

Figure 3. The location of the three ADCPs, top view. Flow direction is from right to left.

Power supply and data transfer were ensured through a cable connection to each of the instruments. The data were stored by a laptop located in a switchboard in the nearby access tunnel. Since the data from the vertical beam had no value this sensor was muted, which increased the transfer capacity. The cross section form and area of an unlined rock-blasted tunnel varies (Rønn & Skog, 1997). A 3D laser scan was performed to obtain correct geometry input for the CFD simulation. The details regarding scanning and generation of the surface mesh are presented by Bråtveit, Lia and Olsen (2012).

2.2. *Field measurements*

The ADCP measurements and the turbine efficiency measurement were conducted simultaneously at the fifth unit at Tonstad Hydro Power plant. The fifth turbine unit is located in one of the three branches of the conveyance tunnel system, i.e., downstream from the location of the three ADCPs, as seen in Figure 2. A standard procedure for turbine efficiency measurement was applied, and the unit was regulated stepwise at different outputs. For each step the production was kept constant for an adequate time period to ensure the flow had stabilized, and the transients were damped out. The four other units, two adjacent to each of the other two pressure shafts, were contra operated striving to maintain constant pressure and head loss in the system. The measuring program is presented in Table 1.

2.2.1. *Thermodynamic efficiency measurement*

During performance of a conventional TEM the turbine discharge has to be determined for each output. This type of measurement requires several instruments, such as thermometers, rotameters, pressure sensors and pulse counters. The thermometers and the pressure transducers are the most essential instruments regarding accurate determination of the discharge. Each of the measurements taken during the test included uncertainties, due to inaccuracy within the instrument used and random variations of the measured value. The individual biases were combined by the law of uncertainty propagation, and it was reported ± 1.0% of the actual discharge. The resulting discharges are provided in Table 1. Further details regarding the TEM can be found in Brænd (2013).

Table 1. Measuring procedure and discharges obtained by TEM.

No.	Hour	Turbine Output [MW]	Q_{TEM} [m^3/s]
1	10:15	283,0	70,17
2	11:00	326,77	81,53
3	11:45	340,51	84,96
4	13:05	149,04	39,34
5	13:35	187,08	47,77
6	14:20	206,52	52,03
7	14:50	233,43	58,26

2.2.2. *Velocity measurements by ADCP*

The ADCP measurements were performed parallel to the TEM measurements. The water velocities were measured at each of the three locations by utilizing equal configuration for all sensors. An overview of the configuration used can be found in Table 2.

ADCP discharge measurements contain systematic and random errors, as in any other measurement. Main systematic errors arise due to biases in ADCP water velocity measurements, while the random errors are largely due to Doppler noise, or flow pulsations caused by turbulence (Huang, 2008). Random errors due to spatial resolution occur, since the velocities are measured radially, before being transformed into an orthogonal coordinate system by assuming that the flow between the two beams is horizontally homogeneous. Since the resulting velocities are spatially averaged, they are suspected to yield biases estimated of mean flow for highly three-dimensional flows (Nystrom, Oberg, & Rehmann, 2002). However, averaging the velocity data over a longer time period will reduce the random errors (Simpson, 2001). By choosing a long sampling time, the turbulence flow pattern will be averaged out (Muste, Yu, Pratt, & Abraham, 2002; Szupiany, Amsler, Best, & Parsons, 2007). The average interval in this study

Table 2. Set up configuration of ADCPs.

Specification	Quantity
Number of cells	50
Pings per ensemble	12
Depth cell length [cm]	25
Blank after transmit [cm]	10
Time between ensemble [s]	6

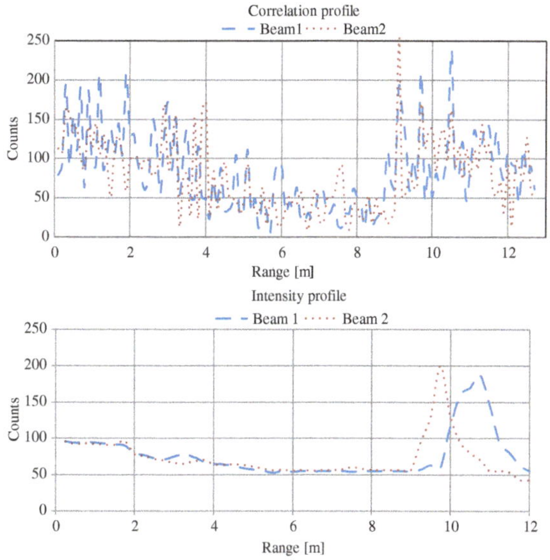

Figure 4. (a) Correlation profile and (b) Intensity profile form ADCP I at discharge level four.

was therefore set to 10 min, in order to reduce the uncertainty in the velocity data due to spatial resolution errors. The production output data were evaluated, and ADCP data were extracted for periods with stable production.

The quality of the recorded velocity data had to be evaluated to ensure that invalid data were excluded. The first limitation to consider was the width of the tunnel. Signals contaminated due to side lobbing from the opposite wall had to be excluded. For each of the three ADCPs, the intensity profile and correlation profile were evaluated. In Figure 4, two representative examples of these charts are presented by plots of values obtained from ADCP I. The location of the two peaks in the intensity profile, shown in Figure 4b), reveals where the signals were reflected by the wall. Due to the roughness of the tunnel wall, the signals from the two beams hit the wall at different distances. Data that originated from bins located after the peak were removed. In addition this width was cut off by 6%, plus one bin size.

The second limitation considered was the correlation profile, as shown in Figure 4a). The correlation count describes how well the returned signal corresponds to its original code tag. The data were only used if the data had a correlation factor above 72 (Teledyne, 2010). Finally, the echo intensity of the data was evaluated. Due to clear water the echo intensity profiles had very low values, as seen in Figure 4b). The echo intensity profiles were obtained from the backscatter of the signal. A high-intensity count value indicates high sediment concentration (Moore, Coz, Hurther, & Paquier, 2011). It is known that the raw data include contamination caused by background or instrument noise (Levesque & Osberg, 2012). The Channel Masters noise

Table 3. Mean velocities and standard deviations obtained from ADCP measurements.

		ADCP					
		I		II		III	
NO.	Hour	$\bar{v}$ [m/s]	σ_v	$\bar{v}$[m/s]	σ_v	$\bar{v}$[m/s]	σ_v
1	10:15	1.2	0.56	0.56	0.10	0.51	0.09
2	11:00	1.3	0.58	0.66	0.11	0.59	0.09
3	11:45	1.2	0.69	0.69	0.12	0.64	0.10
4	13:05	0.7	0.31	0.30	0.06	0.28	0.06
5	13:35	0.7	0.41	0.36	0.08	0.33	0.07
6	14:20	1.0	0.36	0.42	0.08	0.39	0.07
7	14:50	1.1	0.47	0.47	0.09	0.44	0.08

level is assumed by RDI Teledyne to be around 40 counts. As seen in Figure 4(a) and 4(b) the intensity and correlation profiles have low counts values, and are decreasing into the expected noise level range of the instrument. Therefore, as a conservative assumption, the ADCP's velocity profiles are only considered valid up to five meters from the sensor, when operated in such clear water. The accuracy of the data was considered to be too low after five meters, due to the fact that the measured values were in the range of the background noise level. The resulting mean velocities for each output level, and the corresponding standard deviation are presented in Table 3.

2.3. Numerical modeling by CFD

The CFD simulations were performed by the commercial software STAR-CCM+, which is a CFD-package applied to a broad specter of physics. In this study the software was used to solve the Navier-Stokes equations in three dimensions. The boundary conditions of the model were set according to data obtained by field measurements, production rates, pressure, and temperature data were available. The computational domain was extended upstream of the inlet, in order to create a correct velocity distribution over the inlet. Though, after testing several different domains for the inlet structure the velocity distributions downstream the inlet remained nearly identical. It seems to be governed by the flow separation caused by the rapid expansion 11.2 m downstream the inlet. Both first- and second-order discretization schemes were tested, but little variance in the results was found. Therefore a first order discretization scheme was used for the convective terms. For each output level several CFD simulations were conducted.

A turbulent flow is characterized by a large variation in scales of motions, and an accurate numerical approximation of a turbulent flow field involves resolving these equations for all scales of motion. The optimum way to do that is to perform a Direct Numerical Simulation (DNS) (Wissink, 1995). However, for flows with high Reynolds numbers a DNS simulation requires that the size of the numerical mesh is smaller than the size of small-turbulent motions, where dissipation takes place. As a consequent, the number of grid points are needed for most practical hydraulic engineering problems becomes so large that computational resources necessary exceed the capacity of available computers (Rodi, Conatantinescu, & Stoesser, 2013). A common way to avoid massive computations is to use a turbulence model based on the Reynolds average Navier-Stokes Equations – a so called RANS model. A RANS model is based on averaging the NS equations in time, meaning that this statistical model accounts for the entire effect of the turbulence motions. In the last decades the computer power has kept increasing, and this has revealed that RANS models have limitations, especially when applied for coherent structures. Recent development has provided methods such as Large Eddy Simulations (LES), where the large scales are resolved and the small scales are model (Rodi, Conatantinescu, & Stoesser, 2013). Another recent method is the Detached Eddy Simulation (DES), which is a hybrid modeling approach where RANS closure is applied in some parts of the flow field and LES in others (CD-adapco, 2009). However, LES and DES methods demand a highly dense grid to correctly resolve the dissipation, while RANS simulations can be computed based on coarser grids. Hence, RANS models are most applicable for practical employment, because they are economical in terms of computational and memory resources. Though, one should be aware that RANS simulations have limitations, especially when dealing with large-scale turbulence structures (Rodi, Conatantinescu, & Stoesser, 2013).

The calibration method suggested here is based upon utilizing a RANS model applying a standard k-ε turbulence model with a high Reynolds number wall treatment. A k-ε model is a two-equation model used to estimate the eddy viscosity by relating the mean-flow quantities to the time scales of turbulence. Further details of this study can be found in Bråtveit, Brevik, & Olsen (2013).

Verification of a CFD simulation involves a systematic mesh refinement study (Versteeg & Malalasekera, 2007). Hence, several types of nonorthogonal unstructured grids were generated from the surface mesh. Grids based on both orthogonal hexahedral and polyhedral cells were applied, and the grids were refined from the wall to the centre of the tunnel. Grid-dependency tests were performed throughout, including refinements of the grids in the expansion zone downstream the inlet. Once a grid independent solution was obtained, a final test of CFD model was performed by comparing the output to experimental data (Versteeg & Malalasekera, 2007). The final grid chosen for calibration consisted of 849 627 hexahedral cells, which represents a

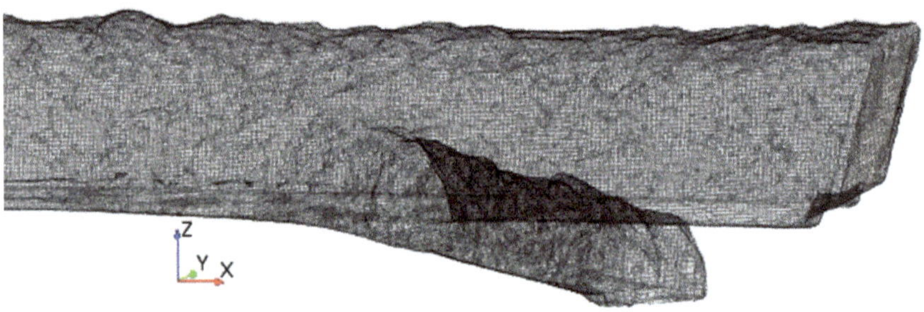

Figure 5. The volume mesh consisted of trimmed cells.

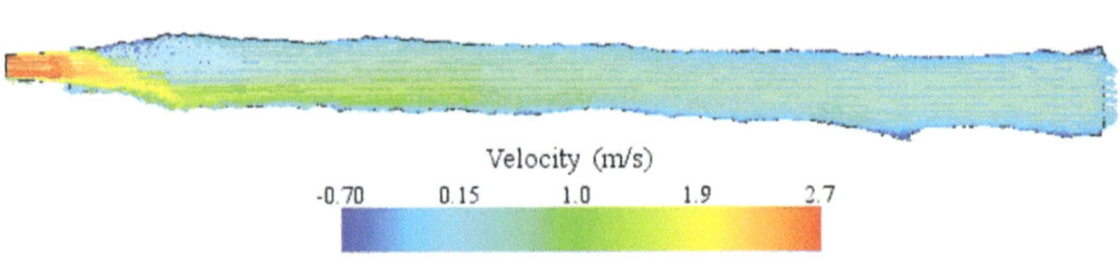

Figure 6. Streamwise velocities plotted for the central plan of the structure.

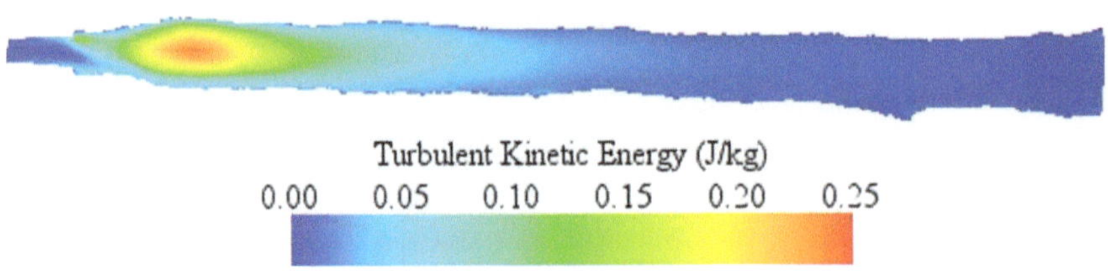

Figure 7. TKE values from simulation of discharge level one.

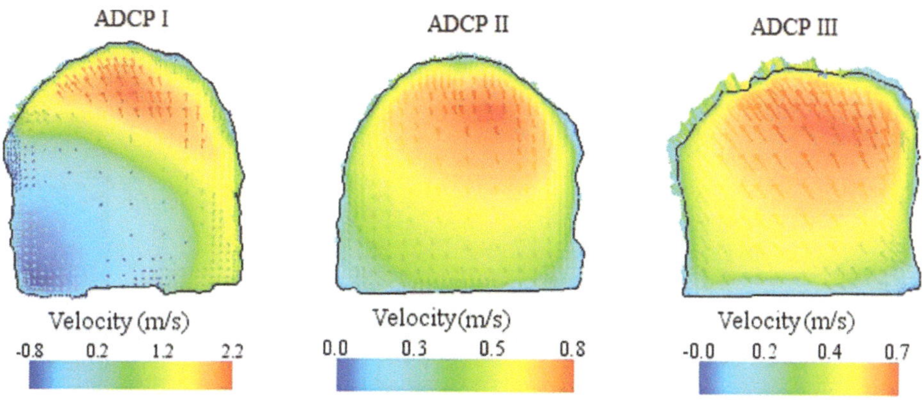

Figure 8. Velocity distributions at different cross sections.

volume mesh of intermediate size, possible to run on an ordinary off-the-shelf computer. A plot of this relatively coarse mesh is provided in Figure 5.

3. Results and discussion

The model evaluation techniques applied in this study include both graphical and statistical methods. First, the flow pattern occurring in the structure was assessed graphically. This is useful because it allows us to evaluate the impact of the geometry on the flow. Figure 6 shows a representative plot of the flow field for the first discharge level in the stream-wise direction for the central plane of the structure. In Figure 7 the corresponding turbulent kinetic energy (TKE) values for the same plane are shown, while plots of the velocity distribution over the cross sections where the instruments are located are provided in Figure 8. The plots indicate that a recirculation zone develops downstream the rapid expansion of the inlet structure, causing increased TKE values in the first part of the tunnel. The plots of the velocity distributions for the cross sections show that the flow field changes from skewed at ADCP I toward more uniform at ADCP II and ADCP III.

The velocity profiles obtained by CFD were graphically compared to the ADCP profiles. One chart was created for each sensor at each discharge level. An example of these plots, obtained from the iteration process for output level one, can be seen in Figure 9. The filtering process previously executed indicated that only the first five meters of the measured profiles are valid, and therefore used for comparison. The graphical evaluation of the numerical simulations reveals that the CFD simulation fails to predict the flow pattern correctly for the first part of the structure, i.e., the simulated velocity profiles obtained for ADCP I do not resemble the profiles from the field measurements. This ADCP is located shortly downstream from an expansion

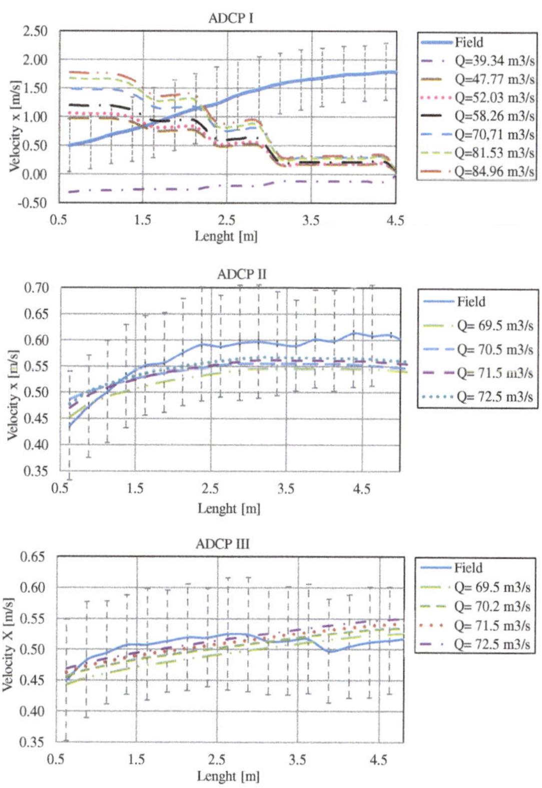

Figure 9. Velocity plots obtained by field measurements and simulation (a) ADCP I; (b) ADCP II; (c) ADCP III.

Table 4. Discharges and statistics obtained during the iteration process of discharge level one for ADCP III.

	Profile III			
Iteration	Q_i [m^3/s]	SSE	RMSE	PBIAS
1	69.5	0.01	0.03	2.1
2	71.5	0.01	0.02	– 1.3
3	72.5	0.01	0.03	– 2.6
4	70.16	0.01	0.02	0.0

Table 5. Results from the statistical evaluation.

	Profile II				Profile III			
No.	Qi [m^3/s]	SSE	RMSE	PBIAS	Qi [m^3/s]	SSE	RMSE	PBIAS
1	73.91	0.01	0.02	0.7	70.16	0.01	0.02	0.0
2	87.75	0.01	0.02	0.5	82.39	0.15	0.02	0.4
3	91.69	0.03	0.04	0.05	89.63	0.01	0.02	0.0
4	41.00	0.00	0.01	0.0	38.83	1.85	0.31	0.1
5	48.90	0.00	0.010	0.3	46.70	0.00	0.01	0.0
6	56.43	0.01	0.02	0.0	55.39	0.00	0.01	0.0
7	63.96	0.00	0.02	0.0	61.29	0.01	0.02	0.0

Table 6. Comparison of the discharge values.

No.	Q_{TEM} [m^3/s]	Q_{IICFD} [m^3/s]	Q_{IIICFD} [m^3/s]	Error Q_{II} [%]	Error Q_{III}[%]	TKE II [J/kg]	TKE III [J/kg]
1	70.17	75	70.16	6.9	0.0	0.014	0.007
2	81.53	87.75	82.39	7.6	1.1	0.022	0.010
3	84.96	91.69	89.63	8.0	5.6	0.026	0.011
4	39.34	41	38.83	4.2	− 1.3	0.004	0.002
5	47.77	48.9	46.7	2.4	− 2.2	0.005	0.002
6	52.03	56.43	55.39	8.5	6.5	0.008	0.004
7	58.26	63.96	61.29	9.8	5.2	0.010	0.004

zone, where the flow is highly unsteady. Figure 9(a) shows that two RANS simulations with nearly identical boundary conditions can yield recirculation zones occurring on opposite side downstream from a rapid expansion. In this case the recirculation zones occurred randomly on the left or the right side of the structure. The velocity profiles for ADCP II and ADCP III fit relatively well, indicating that the CFD simulations were able to estimate the flow conditions fairly well.

Several commonly used statistical error indicators were applied in order to evaluate the fit between the simulated and the calculated velocities during the iteration process. The statistical evaluation includes comparing means, standard square error (SSE), root mean square error (RMSE) and percent bias (PBIAS) (Moriasi et al., 2007). The equation for SSE, RMSE and PBIAS are presented below. Only ADCP II and ADCP III were statistically evaluated, since the graphical evaluation revealed unsatisfactory correlation for ADCP I.

$$SSE = \sum_{i=1}^{n} (Y_i^{obs} - Y_i^{sim})^2 \tag{10}$$

$$RMSE = \sqrt{\frac{\sum_{i=1}^{n} (x_{obs,i} - x_{sim,i})^2}{n}} \tag{11}$$

$$PBIAS = \left[\frac{\sum_{i=1}^{n}(Y_i^{obs} - Y_i^{sim}) \times (100)}{\sum_{i=1}^{n}(Y_i^{obs})}\right] \tag{12}$$

The various discharge values obtained during the iteration process for ADCP III at output level one are presented in Table 4. Similar iteration processes have been conducted for each of the seven discharge levels for both ADCP II and ADCP III. The final results from these iterations are provided in Table 5. It can be observed that the SSE, RMSE and PBIAS values in general are low, but they increase for higher discharges. The discharge values obtained by CFD, Q_{CFD}, are compared against the TEM discharges, and the percent errors between the values were calculated. These results are listed in Table 6. The iteration process shows that the PBIAS values and the errors are lowest for ADCP III, where the TKE values are lowest. The results reveal that the accuracy of the discharges obtained by CFD are restricted by the levels of turbulence arising at the specific site of installation, and that a RANS simulation is not able to predict the flow field close to a rapid expansion correctly.

Table 7. Resulting fit obtained regression analysis.

ADCP No.	R^2 value A)	R^2 value B)
I	0,9609	—
II	0,9735	0,9706
III	0,9900	0,9900

Finally, the index velocity ratings for each of the sensors were established by regression. Two types of regressions were performed, (A) between Q_{TEM} values and the corresponding mean ADCP velocities, and (B) between Q_{CFD} values and the corresponding mean ADCP velocities. The R squared values are presented in Table 7. It can be observed that coefficients of determinations are higher for ADCP III than for ADCP II.

4. Conclusion

In this study, a calibration method for rigidly mounted H-ADCPs based on results from 3D CFD simulations is evaluated. The study shows that detailed CFD simulations provide valuable insight into the flow pattern and the turbulence level arising in different parts of the structure. This is essential information, not only to ensure accurate calibration, but also for deciding the best site of installation for the ADCPs.

One of the well-known limitations of H-ADCP is that the instruments yield only a narrow representative line of the cross section's velocity distribution. Hence, it becomes crucial to apply the instrument so that the measured line captures the changes in the discharge. Locations where the cross sectional velocity distribution is fairly uniform, and not largely influenced by secondary currents, are therefore preferred. In future work, it is of interest to investigate the accuracy of instruments installed at different locations in the same cross-section. For example, comparing results obtained from vertical compared to horizontal position sensors.

A RANS model is economical in terms of memory and computational cost, but has limitations when large-scale turbulence structures dominate the flow. This study shows that a RANS simulation with a standard k-ε model was not able to correctly resolve the flow pattern close to a rapid expansion. Though, it was found that the precision of the flow field increased parallel to the reduction in turbulent kinetic energy. Hence, the accuracy of the calibration method was found confined by the level of turbulence occurring at the locations of the sensors, and the CFD's code capability to correctly resolve the flow pattern.

The simulated discharges were compared to values obtained by TEM. An average error of $\pm 2\%$ was found for the ADCP applied at the tunnel stretch with the most uniform flow condition. This error is in the range of the accuracy of the instruments.

This study shows that H-ADCP calibrated with CFD simulations can yield accurate and continual discharge measurements for pressurized hydro tunnels. Prediction of the flow field can be established by performing economical RANS simulations with standard k- ε models. Though, it should be kept in mind that RANS simulations have difficulties in coping adequately with complex flow phenomena, such as large-scale turbulence structures Rodi, Conatantinescu, & Stoesser, 2013), and one should strive to avoid installing the instrument in or close to recirculation zones.

Acknowledgements

This study is funded by the Centre for Environmental Design of Renewable Energy (CEDREN) in Norway. The writers are grateful to Sira-Kvina Kraftselskap A/S for technical support, facilitating access and willingness in cooperation. A. Sæterdal, K. Grangier, G. Brænd and Å. Killingtveit have provided useful support. We also want to thank NOTUR for providing computational time on the Vilje HPC cluster.

References

Andersson, A. G, Andreasson, P., & Lundström, T. S. (2013). CFD-modelling and validation of free surface flow during spilling of reservoir in down-scale model. *Engineering Applications of Computational Fluid Mechanics*, *7*(1), 159–167.

Baghalian, S., Bonakdari, H., Nazari, F., & Fazil, M. (2012). Closed-form solution for flow field in curved channels in comparison with experimental and numerical analyses and artificial neural network. *Engineering Applications of Computational Fluid Mechanics*, *6*(4), 514–526.

Baranya, S., Olsen, N. R. B., Stoesser, T., & Sturm, T. (2012). Three-dimensional rans modeling of flow around circular piers using nested grids. *Engineering Applications of Computational Fluid Mechanics*, *6*(4), 648–662.

Brænd, G. (2013). *Tonstad kraftverk Virkingsgradsmålinger turbin 5*. Norway: Report, SWECO.

Bråtveit, K., Brevik, O., & Olsen, N. R. B. (2013). *Three dimensional numerical study of an unlined pressurized rock-blasted sand trap*. 35th IAHR World Congress, Chengdu, China.

Bråtveit, K., Lia, L., & Olsen, N. R. B. (2012). An efficient method to describe the geometry and the roughness of an existing unlined hydro power tunnel. *Energy Procedia*, *20*, 200–206.

CD-adapco. (2009). *User Guide STAR-CCM+ Version 4.06.011*. CD-adapco.

Chau, K. W., & Jiang, Y. W. (2001). 3D numerical model for Pearl River Estuary. *Journal of Hydraulic Engineering*, *127*(1), 72–82.

Chau, K. W., & Jiang, Y. W. (2004). A three-dimensional pollutant transport model in orthogonal curvilinear and sigma coordinate system for Pearl river estuary. *International Journal of Environment and Pollution*, *21*(2), 188–198.

Cheng, C. T., Shen, J. J., Wu, X. Y., & Chau, K. W. (2012). Operation challenges for fast-growing China's hydropower systems and respondence to energy saving and emission reduction. *Renewable and Sustainable Energy Reviews*, *16*(5), 2386–2393.

Erduran, K. S., Seckin, G., Kocaman, S., & Atabay, S. (2012). 3D numerical modelling of flow around skewed bridge crossing. *Engineering Applications of Computational Fluid Mechanics*, *6*(3), 475–489.

Gandhi, B. K., Verma, H. K., & Patnaik, S. S. (2008). *Discharge measurement in small hydropower stations using Acoucstic Doppler Current Profiler*. 7th International Group of Hydraulic Efficiency Measurement, Milano, Italy.

Haun, S., Olsen, N. R. B., & Feurich, R. (2011). Numerical modeling of flow over trapezoidal broad-crested weir. *Engineering Applications of Computational Fluid Mechanics*, *5*(3), 397–405.

Huang, H. (2008). *Estimating precision of moving boat ADCP discharge measurement*. Australian Hydrographers Association Conference 2008; Canberra, Austraila.

Levesque, V. A., & Osberg, K. A. (2012). *Computing discharge using the Index Velocity method, techniques and methods 3-A23*. Reston, VA: U.S. Department of The Interior, United States Geological Survey.

McMillan, H., Freer, J., Pappenberger, F., Krueger, T., & Clark, M. (2010). Impacts of uncertain river flow data on rainfall-runoff model calibration and discharge predictions. *Hydrological Processes*, *24*(10), 1270–1284.

Moore, S. A., Coz, L. J., Hurther, D., & Paquier, A. (2011). *On the use of Horizontal-ADCPs for Sediment Flux Measurements in Rivers*. 34th IAHR World Congress, Brisbane, Australia.

Moriasi, D. N., Arnold, J. G., Van Liew, M. W., Bingner, R. L., Harmel, R. D., & Veith, T. L. (2007). Model Evaluation guidelines for systematic quantification of accuracy in watershed simulations. *American Society of Agricultural and Biological Engineers*, *50*(3), 885–900.

Muste, M., Yu, K., Pratt, T. C., & Abraham, D. (2002). ADCP measurements at fixed river locations. In *Hydraulic measurements and experimental methods 2002(CD-ROM) ASCE* (pp. 622–633). Estes Park, CO.

Nihei, Y., & Kimizu, A. (2006). Evaluation of river velocity and discharge with a new assimilated method. *International Journal of River Basin Management*, *4*(1), 27–30.

Nihei, Y., & Kimizu, A. (2008). A new monitoring system for river discharge with horizontal acoustic Doppler current profiler measurements and river flow simulation. *Water Resources Research*, *44*(4), W00D20.

Nystrom, E. A., Oberg, K. A., & Rehmann, C. R. (2002). Measurement of turbulence with acoustic Doppler current profilers—Sources of error and laboratory results. In *Hydraulic Measurements & Experimental Methods 2002 (CD-ROM) ASCE* (pp. 346–355). Estes Park, CO.

Olsen, N. R. B., & Kjellesvig, H. M. (1999). Three-dimensional numerical modelling of bed changes in a sand trap. *Journal of Hydraulic Research*, *37*(2), 189–198.

Rodi, W., Conatantinescu, G., & Stoesser, T. (2013). *Large Eddy simulation in hydraulics*. Nederlands: Book, CRC Press/Balkema.

Rønn, P. E., & Skog, M. (1997). *New method for estimation of head loss in unlined water tunnels*. Proceedings of the 3th International Conference on Hydro power Hydropower'97, Trondheim, Norway.

Sassi, M. G., Hoitink, A. J. F., Vermeulen, B., & Hidayat (2011). *Discharge estimation from H-ADCP measurements in a tidal river subject to sidewall effects and a mobile bed. Water Resources Research*, *47*(6), W06504.

Simpson, M. R. (2001). *Discharge measurements using a broad-band Acoustic Doppler Current Profile* (Open-file report 01-1). Sacramento, CA: United States Geological Survey.

Sulzer, S., Rutschmann, P., & Kinzelbach, W. (2002). Flood discharge prediction using two-dimensional inverse modeling. *Journal of Hydraulic Engineering*, *128*(1), 46–54.

Szupiany, R. N., Amsler, M. L., Best, J. L., & Parsons, D. R. (2007). Comparison of fixed- and moving-vessel flow measurements with an aDp in a large river. *Journal of Hydraulic Engineering*, *133*(12), 1299–1309.

Teledyne, R. D. I. (2010). *ADCP coordinate transformation, formulas and calculations*. User's manual, RDI Teledyne, San Diego, USA.

Versteeg, H. K., & Malalasekera, W. (2007). *An introduction to computational fluid dynamics. The finite volume method* (2nd ed.). Essex, UK: Person Education.

Wissink, J. G. (1995). *Direct numerical simulations of turbulence*. Nederlands: Proefschrift, Rijksuniversiteit Groningen.

13

Evaluation of RANS turbulence models in simulating the corner separation of a high-speed compressor cascade

Haideng Zhang[a,b], Yun Wu[a]* and Yinghong Li[a]

[a]*College of Aeronautical and Aerospace Engineering, Airforce Engineering University, Xi'an, China;* [b]*College of Energy and Propulsion Engineering, Beihang University, Beijing, China*

A typical high-speed compressor cascade flow is simulated using an eddy viscosity model and a Reynolds stress model. Based on comparisons with experimental results, the simulation accuracies of the two turbulence models are evaluated, and some recommendations are provided for further research. Inside the cascade flow passage, the simulation accuracies of both turbulence models decrease because of the corner effect. Downstream of the blade, the wake mixing influence, and thus the flow loss, is under-predicted by both turbulence models. Compared with the linear eddy viscosity model, the modification of the simulation accuracy with regard to endwall flow prediction using the Reynolds stress model is limited, which only occurs at a highly positive incidence angle. The corner separation strength and wake mixing influence predicted by the eddy viscosity model are surpassed by those of the Reynolds stress model. At a negative incidence angle, the simulation result of the Reynolds stress model is proved to be physically unreasonable according to the flow loss prediction.

Keywords: turbulence model; compressor cascade; corner separation; endwall flow; secondary flow

1. Introduction

High axial compressor performance and operability are key factors and considerations in the design process of high-performance aero-engines. Experimental testing of three-dimensional turbulent flows in high-speed and highly-loaded axial compressors is rather expensive (Place, 1997). Therefore, the reliance of axial compressor design processes and research on computational fluid dynamics (CFD) is ever increasing. As noted by Denton (2010) and Menter, Langtry, and Hansen (2004), the reliable prediction of turbulence is critical for turbomachinery aerothermal design. The state of the art for the development and challenges of turbulent flow CFD in turbomachinery is summarized by Tucker (2013). Although the Reynolds-averaged Navier-Stokes (RANS) method has proven to be rather insufficient in modeling complex turbulent flow in turbomachinery, its future in engineering projects of turbomachinery aerodynamics is still promising because of its robustness and relatively low computational resource consumption (Tucker, 2011, 2013).

In axial compressor stator blade flow passages, cross flow on the shroud/hub gives rise to the accumulation of low-energy fluid at conjunctions between the blade suction surface and shroud/hub endwall, which causes a three-dimensional corner separation or even corner stall under the influence of adverse pressure gradients (Dong, Gallimore, & Hodson, 1987; Joslyn & Dring, 1985; Lei, Spakovszky, & Greitzer, 2008; Schulz, Gallus, & Lakshminarayana, 1990). Because of its dramatic impact on compressor performance, corner separation has been studied both experimentally and numerically with the progress of axial compressor design techniques using compressor cascade models (Chen, Qiao, Liesner, & Meyer, 2014; Evans, Hodson, Hynes, & Wakelam, 2010; Gbadebo, Cumpsty, & Hynes, 2005, 2008; Zachos, Grech, Charnley, Pachidis, & Singh, 2011).

With the enormous development in computer science, the large eddy simulation (LES) method has been used in compressor cascade flow simulation in recent years, utilizing a large amount of computer resources (Matsuura & Kato, 2007; McMullan & Page, 2011; Teramoto, 2005; Zaki, Wissink, Durbin, & Rodi, 2009). Modeling three-dimensional corner separation requires the CFD code to be very robust. Therefore, most of the LES work is typically focused on the two-dimensional flow characteristics in compressor cascade flow passages, and the relatively more influential three-dimensional corner separation has been neglected. Recently, the three-dimensional flow structures in a compressor cascade were simulated using the detached eddy simulation (DES) method (Z. Wang & Yuan, 2013), but the blade airfoil NACA-0009 studied is lightly loaded, which is not representative of modern highly loaded axial compressors. Therefore, the RANS method is widely used in detailed numerical work on

*Corresponding author. Email: wuyun1223@126.com

three-dimensional corner separation. In order to improve the simulation accuracy of the RANS method in corner separation modeling, modification of current RANS models was performed (Liu, Lu, Fang, & Gao, 2011; D. Wang, Lu, & Li, 2009). This, in turn, extended the application of RANS methods in the research on axial compressor corner separations.

Most of the past studies on compressor cascade corner separation were performed in the low-speed regime (with inflow Mach numbers smaller than 0.3) to reveal the flow separation mechanisms and determine an efficient control method to suppress the flow separation. However, it is not always easy to transfer these research results to engineering applications of high-speed flow (Tiedemann, Heinrich, & Peitsch, 2012). CFD application to a high-speed compressor cascade is more difficult because of the more severe corner separations and fully developed flow unsteadiness. Even so, studies on high-speed compressor cascades have greatly benefited from the robust RANS methods (Chen et al., 2014; Gmelin, Thiele, & Liesner, 2011; Hergt, Meyer, & Engel, 2013; Hergt, Meyer, Liesner, & Nicke, 2011; Weber, Schreiber, Fuchs, & Steinert, 2002). The turbulence models applied in the previous high-speed compressor cascade studies were carefully selected eddy viscosity models, which can easily diverge. However, by neglecting the anisotropic nature of the turbulence, the linear eddy hypothesis produces apparent deviations between simulations and experiments. Hence, to investigate high-speed compressor cascade flows, improving the simulation accuracy of the eddy viscosity model is necessary and meaningful. The efforts of D. Wang et al. (2009) and Liu et al. (2011) provided promising guidelines for modifications to eddy viscosity models to better simulate three-dimensional corner separation. However, the present encouraging results are obtained in the low-flow regime and constrained to the one-equation Spalart-Allmaras model. Non-linear eddy viscosity models have also been used, with better simulation accuracy (Balabel & El-Askary, 2011; Craft, Launder, & Suga, 1996; Gatski & Speziale, 1993; Kulisa & Dano, 2006; Mani, Babcock, Winkler, & Spalart, 2013; Mehdizadeh, Temmerman, Tartinville, & Hirsch, 2012). However, the shortcomings of non-linear eddy viscosity models are also unavoidable (Tucker, 2013). Through a careful numerical investigation, it is concluded that a linear turbulence model could provide good results in a simple jet flow, whereas in a complicated jet flow, a non-linear model would be more favorable (Balabel & El-Askary, 2011). In contrast, compared with linear turbulence models, it has also been pointed out that no significant modifications are observed when using the non-linear model to simulate a turbine flow (Kulisa et al., 2006). Thus, it is indeed necessary to use the stress-strain relationship provided by a full Reynolds stress model to obtain accurate predictions. Although the Reynolds stress model is always expected to give improved accuracy, this is not always the case (Tucker, 2013). To achieve closure of equations, the Reynolds stress model, in practice, contains substantial empiricism. Therefore, to apply a Reynolds stress model to high-speed compressor cascade CFD research, careful calibrations with experiments must be applied first.

As previously discussed, to study the corner separation of axial compressors using RANS turbulence models, as a first step, it is essential to evaluate the accuracy of RANS turbulence models in simulating three-dimensional corner separation, especially in the high-speed flow regime. However, because experimental tests of corner separations in the high-speed flow regime are very expensive, researchers are seldom able to perform such evaluations. Therefore, few references are available in this research field.

In this paper, an eddy viscosity model is applied to simulate a high-speed compressor cascade flow. The accuracy of the eddy viscosity model is discussed in detail. Because of the unavoidable deviations between the simulation results of the eddy viscosity model and the results of experiments, a Reynolds stress model is introduced with the expectation of higher simulation accuracy. A better understanding of the advantages and disadvantages of the eddy viscosity model and Reynolds stress model in simulating three-dimensional corner separation is obtained, which is useful for promoting axial compressor corner separation CFD research, especially in the high-speed flow regime.

2. RANS equations

With the introduction of averaged and fluctuating quantities in the original unsteady Navier-Stokes equations, the RANS equations are produced as shown in Equation (1):

$$\frac{\partial \rho U_i}{\partial t} + \frac{\partial}{\partial x_j}(\rho U_i U_j) = -\frac{\partial p}{\partial x_i} + \frac{\partial}{\partial x_j}(\tau_{ij} - \rho \overline{u_i u_j}) + S_M \tag{1}$$

Compared to the direct numerical simulation (DNS) method, computational resources are greatly reduced by the simulation of RANS equations. Therefore, RANS equations have been widely adopted for practical engineering applications. However, the averaging procedure introduces additional unknown terms ($\rho\overline{u_i u_j}$), which contains products of the fluctuating quantities (u_i). The unknown terms $\rho\overline{u_i u_j}$, referred to as Reynolds stress, act like additional stresses in the fluid and are difficult to determine directly. So additional equations are necessary to model the Reynolds stress with known quantities in order to achieve 'closure' and the turbulence model type is defined by the equations used to close the system.

2.1. *Eddy viscosity model*

The eddy viscosity model is based on the hypothesis that the Reynolds stress can be related to the mean velocity gradients and eddy viscosity by the gradient diffusion hypothesis, in a manner similar to the relationship between stress and strain tensors in laminar Newtonian flow:

$$-\rho\overline{u_iu_j}=\mu_t\left(\frac{\partial U_i}{\partial x_j}+\frac{\partial U_j}{\partial x_i}\right)-\frac{2}{3}\delta_{ij}\left(\rho k+\mu_t\frac{\partial U_k}{\partial x_k}\right) \quad (2)$$

where μ_t is the turbulent viscosity and k is the kinetic energy.

Inspired by previous work on high-speed compressor cascades (Chen et al., 2014; Hergt et al., 2011; Hergt, Meyer, & Engel, 2006), the Wilcox k-omega model (Wilcox, 1988) is applied in this paper. The turbulent viscosity μ_t is linked to the turbulence kinetic energy k and turbulent frequency ω via the relation

$$\mu_t=\rho\frac{k}{\omega} \quad (3)$$

For the two-equation Wilcox k-omega model, transport equations for turbulent kinetic energy k and turbulent frequency ω are as follows:

k equation:

$$\frac{\partial(\rho k)}{\partial t}+\frac{\partial}{\partial x_j}(\rho U_j k)=\frac{\partial}{\partial x_j}\left[\left(\mu+\frac{\mu_t}{\sigma_k}\right)\frac{\partial k}{\partial x_j}\right]+P_k-\beta'\rho k\omega \quad (4)$$

ω equation:

$$\frac{\partial(\rho\omega)}{\partial t}+\frac{\partial}{\partial x_j}(\rho U_j\omega)=\frac{\partial}{\partial x_j}\left[\left(\mu+\frac{\mu_t}{\sigma_k}\right)\frac{\partial \omega}{\partial x_j}\right]+\alpha\frac{\omega}{k}P_k-\beta\rho\omega^2 \quad (5)$$

In Equation (5), P_k is the turbulence production due to viscous forces, and is modeled with Equation (6):

$$P_k=\mu_t\left(\frac{\partial U_i}{\partial x_j}+\frac{\partial U_j}{\partial x_i}\right)\frac{\partial U_i}{\partial x_j}-\frac{2}{3}\frac{\partial U_k}{\partial x_k}\left(3\mu_t\frac{\partial U_k}{\partial x_k}+\rho k\right) \quad (6)$$

For a special simulating case, definition of the modeled constants should be based on rich experience. In this paper, having referred to the advice of commonly used CFD software, the modeled constants applied are set as below:

Constants	β'	α	β	σ_k	σ_ω
Value	0.09	5/9	0.075	2	2

2.2. *Reynolds stress model*

Unlike the eddy viscosity model, the Reynolds stress model (Launder, Reece, & Rodi, 1975; Speziale, Sarkar, & Gatski, 1991) is based on transport equations for all components of the Reynolds stress tensor and the dissipation rate. A separate transport equation must be solved for each of the six Reynolds stress components $\rho\overline{u_iu_j}$.

The omega-based Reynolds stress model is applied in this paper and the modeled equations for Reynolds stress can be written as follows:

$$\frac{\partial\rho\overline{u_iu_j}}{\partial t}+\frac{\partial}{\partial x_k}(U_k\rho\overline{u_iu_j})=P_{ij}-\frac{2}{3}\beta'\rho\omega k\delta_{ij}+\Phi_{ij}+\frac{\partial}{\partial x_k}\left((\mu+\mu_t)\frac{\partial\overline{u_iu_j}}{\partial x_k}\right) \quad (7)$$

3. Experimental procedure

The typical high-speed cascade NACA65-K48 (Chen et al., 2014; Hergt et al., 2006, 2011, 2013) was studied. The design Mach number is 0.67, with a design Reynolds number of 4.5×10^5 based on the blade chord length, and the pitch to chord ratio is 0.55. Its main geometry parameters are shown in Table 1.

The experimental results in this paper were obtained in the Dalian Maritime University high-speed linear cascade wind tunnel. The highest inlet Mach number reaches about 1.1, thus being qualified for high subsonic compressor cascade experiments. During the experiment, the inlet Mach number was set as 0.7 and the blades were confined to a rotary plate. Through rotating the plate, the inlet flow incidence can be changed continuously. To keep periodicity, seven blades were utilized in the experiment. Through measuring the transverse distribution of the total pressure loss coefficient downstream of the central three blades, periodicity of the flow field is testified (as shown in Figure 1).

The measurement system employed in the experiments consisted of dynamic probes, a high-frequency pressure transducer, a high-accuracy pressure scanning module (DAS3217) and a high-accuracy thermocouple temperature measurement module (DTS 3250).

4. Numerical method

To reduce the numerical errors, a second-order accurate approximation was applied in the discretion processes, and the robustness was ensured using high-quality calculation

Table 1. Main parameters of NACA65-K48.

Geometric parameters	Value
Chord (c)	60 mm
Span (h)	100 mm
Turning angle (θ)	42°
Inlet flow angle (β_1)	132°
Outlet flow angle (β_2)	0°
Stagger angle (β_3)	112.5°

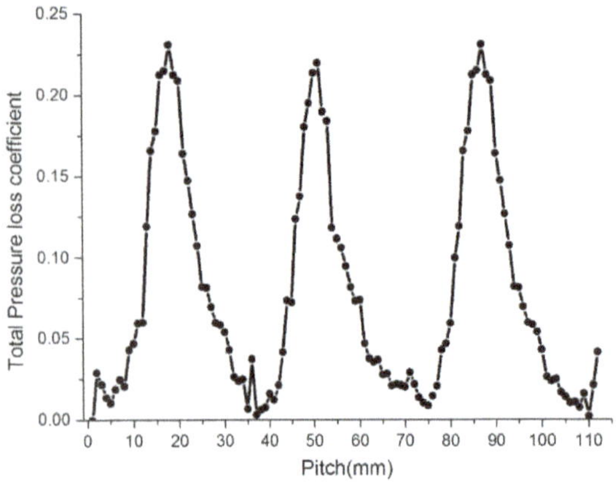

Figure 1. Periodicity of the flow field.

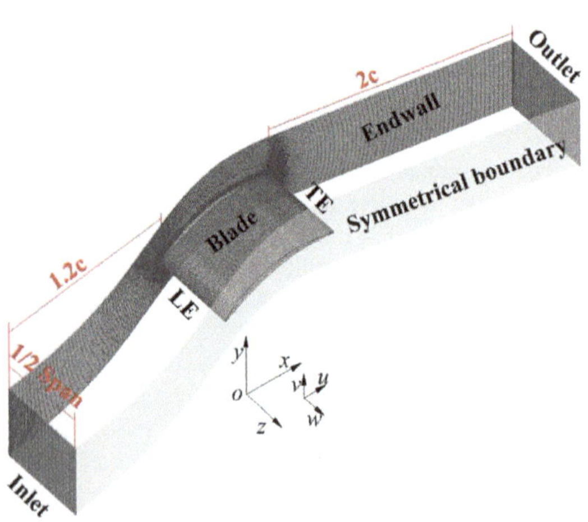

Figure 2. Calculation grid model.

grids. The time step in the simulation was set as 5e-5s, with which fluid is expected to flow pass the cascade blade in about 5 steps.

The calculation grid model is shown in Figure 2. With a symmetrical boundary at mid-span, the cascade flow passage of half span was modeled to reduce the computation cost. The grid element number was varied to test the corresponding influences on the simulation results in preparation for the formal work of this paper. Finally, a grid model of 1.5 million was chosen as optimal, with high numerical accuracy being the first optimizing goal and low computational cost being the second. Near solid walls, grids were refined and the dimensionless variable y^+ is restricted under 1, as the low-Re treatment is applied for near wall turbulence.

The inlet boundary was set 1.2 chord-lengths upstream of the blade leading edge, and the outlet boundary was placed 2 chord-lengths downstream of the blade trailing edge. At the inlet, total pressure distributions in the spanwise direction together with a fixed total temperature from experiments were imposed. For the outlet, an average static pressure boundary was imposed. During the simulations, solid walls were assumed to be adiabatic and non-slipping.

5. Endwall flow

The static pressure coefficient Cp is defined as

$$Cp = \frac{p_{2s} - p_{1s}}{p_1 - p_{1s}},$$

where p_1 and p_{1s} are the total pressure and static pressure of inlet flow, and p_{2s} is the local static pressure. The distribution of the static pressure measurement points in the experiments is shown in Figure 3. Experimentally obtained and simulated static pressure distributions from locations L1 to L12 are compared in Figure 4 to evaluate the simulation accuracy of the turbulence models.

According to Figure 4, in both the streamwise and pitchwise directions, the general variance law of the endwall static pressure is well predicted by Reynolds-averaged turbulence models. Along the streamwise direction, the static pressure increases first and then decreases. Along the pitchwise direction, the static pressure increases from the suction side to the pressure side. Near the blade suction surface, the onset of corner separation is denoted by the constant static pressure distribution along the pitchwise direction in the separation region. During the experimental process, measurement points were located several millimeters away from the blade suction surface, and the onset of corner separation was not observed. For the design and positive incidence angles, when corner separation is fully developed, constant static pressure distribution regions were measured by experiments at locations L8, L7 and L6 respectively, as shown in Figures 4(a)–4(c). Based on comparisons between simulated and experimentally obtained

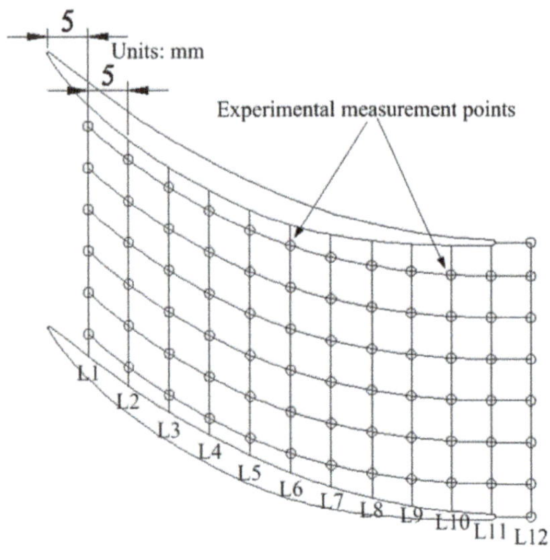

Figure 3. Distribution of static pressure measurement points.

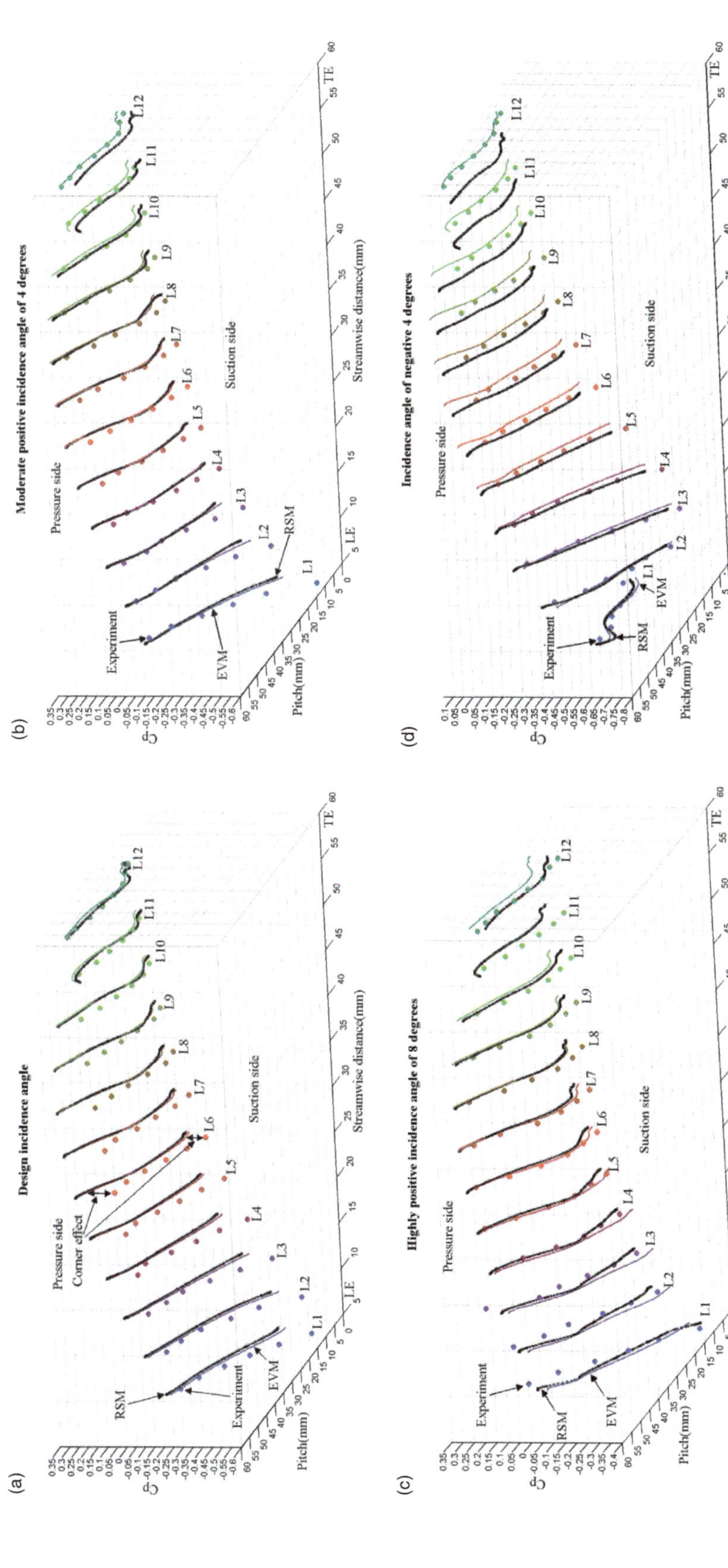

Figure 4. Endwall static pressure distributions.

static pressure variance laws along the pitchwise direction at these locations, the onset of corner separation is inferred to be reasonably predicted by the Reynolds-averaged turbulence models. For a negative incidence angle, as shown in Figure 4(d), the corner separation is constrained to a small scale and was not measured in the experiments.

Inside the cascade flow passage, a great challenge to the accuracy of turbulence model simulations is imposed by the corner effect, as illustrated in Figure 4(a). At conjunctions between the blade surfaces and endwall, the turbulence anisotropy, non-equilibrium, and energy backscattering are enhanced by the transport and accumulation of the boundary layer, as well as the emergence of corner separation. Thus, both turbulence models fail to predict the fluid state variance properly, as shown by the results at locations from L1 to L10. Compared with the results of experiments, the most severe deviations of the simulated static pressure distributions around the corner are observed at the design incidence angle, as shown in Figure 4(a). However, according to Figures 4(b)–4(d), at both positive and negative incidence angles, the influences of the corner effect are weakened, as indicated by smaller deviations between the turbulence model predictions and experiments, especially for the highly positive incidence angle. Near the blade trailing edge, at locations L11 and L12, the corner effects are almost wiped out, and the static pressure distributions in the experiments are well predicted by the eddy viscosity model, except at the highly positive incidence angle condition, as shown in Figure 4(c). A limited improvement in the endwall flow prediction accuracy is achieved by the Reynolds stress model, although obvious differences between the simulation results with the eddy viscosity model and the Reynolds stress model are observed at a negative incidence angle, as shown in Figure 4(d), and a highly positive incidence angle, as shown in Figure 4(c).

At the design incidence angle, as shown in Figure 4(a), the simulation results with the eddy viscosity model and the Reynolds stress model are almost coincident. Both models behave poorly at the corner inside the cascade flow passage and provide rather satisfying predictions at locations L11 and L12 near the blade trailing edge.

As the incidence angle increases and becomes positive, differences between the simulation results for both turbulence models appear. For a moderately positive incidence angle (shown in Figure 4(b)), differences are mostly observed near the blade trailing edge. Under this circumstance, solving the Reynolds stress transport equations is inferior to choosing the eddy viscosity hypothesis, because according to Figure 4(b), at locations L11 and L12, the static pressure distributions are better predicted by the eddy viscosity model. According to Figure 4(c), at a highly positive incidence angle, the corner effect greatly diminishes inside the cascade flow passage, and obvious differences between simulation results for the two turbulence models are found at both the front part and outside of the cascade flow passage at L12. At the front part, the simulation results of the Reynolds stress model are a little closer to the experimental results, but not apparently so. Outside the cascade flow passage at L12, when the flow is totally free of the blade restriction, the static pressure is more accurately predicted by the Reynolds stress model.

From Figure 4(d), at a negative incidence angle, the static pressure predicted by Reynolds stress model is smaller than that of the eddy viscosity model. At locations L1, L2, L3, L4, L5, L6, and L7, the simulation results of the Reynolds stress model are closer to the results of the experiments. In contrast, at locations L9, L10, L11 and L12, the simulation results of the Reynolds stress model become much smaller than the experimental results, which agree with the simulation results of the eddy viscosity model.

6. Downstream flow loss

To evaluate the flow loss, the total pressure loss coefficient ζ is defined as

$$\zeta = \frac{p_1 - p_2}{p_1 - p_{1s}},$$

where, p_2 is the local total pressure.

To better evaluate the flow loss at different blade heights, the pitch-averaged total pressure loss coefficient ζ_t is defined as

$$\zeta_t = \sum_{i=1}^{n} \frac{(\zeta_i \cdot \Delta t_i)}{t}.$$

Figure 5 shows comparisons between the flow loss predictions by the turbulence models and the experimental results. Data were obtained at 60% of the chord length downstream of the blade trailing edge. The variance law of flow loss along the spanwise direction predicted by the turbulence models agrees very well with the experimental results, which accounts for the popularity of turbulence models in engineering applications.

In order to evaluate the simulation accuracy of the two turbulence models, the cascade flow passage is divided into three zones along the spanwise direction – the endwall zone (E-zone), the mid-span zone (M-zone) and the intermediary zone (I-zone), between the E-zone and M-zone.

At the design and positive incidence angles, as shown in Figures 5(a)–5(c), in the E-zone, the increase and decrease process of the flow loss along the spanwise direction is well predicted by both turbulence models. However, the flow loss is obviously under-predicted by the eddy viscosity model while the Reynolds stress model provides flow loss predictions with a higher accuracy, especially for the regions closest to the endwall. At a negative incidence angle, as shown in Figure 5(d), in the E-zone, the flow loss predictions by the Reynolds stress model

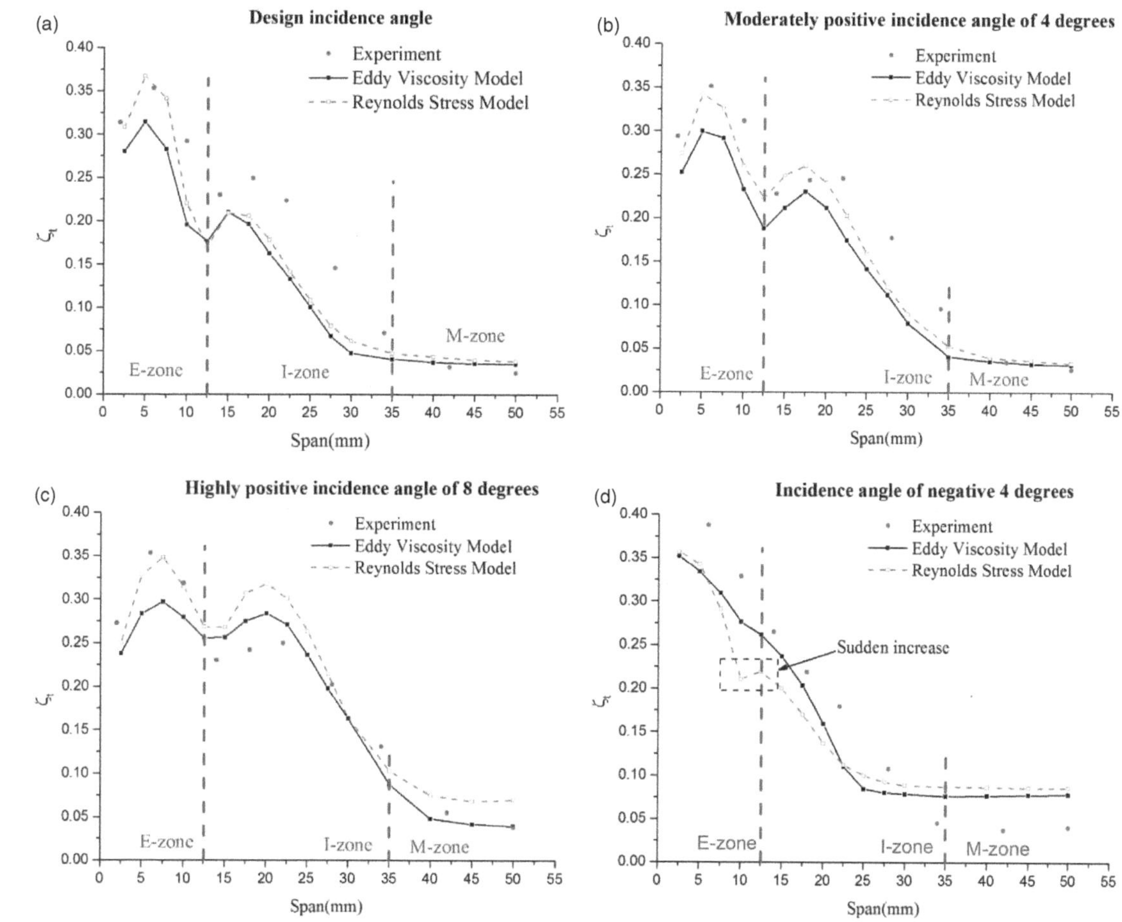

Figure 5. Flow loss distributions along span.

are not physically reasonable because of the sudden flow loss increase phenomenon. Although the flow loss is still under-predicted by the eddy viscosity model, the decrease process of flow loss along the spanwise direction predicted agrees very well with the results of experiment.

As shown in Figure 5(a), at the design incidence angle, the flow loss is under-predicted by both turbulence models in the I-zone. According to Figure 5(b), at a moderately positive incidence angle, flow loss over-prediction by the Reynolds stress model is partly observed in the I-zone. For a highly positive incidence angle, as shown in Figure 4(c), the I-zone flow loss is mostly over-predicted by the Reynolds stress model and partly over-predicted by the eddy viscosity model. However, the process whereby the flow loss first increases and then decreases in the spanwise direction is again well predicted by both turbulence models, as shown in Figures 5(a)–5(c). Similar to the E-zone, at a negative incidence angle, the simulation results of the eddy viscosity model in the I-zone are closer to the experimental results compared with those of the Reynolds stress model.

In the M-zone, as shown in Figures 5(a) and 5(b), at the design and moderately positive incidence angles, the flow loss predicted by both turbulence models agrees very well with the experimental results. At a highly positive incidence angle, as shown in Figure 5(c), the M-zone flow loss is greatly over-predicted by the Reynolds stress model but properly predicted by the eddy viscosity model. At a negative incidence angle, as shown in Figure 5(d), the flow loss is obviously over-predicted by both turbulence models, but the eddy viscosity model is slightly more accurate.

The M-zone flow is almost undisturbed by the three-dimensional corner separation, and mainly behaves in a two-dimensional way. Thus, the anisotropic nature of the turbulence is not dominant, which results in the relatively high prediction accuracy for both turbulence models, except for negative and highly positive incidence angle conditions, where the two-dimensional separation is further deteriorated. For both the I-zone and E-zone, flow mixing and three-dimensional corner separation coexist. Different scales of vortexes and turbulence structures give rise to prominent anisotropy, non-equilibrium, and energy backscattering in flow interactions. Under this circumstance, both turbulence models fail to predict the flow structures accurately, and the flow loss under-prediction at negative, design, and moderately positive incidence angles, as well as the flow loss over-prediction at highly positive incidence angles, was also observed in previous studies (Kožulovic & Röber, 2006; Lewin, Kožulovic, & Stark, 2010). Through solving the non-linear Reynolds stress transport equations, flow loss prediction accuracy is marginally enhanced by the Reynolds stress model at design and positive incidence angles and even reduced at negative incidence angles. Therefore, the gains are far from satisfying in consideration of the triple computational cost compared to the eddy viscosity model.

7. Secondary flow

7.1. Secondary flow at the design incidence angle

The secondary flow structuresat the design incidence angle obtained from the experiments and the simulations are shown in Figures 6 and 7, respectively. According to Figures 6(a) and 7(a), inside the cascade flow passage, at the 70% chord length transverse plane, the passage vortex (PV) first originates from the E-zone, which is observed in the experiment as well as predicted by both turbulence models. Because of geometric restrictions from the blades in the cascade flow passage, the region close to the blade surfaces is not measured in the experiment. As shown in Figure 6(b), at the 90% chord length plane, in contrast with the simulation results, shown in Figure 7(b), additional topological structures (A-TSs) associated with blade suction side separation are observed in the experiment, and the small-scale corner separation vortex (CV) is only observed in the simulation.

As shownin Figures 6(c), 6(d) and 7(c), downstream of the blade, wake mixing of the pressure side and suction side flow interacts with the passage vortex and makes the secondary flow structures more complicated. Compared with the simulation results, the influences of wake mixing (WM-I) on the secondary flow structures are more prominent in the experiments. At the 140% chord length transverse plane, as shown in Figure 6(d), topological structures corresponding to wake mixing are still observable in the experiment. In addition, further downstream at the 160% chord length transverse plane, as shown in Figure 6(e), only the passage vortex is measureable, which testifies to the almost complete dissipation of wake mixing influences. However, according to the simulation results, the obvious influences of wake mixing are only visible at the 110% chord length transverse plane (shown in Figure 7(c)) for both turbulence models, and the topological structures corresponding to passage vortex and wake mixing interactions are also different from those in the experiments. The fast dissipation of wake mixing during the simulation processes results in the rapid disappearance of its influence on the flow field. As shown in Figure 7(d), at the 120% chord length transverse plane, wake mixing is predicted to dissipate almost completely by both turbulence models. Thus, downstream of the cascade blade, the differences between the simulated and experimentally measured secondary flow structures are dramatic in relation to the wake mixing factor. At the 160% chord length transverse plane (shown in Figures 6(e) and 7(f)), where the wake mixing influences are also wiped out in the experiments, the secondary flow structures predicted by the turbulence models agree well with those in the experiments, just like inside the cascade flow passage.

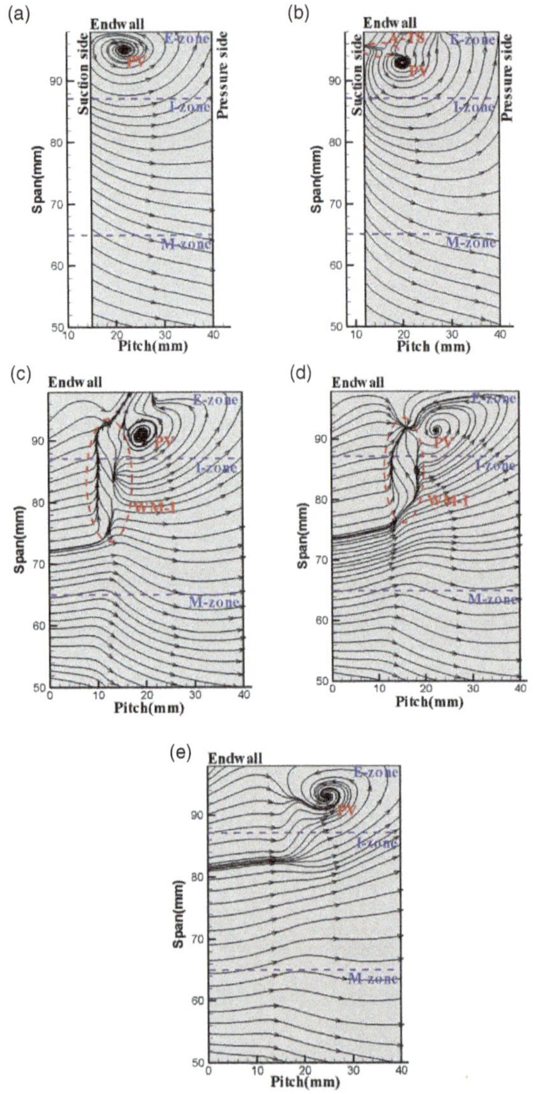

Figure 6. Measured secondary flow structure at design incidence angle.

Although turbulence models are capable of predicting the existence of main corner separation vortexes, predicting the interactions between different flow structures

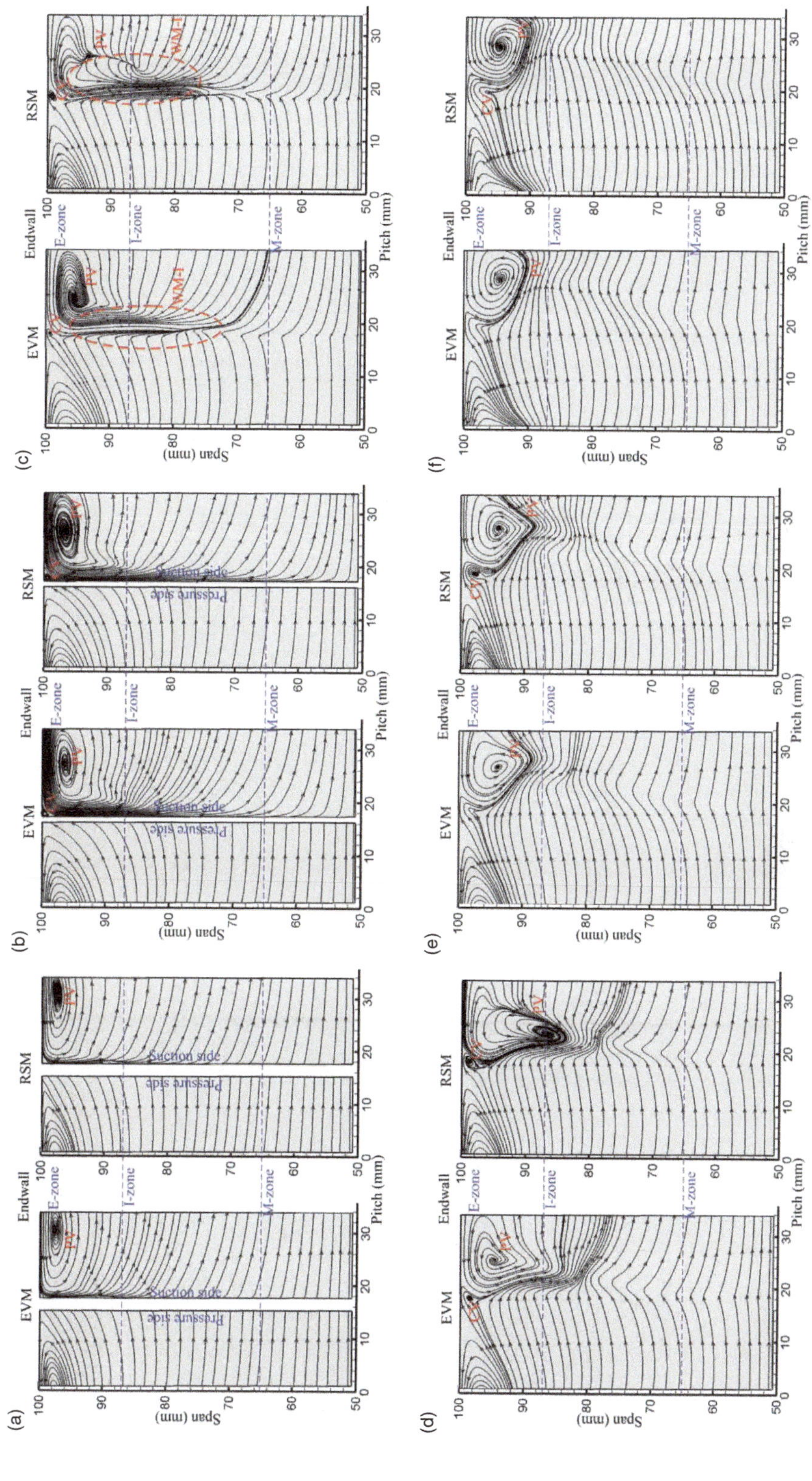

Figure 7. Simulated secondary flow structure at design incidence angle.

is out of their reach. Under the design incidence angle condition, the corner separation and wake mixing influences are both under-predicted by the eddy viscosity model. As a result, the flow losses of both the E-zone and I-zone are also under-predicted. Near the endwall, the Reynolds stress model behaves slightly better in accurately predicting the flow loss, which is in accordance with the more properly predicted corner separation strength. However, the non-linear characteristics of turbulence are greatly enhanced by the introduction of wake mixing, which imposes some unmet challenges on the empirically established Reynolds stress transportation equations. For this reason, the Reynolds stress model behaves poorly in predicting the flow loss at the edge of the E-zone and in the I-zone, where wake mixing interacts severely with the passage vortex.

7.2. Secondary flow at off-design incidence angles

Secondary flow structures at the 160% chord length transverse plane from the experiments and simulations at off-design incidence angles are shown in Figures 8–10. From Figure 8, with an incidence angle of 4°, the wake mixing influence is still under-predicted by both turbulence models. At the 160% chord length transverse plane, where the influence of wake mixing is still obvious according to the experimental results, only the existence of corner vortexes is predicted by both the eddy viscosity model and the Reynolds stress model. Thus, as shown in Figure 5(b), the flow loss in the I-zone is still mostly under-predicted. On the other hand, apart from the passage vortex, the corner separation vortex is only found at the 160% chord length transverse plane in the simulation results for the Reynolds stress model, which indicates a stronger corner separation compared with the simulation results of the eddy viscosity model and the experimental results. For this reason, the flow loss in the I-zone is partly over-predicted by the Reynolds stress model, as shown in Figure 5(b).

For highly positive incidence angles, the corner separation strength predicted by the turbulence models is greatly enhanced, especially for the Reynolds stress model (shown in Figure 9) and, as shown in Figure 5(c), over-prediction

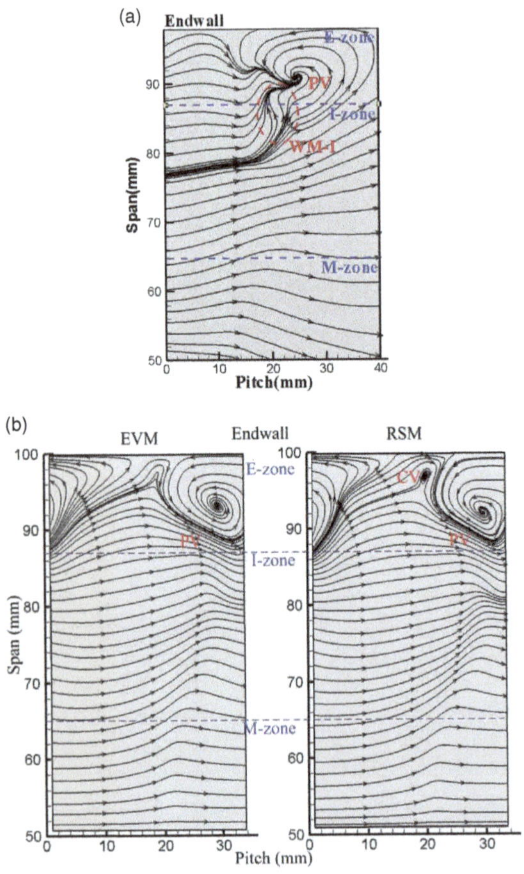

Figure 8. Secondary flow structure on 160% chord length transverse plane at a moderately positive incidence angle of 4°.

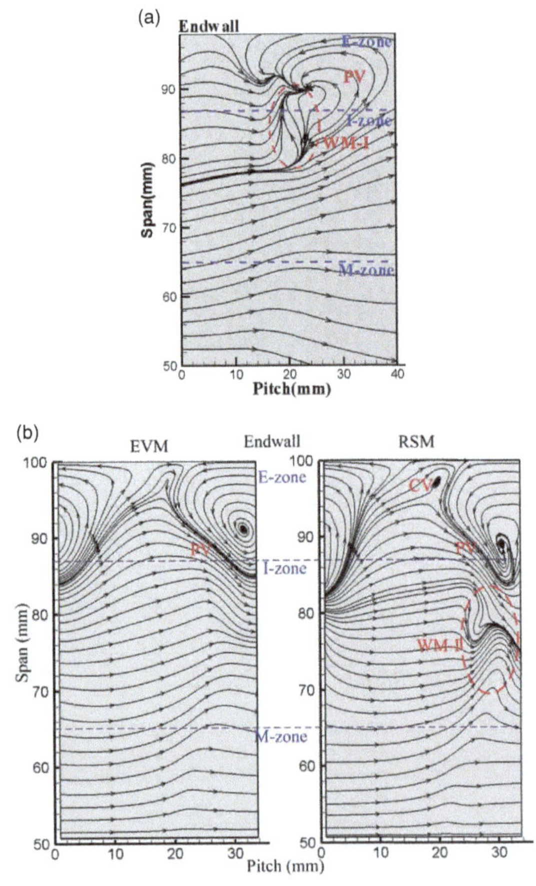

Figure 9. Secondary flow structure on 160% chord length transverse plane at a highly positive incidence angle of 8°.

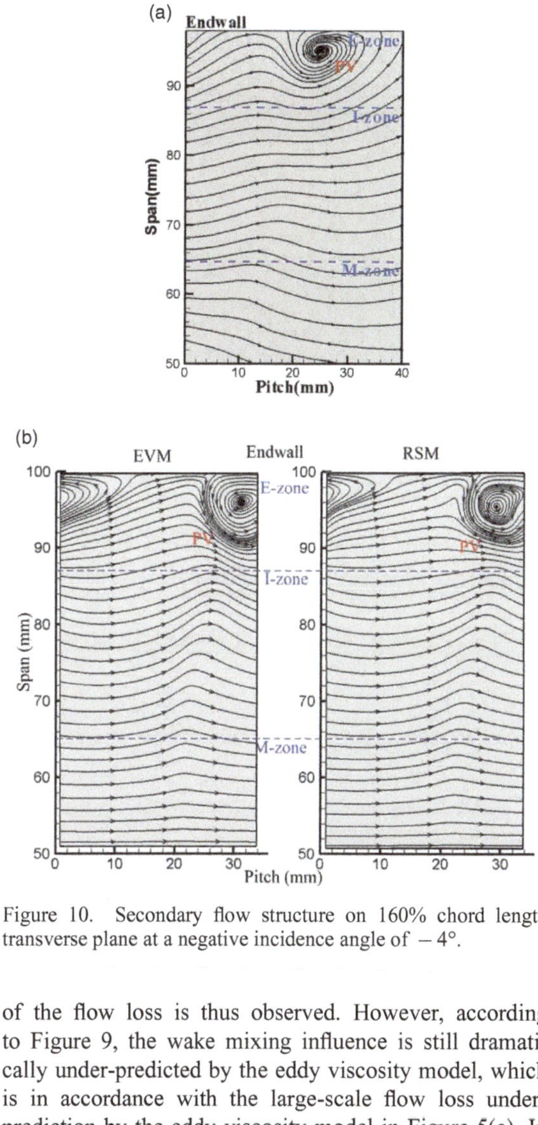

Figure 10. Secondary flow structure on 160% chord length transverse plane at a negative incidence angle of $-4°$.

of the flow loss is thus observed. However, according to Figure 9, the wake mixing influence is still dramatically under-predicted by the eddy viscosity model, which is in accordance with the large-scale flow loss under-prediction by the eddy viscosity model in Figure 5(c). In contrast, wake mixing plays a more important role in the secondary flow structure shown by the simulation results of the Reynolds stress model, and as a result, the under-prediction of the flow loss in Figure 5(c) is very minor. As shown in Figure 10, for negative incidence angles, the wake mixing influence is wiped out both in the experiment and the simulations, and only the passage vortex can be observed at the 160% chord length transverse plane. However, according to Figure 5(d), the simulation results of the eddy viscosity model are more physically reasonable.

8. Conclusions

The main characteristics of endwall flow are well predicted by both the eddy viscosity model and the Reynolds stress model. Inside the cascade flow passage, the corner effect is aroused at solid surface conjunctions, and the simulation accuracies of the turbulence models are greatly reduced. Compared with the design incidence angle, at negative or positive incidence angles, the corner effect is weakened and the simulation accuracy of the turbulence models increases in terms of predicting the endwall flow. However, in spite of the corner effect, the onset of corner separation inside the cascade flow passage is still reasonably predicted by both turbulence models. Outside the cascade flow passage and with the absence of the corner effect, a relatively high prediction accuracy for the endwall flow is achieved by the turbulence models at the design and moderately positive incidence angles. At negative or highly positive incidence angles, there are still apparent deviations between the simulation and experimental results. In a test involving the solution of the non-linear Reynolds stress transport equations, the simulation accuracy in predicting the endwall flow is marginally improved by the Reynolds stress model, but only at highly positive incidence angles.

Downstream of the blade, at the design and positive incidence angles, the wake mixing influence is mostly under-predicted by the turbulence models. As a result, under-prediction of the flow loss is present. With an increase in the incidence angle, the corner separation strength predicted by the turbulence models is increased, and some over-prediction of the flow loss is observed, in spite of the wake mixing influence under-prediction. Generally, at the design and positive incidence angles, the corner separation strength and wake mixing influence predicted by the eddy viscosity model are surpassed by those of the Reynolds stress model. When the incidence angle is increased to be highly positive, obvious over-prediction of flow loss by the Reynolds stress model is observed, while the flow loss is still mostly under-predicted by the eddy viscosity model. At negative incidence angles, the wake mixing influence is reduced and the secondary flow structure is more properly predicted by the Reynolds-averaged turbulence model. However, the simulation results of the Reynolds stress model are proved to be physically unreasonable according to the flow loss prediction.

9. Recommendation and future work

In this paper, to evaluate the accuracy of RANS turbulence models in simulating the corner separation of a high-speed compressor cascade, the corresponding simulation and experimental results were analyzed and compared in detail.

Based on the conclusion of this paper, the following suggestions are provided to researchers carrying out similar projects:

(1) Commonly used RANS methods can be treated as useful tools in simulating the main structures of high-speed compressor cascade corner separation, which are rather unstable.

(2) To simulate a high-speed compressor cascade flow, special attention should be paid to the corner effect inside the cascade passage and wake mixing downstream of the blades.

(3) The Reynolds stress model is not superior to the eddy viscosity model in simulating high-speed compressor cascade corner separations of moderate strength. In simulating high-speed compressor cascade flows under near-stall conditions, the Reynolds stress model may be an alternative choice, but the simulation results should be carefully evaluated in order to avoid physically unreasonable results.

However, it should be mentioned that the experimental methods discussed in this paper are incapable of capturing small-scale and transient high-speed compressor cascade flow structures. Thus, in the future, in order to perform finer evaluations of the simulation accuracy, more advanced experimental methods such as particle image velocimetry (PIV) should be applied. On the other hand, many factors that influence the simulation accuracy, such as the numerical scheme, should be considered in future studies – and with the development of computer technology, some LES simulation work will be quite helpful in studying three-dimensional corner separations in the high-speed regime.

Acknowledgements

The authors thank Prof. Meyer of DLR for the airfoil data.

Disclosure statement

No potential conflict of interest was reported by the authors.

Funding

This work was supported by the National Natural Science Foundation of China [51336011, 50906100], the Science Foundation for the Author of National Excellent Doctoral Dissertation of China, and China Scholarship Council.

References

Balabel, A., & El-Askary, W. A. (2011). On the performance of linear and nonlinear k-ε turbulence models in various jet flow applications. *European Journal of Mechanics - B/Fluids*, *30*, 325–340.

Chen, P. P., Qiao, W. Y., Liesner, K., & Meyer, R. (2014). *Location effect of boundary layer suction on compressor hub-corner separation* (ASME Paper No. GT 2014-25043).

Craft, T. J., Launder, B. E., & Suga, K. (1996). Development and application of a cubic eddy-viscosity model of turbulence. *International Journal of Heat and Fluid Flow*, *17*(2), 108–115.

Denton, J. (2010). *Some limitations of turbomachinery CFD* (ASME Paper No. GT2010-22540).

Dong, Y., Gallimore, S. J., & Hodson, H. P. (1987). Three-dimensional flows and loss reduction in axial compressors. *Journal of Turbomachinery*, *109*, 354–361.

Evans, S., Hodson, H., Hynes, T. P., & Wakelam, C. (2010). Flow control in a compressor cascade at high incidence. *Journal of Propulsion and Power*, *26*(4), 828–836.

Gatski, T. B., & Speziale, C. G. (1993). On explicit algebraic stress models for complex turbulent flows. *Journal of Fluid Mechanics*, *254*(1), 59–78.

Gbadebo, S. A., Cumpsty, N. A., & Hynes, T. P. (2005). Three-dimensional separations in axial compressors. *Journal of Turbomachinery*, *127*, 331–339.

Gbadebo, S. A., Cumpsty, N. A., & Hynes, T. P. (2008). Control of three-dimensional separations in axial compressors by tailored boundary layer suction. *Journal of Turbomachinery*, *130*, 011004–011004-8.

Gmelin, C., Thiele, F., & Liesner, K. (2011). *Investigations of secondary flow suction in a high speed compressor cascade* (ASME Paper No. GT2011-46479).

Hergt, A., Meyer, R., & Engel, K. (2006). *Experimental investigation of flow control in compressor cascades* (ASME Paper No. GT 2006-90415).

Hergt, A., Meyer, R., & Engel, K. (2013). Effects of vortex generator application on the performance of a compressor cascade. *Journal of Turbomachinery*, *135*, 021026–021026-10.

Hergt, A., Meyer, R., Liesner, K., & Nicke, E. (2011). *A new approach for compressor endwall contouring* (ASME Paper No. GT2011-45858).

Joslyn, H. D., & Dring, R. P. (1985). Axial compressor stator aerodynamics. *Journal of Engineering for Gas Turbines and Power*, *107*, 485–493.

Kožulovic, D., & Röber, T. (2006). *Modelling the streamline curvature effects in turbomachinery flows* (ASME Paper No. GT2006-90265).

Kulisa, P., & Dano, C. (2006). Assessment of linear and nonlinear two-equation turbulence models for aerothermal turbomachinery flows. *Journal of Thermal Science*, *15*(1), 14–26.

Launder, B. E., Reece, G. J., & Rodi, W. (1975). Progress in the developments of a Reynolds-stress turbulence closure. *Journal of Fluid Mechanics*, *68*, 537–566.

Lei, V. M., Spakovszky, Z. S., & Greitzer, E. M. (2008). A criterion for axial compressor hub-corner stall. *Journal of Turbomachinery*, *130*(3), 031006–031006-10.

Lewin, E., Kožulovic, D., & Stark, U. (2010). *Experimental and numerical analysis of hub-corner stall in compressor cascades* (ASME Paper No. GT 2010-22704).

Liu, Y. W., Lu, L. P., Fang, L., & Gao, F. (2011). Modification of Spalart-Allmaras model with consideration of turbulence energy backscatter using velocity helicity. *Physics Letter A*, *375*, 2377–2381.

Mani, M., Babcock, D. A., Winkler, C. M., & Spalart, P. R. (2013). *Predictions of a supersonic turbulent flow in a square duct* (AIAA Paper No. 2013-0860).

Matsuura, K., & Kato, C. (2007). Large Eddy Simulation of compressible transitional cascade flows. *Journal of Fluid Science and Technology*, *2*, 558–569.

McMullan, W. A., & Page, G. J. (2011). Large eddy simulation of a controlled diffusion compressor cascade. *Flow, Turbulence and Combustion*, *86*, 207–230.

Mehdizadeh, O. Z., Temmerman, L., Tartinville, B., & Hirsch, C. (2012). *Applications of EARSM turbulence models to internal flows* (ASME Paper No. GT2012-68886).

Menter, F. R., Langtry, R., & Hansen, T. (2004). *CFD simulation of turbomachinery flows - verification, validation, and modelling*. Proceedings of ECCOMAS, Juväs- kylä.

Place, J. M. M. (1997). *Three-dimensional flow in core compressors* (PhD thesis). University of Cambridge.

Schulz, H. D., Gallus, H. E., & Lakshminarayana, B. (1990). Three-dimensional separated flow field in the endwall region of an annular compressor cascade in the presence of rotor-stator interaction: Part 1-quasi-steady flow field and comparison with steady-state data. *Journal of Turbomachinery*, *112*(4), 669–678.

Speziale, C. G., Sarkar, S., & Gatski, T. B. (1991). Modelling the pressure-strain correlation of turbulence: An invariant dynamical systems approach. *Journal of Fluid Mechanics*, *227*, 245–272.

Teramoto, S. (2005). *Large-eddy simulation of a transonic compressor cascade with boundary layer Transition* (AIAA Paper No. 2005-3639).

Tiedemann, C., Heinrich, A., & Peitsch, D. (2012). *A new linear high speed compressor stator cascade for active flow control investigations* (AIAA Paper No. 2012-3251).

Tucker, P. G. (2011). Computation of unsteady turbomachinery flows: Part 1-progress and challenges. *Progress in Aerospace Sciences*, *47*, 522–545.

Tucker, P. G. (2013). Trends in turbomachinery turbulence treatments. *Progress in Aerospace Sciences*, *63*, 1–32.

Wang, D. H., Lu, L. P., & Li, Q. S. (2009). *Improvement on S-A model for predicting corner separation based on turbulence transport nature* (AIAA Paper No. 2009-4931).

Wang, Z. N., & Yuan, X. (2013). *Unsteady mechanisms of compressor corner separation over a range of incidences based on hybrid Les/RANs* (ASME Paper No. GT 2013-95300).

Weber, A., Schreiber, H.-A., Fuchs, R., & Steinert, W. (2002). 3-D Transonic flow in a compressor cascade with shock-induced corner stall. *Journal of Turbomachinery*, *124*, 358–366.

Wilcox, D. C. (1988). Multiscale model for turbulent flows. *AIAA Journal*, *26*(11), 1311–1320.

Zachos, P. K., Grech, N., Charnley, B., Pachidis, V., & Singh, R. (2011). Experimental and numerical investigation of a compressor cascade at highly negative incidence. *Engineering Applications of Computational Fluid Mechanics*, *5*(1), 26–36.

Zaki, T. A., Wissink, J. G., Durbin, P. A., & Rodi, W. (2009). Direct computations of boundary layers distorted by migrating wakes in a linear compressor cascade. *Flow, Turbulence and Combustion*, *83*(3), 307–322.

14

Implicit scheme for meshless compressible Euler solver

Manish K. Singh[a,b]*, V. Ramesh[a] and N. Balakrishnan[b]

[a]*CTFD Division, CSIR-NAL, Bangalore, India;* [b]*Aerospace Engineering Department, Indian Institute of Science, Bangalore, India*

In this paper, an implicit scheme is presented for a meshless compressible Euler solver based on the Least Square Kinetic Upwind Method (LSKUM). The Jameson and Yoon's split flux Jacobians formulation is very popular in finite volume methodology, which leads to a scalar diagonal dominant matrix for an efficient implicit procedure (Jameson & Yoon, 1987). However, this approach leads to a block diagonal matrix when applied to the LSKUM meshless method. The above split flux Jacobian formulation, along with a matrix-free approach, has been adopted to obtain a diagonally dominant, robust and cheap implicit time integration scheme. The efficacy of the scheme is demonstrated by computing 2D flow past a NACA 0012 airfoil under subsonic, transonic and supersonic flow conditions. The results obtained are compared with available experiments and other reliable computational fluid dynamics (CFD) results. The present implicit formulation shows good convergence acceleration over the RK4 explicit procedure. Further, the accuracy and robustness of the scheme in 3D is demonstrated by computing the flow past an ONERA M6 wing and a clipped delta wing with aileron deflection. The computed results show good agreement with wind tunnel experiments and other CFD computations.

Keywords: implicit scheme; meshless Euler solver; LSKUM; matrix-free formulation; NACA 0012 airfoil; ONERA M6 wing; clipped delta wind; aileron deflection

1. Introduction

In the past decade a lot of work has been done towards the development of meshless solvers. The basic idea of such a solver is to develop a method which does not require the process of breaking up the physical domain in terms of smaller hexahedral cells (structured grids) or tetrahedral cells (unstructured grids). The meshless solver only needs a cloud of points to discretize the governing equations. There is no requirement for points to connect in order to form the grids, which are used in the conventional finite volume and finite element methods. This type of solver is able to operate on any arbitrary distribution of points, thus mitigating the problem of generating an entire complex grid. For a complex configuration, with such a solver, we can generate a cloud of points over the entire geometry with relative ease. Individual clouds are generated over each component independently and then merged to form a single large cloud. A suitable deletion operation is carried out to remove hidden points.

One of the earliest meshless methods is called Smoothed Particle Hydrodynamics (SPH; Gingold & Monaghan, 1977; Lucy, 1977). This method was originally developed for astrophysical applications and later used by many researchers for problems in solid mechanics and fluid mechanics (Monaco et al., 2011; Pu, Shao, Huang, & Hussain, 2013). Liszka has developed a generalized finite difference method (meshless) which uses the moving least square approach for derivative approximation (Demkowikz, Karafiat, & Liszka, 1984; Liszka & Orkisz, 1980). The early work on the use of meshless methods for fluid dynamics applications can be traced back to the work of Deshpande (1986a, 1986b) and Batina (1993). Both of them have used the least square procedure to approximate spatial derivatives. Deshpande and colleagues have developed a meshless Euler solver called the Least Square Kinetic Upwind Method (LSKUM; Deshpande, Anandhanarayanan, Praveen, & Ramesh, 2002; Deshpande, Kulkarni, & Ghosh, 1998; Ghosh & Deshpande, 1995; Ramesh, 2001; Ramesh & Deshpande, 2001, 2004, 2007; Ramesh, Mathur, & Deshpande, 1997). Sridar and Balakrishnan (2003, 2006) have developed a meshless solver based on the finite difference approach called the Least Square Upwind Finite Difference Method (LSFD-U). In this approach, the mid-point between a given node and its neighbor is used to determine the upwind fluxes (Balakrishnan & Sridar, 2001). This scheme has the flexibility to choose between different flux formulations such as the Roe, Van Leer, Kinetic Flux Vector Split (KFVS) and the Advection Upstream Splitting Method (AUSM). The capabilities of the LSFD-U were further enhanced by including viscous terms (Munikrishna, 2007; Munikrishna & Balakrishnan, 2007; Ninawe, Munikrishna, & Balakrishnan, 2006). Katz and Jameson (2009a, 2009b, 2009c) have developed an edge-based meshless Euler solver called the 'meshless volume scheme'. They have demonstrated this method along with a multi-grid convergence acceleration

*Corresponding author. Email: mksingh@ctfd.cmmacs.ernet.in

technique, which they named the 'multi-cloud approach', using the NACA 0012 and RAE 2822 airfoils. Zhau and Xu (2010) used the meshless method to compute an unsteady flow field over the NACA 0012 and NACA 0008 airfoils. Mahindra, Singh, and Gouthaman (2011) used an LSKUM-based solver to compute the viscous rotating flow. They observed the ill-conditioning of a least square problem when an explicit LSKUM solver operates on the stretched distribution of points near a boundary, which is required to capture viscous flow features. They chose split stencils (a set of neighbors) selectively to avoid the ill-conditioning of the least square procedure. The present work is focused on in-viscid flow, which does not require the stretched distribution of points as required for viscous solvers. Such difficulties are more pronounced for explicit scheme-based solvers. Moreover, implicit schemes are more stable than explicit schemes.

It is difficult to show mathematically that the above meshless schemes are formally conservative. However, it is shown through numerical experiments that these schemes have mass conservation properties (Sridar & Balakrishanan, 2003). Recently, Chiu, Wang, and Jameson (2011) have developed a finite volume-like conservative meshless scheme in which the discrete coefficients are determined by an algorithm based on minimum-norm solutions such that the scheme is locally conservative.

Many of the above solvers have used explicit time stepping. Generally, computational time per iteration for meshless solvers is more than their finite volume counterpart. So there is interest in convergence acceleration techniques such as implicit methodology or the multi-grid approach. Katz and Jameson (2009c) have reported a convergence acceleration of about 20 using a multi-cloud approach on the NACA 0012 airfoil. The implicit Lower-Upper Symmetric Gauss-Seidel (LU-SGS) scheme is very attractive for convergence acceleration due to its low numerical complexity, moderate memory requirement and unconditional stability for linear wave equations. LU-SGS has been used widely for many Finite Volume Method-based solvers (Shende, 2005; Shende & Balakrishnan, 2004). Sridar and Balakrishnan (2006) have implemented a LU-SGS implicit procedure along with Symmetric Gauss-Seidel (SGS) and modified SGS (SGS-M) implicit procedures on the finite difference-based meshless Euler solver LSFD-U. These implicit procedures were tested on subsonic and transonic flow over a NACA 0012 airfoil, supersonic flow past the ramp in channel and high supersonic flow past the half cylinder. It was found that the LU-SGS implicit procedure performs better than the SGS implicit procedure for flow past the NACA 0012 airfoil (both subsonic and transonic) and supersonic flow past the ramp in channel. The modified SGS procedure gives the best performance. For the modified SGS implicit procedure, a convergence speed-up of at least 20 is observed as compared to the four-stage Range-Kutta (RK4) explicit scheme. Anandhanarayanan (2003) implemented a LU-SGS implicit procedure in a LSKUM meshless Euler solver and obtained a maximum convergence speed close to 4, on flow over bump.

The implicit version of flux vector splitting schemes requires the computation of split flux Jacobians, which is expensive. To avoid this, a matrix-free approach is implemented in finite volume methods for off diagonal terms of the implicit operator (Luo, Baum, & Lohner, 1998). The split flux Jacobians are very popular for diagonal terms with regard to retaining diagonal dominancy (Jameson & Yoon, 1987; Yoon & Jameson, 1986, 1987). In the finite volume or finite deference method this leads to a scalar diagonal matrix. However, this procedure leads to a block diagonal matrix, which again requires expensive inversion when applied to a meshless solver. Anandhanarayanan (2003) has approximated a split flux Jacobian to be half of the spectral radius in order to avoid a block diagonal matrix. Conventional LSKUM solvers achieve second-order accuracy via a two-step defect-correction procedure. The present solver uses modified Courant-Isaacson-Rees (CIR) splitting. We have called this an LSKUM-MCIR or simply MCIR scheme. This enables us to get a single-step higher-order meshless method. It is in this respect that the present solver is distinctly different from the work of Mahindra et al. (2011) and Anandhanarayanan, Nagarathinam, and Deshpande (2004).

In the present work we have developed a modified implicit procedure using a single-step higher-order scheme which can be considered an extension of earlier work on LSKUM-based solvers. We use Jameson and Yoon's split flux Jacobian formulation for diagonal terms in order to retain diagonal dominancy for an efficient implicit procedure. However, a matrix-free approach is applied for diagonal as well as off-diagonal terms of the implicit operator as against the finite volume or finite difference method where it is required to be applied for off-diagonal terms only. This modification leads to a scalar diagonal matrix using a meshless method. The present implicit procedure is tested on a NACA 0012 airfoil, an ONERA M6 wing and a clipped delta wing with aileron deflection. The results are compared with those of the RK4 explicit meshless method and available wind tunnel experiments.

The paper outline is as follows. The LSKUM is presented in section 2. Time integration schemes are presented in section 3 with special emphasis on the adaptation aspect of the LU-SGS implicit scheme for a meshless solver. 2D and 3D results are presented in section 4, followed by the conclusion in section 5.

2. Least Square Kinetic Upwind Method (LSKUM)

2.1. *Formulation*

The Boltzmann equation in two dimensions for a Maxwellian velocity distribution function F (Deshpande,

1986a, 1986b) is given by

$$\frac{\partial F}{\partial t} + v_1 \frac{\partial F}{\partial x} + v_2 \frac{\partial F}{\partial y} = 0, \quad (1)$$

where F is given by

$$F = \frac{\rho}{I_0}\left(\frac{\beta}{\pi}\right).e^{[-\beta(v_1-u_1)^2-\beta(v_2-u_2)^2-(I/I_0)]},$$

where ρ is the fluid density, $\beta = 1/(2RT)$, and I_0 is the internal energy due to non-translational degrees of freedom. $I_0 = (2-\gamma)\,RT/(\gamma-1)$ and u_1 and u_2 are Cartesian components of the fluid velocity, and v_1 and v_2 are Cartesian components of the molecular velocity. R is the gas constant and T is the absolute temperature of the fluid.

Thus we define the Moment Vector, $\psi \equiv [1, v_1, v_2, I + (v_1{}^2 + v_2^2/2)]^T$,

$$\Psi \text{ Moment of } \mathbf{F}, <\psi, F> \equiv \int_0^\infty dI \int_{-\infty}^{\infty} dv_1 \int_{-\infty}^{\infty} dv_2 \psi F.$$

Then the Euler equation is obtained by taking the ψ moment of the Boltzmann equation (1),

$$\langle \psi, \frac{\partial F}{\partial t} + v_1\frac{\partial F}{\partial x} + v_2\frac{\partial F}{\partial y}\rangle = \frac{\partial U}{\partial t} + \frac{\partial}{\partial x}(GX) + \frac{\partial}{\partial y}(GY) = 0, \quad (2)$$

where U represents the conserved vector, GX is the X-component of the flux vector, and GY is the Y-component of the flux vector:

$$\begin{aligned} U &= [\rho, \rho u_1, \rho u_2 \rho e]^T \\ GX &= [\rho u_1, p + \rho u_1^2, \rho u_1 u_2, u_1(p+\rho e)]^T \\ GY &= [\rho u_2, \rho u_1 u_2, p + \rho u_2^2, u_2(p+\rho e)]^T \\ e &= \frac{RT}{\gamma - 1} + \frac{1}{2}(u_1^2 + u_2^2) \\ p &= \rho RT \end{aligned}$$

where, p is pressure in the above expression. The CIR splitting of the molecular velocity leads to an upwind scheme at Euler level through the moment method strategy

$$\begin{aligned} &\left\langle \psi, \frac{\partial F}{\partial t} + \frac{v_1+|v_1|}{2}\frac{\partial F}{\partial x} + \frac{v_1-|v_1|}{2}\frac{\partial F}{\partial x} + \frac{v_2+|v_2|}{2}\frac{\partial F}{\partial y} + \frac{v_2-|v_2|}{2}\frac{\partial F}{\partial y}\right\rangle \\ &= \frac{\partial U}{\partial t} + \frac{\partial}{\partial x}(GX^+)\Big|_{cx^+} + \frac{\partial}{\partial x}(GX^-)\Big|_{cx^-} \\ &+ \frac{\partial}{\partial y}(GY^+)\Big|_{cy^+} + \frac{\partial}{\partial y}(GY^-)\Big|_{cy^-} = 0 \end{aligned} \quad (3)$$

where $GX^\pm$ and $GY^\pm$are KFVS split fluxes (Ramesh, 2001).

The approximations for spatial derivatives in equation (3) for split fluxes GX^+, GX^-, GY^+ and GY^- are obtained by the least square procedure and using the split stencils cx^+, cx^-, cy^+ and cy^-, respectively. The split stencils cx^+, cx^-, cy^+ and cy^- are shown as shaded areas in Figures 1(a–d), respectively. Let 'i' be the point at which we want to calculate the spatial derivative and point 'k' be a point in the neighborhood of point 'i'. Then the approximation for the spatial derivatives at point 'i' is given by

$$\begin{bmatrix} \sum \Delta x_k^2 & \sum \Delta x_k \Delta y_k \\ \sum \Delta x_k \Delta y_k & \sum \Delta y_k^2 \end{bmatrix} \begin{bmatrix} F_{x_i} \\ F_{y_i} \end{bmatrix} = \begin{bmatrix} \sum \Delta F_k \Delta x_k \\ \sum \Delta F_k \Delta y_k \end{bmatrix}, \quad (4)$$

where $\Delta(.)_k = (.)_k - (.)_i$.

The space derivative approximation given by equation (4) is first-order accurate. Second-order accuracy can be obtained by using the two-step defect-correction method of Ghosh and Deshpande (1995). But, in the present work, the single-step MCIR scheme proposed by Ramesh and Deshpande (2004) has been used to obtain higher-order approximations for the spatial derivative. In this scheme, molecular velocity is written as

$$\begin{aligned} v_1 &= \frac{v_1 + |v_1|\varphi_1}{2} + \frac{v_1 - |v_1|\varphi_1}{2}, \\ v_2 &= \frac{v_2 + |v_2|\varphi_2}{2} + \frac{v_2 - |v_2|\varphi_2}{2}; \end{aligned}$$

then the Boltzmann equation can be written as

$$\begin{aligned} &\frac{\partial F}{\partial t} + \frac{v_1 + |v_1|\varphi_1}{2}\frac{\partial F}{\partial x} + \frac{v_1 - |v_1|\varphi_1}{2}\frac{\partial F}{\partial x} \\ &+ \frac{v_2 + |v_2|\varphi_2}{2}\frac{\partial F}{\partial y} + \frac{v_2 - |v_2|\varphi_2}{2}\frac{\partial F}{\partial y} = 0, \end{aligned} \quad (5)$$

where φ_1 and φ_2 represents the dissipation control parameters corresponding to two components of the molecular velocity. Now, several choices are possible for φ_1 and φ_2. Dominic (2004) had chosen dissipation control parameters such that modified split-fluxes satisfy the Rankine-Hugoniot condition across the shock. Anil (2008) has taken the dissipation control parameter as a function of molecular velocity. In the present work, the dissipation control parameters φ_1 and φ_2 are chosen as $\varphi_1, \varphi_2 = (\Delta r)^p$, where $0 < p < 1$ and Δr is the distance between any point and its nearest neighborhood. In numerical experiments it has been observed that when p reaches close to 1, wiggles appear in the solution. The optimum value of p was found to be $.25 < p < .7$. Further details about the MCIR scheme can be found in Ramesh and Deshpande (2007).

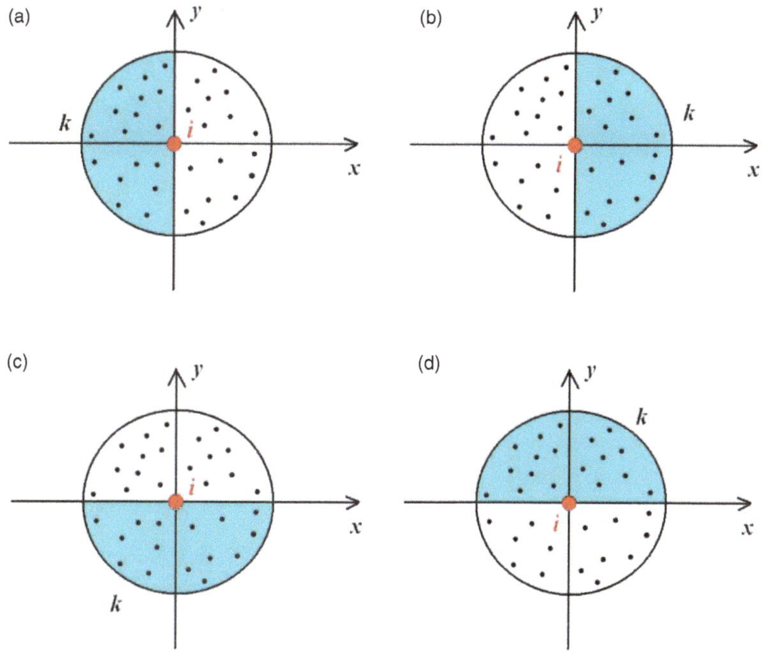

Figure 1. Split stencils: (a) cx$^+$; (b) cx$^-$; (c) cy$^+$; and (d) cy$^-$.

The ψ moment of equation (5) will lead to the Modified Kinetic split flux Euler equation

$$\frac{\partial U}{\partial t}+\frac{\partial}{\partial x}(\mathrm{GX}_M^+)+\frac{\partial}{\partial x}(\mathrm{GX}_M^-)+\frac{\partial}{\partial y}(\mathrm{GY}_M^+)+\frac{\partial}{\partial y}(\mathrm{GY}_M^-)=0, \tag{6}$$

where $\mathrm{GX}_M^{\pm}$ and $\mathrm{GY}_M^{\pm}$ are modified KFVS split fluxes. These can be represented in terms of the usual KFVS fluxes as

$$\mathrm{GX}_M^{\pm}=\frac{1}{2}\{GX\pm\varphi_1(\mathrm{GX}^+-\mathrm{GX}^-)\},$$
$$GY_M^{\pm}=\frac{1}{2}\{\mathrm{GY}\pm\varphi_2(\mathrm{GY}^+-\mathrm{GY}^-)\}.$$

2.2. *Boundary condition*

The boundary conditions (wall or far-field) are implemented in a local Cartesian coordinate system. The treatment of the boundary condition is developed at a molecular level and by taking suitable moments; we obtain the boundary condition at the macroscopic (Euler) level. This procedure is called the 'kinetic treatment of the boundary condition'. To implement a wall boundary condition (surface tangency condition), the Maxwellian velocity distribution function F, is divided into two parts as

$$F=F_I\cup F_R,$$

where F_I and F_R correspond to the incident and reflected molecules, respectively. Following the principle of the specular reflection of kinetic theory, F_R is reconstructed from F_I by reversing the component of velocity in the direction normal to the surface. Referring to Figure 2(a) for the wall point 'i',

$$F_I=F_I(v_1,v_2)\quad\text{for}\quad v_2<0,$$
$$F_R=F_I(v_1,-v_2)\quad\text{for}\quad v_2>0,$$

where, v_1 and v_2 are the molecular velocity in the local coordinate system along the x and y directions, respectively.

Similarly, to implementthe far-field boundary condition, the Maxwellian velocity distribution function F, at any point 'i' lying on the outer boundary surface (Figure 2(b)), is constructed as the union of two half Maxwellians, corresponding to the incoming and outgoing particles. The incoming particles have a negative velocity with respect to the direction normal to the boundary. The velocity distribution function F_{in} for this part corresponds to the free stream conditions. The outgoing particles have a positive velocity in the normal direction. For the outgoing particles the velocity distribution function F_{out} is constructed from the interior region.

$$F_i=F_{in}\cup F_{out}.$$

Details about the boundary conditions can be found in the work of Mandal and Deshpande (1994) and Ramesh and Deshpande (2001).

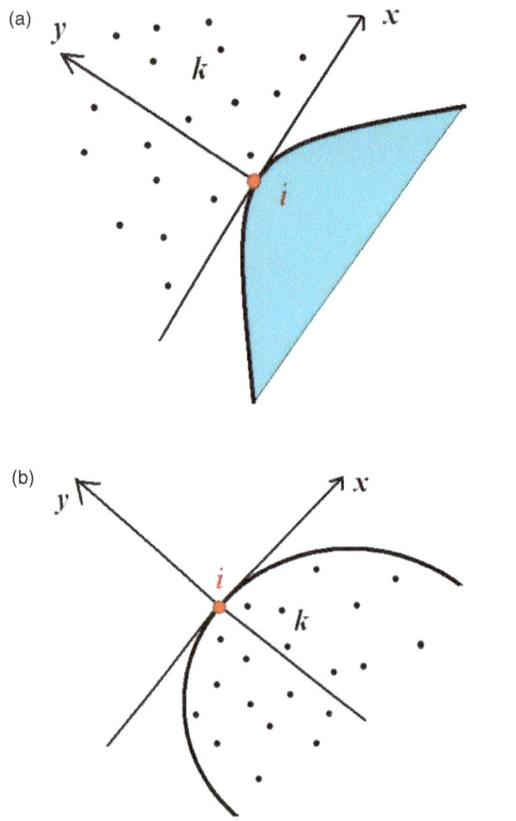

Figure 2. A typical 2D (a) wall boundary and (b) far-field boundary.

3. Implicit Time Integration Scheme

A 2D Euler equation in implicit form using split fluxes can be written as

$$\left(\frac{\partial U}{\partial t}\right)^n + \frac{\partial (\mathrm{GX}^+)^{n+1}}{\partial X} + \frac{\partial (\mathrm{GX}^-)^{n+1}}{\partial X} + \frac{\partial (\mathrm{GY}^+)^{n+1}}{\partial Y} + \frac{\partial (\mathrm{GY}^-)^{n+1}}{\partial Y} = 0, \quad (7)$$

where U, $\mathrm{GX}^\pm$ and $\mathrm{GY}^\pm$ are vectors of conserved variables and in-viscid split flux vectors in the x and y directions, respectively. Using least square approximation, the flux derivative for point 'i' can be written in compact form as

$$\frac{\partial (\mathrm{GX}^\pm)^{n+1}}{\partial x} = p_1^\pm \sum_{k \in cx^\pm} (\Delta \mathrm{GX}_k^\pm)^{n+1}.\Delta x_k + p_2^\pm \sum_{k \in cx^\pm} (\Delta \mathrm{GX}_k^\pm)^{n+1}.\Delta y_k, \quad (8)$$

$$\frac{\partial (GY^\pm)^{n+1}}{\partial y} = q_1^\pm \sum_{k \in cy^\pm} (\Delta GY_k^\pm)^{n+1}.\Delta x_k + q_2^\pm \sum_{k \in cy^\pm} (\Delta \mathrm{GY}_k^\pm)^{n+1}.\Delta y_k, \quad (9)$$

where p_1, p_2, q_1 and q_2 are geometrical constants (see Appendix A) and $\Delta(.)_k = (.)_k - (.)_i$. The superscript '$+$' and '$-$' indicate that split stencils are used to evaluate corresponding constants (see Figures 1(a–d)). For simplicity, we take the first term of the spatial derivative of equation (7),

$$\frac{\partial (\mathrm{GX}^+)^{n+1}}{\partial X} = {p_1}^+ \sum_k (\mathrm{GX}_k^+ - \mathrm{GX}_i^+)^{n+1}.\Delta x_k + {p_2}^+ \sum_k (\mathrm{GX}_k^+ - \mathrm{GX}_i^+)^{n+1}.\Delta y_k. \quad (10)$$

After temporal linearization for the current time level n,

$$(\mathrm{GX}^\pm)^{n+1} = (\mathrm{GX}^\pm)^n + A^\pm.\Delta_t U \quad (11)$$

where $A^\pm$ is split flux Jacobian for the x direction and $\Delta_t U = U^{n+1} - U^n$. From equation (10) and equation (11), we have

$$\frac{\partial (\mathrm{GX}^+)^{n+1}}{\partial X} = \overbrace{\left[\begin{array}{l} {p_1}^+ \sum_k (A_k^+.\Delta_t U_k - A_i^+.\Delta_t U_i).\Delta x_k \\ +{p_2}^+ \sum_k (A_k^+.\Delta_t U_k - A_i^+.\Delta_t U_i).\Delta y_k \end{array}\right]}^{\mathrm{Im}(R)} + \underbrace{\left[p_1^+ \sum_k \Delta \mathrm{GX}_k^+ \Delta x_k + p_2^+ \sum_k \Delta \mathrm{GX}_k^+ \Delta y_k\right]}_{R^n} \quad (12)$$

The terms in the first set of square brackets of equation (12) are termed as the implicit operator Im(R), while the terms in the second set of square brackets are termed as the explicit operator R^n (or the explicit residual). The implicit operator requires computation of split flux Jacobians. We make use of a matrix-free approach (Luo et al., 1998), which is very economical as compared to the expensive computation of split flux Jacobians. In this approach, the product of split flux Jacobians (or flux Jacobians) with incremental conserved variable vector is approximated by an incremental change in the split flux vector in time:

$$A_k^\pm \Delta_t U_k = \frac{\partial (\mathrm{GX}_k^\pm)}{\partial U_k} \Delta_t U_k = \mathrm{GX}_k^\pm (U + \Delta U) - \mathrm{GX}_k^\pm (U) = \Delta_t \mathrm{GX}_k^\pm.$$

With this, the implicit operator term of equation (12) becomes

$$\mathrm{Im}(R) = \left({p_1}^+ \sum_k \Delta_t GX_k^+.\Delta x_k + {p_2}^+ \sum_k \Delta_t GX_k^+.\Delta y_k\right) - A_i^+.\Delta_t U_i \left(p_1^+ \sum_k \Delta x_k + p_2^+ \sum_k \Delta y_k\right). \quad (13)$$

The terms in the first set of square brackets of equation (13) correspond to the lower and upper matrix of the implicit operator, which is clubbed along with the explicit operator. The terms in the second set of square brackets correspond to the diagonal term of the implicit operator. Likewise, working with the other spatial derivative terms of split fluxes from equation (7), we arrive at a discretized equation corresponding to equation (7) as

$$\underbrace{\left[\frac{I}{\Delta t} - \left\{\begin{matrix} p_1^+ \sum_k A_i^+ \Delta x_k + p_2^+ \sum_k A_i^+ \Delta y_k \\ + p_1^- \sum_k A_i^- \Delta x_k + p_2^- \sum_k A_i^- \Delta y_k \end{matrix}\right\} - \left\{\begin{matrix} q_1^+ \sum_k B_i^+ \Delta x_k + q_2^+ \sum_k B_i^+ \Delta y_k \\ + q_1^- \sum_k B_i^- \Delta x_k + q_2^- \sum_k B_i^- \Delta y_k \end{matrix}\right\}\right]}_{D_i}$$

$$\Delta_t U_i = -R_i^n - (\mathbf{E} + \mathbf{C}), \tag{14}$$

where, $A_i^{\pm}$ and $B_i^{\pm}$ are split flux Jacobians in the x and y directions, respectively. R_i^n is the explicit residual (or explicit operator) and $E + C$ are terms which correspond to the lower and upper matrix of the implicit operator (see Appendix B). Following Jameson and Yoon's split flux Jacobians approximation based on the spectral radius,

$$A_i^{\pm} = \frac{A_i \pm \rho_A I}{2}, \quad B_i^{\pm} = \frac{B_i \pm \rho_B I}{2},$$

where ρ_A and ρ_B are the spectral radius in the x and y directions, respectively.

The left side of equation (14) is given by

$$D_i.\Delta_t U_i = \overbrace{\left[\frac{I}{\Delta t} - \frac{\rho_A}{2}\left\{\begin{matrix} p_1^+ \sum_k \Delta x_k + p_2^+ \sum_k \Delta y_k \\ - p_1^- \sum_k \Delta x_k - p_2^- \sum_k \Delta y_k \end{matrix}\right\} - \frac{\rho_B}{2}\left\{\begin{matrix} q_1^+ \sum_k \Delta x_k + q_2^+ \sum_k \Delta y_k \\ - q_1^- \sum_k \Delta x_k - q_2^- \sum_k \Delta y_k \end{matrix}\right\}\right]}^{D_{ii}} \Delta_t U_i$$
$$- \underbrace{\left[\frac{A_i \Delta_t U_i}{2}\left\{\begin{matrix} p_1^+ \sum_k \Delta x_k + p_2^+ \sum_k \Delta y_k \\ + p_1^- \sum_k \Delta x_k + p_2^- \sum_k \Delta y_k \end{matrix}\right\} + \frac{B_i \Delta_t U_i}{2}\left\{\begin{matrix} q_1^+ \sum_k \Delta x_k + q_2^+ \sum_k \Delta y_k \\ + q_1^- \sum_k \Delta x_k + q_2^- \sum_k \Delta y_k \end{matrix}\right\}\right]}_{S} \tag{15}$$

Generally, the term in the second set of square brackets of equation (15) will not equal zero. Hence, the above is a block diagonal matrix. In the case of the finite difference method,

$$p_1^+ \sum_k \Delta x_k + p_2^+ \sum_k \Delta y_k + p_1^- \sum_k \Delta x_k + p_2^- \sum_k \Delta y_k = 0;$$

$$q_1^+ \sum_k \Delta x_k + q_2^+ \sum_k \Delta y_k + q_1^- \sum_k \Delta x_k + q_2^- \sum_k \Delta y_k = 0.$$

This leads to a scalar diagonal matrix. The finite volume method uses the integral form of the governing equation, which again leads to a scalar diagonal matrix. If we employ a split flux Jacobians approximation as

$$A_i^{\pm} = \frac{\pm \rho_A I}{2}, \quad B_i^{\pm} = \frac{\pm \rho_B I}{2},$$

then the terms in the second set of square brackets of equation (15) will not appear and we get a scalar diagonal matrix. This approach has been followed in Anandhanarayanan (2003) and Anandhanarayanan et al. (2004). In the present work, we stick to Jameson and Yoon's split flux Jacobians in the diagonal term and employ the matrix-free approach again as

$$\begin{aligned} A_i \Delta_t U_i &= \frac{\partial (\mathrm{GX}_i)}{\partial U_i} \Delta_t U_i = \mathrm{GX_i}(U + \Delta U) \\ &\quad - \mathrm{GX}_i(U) = \Delta_t \mathrm{GX_i} \\ B_i \Delta_t U_i &= \frac{\partial (\mathrm{GY}_i)}{\partial U_i} \Delta_t U_i = \mathrm{GY}_i(U + \Delta U) \\ &\quad - \mathrm{GY}_i(U) = \Delta_t \mathrm{GY_i} \end{aligned}$$

We club the term in the second set of square brackets of equation (15) along with the explicit operator. Denoting this term as S, equation (14) can be written as

$$\mathbf{D_{ii}}.\Delta_t U_i = -R_i^n - (\mathbf{E} + \mathbf{C}) + \mathbf{S} \tag{16}$$

It is to be noted that $\mathbf{D_{ii}}$ is a scalar diagonal matrix, which depends on geometrical constants and the spectral radius of flux Jacobians. The expressions for S and $\mathbf{D_{ii}}$ are given in Appendix B. All the terms corresponding to fluxes are on the right side of equation (16). Then a forward and backward sweep of the implicit scheme is given by:

1. Forward sweep:

$$\Delta U^* = \mathbf{D_{ii}^{-1}} \left[-R_i^n - \underbrace{(\mathbf{E} + \mathbf{C})}_{k<i} + \mathbf{S} \right]; \tag{17a}$$

2. Backward sweep:

$$\Delta U^{n+1} = \mathbf{D_{ii}^{-1}} \left[-R_i^n - \underbrace{(\mathbf{E} + \mathbf{C})}_{k<i} - \underbrace{(\mathbf{E} + \mathbf{C})}_{k>i} + \mathbf{S} \right]. \tag{17b}$$

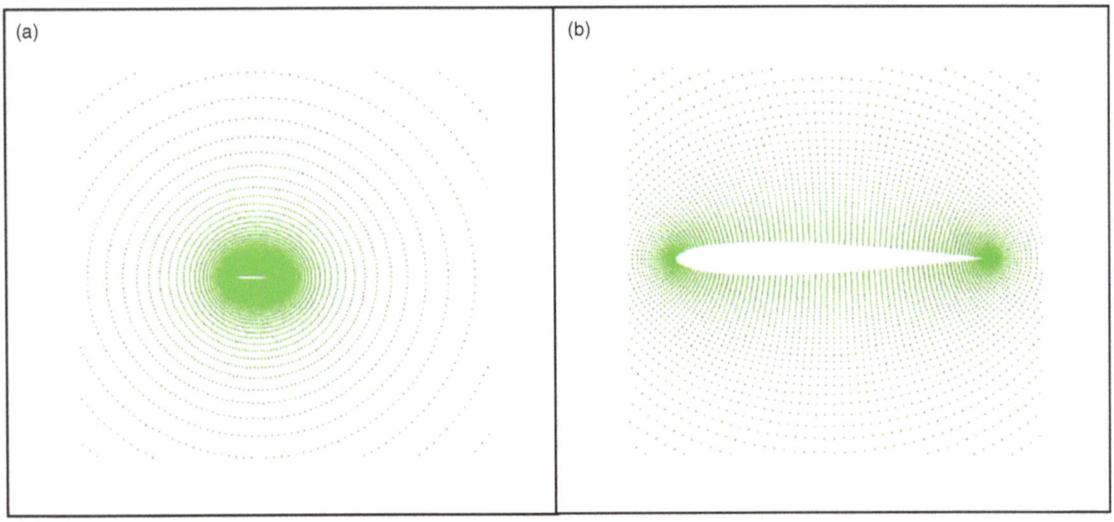

Figure 3. (a) A structured point distribution over a NACA 0012 airfoil and (b) a close-up view.

It is to be noted that $\mathbf{D_{ii}}$ is a scalar diagonal matrix for which $\mathbf{D_{ii}^{-1}} = 1/\mathbf{D_{ii}}$. The spectral radius is taken as $\rho_A = |u| + c, \rho_B = |v| + c,$
where c is the speed of sound and u and v are the fluid velocities in the x and y directions, respectively (for more details, see Singh, 2010; Singh, Ramesh, & Balakrishnan, 2010, 2012).

4. Results and discussion

4.1. 2D test cases

To test the implicit and explicit solver on the NACA 0012 airfoil, a point cloud consisting of 9600 points is created (see Figure 3). There are a total of 160 points on the surface of the airfoil. The far-field distance is taken as 10 times the chord length. All the computations are carried out on Silicon Graphics International (SGI) Altix ICE clusters with a clock speed of 2.93 GHz. The solution is assumed to have converged after 5 decade density residue fall. The Courant–Friedrichs–Lewy (CFL) number for explicit schemes is limited by stability considerations. In the present explicit scheme, the CFL number is taken as 0.8. A larger CFL number can be used in the case of implicit schemes. Figure 4 shows the effect of the CFL number on the convergence of the present implicit scheme for flow over a NACA 0012 airfoil at Mach 0.65 with an angle of attack of 1.86°. It is observed that the convergence speed-up increases with the CFL number up to 1000. Increasing the CFL number beyond 1000 does not have any effect on the convergence. In the present implicit scheme, the CFL number is kept at 1000.

4.1.1. Subsonic case

The subsonic test case is the flow over the NACA 0012 airfoil at Mach 0.65 with an angle of attack of 1.86°. Wind tunnel experimental data is available in Harris (1981). The computed surface pressure coefficient (Cp) distribution along the chord line is compared with the wind tunnel data in Figure 5(a). It is evident that explicit and implicit meshless computed data are in good agreement with the wind tunnel data. The iso-pressure contour plots are smooth and identical for the explicit and implicit schemes (Figure 5(b)). The convergence history for the density residue and lift coefficient are given in Figures 6(a and b), respectively. It is found that the present implicit scheme takes about 80 seconds to converge, as opposed to 1092 seconds in the RK4 explicit scheme. So, in this case, the implicit scheme is 13.6 times faster than the RK4 explicit scheme. Table 1 summarizes this test case.

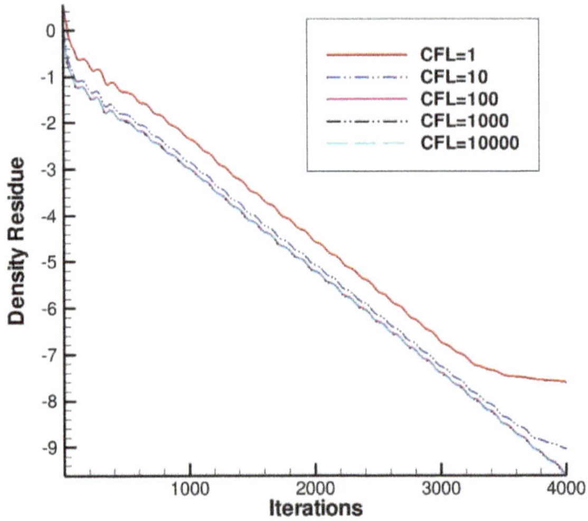

Figure 4. The effect of CFL on the convergence of an implicit scheme for flow over a NACA 0012 airfoil at $M_\infty = 0.65$, $\alpha = 1.86°$.

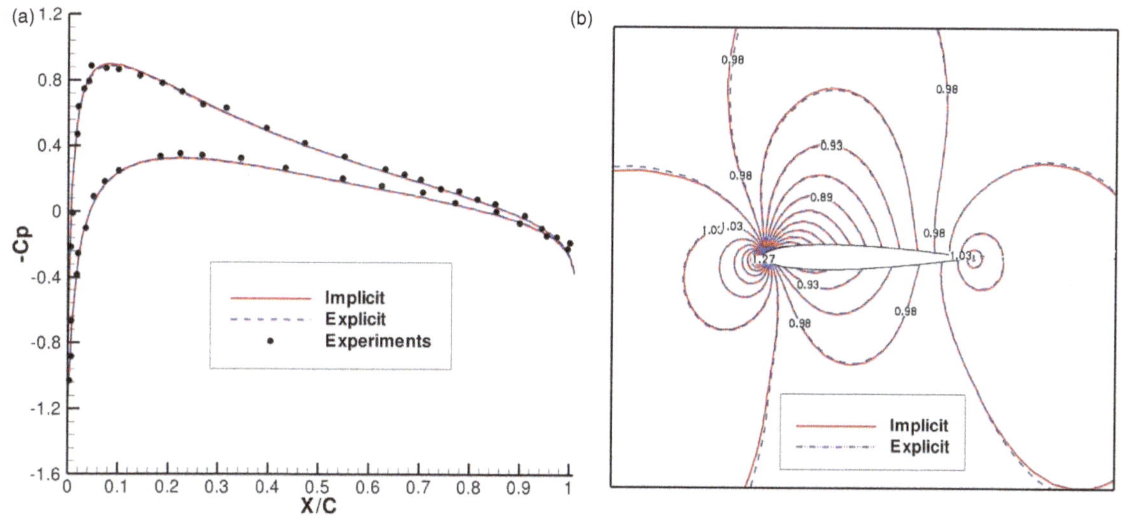

Figure 5. A NACA 0012 airfoil at $M_\infty = 0.65, \alpha = 1.86°$: (a) the surface pressure coefficient distribution along the chord line and (b) the normalized pressure contour.

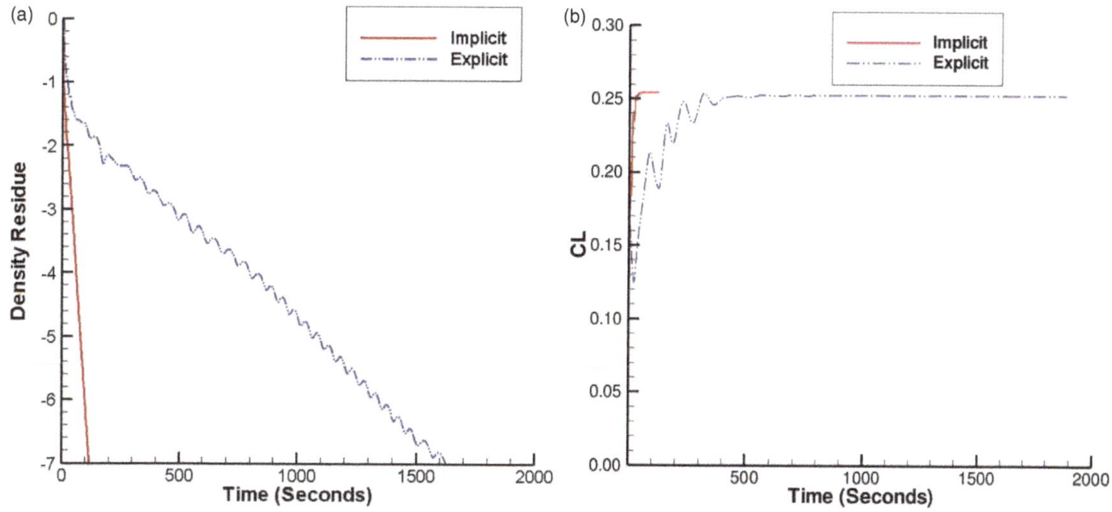

Figure 6. A NACA 0012 airfoil at $M_\infty = 0.65, \alpha = 1.86°$: the convergence history for (a) density residue vs time and (b) lift coefficient (CL) vs time.

Table 1. Subsonic test case ($M_\infty = 0.65, \alpha = 1.86°$).

	Explicit	Implicit	Exp. (Harris, 1981)
Iterations	13,920	1911	
Time (seconds)	1092.5	80.1	
Lift coefficient	.253	.255	.233
Speed-up	13.6		

4.1.2. *Transonic case*

The standard AGARD test case (Viviand, 1985) of transonic flow past a NACA 0012 airfoil at Mach 0.85 with an angle of attack of 1° was chosen for demonstrating the present methodology under the transonic regime. In this test case, shocks appear on both the lower and upper surfaces (Figure 7(a)), which are captured by the present methodology reasonably well. The pressure contour plot is shown in Figure 7(b). The density convergence history is plotted in Figures 8(a and b) against the time and number of iterations, respectively. A speed-up of 7.2 is obtained by using the implicit procedure as compared to the RK4 explicit procedure (see Table 2). Table 2 also compares the lift and drag coefficients with the AGARD results. The explicit and implicit meshless predictions are well within the range of the AGARD data.

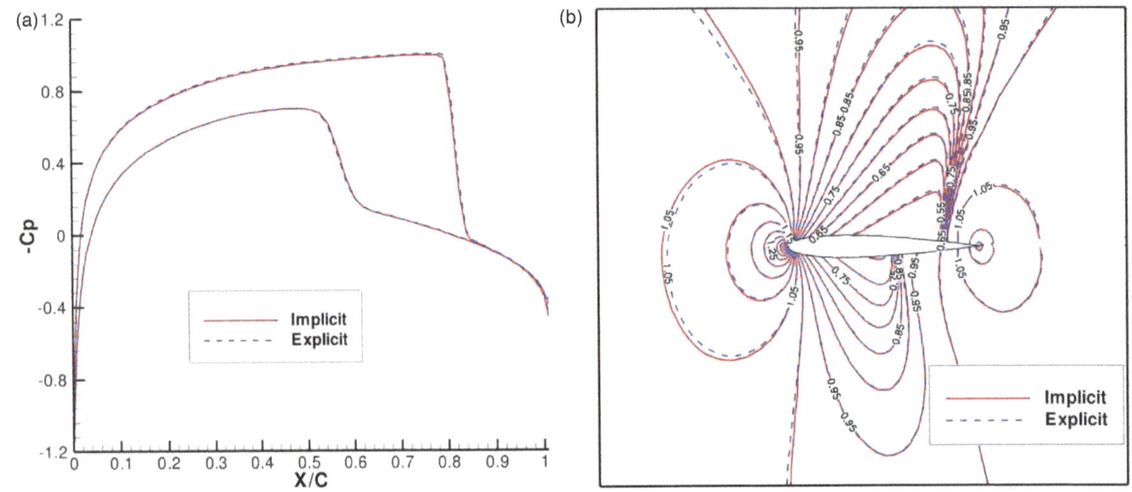

Figure 7. A NACA 0012 airfoil at $M_\infty = 0.85, \alpha = 1.0°$: (a) the surface pressure coefficient distribution along the chord line and (b) the normalized pressure contour.

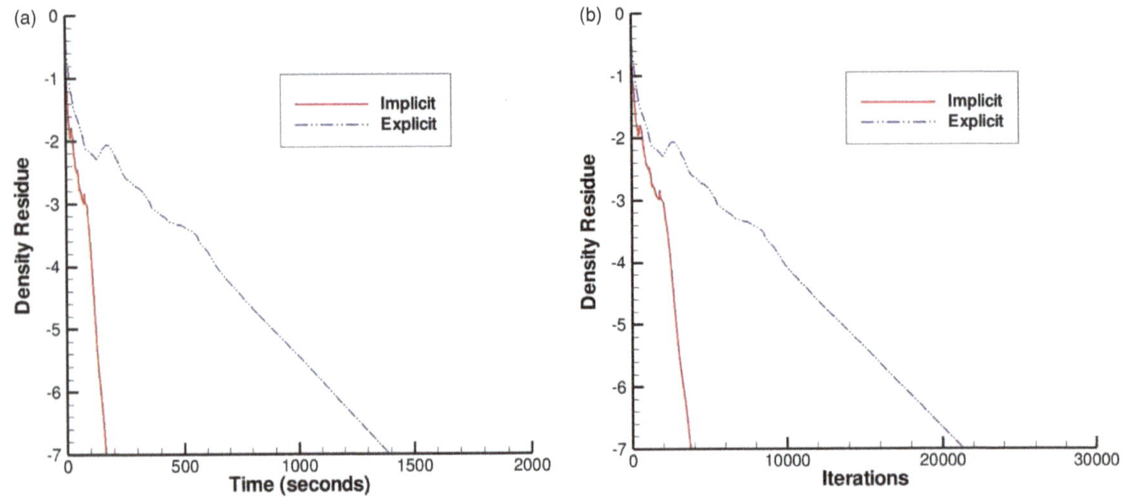

Figure 8. A NACA 0012 airfoil at $M_\infty = 0.85, \alpha = 1.0°$: the convergence history for (a) density residue vs time and (b) density residue vs iterations.

Table 2. Transonic test case ($M_\infty = 0.85, \alpha = 1.0°$).

	Explicit	Implicit	AGARD AR-211
Iterations	13,540	2824	
Time (seconds)	885.0	123.2	
Lift coefficient	.368	.361	.330–.389
Drag coefficient	.0515	.0513	.464–.0590
Speed-up	7.2		

4.1.3. *Supersonic case*

The test case here is the flow past a NACA 0012 airfoil at Mach 1.2 and with an angle of attack of 7°. In Figure 9(a), the computed surface pressure coefficient distribution along the chord line shows good agreement with Schmidt and Jameson's computations from the AGARD report (Viviand, 1985). Figure 9(b) shows the pressure contour in the flow field. The bow shock ahead of the leading edge is captured reasonably well. The convergence of the lift and drag coefficients is shown in Figures 10(a and b), respectively. Table 3 compares the lift and drag coefficients with the AGARD results.

4.2. *Convergence speed vs memory*

One of the key attractions of implicit solvers is the convergence speed-up, which varies according to the Mach number. The convergence speed-up obtained for subsonic

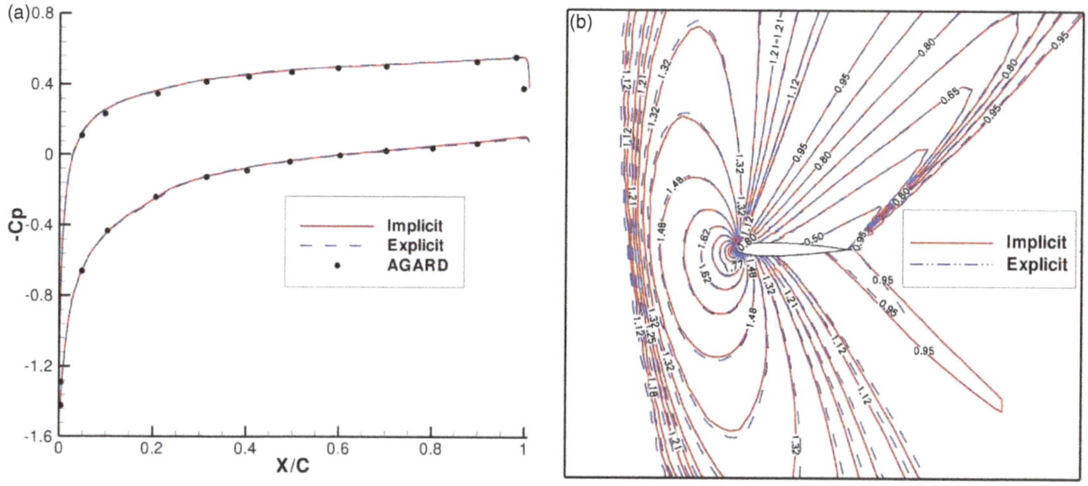

Figure 9. A NACA 0012 airfoil at $M_{\infty} = 1.2$, $\alpha = 7.0°$: (a) the surface pressure coefficient distribution along the chord line and (b) the normalized pressure contour.

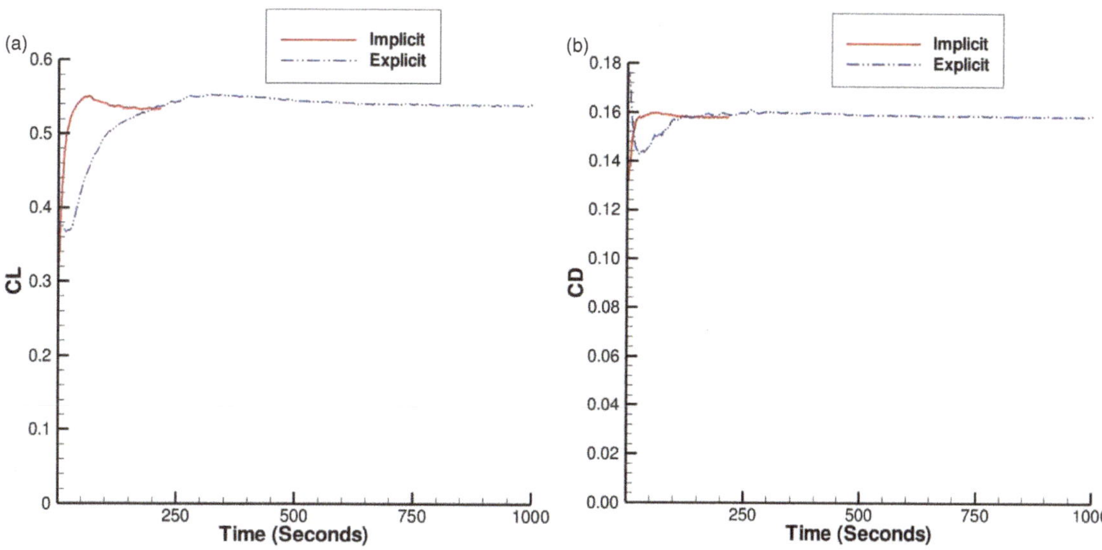

Figure 10. A NACA 0012 airfoil at $M_{\infty} = 1.2$, $\alpha = 7.0°$: the convergence history for (a) the lift coefficient (CL) vs time and (b) the drag coefficient (CD) vs time.

Table 3. Supersonic test case ($M_{\infty} = 1.20$, $\alpha = 7.0°$).

	Explicit	Implicit	AGARD AR-211
Lift coefficient	.5399	.5342	.5138–.5280
Drag coefficient	.1586	.1581	.1531–.1543

flow is close to 14, as compared to 7 for transonic flow. However, the price paid for this convergence acceleration is an increase in memory requirement. Implicit solvers need to store more parameters than with the explicit counterpart. It is observed that for all 2D test cases mentioned above, the implicit solver requires approximately 80% more memory than the explicit solver.

4.3. *3D test case*

The performance of the implicit 3D solvers is tested for transonic flow on an ONERA M6 wing. At Mach 0.84 and an angle of attack 0f 3.06°, lambda shock occurs on the upper surface of the ONERA wing, and this is the standard test case for 3D CFD codes. The experimental results are available in Schmitt and Charpin (1979). The clipped delta wing with deflected aileron is selected to demonstrate the capability of a meshless solver to handle the small gaps effectively. The experimental data is taken from Hess, Cazier, and Wyne (1986). All of these computations are carried out on Centre for Mathematical Modeling and Computer Simulations (CMMACS) SGI clusters (Altix system).

4.3.1. ONERA M6 Wing

The ONERA M6 wing is a swept, semi-span wing with no twist. It uses a symmetric airfoil using the ONERA D section. The ONERA M6 wing is a standard CFD validation case because of its simple geometry combined with a complex transonic flow. A large number of CFD codes have been validated against this test case over the years. Euler simulation of flow over ONERA M6 wing is performed using explicit and implicit meshless solvers. To obtain the point distribution, an unstructured grid is created in the flow domain and only point coordinate information is retained for the meshless solver. Details about flow and point distribution are given in Table 4. The point distribution around the wing is shown in Figure 11. For a higher order of accuracy, the MCIR factor p-value should be as close to unity as possible ($p < 1$). For the ONERA M6 wing, a maximum p-value of only .56 is achieved using an explicit scheme, as compared to .63 for an implicit scheme. Each iteration of the RK4 explicit scheme is about 1.6 times costlier than the implicit scheme. It is found that the implicit scheme converges in 1000 iterations against 4500 iterations in the explicit scheme. So, the speed-up obtained is more than 7 in this case. However, only the implicit results are presented here because of the better accuracy achieved through the higher p-value.

The comparison of the pressure coefficient (Cp) distribution along a local chord line (in a fraction of local chord length, X/C) between computations and experiments are shown in Figures 12 and 13 for different span wise locations ranging from 20% span to 95% span. Figures 12(a and b) show Cp comparison plots at the locations of 20% and 44% span, respectively. It is observed that implicit computation is able to accurately capture both the pressure peak and the shock location quite satisfactorily.

Figures 13(a–c) showCp comparison plots at the locations of 65%, 80% and 90% span, respectively. It is observed that implicit computation can capture the pressure peak and second shock location very accurately. At

Table 4. Flow and point distribution parameter for an ONERA M6 wing.

Parameter	Value
Mach number (M_∞)	0.84
Angle of attack (α)	3.06°
Total number of points	632,164
Points on wing surface	104,618
Far-field along span	5 (Span length)
Far-field upstream and downstream	10 (Root chord length)

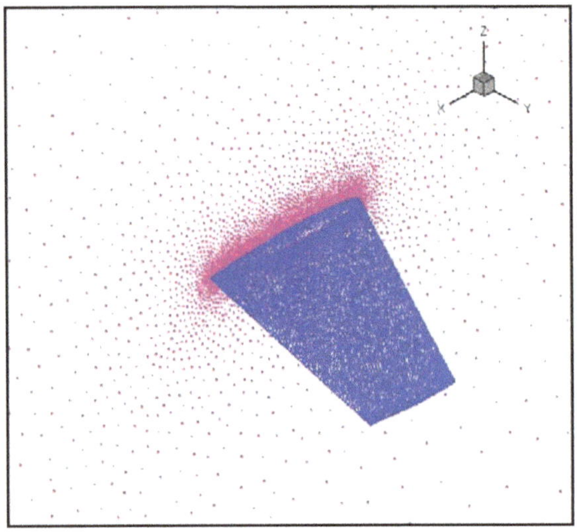

Figure 11. Point distribution around an ONERA M6 wing.

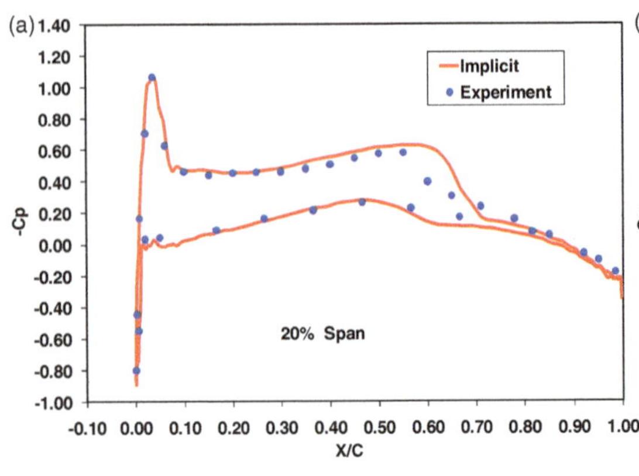

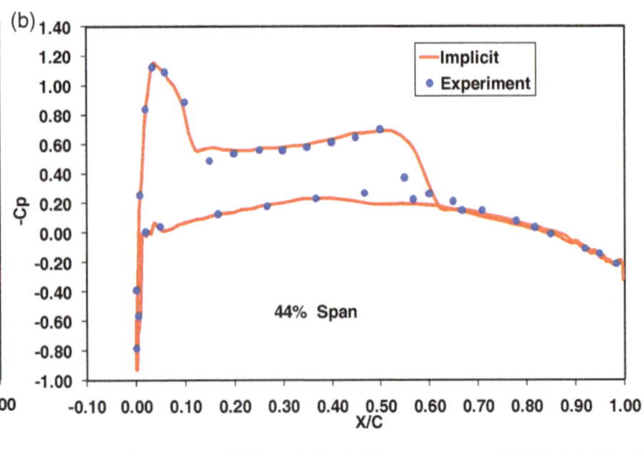

Figure 12. The surface pressure coefficient distribution along the local chord line at (a) 20% and (b) 44% span on an ONERA M6 wing at $M_\infty = 0.84$, $\alpha = 3.06°$.

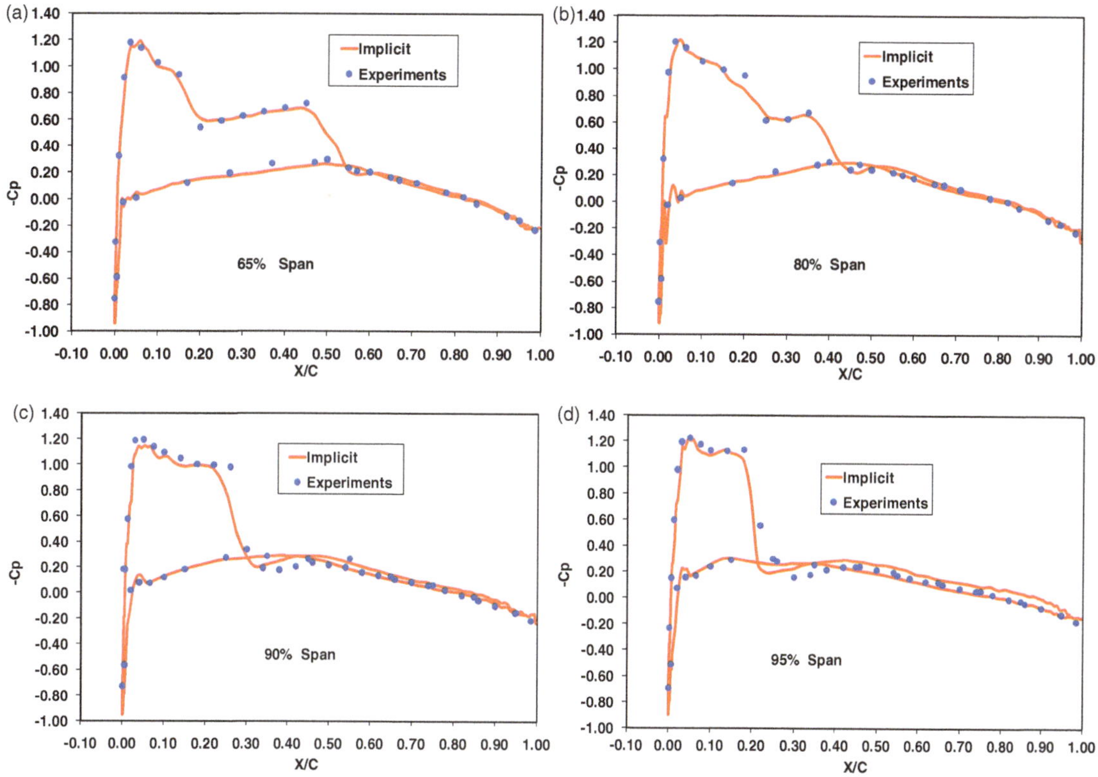

Figure 13. The surface pressure coefficient distribution along the local chord line at (a) 65%, (b) 80%, (c) 90% and (d) 95% span on an ONERA M6 wing at $M_{\infty} = 0.84, \alpha = 3.06°$.

the 90% span location, there is a single shock which is captured very well by computation. Figure 13(d) shows a Cp comparison plot at the location of 95% span. This section also has a single shock similar to the 90% span location. The computed results predicted the shock location and pressure peaks very well. Finally, a normalized pressure (with respect to free stream pressure) contour plot is given in Figure 14. The lambda-shape shock can be noticed very clearly in this figure. Present computation shows that LSKUM with MCIR splitting can capture all the flow features very well for the case of transonic flow past an ONERA M6 wing.

4.3.2. Clipped delta wing with aileron deflection of 6°

Figure 15(a) shows the plan-form of the cropped delta wing. The leading edge backward sweep is 50.4°. The aileron is hinged from 56.5% semi-span to 82.9% of semi-span and at 80% of the local chord from the leading edge. The aileron deflection is 6°. The slot gap between the deflected aileron and the main wing is modeled as it is. The present computation uses 601,359 points (for details of the flow and point distribution, see Table 5). The information about five sectional locations along the span, where experimental Cp data is available, is given in Figure 15(b) and Table 6. At location D, which lies in the middle of

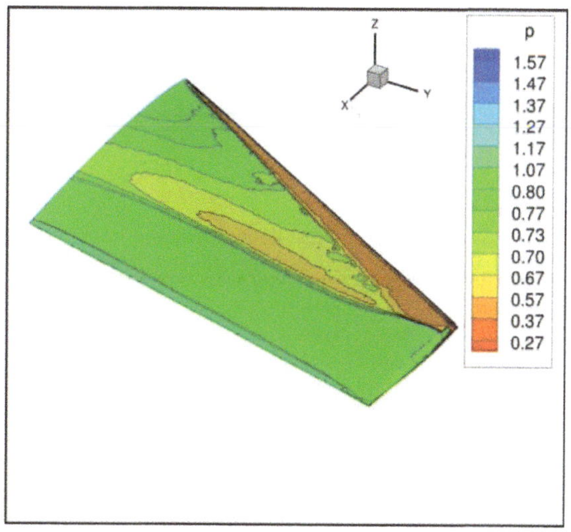

Figure 14. Normalized pressure contours on an ONERA M6 wing at $M_{\infty} = 0.84, \alpha = 3.06°$.

the deflected aileron, present Euler computations are also compared with the Euler and Reynolds-Averaged Navier-Stokes (RANS) of Li, Zhu, Chen, and Li (1999) using finite volume code. The finite volume code of Li et al.

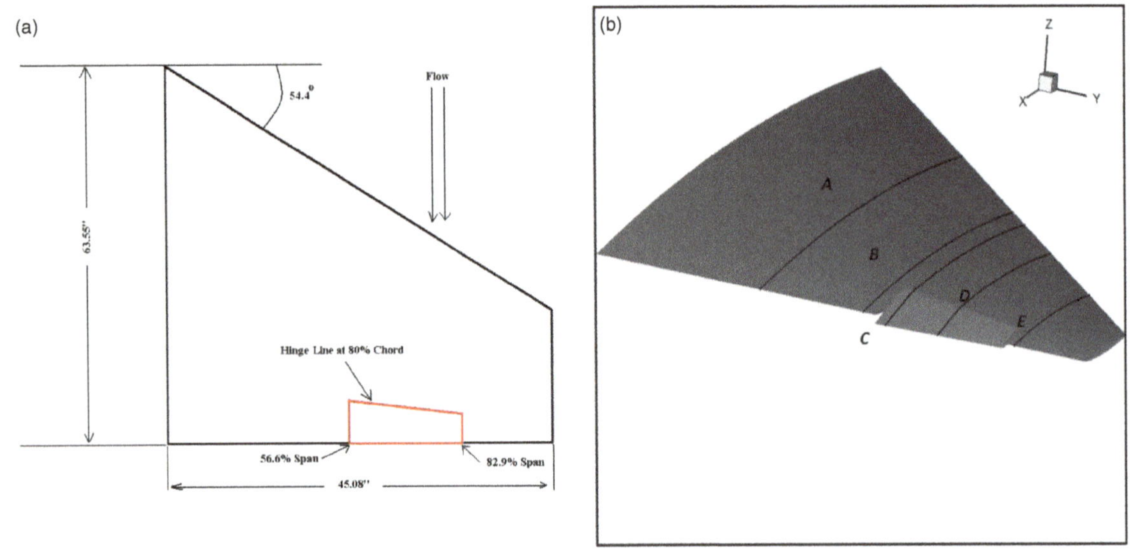

Figure 15. Clipped delta wing with aileron deflection: (a) plan-form; and (b) location of various sections of the clipped delta wing.

Table 5. Flow and point distribution parameter for a clipped delta wing.

Parameter	Value
Mach number (M_∞)	0.90
Angle of attack (α)	1°, 3°
Total number of points	601,359
Points on wing surface	143,413
Far-field along span	5 (Span length)
Far-field upstream and downstream	10 (Root chord length)

Table 6. Span-wise section locations.

Location	% of span	Comments
A	33.2	Section on main wing
B	54.1	On the main wing but close to the deflected aileron
C	58.7	On deflected aileron
D	69.4	Middle of the deflected aileron
E	85.1	On main wing but close to the deflected aileron

(1999) uses a Van Leer flux vector splitting scheme for the discretization of in-viscid fluxes. This code uses 120,000 and 506,250 grid cells for the Euler and NS computations, respectively.

The Cp comparisons between the present meshless Euler computations, the wind tunnel experiments (Hess et al., 1986) and the earlier Euler and RANS computations (Li et al., 1999) are presented in Figures 16(a and b) for angles of attack of 1° and 3°, respectively at section D (see Figure 15(b)). Both of the Euler computations over-predict the suction pressure at the deflection location. However, the RANS computation of Li et al. (1999), which is closer

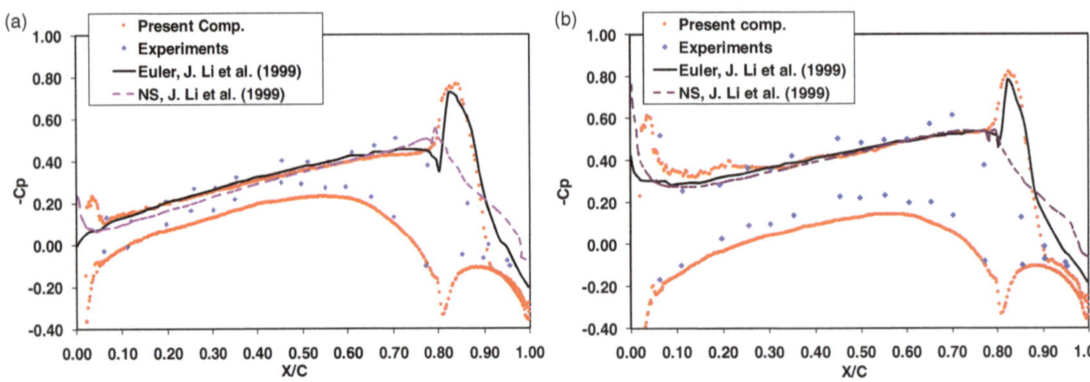

Figure 16. Pressure coefficient distribution comparison along section 'D' on the surface of a clipped delta wing with aileron deflection of $\delta = 6°$ at $M_\infty = 0.9$ with (a) $\alpha = 1°$ and (b) $\alpha = 3°$.

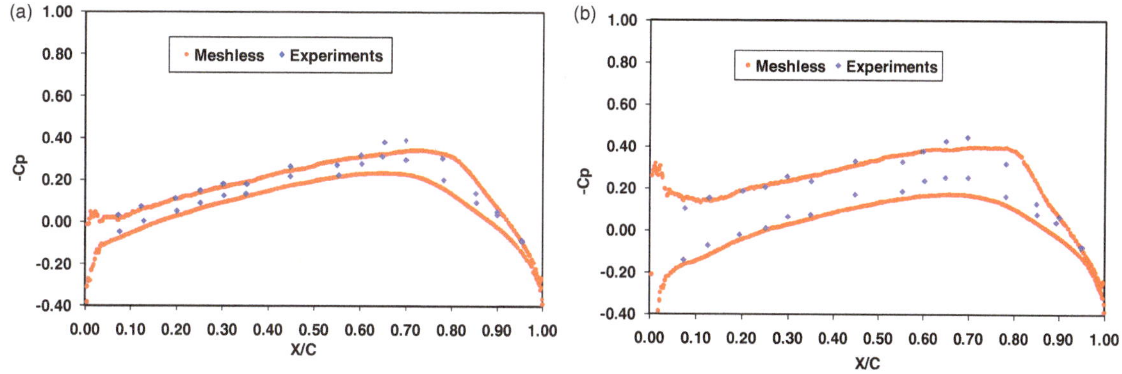

Figure 17. Pressure coefficient distribution comparison along section 'A' on the surface of a clipped delta wing with aileron deflection of $\delta = 6°$ at $M_{\infty} = 0.9$ with (a) $\alpha = 1°$ and (b) $\alpha = 3°$.

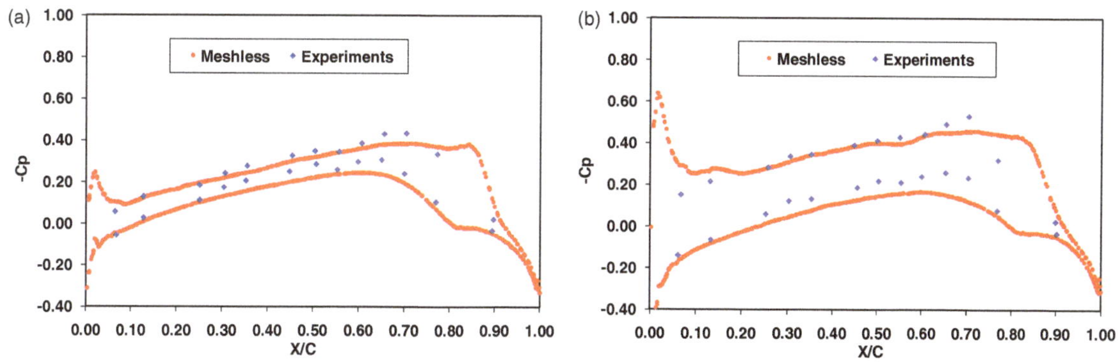

Figure 18. Pressure coefficient distribution comparison along section 'B' on the surface of a clipped delta wing with aileron deflection of $\delta = 6°$ at $M_{\infty} = 0.9$ with (a) $\alpha = 1°$ and (b) $\alpha = 3°$.

to the experiments, does not show pressure suction to this extent at this location. The present meshless Euler computations capture the leading edge suction peak quite well.

The Cp comparison between the implicit computations and experiments, for locations A, B, C and E are presented in Figures 17a, 18a, 19a and 20a, respectively for an angle of attack of 1°, and in Figures 17b, 18b, 19b and 20b for an angle of attack of 3°. The computed Cp matches reasonably well with the experiments except at location C (Figures 19(a and b)), which is on the deflected portion of the aileron (see Figure 15(b)). The present Euler computations predict an increase in suction pressure at the deflected location for both angles of attack. This trend is similar to

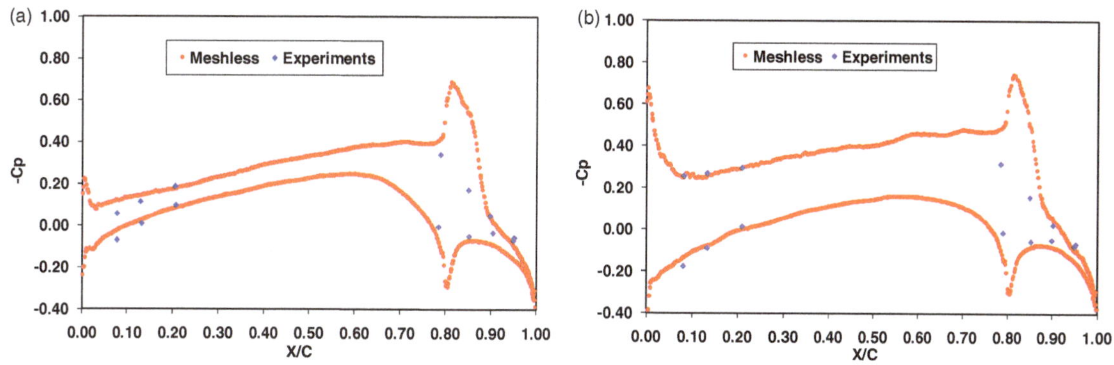

Figure 19. Pressure coefficient distribution comparison along section 'C' on the surface of a clipped delta wing with aileron deflection of $\delta = 6°$ at $M_{\infty} = 0.9$ with (a) $\alpha = 1°$ and (b) $\alpha = 3°$.

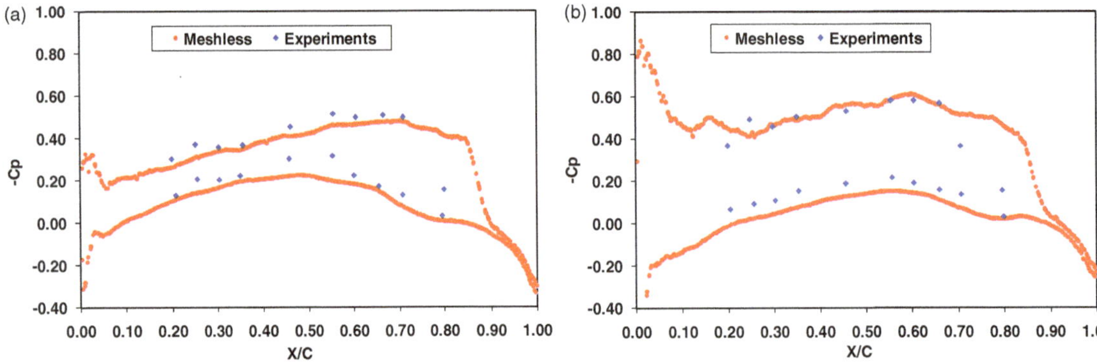

Figure 20. Pressure coefficient distribution comparison along section 'E' on the surface of a clipped delta wing with aileron deflection of $\delta = 6°$ at $M_\infty = 0.9$ with (a) $\alpha = 1°$ and (b) $\alpha = 3°$.

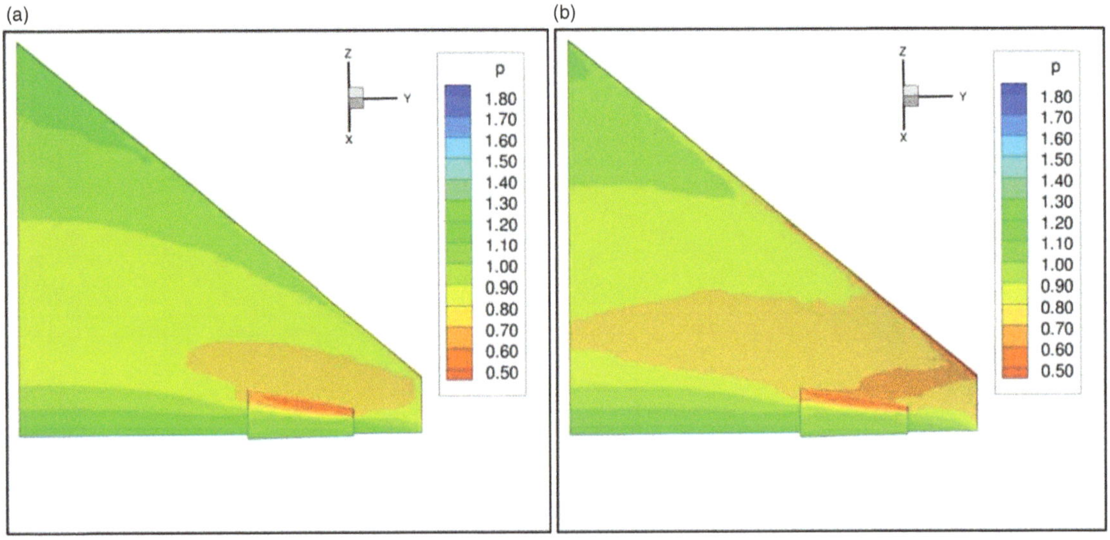

Figure 21. Normalized pressure contour plot on the surface of a clipped delta wing with aileron deflection of $\delta = 6°$ at $M_\infty = 0.9$ with (a) $\alpha = 1°$ and (b) $\alpha = 3°$.

what has been observed at location D. For all locations, there is not enough experimental data available between 80 and 90% of the local chord length. The leading edge suction pressure is more for $\alpha = 3°$, in comparison with $\alpha = 1°$ at all sections. Normalized pressure (with respect to free stream pressure) contour plots are given in Figures 21(a and b) for $\alpha = 1°$ and $\alpha = 3°$, respectively. On the upper surface of the wing, a relatively lower pressure region is observed in a larger area at $\alpha = 3°$.

5. Conclusion

An implicit scheme for a meshless solver based on the kinetic upwind method is presented. Advances made in the implicit finite volume methodology, like the matrix-free approach and Jameson and Yoon's split flux Jacobian for the diagonal-dominant implicit procedure are adopted for the meshless framework. The capability of the implicit scheme is demonstrated by the computing flow past a NACA 0012 airfoil under subsonic, transonic and supersonic regimes. A convergence acceleration up to 14 is obtained over the RK4 explicit scheme. The results obtained are in good agreement with the wind tunnel experimental data and other reliable CFD computations. However, the memory requirement for the implicit solver is approximately 80% more than that of the RK4 explicit solver. Further, the accuracy and robustness of the scheme in 3D is demonstrated by computing the flow past an ONERA M6 wing and a clipped delta wing with an aileron deflection of 6°. The computed results show good agreement with the wind tunnel experimental data and other CFD computations.

At present, this study is limited to a meshless in-viscid Euler solver. We plan to implement this methodology in a viscous meshless solver as well. Reducing the memory requirement of the implicit solver is also one of the priorities.

Acknowledgment

The authors would like to thank Mr Richard A. Evans, College of Engineering, Swansea University, Swansea, United Kingdom for editing the manuscript and suggesting corrections and improvements.

Disclosure statement

No potential conflict of interest was reported by the authors.

References

Anandhanarayanan, K. (2003). *Development and application of a Gridfree kinetic upwind solver to multibody configurations* (Ph.D. Thesis), Dept. of Aerospace Engg., Indian Institute of Science, Bangalore, India.

Anandhanarayanan, K., Nagarathinam, M., & Deshpande, S. M. (2004). An entropy variable based grid-free solver with LU-SGS accelerator. *24th Intl. Congress of Aeronautical Sciences, Yakohama, Japan*. ICAS2004-2.4.3.

Anil, N. (2008). *Optimal control of numerical dissipation in modified KFVS (m-KFVS) using Discrete Adjoint Method* (Ph.D. Thesis). Dept. of Aerospace Engg., Indian Institute of Science, Bangalore, India.

Balakrishnan, N., & Sridar, D. (2001). On a generalised finite difference framework and its order of accuracy. *Computational Fluid Dynamics 2000*, Springer New York: 797–798.

Batina, J. T. (1993). A gridless Euler/Navier–Stokes solution algorithm for complex aircraft applications. AIAA *Paper 1993-0333*.

Chiu, E. K., Wang, Q., & Jameson, A. (2011). A conservative meshless scheme: General order formulation and application to euler equations. *AIAA Paper 2011-0651*.

Demkowikz, L., Karafiat, A., & Liszka, T. (1984). On some convergence results for FDM with irregular mesh. *Computer Methods in Applied Mechanics and Engineering*, *42*, 343–355.

Deshpande, S. M. (1986a). On the Maxwellian distribution, symmetric form and entropy conservation for the euler equations. *NASA TP 2583*.

Deshpande, S. M. (1986b). A second order accurate kinetic theory based method for inviscid compressible flow. *NASA TP 2613*.

Deshpande, S. M., Anandhanarayanan, K., Praveen, C., & Ramesh, V. (2002). Theory and application of 3-D LSKUM based on entropy variables. *International Journal for Numerical Methods in Fluids*, *40*, 47–62.

Deshpande, S. M., Kulkarni, P. S., & Ghosh, A. K. (1998). New developments in kinetic schemes. *Journal of Computers and Mathematics with Applications*, *35*(1/2), 75–93.

Dominic, D. J. C. (2004). *A one point shock capturing kinetic scheme for hyperbolic conservation laws* (M.E. Thesis). Dept. of Aerospace Engg., IISc, Bangalore, India.

Ghosh, A. K., & Deshpande, S. M. (1995). Least Squares Kinetic Upwind Method for inviscid compressible flows. *AIAA Paper 1995–1735*.

Gingold, R. A., & Monaghan, J. J. (1977). Smoothed particle hydrodynamics - Theory and applications to non-spherical stars. *Monthly Notices of the Royal Astronomical Society*, *181*, 375–389.

Harris, C. D. (1981). Two-dimensional aerodynamic characteristics of the NACA 0012 Airfoil in the Langley 7-foot transonic pressure tunnel. *NASA Technical Memorandum 81927*.

Hess, R. W., Cazier, F. W., & Wyne, E. C. (1986). Steady and unsteady transonic pressure measurements on a clipped delta wing for pitching and control-surface oscillations. *NASA TP 2594*.

Jameson, A., & Yoon, S. (1987). Lower-upper implicit schemes with multiple grids for the euler equations. *AIAA Journal*, *25*(7), 929–935.

Katz, A., & Jameson, A. (2009a). A comparison of various meshless schemes within a unified algorithm. *AIAA paper 2009-596*.

Katz, A., & Jameson, A. (2009b). A meshless volume scheme. *AIAA Paper 2009–3534*.

Katz, A., & Jameson, A. (2009c). Multicloud: multigrid convergence with a meshless operator. *Journal of Computational Physics*, *228*, 5237–5250.

Li, J., Zhu, Z. Q., Chen, Z. M., & Li, H. M. (1999). The Euler and Navier-Stokes solutions of a 3-D wing with aileron. *Acta Mechanica*, *138*, 51–59.

Liszka, T., & Orkisz, J. (1980). The finite difference method at arbitrary irregular grids and its application in applied mechanics. *Computers and Structures*, *11*(1/2), 83–95.

Lucy, L. B. (1977). A numerical approach to the testing of fission hypothesis. *The Astronomical Journal*, *82*, 1013–1024.

Luo, H., Baum, J. D., & Lohner, R. (1998). A fast, matrix-free implicit method for compressible flows on unstructured grid. *Journal of Computational Physics*, *146*, 664–690.

Mahindra, A. K., Singh, R. K., & Gouthaman, G. (2011). Meshless kinetic upwind method for compressible, viscous rotating flows. *Computers and Fluids*, *46*(1), 325–332.

Mandal, J. C., & Deshpande, S. M. (1994). Kinetic flux vector splitting method for euler equations. *Computers and Fluids*, *23*(2), 447–478.

Monaco, A. D., Manenti, S., Gallati, M., Sibilla, S., Agate, G., & Guandalini, R. (2011). SPH modeling of solid boundaries through a semi-analytic approach. *Engineering Applications of Computational Fluid Mechanics*, *5*(1), 1–15.

Munikrishna, N. (2007). *On viscous flux discretization procedures for finite volume and meshless solvers*. Ph. D. Thesis, Dept. of Aerospace Engg., Indian Institute of Science, Bangalore, India.

Munikrishna, N., & Balakrishnan, N. (2007). Turbulent Flow Computations using Meshless Solver LSFD-U. *Proceedings of Seventh Asian Computational Fluid Dynamics Conference (ACFD7) Bangalore*.

Ninawe, A., Munikrishna, N., & Balakrishnan, N. (2006). Viscous Flow Computations Using a Meshless Solver, LSFD–U. *Computational Fluid Dynamics 2004*, Springer–Verlag, Berlin: 509–514.

Pu, J. H., Shao, S., Huang, Y., & Hussain, K. (2013). Evaluations of SWEs and SPH numerical modelling techniques for dam break flows. *Engineering Applications of Computational Fluid Mechanics*, *7*(4), 544–563.

Ramesh, V. (2001). *Least Squares Grid Free Kinetic Upwind Method* (Ph.D. Thesis). Dept of Aerospace Engg., Indian Institute of Science, Bangalore, India.

Ramesh, V., & Deshpande, S. M. (2001). Least squares kinetic upwind method on moving grids for unsteady euler computations. *Computers and Fluids*, *30*, 621–641.

Ramesh, V., & Deshpande, S. M. (2004). Low dissipation grid free upwind kinetic scheme with modified CIR splitting. *Fluid Mechanics report, FM*-20, Dept of Aerospace Engg., Indian Institute of Science, Bangalore, India.

Ramesh, V., & Deshpande, S. M. (2007). Unsteady flow computations for flow past multiple moving boundaries using LSKUM. *Computers and Fluids*, *36*(10), 1592–1608.

Ramesh, V., Mathur, J. S., & Deshpande, S. M. (1997). Kinetic treatment of the far-field boundary condition. *Fluid Mechanics report FM 2*, Dept of Aerospace Engg., Indian Institute of Science Bangalore India.

Schmitt, V., & Charpin, F. (1979). Pressure Distribution on the ONERA M6 wing at transonic Mach numbers. *AGARD AR-138*.

Shende, N. (2005). *Development of a general purpose flow solver for Euler equations*. (Ph.D. Thesis), Dept of Aerospace Engg., Indian Institute of Science, Bangalore, India.

Shende, N., & Balakrishnan, N. (2004). A new migratory memory algorithm for implicit finite volume solvers. *AIAA Journal*, *42*(9), 1863–1870.

Singh, M. K. (2010). *LU-SGS implicit scheme for a mesh-less Euler solver*. M.Sc. (Engg.) thesis, Dept. of Aerospace Engineering, Indian Institute of Science, Bangalore, India.

Singh, M. K., Ramesh, V., & Balakrishnan, N. (2010). Convergence acceleration of meshless Euler solver. *12th Annual CFD AeSI Symposium, Bangalore, India*.

Singh, M. K., Ramesh, V., & Balakrishnan, N. (2012). An implicit scheme for a meshless Euler solver. *26th National Convention of Aerospace Engineers: Expanding frontiers in aerospace technologies – challenges and opportunities, Hyderabad, India*.

Sridar, D., & Balakrishnan, N. (2003). An upwind finite difference scheme for mesh-less solvers. *Journal of Computational Physics*, *189*, 1–29.

Sridar, D., & Balakrishnan, N. (2006). Convergence acceleration of an upwind least squares finite difference based meshless solver. *AIAA Journal*, *44*(10), 2189–2196.

Viviand, H. (1985). Test cases for inviscid flow field methods. *AGARD AR-211*.

Yoon, S., & Jameson, A. (1986). A multigrid LU-SSOR scheme for approximate newton-iteration applied to the euler equations. *NASA CR-17954*.

Yoon, S., & Jameson, A. (1987). A lower-upper symmetric-gauss-seidel method for the euler and navier-stokes equations. *AIAA Paper 1987-0600*.

Zhau, X., & Xu, H. (2010). Gridless method for unsteady flow involving moving discrete points and its applications. *Engineering Applications of Computational Fluid Mechanics*, *4*(2), 276–286.

Appendix A

In least square approximation, the flux derivative for point 'i' can be written as

$$\frac{\partial \mathrm{GX}}{\partial x} = p_1 \sum_k \Delta \mathrm{GX}_k.\Delta x_k + p_2 \sum_k \Delta \mathrm{GX}_k.\Delta y_k$$

$$\frac{\partial \mathrm{GY}}{\partial x} = q_1 \sum_k \Delta \mathrm{GY}_k.\Delta x_k + q_2 \sum_k \Delta \mathrm{GY}_k.\Delta y_k$$

where '*k*' is point in connectivity, and p_1, p_2, q_1 and q_2 are the geometrical constants given by

$$p_1 = \frac{-\sum_k \Delta y_k^2}{\left(\sum_k \Delta x_k \Delta y_k\right)^2 - \sum_k \Delta x_k^2 \sum_k \Delta y_k^2},$$

$$p_2 = \frac{\sum_k \Delta x_k \Delta y_k}{\left(\sum_k \Delta x_k \Delta y_k\right)^2 - \sum_k \Delta x_k^2 \sum_k \Delta y_k^2}$$

$$q_1 = \frac{\sum_k \Delta x_k \Delta y_k}{\left(\sum_k \Delta x_k \Delta y_k\right)^2 - \sum_k \Delta x_k^2 \sum_k \Delta y_k^2},$$

$$q_2 = \frac{-\sum_k \Delta y_k^2}{\left(\sum_k \Delta x_k \Delta y_k\right)^2 - \sum_k \Delta x_k^2 \sum_k \Delta y_k^2}$$

Appendix B

The explicit operator is given by

$$R_i^n = \left[\begin{array}{l} \left\{\begin{array}{l} p_1^+ \sum_k \Delta \mathrm{GX}_k^+ \Delta x_k + p_2^+ \sum_k \Delta \mathrm{GX}_k^+ \Delta y_k \\ +p_1^- \sum_k \Delta \mathrm{GX}_k^- \Delta x_k + p_2^- \sum_k \Delta \mathrm{GX}_k^- \Delta y_k \end{array}\right\} \\ + \left\{\begin{array}{l} q_1^+ \sum_k \Delta \mathrm{GY}_k^+ \Delta x_k + q_2^+ \sum_k \Delta \mathrm{GY}_k^+ \Delta y_k \\ +q_1^- \sum_k \Delta \mathrm{GY}_k^- \Delta x_k + q_2^- \sum_k \Delta \mathrm{GY}_k^- \Delta y_k \end{array}\right\} \end{array}\right]$$

The lower and upper implicit operator is given by

$$\mathbf{E} + \mathbf{C} = \left[\begin{array}{l} \left\{\begin{array}{l} p_1^+ \sum_k \Delta_t \mathrm{GX}_k^+ \Delta x_k + p_2^+ \sum_k \Delta_t \mathrm{GX}_k^+ \Delta y_k \\ +p_1^- \sum_k \Delta_t \mathrm{GX}_k^- \Delta x_k + p_2^- \sum_k \Delta_t \mathrm{GX}_k^- \Delta y_k \end{array}\right\} \\ + \left\{\begin{array}{l} q_1^+ \sum_k \Delta_t \mathrm{GY}_k^+ \Delta x_k + q_2^+ \sum_k \Delta_t \mathrm{GY}_k^+ \Delta y_k \\ +q_1^- \sum_k \Delta_t \mathrm{GY}_k^- \Delta x_k + q_2^- \sum_k \Delta_t \mathrm{GY}_k^- \Delta y_k \end{array}\right\} \end{array}\right]$$

The scalar diagonal matrix is given by

$$\mathbf{D}_{\mathbf{ii}} = \left[\begin{array}{l} \frac{I}{\Delta t} - \frac{\rho_A}{2} \left\{\begin{array}{l} p_1^+ \sum_k \Delta x_k + p_2^+ \sum_k \Delta y_k \\ -p_1^- \sum_k \Delta x_k - p_2^- \sum_k \Delta y_k \end{array}\right\} \\ - \frac{\rho_B}{2} \left\{\begin{array}{l} q_1^+ \sum_k \Delta x_k + q_2^+ \sum_k \Delta y_k \\ -q_1^- \sum_k \Delta x_k - q_2^- \sum_k \Delta y_k \end{array}\right\} \end{array}\right]$$

The term '*S*' in equation (16) is given by

$$\mathbf{S} = \left[\begin{array}{l} \frac{\Delta_t GX_i}{2} \left\{\begin{array}{l} p_1^+ \sum_k \Delta x_k + p_2^+ \sum_k \Delta y_k \\ +p_1^- \sum_k \Delta x_k + p_2^- \sum_k \Delta y_k \end{array}\right\} \\ + \frac{\Delta_t GY_i}{2} \left\{\begin{array}{l} q_1^+ \sum_k \Delta x_k + q_2^+ \sum_k \Delta y_k \\ +q_1^- \sum_k \Delta x_k + q_2^- \sum_k \Delta y_k \end{array}\right\} \end{array}\right]$$

In the above expressions, $\Delta(.)_k = (.)_k - (.)_i$ and $\Delta_t(.)_k = (.)_k^{n+1} - (.)_k^n$.

15

Transient numerical simulation for solid-liquid flow in a centrifugal pump by DEM-CFD coupling

Si Huang*, Xianghui Su and Guangqi Qiu

School of Mechanical and Automotive Engineering, South China University of Technology, Guangzhou, China

The two-phase solid-liquid flow in the computational fluid dynamic (CFD) method is generally analyzed by treating the solid particles as a quasi-fluid element, but this does not take into account the effects of the physical features of the solid particles and collisions among the particles. In this study, the discrete element method (DEM) was coupled with the CFD method in order to analyze the transient two-phase solid-liquid flow in a single-stage centrifugal pump with consideration for the particle-particle and particle-structure interactions. The numerical simulation process involving EDEM and FLUENT is described. Simulations were performed for the water at 25âĎČ at the continuous phase together with 15% of the volume as spherical solid particles (density $\rho = 1500$ kg/m^3) with the diameter ranging from 1 to 3 mm in a random order. The results illustrate the motion of the solid particles in the pump and their effect on the flow performance of the pump in terms of time variation in the head. The velocity fields for the two-phase flow together with the volume fraction distributions and trajectories of the solid particles in the centrifugal pump are also presented.

Keywords: DEM-CFD coupling; centrifugal pump; trajectories of particles; transient two-phase flow

1. Introduction

Centrifugal pumps are widely used in various slurry transport systems and are considered to be more cost effective than transportation by truck or conveyor belt. Transportation of slurry mixture via centrifugal pumps is most commonly found in operations to dredge various bodies of water, maintain navigation in harbors and rivers, alter coastlines, and collect material for landfill and construction purposes. As well as the above, the mining, fertilizer and food-processing sectors also employ large-scale slurry transportation via pumps. Moreover, preservation of the environment requires that waste be conveyed to a disposal site via pumps. The operational efficiency of centrifugal pumps employed in slurry transportation strongly depends on the pump design and the characteristics of two-phase slurry flow (Addie, Roudnev, & Sellgren, 2007; Clark & Abbott, 1992; Lyczkowski & Bouillard, 2002; Messa & Malavasi, 2014; Wilson, Addie, Sellgren, & Clift, 2006). A systematic study of two-phase solid-liquid flow in pumps and the effects of solid particle size on flow characteristics is thus essential in order to seek crucial design guidance for enhancing the operational efficiency of the transportation process.

The numerical analyses of two-phase flow in various applications have mostly been conducted using the computational fluid dynamic (CFD) method, sometimes coupled with the discrete element method (DEM). In the CFD method, the solid particles in the mixed flow are typically treated as a pseudo-liquid medium in order to obtain the particles' motion through solutions of the governing equations of multiphase flow models. This method has been widely used in a range of engineering applications due to its global accuracy in two-phase flow prediction, its applicability to a wide range of solid volume fractions and its relatively low computational demand. Simulations of multiphase flow models can be conveniently carried out using commercially available CFD code. This is evident from the various reported studies on two-phase solid-liquid flow in different rotating machinery (Cheng, Li, Gao, & Guo, 2013; Chochua, Koch, & Sorokes, 2005; Fang, Ling, & Sang, 2013; Han, Ma, Li, & Li, 2012; Hua, Yi, & Yu-Liang, 2013; Ma, Wu, Huang, Yu, & Wang, 2013; Pagalthivarthi, Gupta, Tyagi, & Ravi, 2011; Roudnev, Bourgeois, & Kosmicki, 2009; Roudnev & Kosmicki, 2007; Zhang, Li, Cui, Zhu, & Dou, 2013), which have substantially contributed to the design of more efficient rotating machines.

The CFD method, however, exhibits inherent limitations due to the pseudo-fluid approximation of the solid particles, primarily due to discontinuities caused by the motion of the solid particles. The effects of various particle features such as material properties, shape and size cannot be accurately evaluated using the CFD method. Moreover, the CFD method does not provide an accurate assessment

*Corresponding author. Email: huangsi@scut.edu.cn

of the effects of the collision and wear on the two-phase flow characteristics caused by the solid particles. Alternatively, the discrete element method (DEM), proposed by Cundall (1971), could yield an accurate analysis of solid particle motion in two-phase flow by considering the essential solid particle features. The proposed method provides an accurate description of the particles' dynamic motions (position and velocity) in accordance with Newton's second law of motion, factoring in the shape, size and material properties of the particles. The DEM coupled with the CFD method has also been proposed for effectively studying the liquid-solid particle interactions and particle motion in two-phase flow (Kafui, Thornton, & Adams, 2002; Tsuji, Kawaguchi, & Tanaka, 1993). This approach uses the DEM and CFD method to iteratively solve for the particle motion and the fluid flow field so as to achieve momentum balance between the liquid and solid phases, while considering the solid particle concentration.

In this paper, the coupled DEM-CFD method was used to investigate the interactions between the solid particles and the liquid flows through an analysis of the two-phase flow in a centrifugal pump. The simulations were performed using the FLUENT software (FLUENT 16.0) together with the EDEM (DEM, 2010), and factoring in the size and material properties of the solid particles. The results illustrate the dynamic motion of the solid particles as a function of the material property and particle size and their impact on pump performance in terms of energy consumption and wear. The results are used to highlight design guidance for reducing energy demand and component wear.

2. Computational model and method

2.1. *EDEM-FLUENT coupling*

The Euler-Lagrange method was used to formulate the computational model of the two-phase solid-liquid flow in a centrifugal pump. The computational model employs DEM for simulating the solid particle dynamics in the two-phase flow (DEM, 2010) in conjunction with the CFD method, which uses FLUENT code. Figure 1 illustrates the EDEM-FLUENT coupling process. The transient fluid phase flow is initially analyzed using CFD simulations through solutions of the transient Reynolds-averaged Navier-Stokes (RANS) equation and the standard k-epsilon turbulence model in the FLUENT platform. The flow simulation is iterated until convergence is achieved for a given time step. The DEM simulations are subsequently initiated to compute the instantaneous drag force acting on each DEM particle from the transient fluid conditions of the mesh cell containing the particular particle. The simulations are continued using the DEM to determine the instantaneous position, velocity and forces developed by the particle. The coordinates of the particles are subsequently updated and transferred to the CFD solver, together with the inertia forces due to the particles. For this purpose a momentum source is introduced to each mesh cell to represent the effect of energy transfer from each DEM particle, taking into account the volume fraction of the CFD mesh cell occupied by the solid particle.

In the DEM-CFD coupling process, the DEM iterations are performed in an independent integration time step following the convergence corresponding to each CFD simulation time step. The total simulation time is thus significantly higher, especially when the dynamic motion of a large number of solid particles is tracked. The total simulation time for a single iteration, however, is affected not only by the number of particle elements but also by the surface and geometric properties of the elements, and the contact conditions. EDEM simulations are nearly linearly dependent on the number of elements.

2.2. *Material and interaction data*

The simulations were performed for water at 25°C as the continuous phase, while the solid particles formed the discrete phase. The physical parameters of the solid particles were chosen so as to approximately simulate the conditions encountered in most slurry transport applications (Wilson et al., 2006). The density of the particles was taken as 1500 kg/m^3, while the sizes of the particles were varied from 1.0 to 3.0 mm in diameter in a random manner. The volume fraction of the solid particles was set to 15% at the pump inlet. Some of the material properties of the particles

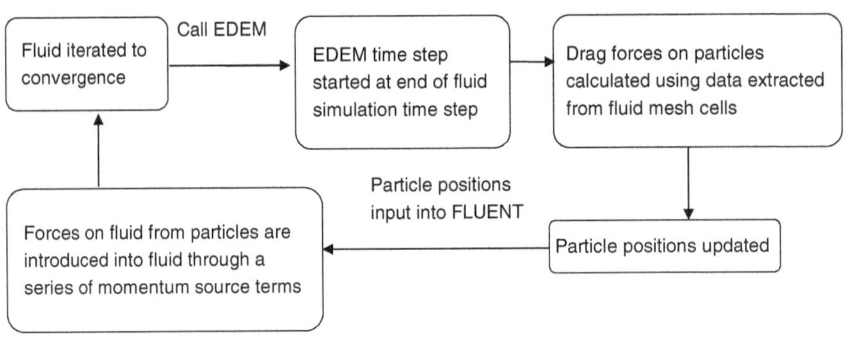

Figure 1. EDEM-FLUENT coupling process.

Table 1. Material properties of the solid particles and the pump structure.

Material	Poisson's ratio	Shear modulus (MPa)	Density (kg/m^3)	Diameter (mm)	Volume fraction at pump inlet (%)
Pump	0.30	70.0	7800	–	–
Particles	0.40	21.3	1500	1.0–3.0	15

Table 2. Model parameters for the particle-particle and particle-pump interactions.

Interaction	Coefficient of restitution	Coefficient of static friction	Coefficient of rolling friction
Particle-particle	0.44	0.27	0.01
Particle-pump	0.50	0.15	0.01

and pump structure are presented in Table 1.The shape of each particle was considered to be spherical in order to simplify the computations. A Hertz-Mindlin non-slip contact model was used for describing the particle-particle and particle-pump interactions. The coefficients of friction and restitution used in the model are summarized in Table 2.

Figure 2. Computational domains and grids.

2.3. *Computational domain and meshing*

The computational domain consisted of the inlet pipe, impeller, volute and outline pipe of a centrifugal pump, as shown in Figure 2. A single stage centrifugal pump was selected for the study of two-phase flow and the interactions of the solid particles. The flow rate capacity of the pump was 88.5 m^3/h with the head being 14.5 m, when operating at 1450 rpm. The impeller comprised a total of six blades with a wrap angle of 100° (inlet diameter $D_0 = 125$ mm; outlet width $b_2 = 26.5$ mm). The hexahedral meshes were created using Integrated Computer Engineering and Manufacturing code (ICEM) and a total of 1,019,538 cells constituted the entire computational domain. Sliding meshes were adopted at the interfaces of the rotating impeller with the stationary components of the pump for the transient analyses. In addition, the acceleration due to gravity ($g = 9.81$ m/s^2) was taken into account and its direction was considered to be opposite to the flow entering the pump inlet. This type of inlet-downward installation is quite often found in slurry pumps used for dredging operations in rivers.

2.4. *Transient analysis*

Both the pump and the fluid medium were considered to be in a static state prior to start-up. Once the pump starts, the impeller rotational speed grows with time until the design value and the flow field inside the pump gradually enter a stable circulating state (Saito, 1982; Tsukamoto & Ohashi, 1982). The simulations were designed to study the transient solid-liquid flow fields but without factoring in the contributions due to the transient response of the impeller, in order to preserve reasonable computational efficiency. A steady impeller speed of 1450 rpm was thus considered in the transient analysis. The initial velocity of the liquid phase was set as zero in all computational domains. Computations were initiated with solid particles being released continuously at the pump inlet, while the volume of solid particles was maintained at 15%. The inlet boundary condition in FLUENT included the design flow rate of the pump, while the outlet boundary was given by the outlet pressure ($p = 10^5$ Pa). The computations in FLUENT were performed using the time step $\Delta t = 60/65nZ \approx 10^{-4}$ s, which represents an impeller rotation of less than one degree. A considerably smaller integration time step, however, is essential for simulations in EDEM software for capturing the contact behavior of the solid particles accurately (DEM, 2010). The time step of 10^{-6} s was thus selected for the EDEM simulations and the ratio of DEM-CFD simulation steps was thus 1:100. The absolute convergence criteria for the calculated residuals were set as 10^{-3} by default, which was monitored in terms of the pump head (total pressure difference between the outlet and the inlet) during the computations. The computations were terminated on the basis of the above convergence condition together with the regular variations in the pump head.

3. Results and discussion

3.1. *Time-dependent performance of the solid-liquid pump*

The computations were performed with the pump design condition taken into account (flowrate, $Q_l = 88.5$ m^3/h; impeller speed, $n = 1450$ rpm). The transient response of the pump was obtained in terms of the head using the FLUENT post-processor. Figure 3 compares the variation in the pump head with time for the two-phase solid-liquid flow and the water-only flow. The results show periodic oscillations in the pump head for both flow conditions. The amplitudes of oscillations in both cases approach nearly steady values as t exceeds three times the fundamental rotation period of the impeller ($T = 0.0414$ s). It should be noted that the settling time of $3T$ for the pump head corresponds to the conditions considered, namely, constant impeller speed and zero initial velocity of the liquid phase in all domains, as stated in section 2.4 above.

The results further show oscillations in the pump head at a frequency of 145 Hz for both the flows, which directly relates to the product of impeller blades and the fundamental frequency of the impeller (24.16 Hz). Within a rotation period T, there are six peak values of H, corresponding to the impeller vane number. The amplitude of oscillations in the pump head for two-phase solid-liquid flow, however, is considerably higher than that for water alone, while the steady-state mean value of the head for the purely liquid phase flow is higher than that for the two-phase flow.

Figure 4 shows the variations in the total particle volume normalized with the total volume in the pump domain, $V'_p = V_p/V$, where Vp and V denote the total volumes of the particle phase and the entire pump domain, respectively. The normalized particle volume V'_p increases nearly linearly with time until $t \approx 0.24$ s and tends to saturate at $t \geq 0.3$ s, suggesting that the total volume of particles inside the pump approaches a state of equilibrium. The results also suggest that the particles entering the pump are more than those discharged during the period. Small amplitude variations in the volume of normalized particles were also observed during the interval $0.28 \leq t \leq 0.35$ s, which is likely due to the clustering of the solid particles during discharge.

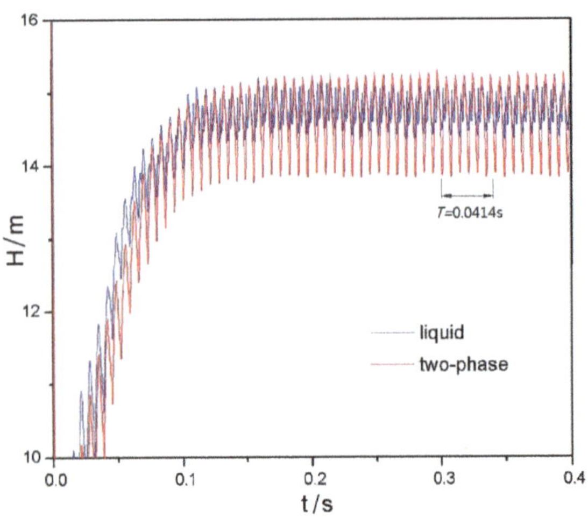

Figure 3. Comparison of variations in the pump head (H) with time for solid-liquid and liquid-only flows.

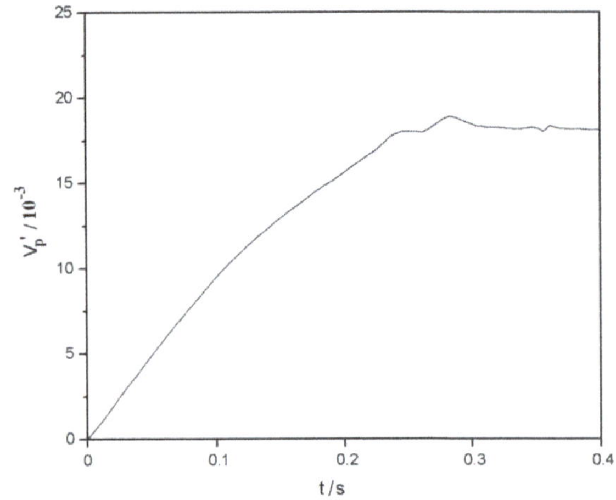

Figure 4. Variations in the normalized solid particle volume (V'_p) with time.

3.2. *Trajectories of the solid particles*

Figure 5 shows the trajectories of the solid particles inside the pump at various instants during the simulation. The computational domains of the inlet and the outlet pipes are omitted so as to clearly observe the distributions of the solid particles phase around the impeller and the volute. The trajectories clearly show a gradually increasing particle volume with time for $t \leq 0.3$ s, as seen in Figure 4. The solid particles tend to distribute more uniformly in the inlet section (Figure 5(a)), and enter the impeller along the pressure sides of the vanes (Figure 5(b–e)). In general, the particles tend to maintain a steady trajectory during flow towards the volute that corresponds with the shape of the impeller vanes, as seen in Figure 5(c). The particles tend to cluster along the volute outside wall and move downstream towards the outlet (Figure 5(d)).

Figure6 further shows the detailed particle trajectories around some of the impeller vanes, which is an enlarged view of the particles' motion shown in Figure 5(f). This enlarged view illustrates the motion of particles of different sizes – the trajectories of different-sized particles could not be distinguished from the figure. This is due to frequent particle-particle and particle-pump interactions. The large particles are presented in warmer colors, while smaller particles are shown in cooler colors. The trajectories further show that the solid particles dominantly move to the pressure sides of the impeller vanes. This tendency has also been observed in an experimental study of flow in a similar specific speed centrifugal pump (Liu, Xu, & Tang, 2008). The particles, however, tend to deviate from the pressure

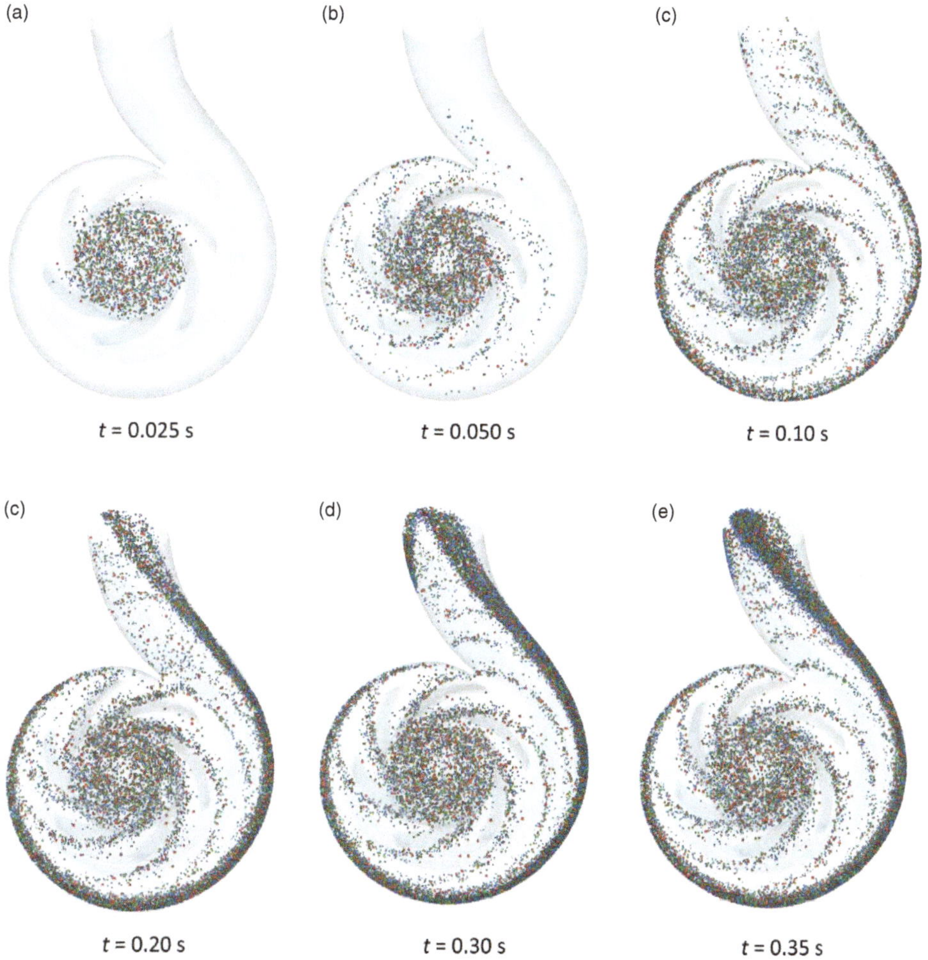

Figure 5. Trajectories of the solid particles inside the pump at different selected instants.

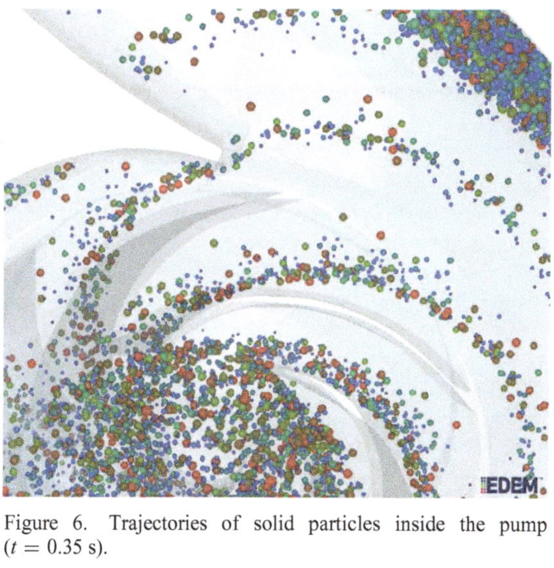

Figure 6. Trajectories of solid particles inside the pump ($t = 0.35$ s).

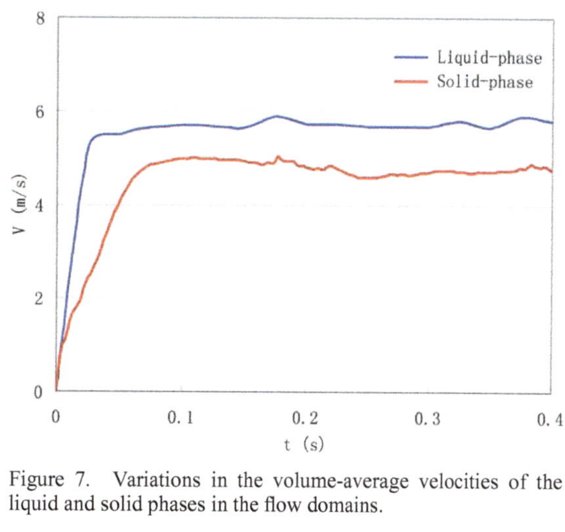

Figure 7. Variations in the volume-average velocities of the liquid and solid phases in the flow domains.

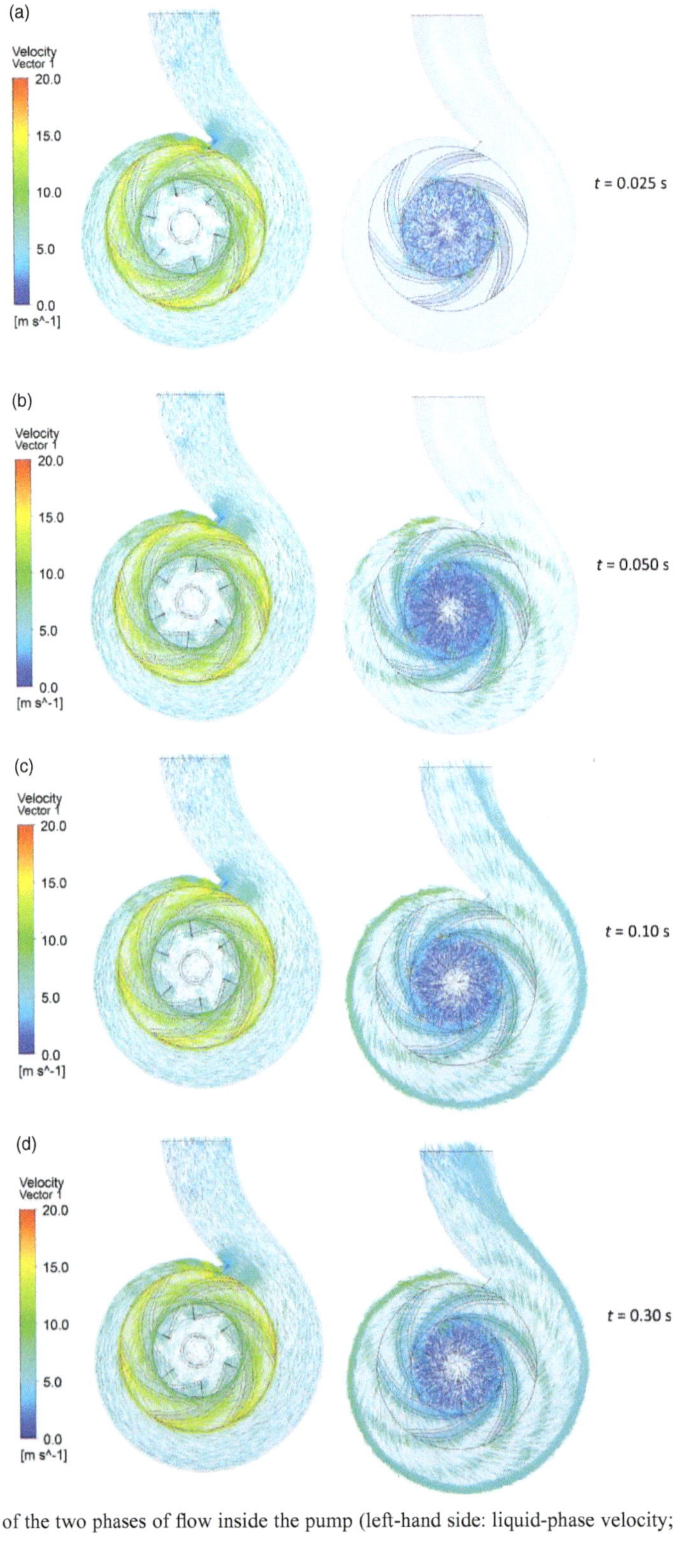

Figure 8. Velocity fields of the two phases of flow inside the pump (left-hand side: liquid-phase velocity; right-hand side: solid-phase velocity).

sides near about one third of the length of the vanes while moving downstream.

3.3. *Velocity fields of the flow phases*

Figure7 shows the variations in the volume-average velocities of both flow phases in the pump domains, which are obtained from the EDEM-FLUENT post-processors. The results show that the average velocity of the liquid phase approaches a nearly steady value of 5.7 m/s, while that of the solid phase approaches a considerably lower value of 4.7 m/s. There are apparently whole slip velocities between the liquid and solid phase flows inside the pump.

Figure 8 shows the velocity fields of the liquid phase (left side) and the particle phase (right side) inside the pump at different selected instants. The figures show quite similar velocity fields for the liquid phase at all of the selected instants. This is due to the fact that the volume-average velocity of the liquid phase in the pump domain rapidly approaches a steady value after $t = 0.025$ s, as seen in Figure 7. The flow velocity magnitude near the impeller outlet tends to be relatively high due to the rotating impeller, while the maximum was obtained near 16.4 m/s. In contrast, the velocities of the solid particles are substantially lower, particularly at the inlet. The particle velocity is only about 2 m/s in the inlet pipeline, while they tend to accelerate in the impeller and discharge from the impeller outlet at a speed of about 10 m/s. From Figure 7, it is evident that the volume-average velocity of the solid particle phase in the pump domains approaches a steady value after $t = 0.08$ s. The velocity fields for the particle phase at the instants $t = 0.10$ s and $t = 0.30$ s thus appear to be quite similar. While the higher velocity of the liquid phase compared to that of the particle phase in the inlet pipeline and the impeller is clearly evident, the velocity slips between both the phases are not obvious in the volute. This is due to the fact that the liquid-phase velocity decreases considerably with increase in the cross-sectional area of the volute, while the solid particles, as the dispersed phase, retain their inertia from the impeller.

3.4. *Solid-phase volume distribution*

The distributions of the particle volume fraction, as well as the trajectories of the solid particles inside the pump, can provide important guidelines for identifying potential locations of wear caused by the solid particles. The contours of the particle volume fraction α inside the pump are thus evaluated from the coupled DEM-CFD simulations and presented in Figure 9. The simulation results revealed a maximum particle volume fraction ($\alpha_{\max}$) in the equilibrium state ($t \geq 0.30$ s) in the order of 0.5 occurring on the outside wall of the volute. More particles tend to cluster together on the inlet-downward side of the volute wall due to the gravity effect, especially in the downstream path of the volute, since the kinetic energy of the two-phase flow decreases gradually. The rate of erosion of the pump wall due to the particles E_r can be calculated as follows (DEM, 2010):

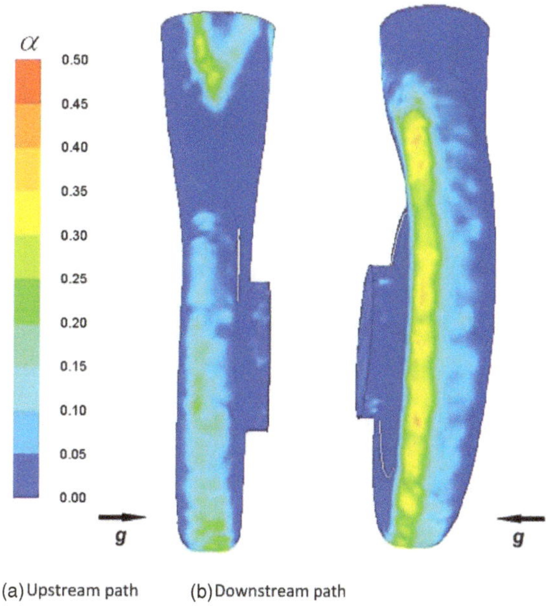

Figure 9. Particle volume fraction in the volute in the equilibrium state ($t = 0.35$ s).

$$E_r = \mathrm{E}\dot{\mathrm{N}}m_p, \tag{1}$$

where m_p is the mass of the particle, $\dot{N}$ is particle number rate and E is a constant, which is a function of the particle impact velocity, the impact angle, and the particle and the wall properties. The software offers two erosion models for computing the constant E, namely, the models reported by Finnie (1960) and Grant and Tabakoff (1975). Further studies of the erosion of the pump components due to interactions with solid particles would be desirable using the results obtained from coupled DEM-CFD simulations accounting for the effects of particle size and volume fractions.

4. Conclusions

The transient two-phase solid-liquid flow in a centrifugal pump was analyzed using the coupled DEM-CFD simulations. The conclusions are as follows:

(1) For the two-phase flow with densely distributed particles (1500 kg/m^3 in density, 1.0–3.0 mm in diameter, 15% volume fraction at the pump inlet), the results show that the pump head in the equilibrium state is considerably lower compared to that for the liquid-only flow. The two-phase flow also results in substantially higher amplitude oscillations in the pump head, an effect which is attributed to collisions among the particles in the densely distributed state. The frequency of the pump head oscillations, however, is identical for both flows,

which could be directly related to the fundamental frequency of the impeller and the number of impeller vanes.

(2) The results further show an increase in the total volume of the particles inside the pump during the start-up period, suggesting that the particles entering the pump are greater than those discharged in the same period. The velocity fields of the flows also reveal whole slip velocities between the two phases inside the pump.

(3) The solid particles are more uniformly distributed in the inlet section, and enter the impeller along the pressure sides of the vanes. The trajectories of the particles are consistent with the shape of the impeller vanes when moving towards the volute. Furthermore, most of the particles along the outside wall of the volute move to the downstream towards the outlet. The effect of particle size on the trajectory, however, could not be distinguished due to frequent particle-particle interactions and higher concentration of the particles. The results also suggest that the contribution due to the effect of gravity cannot be ignored in the analysis.

(4) The results in this study are limited to the solid, liquid and two-phase flows in a pump; it would be desirable to compare these with those obtained from the pseudo-fluid multiphase flow approach so as to assess the relative errors. Further studies on the applications of the results for predicting pump component wear due to interactions with solid particles are also desirable.

Disclosure statement

No potential conflict of interest was reported by the authors.

References

Addie, G. R., Roudnev, A. S., & Sellgren, A. (2007). The new ANSI/HI centrifugal slurry pump standard. *Journal of the South African Institute of Mining and Metallurgy*, *107*(6), 403–409.

Cheng, X. R., Li, R. N., Gao, Y., & Guo, W. L. (2013). *Numerical research on the effects of impeller pump-out vanes on axial force in a solid-liquid screw centrifugal pump*. IOP conference series: Materials science and engineering. The 6th international conference on pumps and fans with compressors and wind turbines (ICPF2013), September 19–22, Tsinghua University, Beijing, China, Paper No. 062008.

Chochua, G., Koch, J. M., & Sorokes, J. M. (2005). *Analytical and computational study of radial loads in volutes and collectors*. American Society of Mechanical Engineers. Reno, Nevada, USA, 871–879.

Clark, D., & Abbott, C. (1992). Improve slurry pump life. *Hydrocarbon Processing*, *71*(7), 53–55.

Cundall, P. A. (1971). *A computer model for simulating progressive, large scale movements in blocky rock system*. Proceedings of the international symposium on rock mechanics. Nancy France, 8–12.

Fang, J., Ling, X., & Sang, Z. F. (2013). Solid suspension in stirred tank equipped with multi-side-entering agitators. *Engineering Applications of Computational Fluid Mechanics*, *7*(2), 282–294.

Finnie, L. (1960). Erosion of surfaces by solid particles. *Wear*, *3*(2), 87–103.

FLUENT (16.0) [Computer software]. http://www.ansys.com/Products/Simulation + Technology/Fluid + Dynamics/Fluid + Dynamics + Products/ANSYS + Fluent.

Grant, G., & Tabakoff, W. (1975). Erosion prediction in turbomachinery resulting from environmental solid particles. *Journal of Aircraft*, *12*(5), 471–478.

Han, W., Ma, W., Li, R., & Li, Q. (2012). The Numerical analysis of radial thrust and axial thrust in the screw centrifugal pump. *Procedia Engineering*, *31*, 176–181.

Hua, T., Yi, L., & Yu-Liang, Z. (2013). Numerical analysis of a prototype centrifugal pump delivering solid-liquid two-phase flow. *Journal of Applied Sciences*, *13*(17), 3416–3420.

Kafui, K. D., Thornton, C., & Adams, M. J. (2002). Discrete particle-continuum fluid modelling of gas-solid fluidized beds. *Chemical Engineering Science*, *57*(13), 2395–2410.

Liu, J., Xu, H., & Tang, S. (2008). Experimental research on the movement rule of solid particles in centrifugal pump. *Journal of Hydroelectric Engineering*, *27*(6), 168–173.

Lyczkowski, R. W., & Bouillard, J. X. (2002). State-of-the-art review of erosion modeling in fluid/solids systems. *Progress in Energy and Combustion Science*, *28*(6), 543–602.

Ma, X. D., Wu, D. Z., Huang, D. S., Yu, H., & Wang, L. Q. (2013). *CFD analysis on a turbulence generator of medium consistency pump*. IOP conference series: Materials science and engineering. The 6th international conference on pumps and fans with compressors and wind turbines (ICPF2013), September 19–22, Tsinghua University, Beijing, China, Paper No.032003.

Messa, G. V., & Malavasi, S. (2014). Numerical prediction of particle distribution of solid-liquid slurries in straight pipes and bends. *Engineering Applications of Computational Fluid Mechanics*, *8*(3), 356–372.

Pagalthivarthi, K. V., Gupta, P. K., Tyagi, V., & Ravi, M. R. (2011). CFD predictions of dense slurry flow in centrifugal pump casings. *World Academy of Science, Engineering and Technology*, *5*(3), 16–28.

Roudnev, A. S., Bourgeois, R. J., & Kosmicki, R. J. (2009). *Slurry pump casing wear prediction using numerical multiphase flow simulation*. ASME 2009 fluids engineering division summer meeting, American Society of Mechanical Engineers, Vail, Colorado, USA, 515–523.

Roudnev, A. S., & Kosmicki, R. J. (2007). *Effects of CFD modeling configuration on centrifugal slurry pump casing wear prediction*. The 17th international conference on the hydraulic transport of solids, 7–11 May 2007, Cape Town, Southern African, 271–280.

Saito, S. (1982). The transient characteristics of a pump during start up. *Bulletin of JSME*, *25*(201), 372–379.

Solutions DEM. (2010). *EDEM 2.3 user guide*. Edinburgh: DEM Solutions.

Tsuji, Y., Kawaguchi, T., & Tanaka, T. (1993). Discrete particle simulation of two-dimensional fluidized bed. *Powder Technology*, *77*(1), 79–87.

Tsukamoto, H., & Ohashi, H. (1982). Transient characteristics of a centrifugal pump during starting period. *ASME Journal of Fluid Engineering*, *104*(1), 6–13.

Wilson, K. C., Addie, G. R., Sellgren, A., & Clift, R. (Eds.) (1997). *Slurry transport using centrifugal pumps*. London: Blackie Academic & Professional.

Zhang, Y., Li, Y., Cui, B., Zhu, Z., & Dou, H. (2013). Numerical simulation and analysis of solid-liquid two-phase flow in centrifugal pump. *Chinese Journal of Mechanical Engineering*, *26*(1), 53–60.

16

Development of new "multivolute casing" geometries for radial force reduction in centrifugal pumps

Hamed Alemi*, Seyyed Ahmad Nourbakhsh, Mehrdad Raisee and Amir Farhad Najafi

Hydraulic Machinery Research Institute, School of Mechanical Engineering, College of Engineering, University of Tehran, Tehran, 1439955961, Iran

Large radial force causes several issues in pumps, such as noise, vibration, and extra load on the bearings. To reduce the radial force, the effects of concentric volute and multivolute geometry on the head, efficiency, and radial force of a low speed centrifugal pump at off-design conditions were investigated. Commercial software with the $k-\omega$ turbulence model and automatic near wall treatment was employed for the prediction of fluid flow inside the pump. Flow simulations for three casings concentric at 180°, 270°, and 360° from the tongue showed that the 270° concentric volute generates the lowest radial force at throughout the entire range of flow rate. The triple-volute and tetravolute casings are also proposed as new volute geometries. The flow analysis of a double-volute, triple-volute, and tetravolute show that the triple-volute is the most appropriate volute geometry at off-design conditions.

Keywords: pump; volute; CFD; radial force; multivolute; concentric volute

1. Introduction

The circumferential pressure distribution around the impeller periphery becomes non-uniform at off-design conditions. This leads to radial force on the pump shaft. A large radial force generates noise, vibration, and extra load on the bearings, thus deteriorating the natural pump performance. The magnitude of the radial forces can be directly measured using the bearing or shaft forces. Moreover, the radial forces can be estimated by static pressure measurement or computation around the impeller periphery. From a computational point of view, the prediction of the internal flow developing in a pump is extremely challenging at off-design conditions due to the presence of rotation, the curvature of the impeller, impeller-volute interaction, and flow separation. Nevertheless, numerous investigations have recently been carried out to predict turbulent flow in centrifugal pumps. Kelder, Dijkers, Vas Esch, and Kruyt (2001) used both experimental and numerical methods to study turbulent flow at a Reynolds number of 1.7 $\times$ 10^6 in the volute of a low specific-speed pump close to the design point. They showed that in such conditions, the core flow characteristics are similar to the potential flow. Thus, they employed an inviscid solver to obtain the flow field. Gonzalez, Fernandez, Blanco, and Santolaria (2002) investigated gap variation for a pump with a specific speed of $N_S = 26$. They found that an increase by about 50% of the maximum pressure amplitude occurs where the gap distance is reduced from 15.8% to 10% of the impeller radius. Spence and Amaral-Texeira (2008) attempted to estimate pressure pulsation in a double-entry, double-volute centrifugal pump with a specific speed of $N_S = 39$ by means of using CFX-TASC, a commercial flow code. Their numerical results were in good agreement with the measurements. Yedidiah (2008) tried to synthesize knowledge between pump experts and CFD specialists with a test case study. Pumps with higher N_s were found to need a larger cutwater gap for pump performance improvement. Feng, Benra, and Dohmen (2010) applied various turbulence models to simulate non-stationary turbulent flow in a radial diffuser pump. The numerical results obtained for the pressure, velocity, and turbulence fields were analyzed and compared with LDV and PIV data. They reported that the choice of turbulence model does not have a large influence on the pressure field predictions. Among the turbulence models examined, the $k-\omega$ turbulence model had superior performance in the prediction of the velocity vector directions and the turbulent kinetic energy levels. Barrio, Fernandez, Blanco, and Parrondo (2011) computed steady and unsteady pressure distributions around an impeller by solving full 3D-URANS equations with the commercial code FLUENT as a function of flow rate between 10% and 130% of the rated conditions. Zhou, Wang, and Yang (2012) numerically analyzed a double-suction pump using SST turbulence models and compared the results with experiment results. They reported that an SST turbulence model can accurately predict pump performance

*Corresponding author. Email: halemi@ut.ac.ir

at different flowrates. Liu, Liu, Dong, Ren, and Du (2012) used commercial CFX software to investigate the effects of mesh size and turbulence models on the numerical simulation of a centrifugal pump at the design point. Their CFD calculation was carried out with five different turbulence models ($k-\varepsilon$, *SSG* Reynolds Stress, $k-\varepsilon$ *EARSM, RNG* $k-\varepsilon$, and $k-\omega$) and six different mesh numbers ranging from 1 million to 25 million. It was confirmed that the $k-\varepsilon$ *EARSM* turbulence model is more robust than the others, and the internal flow was depicted more perfectly with increased mesh numbers. Zhongyong, Junjie, Shuai, and Shouqi (2011) presented a numerical simulation of the steady and unsteady radial force of centrifugal pump impellers equipped with various volutes, including a single volute, double-volute, and a vaned volute. Their results showed that both the double-volute and the vaned volute decreased the radial forces. They concluded that the double-volute is superior, since it significantly decreases the radial force in comparison with the single volute, and it has a more compact radial size than the vaned volute. Asuaje, Bakir, Kouidri, Kenyery, and Rey (2005) analyzed a centrifugal pump with a specific speed of $N_S = 32$ using CFX code. In their study, the three well-known $k-\varepsilon$, $k-\omega$, and *SST* turbulence models were employed. They computed the velocity and pressure fields in the flow domain and the radial force on the pump shaft. Torabi and Nourbakhsh (2011) studied the effects of variation in some of the geometrical factors of volutes on the radial force. They found that the gap between the impeller and tongue severely affects the radial force. More specifically, they found that a 6% reduction in the gap increases the maximum value of the static pressure by 50% round the impeller exit.

As discussed, a number of experimental and computational investigations have been carried out to study the effects of various geometrical parameters on the hydraulic performance and radial force of centrifugal pumps. In this study, multivolute casings are proposed for application in a centrifugal pump for the first time. To the best of our knowledge, no numerical simulation has been performed to investigate various concentric volutes for different angles from the tongue. One of the difficulties of this study was the grid generation of multivolute casings due to the intricate geometry. We overcame this challenge using special tools in ANSIS-Meshing 13.0 software, such as local refinement and inflation.

2. Main pump geometry

In the present study, we used experimental measurements of Kelder et al. (2001) for CFD validation. The details of pump geometry used by Kelder et al. are shown in Figure 1. The investigated centrifugal pumps have a low-N_S of 21. Design parameters of the studied pump are summarized in Table 1.

Table 1. Main parameters of pump.

Parameter	Value
Metric specific speed, N_s	21
Nominal flow coefficient, ϕ_n	0.15
Nominal head coefficient, ψ_n	0.124
Impeller exit diameter, d_2 (mm)	640
Impeller exit width, b_2(mm)	25
Number of Blades, Z	7

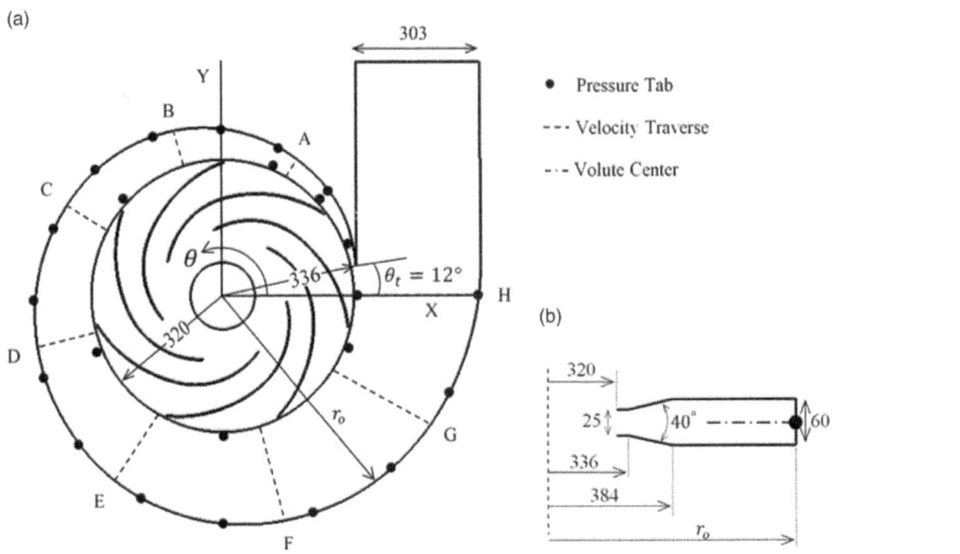

Figure 1. (a) Kelder volute geometry and locations of velocity and static pressure measurement (b) Volute cross section dimensions in millimetres.

Experiments were performed at flow rates of *82.5%*, *100%*, and *112%* of the design flow rate (28.8 m^3/h). An LDV measurement system was employed for velocity measurements. Velocity and static pressure data were collected at the locations shown in Figure 1(a). As shown in Figure 1(a), with the pressure tabs are located at the shroud side just outside the impeller, yet in volute at mid-height of the outer wall. More details of the pump configuration and experimental methods can be found in (Kelder et al., 2001).

3. Numerical method and model description

To solve the full 3D Reynolds Averaged Navier-Stokes (3D-RANS) equations for different working conditions the commercial software CFX-13.0 was used in this study. The solver was a 3D CFD code in which the governing equations are discretized using the finite-element based finite-volume method. Two main advantages of such a hybrid CFD method are geometric flexibility due to the finite-element method and retaining the conservation properties of the finite-volume method which results in low numerical error in the case of non-smooth grids. For space discretization, the second-order upwind scheme was applied to achieve high accuracy and reasonable solution stability.

For grid generation, the pump was divided into four components. As shown in Figure 2(a) these four components are: (i) inlet duct, (ii) impeller, (iii) volute, and (iv) outlet duct. Uniform flow normal to boundary and constant static pressure were implemented for inlet and outlet boundary conditions, respectively. Straight ducts are enough long to eliminate effects of inlet and outlet boundary conditions on flow regime in pump. Spence and Amaral-Texeira (2009) reported that these boundary conditions cause rapid convergence. Figure 2(b) shows the generated grids for the pump impeller and volute. As shown, due to complexity of geometry, hexahedral cells near important regions (with curvature and strong gradient) including blades and volute outer wall, and tetrahedral cells in other regions were used. Mesh clustering was applied close to some effective regions, such as leading and trailing edge of the blades and the volute tongue where flow separation may occur (see Figure 2). During grid generation process, orthogonal quality, aspect ratio, and skewness were assured to be in desirable range. For quasi-steady simulation the grids between the impeller and volute in one hand, and inlet duct and impeller on the other hand, were conjoined by means of a frozen rotor interface. In other words, impeller rotation was considered and the frame of reference was changed but relative orientation of the components across the interface was fixed. Both two frames of reference were connected in such a way that each of them had a fixed relative position throughout the calculation. This was equivalent to taking a snapshot from the flow field in an instance of time. Moreover, general grid interface (GGI) was employed between the volute and outlet duct. Boundary conditions in solution domain are summarized in Table 2.

Grid independency studies were performed using different mesh sizes. As shown in Figure 3, it was found that by increasing the number of cells from 7×10^5 to

Table 2. Boundary conditions of numerical solution domain.

Location	Boundary condition
Inlet duct entrance	Uniform mass flow, normal to boundary
Inlet duct/impeller interface	Frozen rotor interface
Impeller/volute interface	Frozen rotor interface
Volute/outlet duct interface	General grid interface (GGI)
Outlet duct exit	Constant static pressure
Walls	Smooth and No-slip

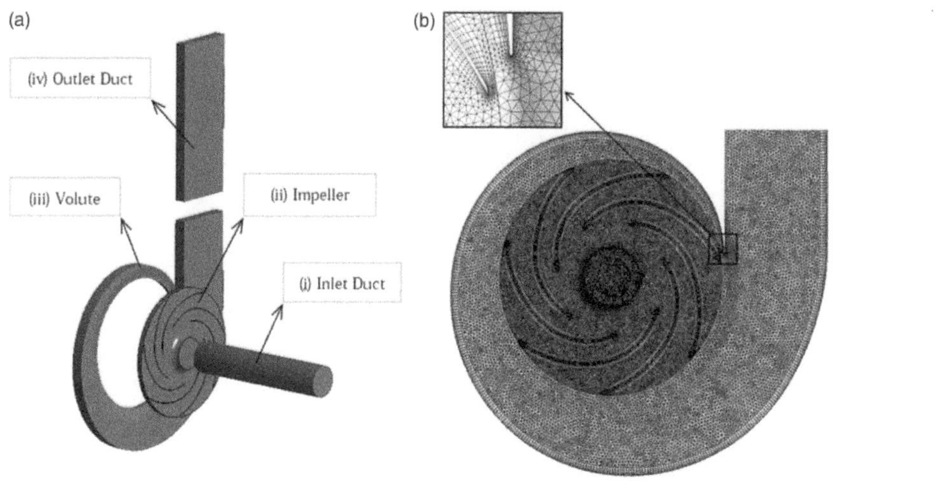

Figure 2. (a). 3D models of total domain including (i) inlet duct, (ii) impeller, (iii) volute, and (iv) outlet duct, (b) Computational mesh.

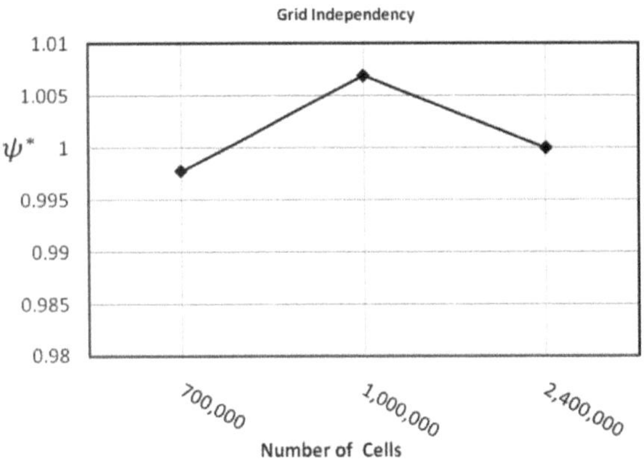

Figure 3. Non-dimensional head versus mesh size.

1×10^6, change in the predicted head is about 0.8%. Further mesh refinement from 1×10^6 cells to 2.4×10^6 cells, decreases the head by about 0.6 %. Since the predicted head did not show a strong sensitivity to the grid refinement, the numerical results can be considered as grid independent. All the numerical results were obtained using the 1×10^6 cells mesh.

It was found that by increasing the turbulent intensity at inlet from 5% to 10% and viscosity ratio from 10 to 100, the predicted head changes less than 0.01%. Thus, turbulent intensity and viscosity ratio were set to 5% and 10%, respectively.

4. Validation of CFD results

Numerical simulations of the present study were validated by comparing the predicted static pressure and velocity components with the experimental data of Kelder et al. at selected locations (see Figure 1). In this paper, three well-known turbulence models were examined; namely, the standard $k-\varepsilon$, the low-Re $k-\omega$, and the *SST*. The standard $k-\varepsilon$, developed by Jones and Launder (1972) is a high-Re turbulence model that is able to predict invisid flow regime properly. So, this model needs wall function approximation to solve viscous region near walls. The low-Re $k-\omega$ turbulence model, developed by Wilcox (1986), could be used for entire flow region even near walls. The low-Re $k-\omega$ model would typically require a near wall resolution of $y^+ < 2$. In turbomachinery flows, even $y^+ < 2$ can not be guaranteed and for this reason, a new near wall treatment was developed for the $k-\omega$ models. It allows for smooth shift from $k-\omega$ model to a wall function formulation regarding local y^+ value. The SST turbulence model, developed by Menter (1993, 1994), combines the $k-\omega$ and the $k-\varepsilon$ turbulence models. It consists of a transformation of the $k-\varepsilon$ model in the outer region to a $k-\omega$ formulation near the walls.

In Figure 4, pressure distributions obtained from three turbulence models are compared with the experimental data of Kelder et al.

It is observed that all three turbulence models produce comparable results in both design and off-design conditions. Despite the presence of local discrepancies between CFD predictions and measurements, both the level and the trend of CFD results are in overall agreement with the measurements. The wavy shape of pressure distribution seen in CFD predictions is due to impeller blades effects. Experimental and numerical results showed that depending on the value of flow rate, the static pressure distribution has different behavior. For the design flow rate, the static pressure remains almost uniform, which leads to minimum radial force. At the off-design conditions corresponding to the low capacity $\phi/\phi_n = 0.825$, the static pressure increases with θ while an opposite trend is observed for $\phi/\phi_n = 1.12$. At overload, the effective cross-sections are small for the flow. As a result, volute acts like a nozzle and the pressure decreases. In contrast, at low capacity, the effective cross-sections are large for the flow; accordingly, the volute acts like a diffuser and the pressure raises.

In Figure 5 the predicted radial and tangential velocity components were compared with the experimental data of Kelder et al. Comparisons were made for three working conditions ($\phi/\phi_n = 0.825$, 1.0, and 1.12) along four traverses (A, D, F and H) in the volute (see Figure 1). More or less, all turbulence models returned similar results consistent with the experimental data of Kelder et al. There are some discrepancies between the simulation results and experimental data in the V_θ on the Traverse A. The Traverse A is close to tongue, thus flow regime induced by blade-tongue interaction is strongly unsteady. Since, using quasi-steady CFD simulation may lead to numerical error increment.

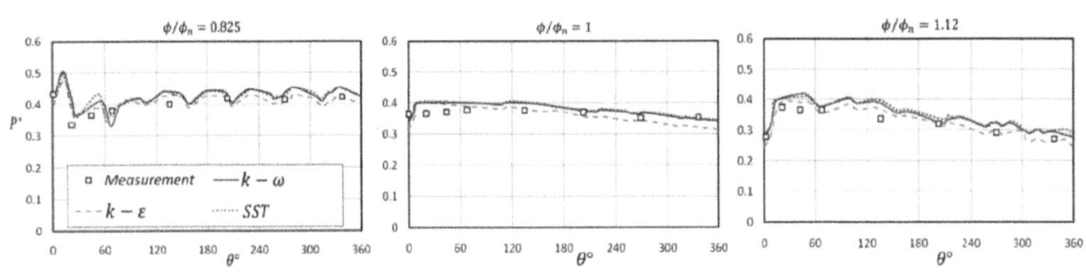

Figure 4. The pressure distributions around the impeller periphery.

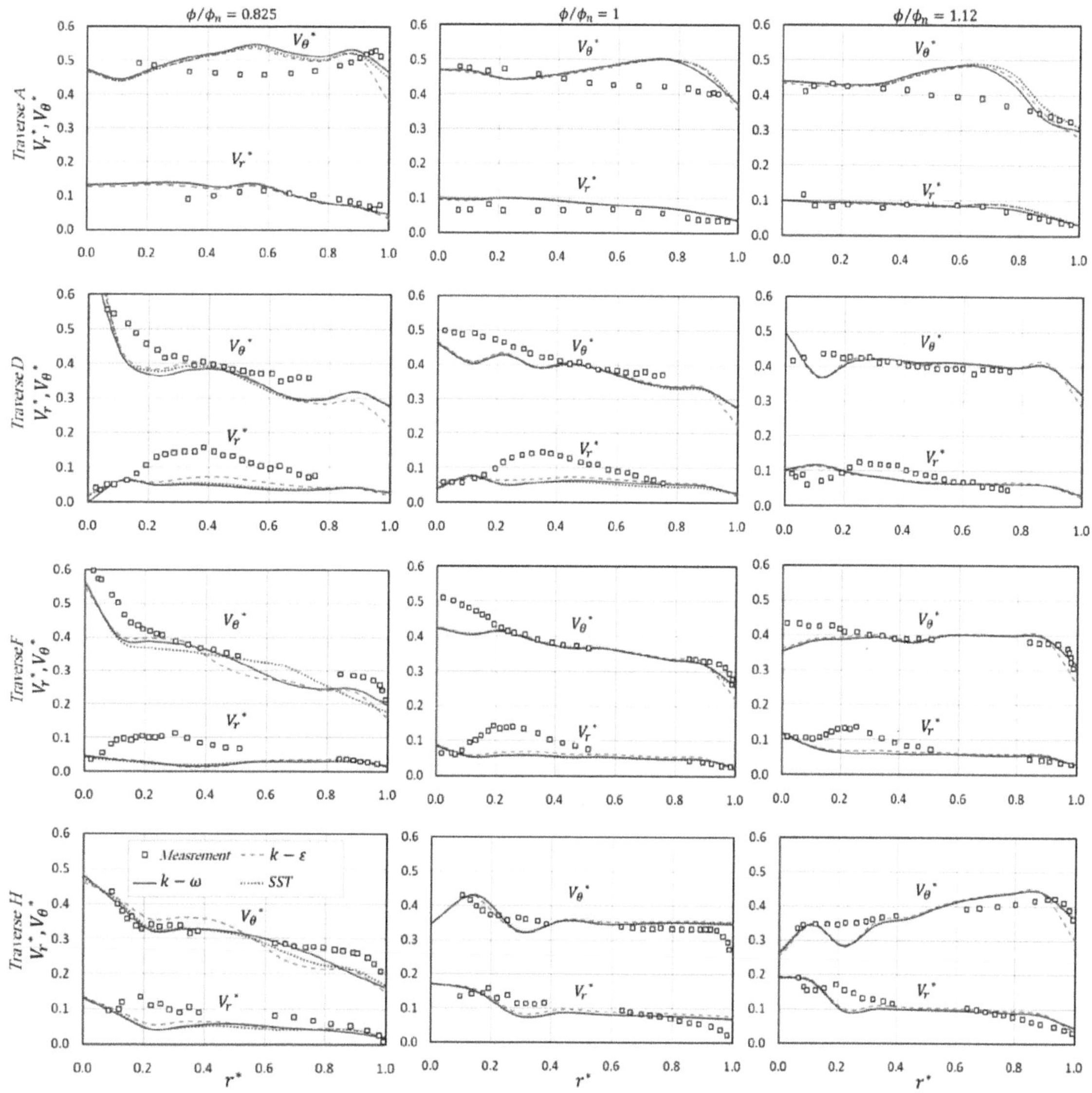

Figure 5. Comparison of predicted and measured non-dimensional radial and circumferential velocities in the volute along different traverses at $\phi/\phi_n = 0.825$, 1, and 1.12.

As mentioned in the introduction, Kelder et al. employed Euler equations in their numerical predictions and found reasonable agreement between computational and experimental data. However, level of agreement with the measured data was not as close as those shown in Figure 5.

The predicted coefficients using three turbulent models with steady simulation and just $k-\omega$ model with unsteady simulation were compared with experimental data for a wide range of flow coefficients; namely, the design point and off-design conditions (see Figure 6). Firstly, in steady simulations, although the $k-\varepsilon$ turbulence model returns more accurate results for the non-dimensional head near the design point, the head predictions of the $k-\omega$ model at off-design conditions are closer to the experimental data, especially at high flow rate (Figure 6(a)). Radial forces were computed by means of a full integration of the pressure and shear stress distribution on the impeller surfaces (blades, Hub, and shroud). The $k-\omega$ turbulence model returns radial force predictions in better agreement with experiments than other models (Figure 6(b)). Secondly, as shown in Figure 6, although unsteady simulation using $k-\omega$ turbulence model yields the closest head coefficients to the measurement data, the estimated radial forces by unsteady simulation are more or less similar to steady simulation. Since, the main focus of the current study is radial

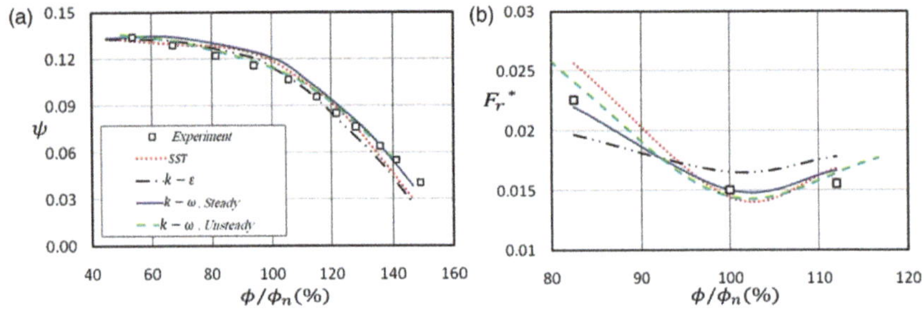

Figure 6. Comparison of predicted head and radial force using various turbulence models with the experimental data.

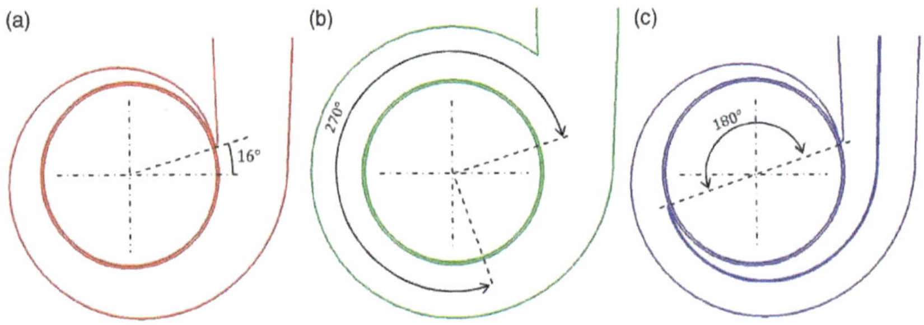

Figure 7. Different volute shapes: (a) standard volute, (b) 270° concentric volute, and (c) double-volute.

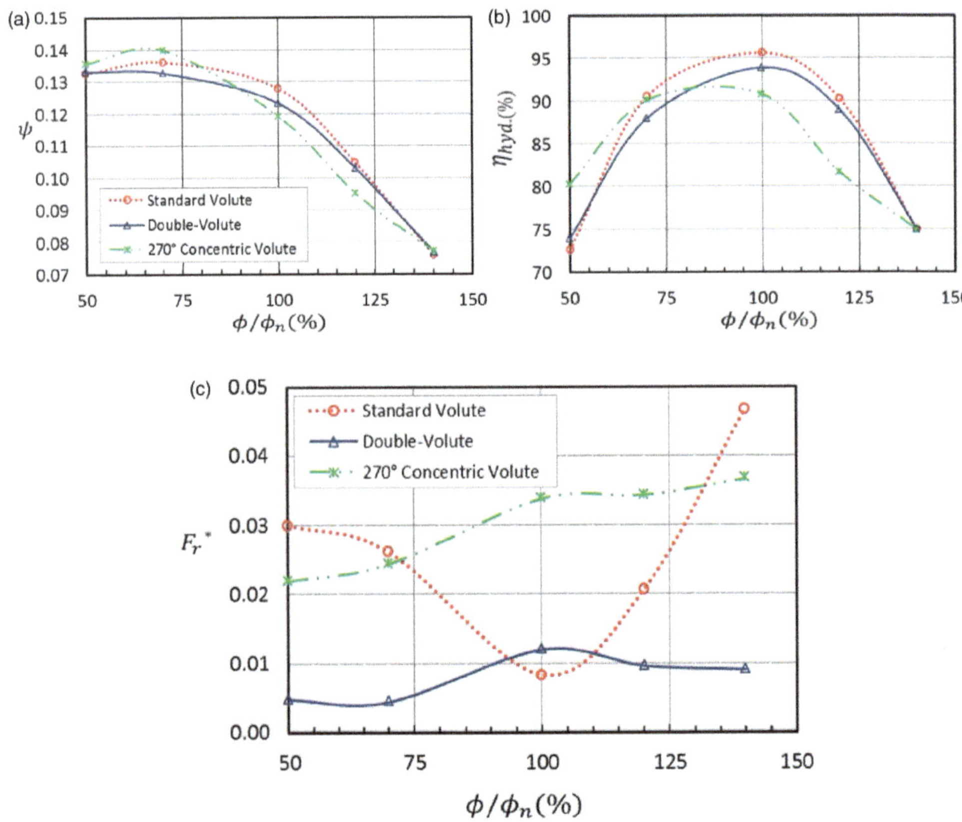

Figure 8. Comparison of effect of three volute shapes; namely, standard, concentric, and double-volute casings on pump characteristics.

force prediction at off-design conditions; the $k-\omega$ turbulence model with automatic near wall treatments was employed for calculations.

5. Results and discussion

To study the effects of volute geometry on the head, efficiency and radial forces in this investigation, seven various volutes were designed. The results obtained for different volutes are discussed in the following sections.

5.1. Volutes with various shapes

One of the main objectives was to minimize the radial force while having reasonable head and efficiency. The high radial force created by single volutes at low capacity is due to the interaction between volute cutwater and impeller outflow, which leads to flow separation and a pressure minimum downstream of the cutwater. This can be clearly observed from the radial force distribution in Figure 8(c). The radial force drops in a slightly downward curve with increasing flow rate until the point of best efficiency. At overload, the cross section is small for the flow. As a result, a low pressure zone is formed in the domain upstream of the discharge nozzle, as shown in Figure 9. Thus, the radial force increases (Figure 8(c)). Double-volute and concentric casings are employed to decrease the radial force, especially at off-design conditions. Generally, in double-volute, a second cutwater is located at an angle of 180° from the

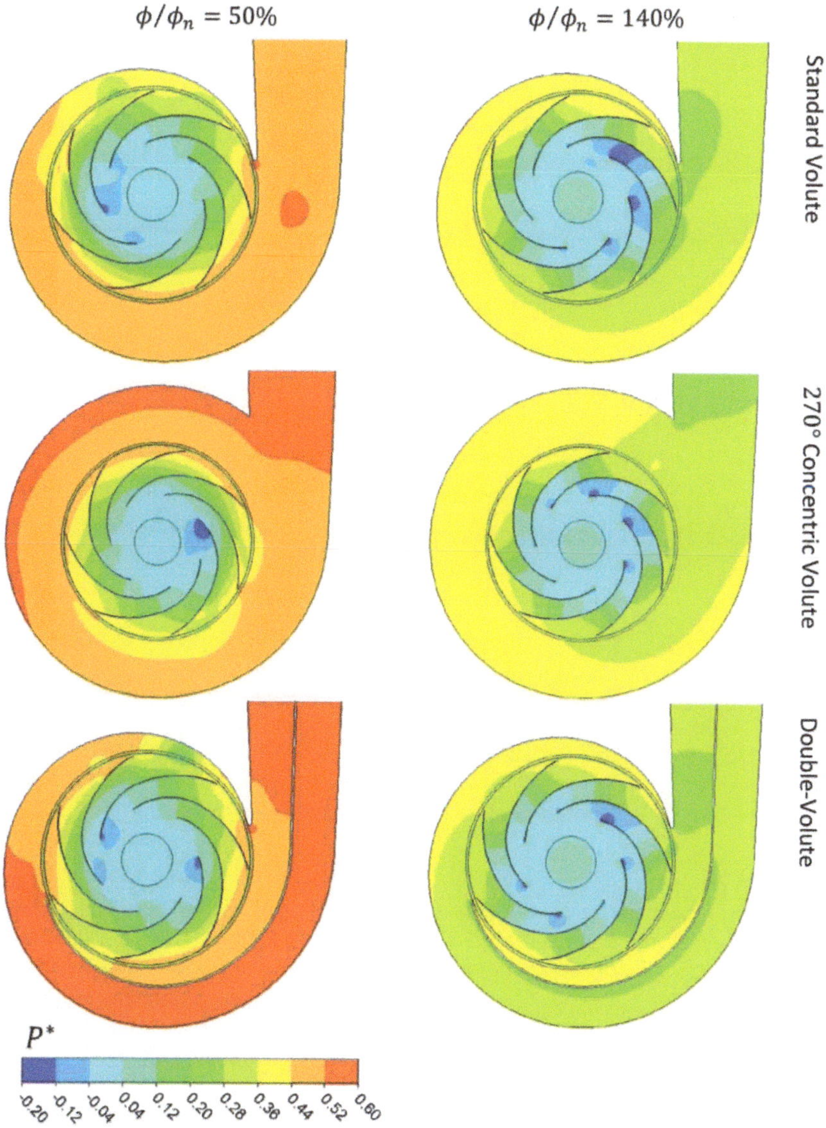

Figure 9. Pressure contours in volutes with various shapes.

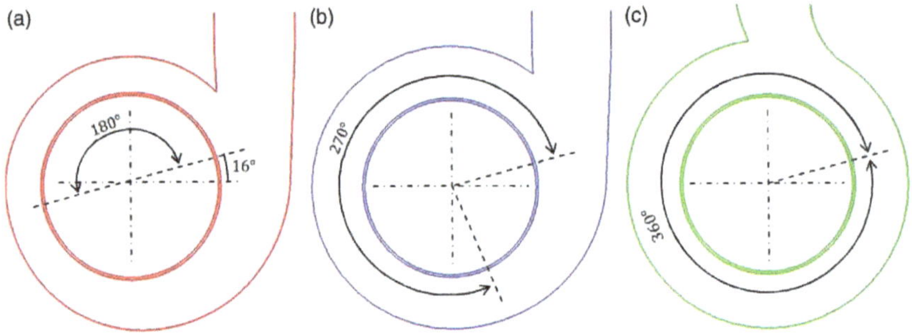

Figure 10. Three casings which were concentric from tongue for (a) 180°, (b) 270°, and (c) 360°.

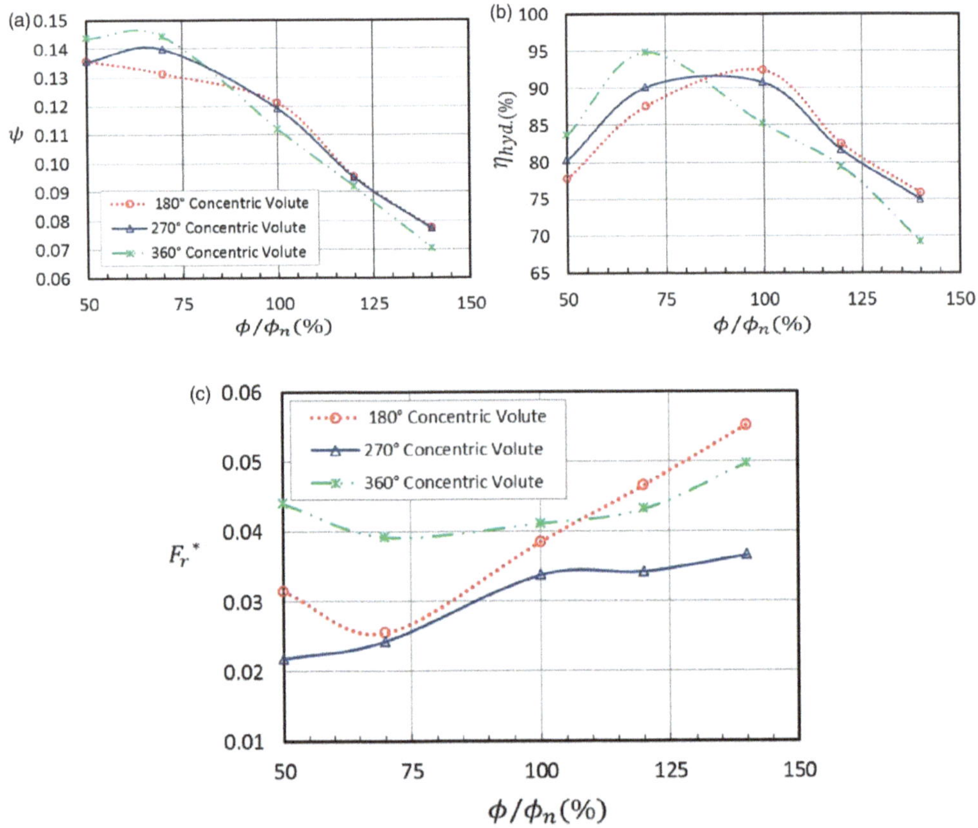

Figure 11. Comparison of effect of three concentric volute designs on pump performance.

first cutwater, as shown in Figure 7 (c). Consequently, as in Figure 9, similar pressure distributions are created in both partial volutes around the impeller periphery. This results in a reasonable reduction of the radial force at off-design conditions.

As shown in Figures 7(a), (b), and (c), three distinct volute shapes have been numerically investigated: a standard volute, concentric volute, and double-volute casing. The predicted output curves for three volutes are displayed in Figure 8. As shown in Figures 8(a) and (b), the concentric volute yields higher head and efficiency at low capacity. As expected, the standard volute has the best performance and the lowest radial force at the design point. The double-volute casing generates the minimum radial force at off-design conditions. Furthermore, the force at the design point is not very different from the standard one.

The 270° volute gives the highest radial force and the lowest efficiency at the design point. This volute generates lower radial force than the standard one at off-design conditions. The observations for the radial force are consistent with Figure 9, which shows that the pressure values at off-design conditions are reasonable equal at opposite positions (with respect to center of rotation) around the impeller in the double-volute. It can be concluded that the radial force in the double-volute depends on the equality of pressure in opposite positions rather than pressure uniformity around the impeller, which is important in a standard volute. This means that the minimum radial force does not necessarily occur at the design point in the double-volute, as shown in Figure 8(c). For a pump which is likely to work at off-design conditions, double-volute could be chosen as the most promising.

5.2. *Various concentric volutes*

As shown in Figure 8, the radial force can be reduced at the low flow rate range by using a concentric casing. Also, these volutes generate considerably more head and efficiency when the pump operates mainly below the best efficiency point, which frequently occurs in process pumps. The large cutwater gap in these volutes reduces interaction effects between the impeller blades and volute tongue. Consequently, the pressure distribution around the impeller periphery becomes more uniform at low flow rate. At the design point, the pressure distribution is not uniform around the impeller due to the constant cross-sectional area

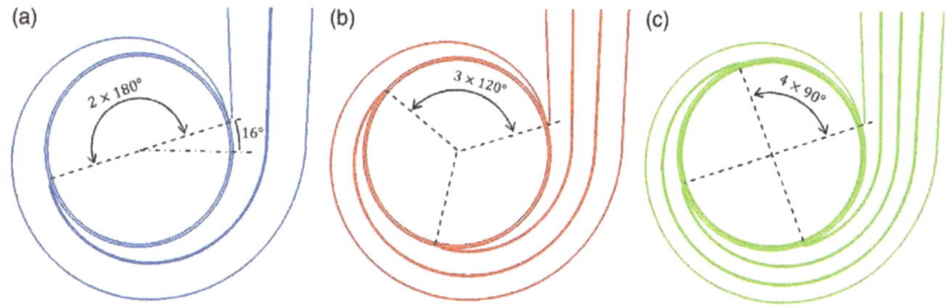

Figure 12. Designs of three multivolute; namely, (a) Double-volute, (b) Triple-volute, and (c) Tetravolute.

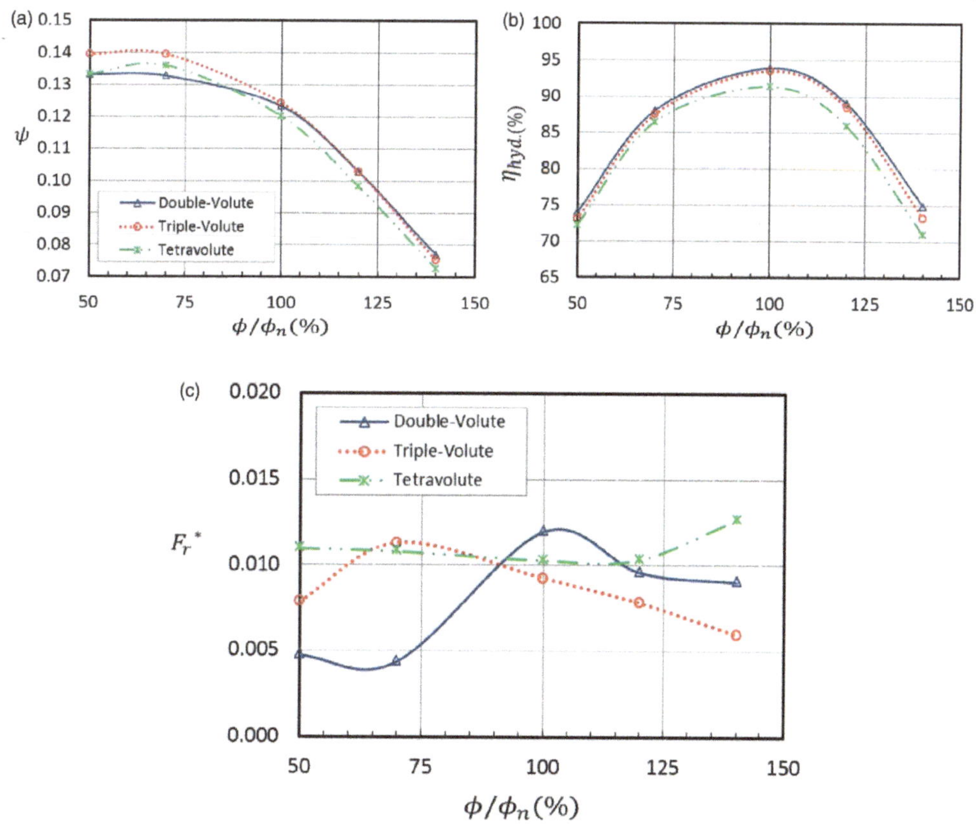

Figure 13. Comparison of effect of three multivolute casing designs on pump performance.

of the volute. It causes the minimum radial force point to be at a low flow rate instead of the design flow rate. The schematics of the three casings concentric at 180°, 270°, and 360° from the tongue are shown in Figure 10. The numerical results in Figure 11 show that the pump with the 180° volute yields higher head and efficiency at the design condition. However, the 360° volute displays more adequate head and efficiency at low flow rate. Interestingly, in the obtained curves, the best efficiency point has been shifted to 70% Q_d in the 360° concentric volutes. In a similar way, the minimum radial force occurred at 70% Q_d in the 180° and 360° concentric volutes. The 270° volute dominantly creates lower radial force at all flow rates and has moderately good head and efficiency. Considering all these aspects, the 270° volute could be selected as the best choice among these three concentric volutes.

5.3. *Multivolute casings*

For the first time, the application of multivolute casings is proposed to improve pump performance, especially at off-design conditions. Thus, as shown in Figure 12, three designs were generated, including a double-volute, triple-volute, and tetravolute casings. CFD simulations of the flow field within the volutes are illustrated in Figure 13.

As shown in Figure 13(a), the triple-volute provides the highest head at low capacity and the lowest radial force at the design and overload points, whereas the double-volute yields the lowest radial force at low capacity. As shown in Figure 13(c), the trend of radial force in the tetravolute is the most uniform at all flow rates. As a result, the triple-volute could be considered as the most appropriate alternative.

6. Conclusion

Quasi-steady flow through a low-N_S pump was numerically simulated by CFX 13.0 commercial code using a $k-\omega$ turbulence model and automatic near wall treatments. The dependence of computational prediction was checked with the grid. Subsequently, numerical results including the pressure distribution and velocity profile were compared with available experimental measurements for output validation. After that, various concentric and multivolute casings were developed to study their hydraulic effects on head, efficiency, and radial reaction of a low-N_S centrifugal pump. As expected, the standard volute provides the most convenient head, efficiency, and radial force at the design point. In comparison with the standard and concentric volutes, the double-volute showed superiority in the radial force at off-design conditions, whereas the concentric volute created the highest head and efficiency at low capacity.

The 180° and 360° concentric volutes produced the highest head and efficiency at low capacity and the design capacity, respectively. The 270° concentric volute gives the lowest radial force at entire flow rate. Interestingly, in the 360° concentric volute, the best efficiency point was shifted to 70% Q_d. The analysis of the various multivolute casings showed that the triple-volute is the most appropriate volute configuration due to its superior overhead, efficient performance at all capacities, and low radial reaction curve at the design flow rate and high flow rate. The absolute minimum radial force was generated by the double-volute at 65% Q_d.

In this study, we did not carry out experimental measurements or unsteady simulation of the new geometry designs due to running time reduction and cost limitations. In future work, preparing the experimental data for new designs of volute geometries and comparing the results with quasi-steady and pure unsteady CFD results could complete this research.

Acknowledgements

The authors gratefully acknowledge the cooperation of Hydraulic Machinery Research Institute of Tehran University.

NOTATION

b_2	Impeller outlet width, mm
d_2	Impeller outlet diameter, mm
r_0	Radius of volute wall midline, mm
r^*	$r - r_4/r_o - r_4$Volute radial distance, non-dimensional
u_2	Blade tip velocity, m/s
V_r	Radial velocity, m/s
V_θ	Circumferential velocity, m/s
F_r	Radial force, N
F_r^*	$Fr/(0.5\rho u_2^2 \pi d_2 b_2)$Radial force coefficient, non-dimensional
g	Gravity, m/s^2
H, H_n	Head, Normal or design head, m
k	Turbulent kinetic energy, m^2/s^2
N	Rotation speed, r/min
N_S	$NQ^0.5H^{-0.75}$ Metric specific speed, [rpm, m^3/s, m]
P	Static pressure, Pa
P_{in}	Inlet static pressure, Pa
P^*	$(P - P_{in})/\rho u_2{}^2$ Pressure coefficient, non-dimensional
Q, Q_n	Flow rate, Normal or design flow rate, m^3/s
$T_{ave.}$	Average pressure and viscous moment on impeller walls, N.m
y^+	Distance from the wall, non-dimensional
Z	Number of blades

Greek Symbols

ε	Dissipation rate, m^2/s^3
$\eta_{hyd.}$	$\rho gQH/(T_{ave.}\Omega)$Hydraulic efficiency, %
ρ	Fluid density, kg/m^3
$\theta, \theta v$	Angular coordinate, Angular position of tongue-tip,°
ω	Specific dissipation rate, 1/s
ϕ, ϕ_n	$Q/(\Omega d_2^2 b_2)$Flow coefficient, Flow coefficient at design flow rate Q_n, non-dimensional
ψ, ψ_n	$gH/(\Omega^2 d^2)$ Head coefficient, Head coefficient at design head H_n, non-dimensional

References

Asuaje, M., Bakir, F., Kouidri, S., Kenyery, F., & Rey, R. (2005). Numerical modelization of the flow in centrifugal pump: volute influence in velocity and pressure fields. *International Journal of Rotating Machinery*, *3*, 244–255.

Barrio, R., Fernandez, J., Blanco, E., & Parrondo, J. (2011). Estimation of radial load in centrifugal pumps using computational fluid dynamics. *European Journal of Machines B/Fluids*, *30*, 316–324.

Feng, J., Benra, F. K., & Dohmen, H. J. (2010). Application of different turbulence models in unsteady flow simulations of radial diffuser pump. *Forsch Ingenieurwes*, *74*, 123–133.

Gonzalez, J. J., Fernandez, J., Blanco, E., & Santolaria, C. (2002). Numerical simulation of the dynamic effects due to impeller volute interaction in a centrifugal pump. *ASME Journal of Fluids Engineering*, *124*, 348–355.

Jones, W. P., & Launder, B. E. (1972). The prediction of laminarization with a two Equation Model of Turbulence. *International Journal of Heat & Mass Transfer*, *15*, 301–314.

Kelder, J. D. H., Dijkers, R. J. H., Vas Esch, B. P. M., & Kruyt, N. P. (2001). Experimental and theoretical study of the flow in the volute of a low specific-speed pump. *Journal of Fluid Dynamic Research*, *28*, 267–280.

Liu, H. L., Liu, M. M., Dong, L., Ren, Y., & Du, H. (2012, August 19–23). *Effects of computational grids and turbulence models on numerical simulation of centrifugal pump with CFD*. Proceeding of 26th IAHR Symposium on Hydraulic Machinery and Systems, Beijing, China.

Menter, F. R. (1993, July 6–8). Multiscale model for turbulent flows. In *24th Fluid Dynamics Conference, Orlando, Florida*. Reston, VA: American Institute of Aeronautics and Astronautics.

Menter, F. R. (1994). Two-equation eddy-viscosity turbulence models for engineering applications. *AIAA Journal*, *32*, 1598–1606.

Spence, R., & Amaral-Texeira, J. (2008). Investigation into pressure pulsations in a centrifugal pump using numerical methods supported by industrial tests. *Computers & Fluids*, *37*, 690–704.

Spence, R., & Amaral-Texeira, J. (2009). A CFD parametric study of geometrical variations on the pressure pulsations and performance characteristics of a centrifugal pump. *Computers & Fluids*, *38*, 1243–1257.

Torabi, R., & Nourbakhsh, S. A. (2011, July 24–29). *Hydrodynamic design of the volute of a centrifugal pump using CFD*. Proceeding of ASME-JSME-KSME Joint Fluids Eng. Conference, Hamamatsu, Shizuoka, Japan.

Yedidiah, S. (2008). A study in the use of CFD in the design of centrifugal pumps. *Engineering Applications of Computational Fluid Mechanics*, *2*, 331–343.

Wilcox, D. C. (1986, January 6–8). Multiscale model for turbulent flows. In *AIAA 24th Aerospace Sciences Meeting, Reno, NV*. Reston, VA: American Institute of Aeronautics and Astronautics.

Zhongyong, P., Junjie, L., Shuai, L., & Shouqi, Y. (2011, July 24–29). *Study on radial forces of centrifugal pumps with various collectors*. Proceeding of ASME-JSME-KSME Joint Fluids Eng. Conference, Hamamatsu, Shizuoka, Japan.

Zhou, P. J., Wang, F. J., & Yang, M. (2012, August 19–23). *Internal flow numerical simulation of double-suction centrifugal pump using DES model*. Proceeding of 26th IAHR Symposium on Hydraulic Machinery and Systems, Beijing, China.

Permissions

All chapters in this book were first published in EACFM, by Taylor & Francis Online; hereby published with permission under the Creative Commons Attribution License or equivalent. Every chapter published in this book has been scrutinized by our experts. Their significance has been extensively debated. The topics covered herein carry significant findings which will fuel the growth of the discipline. They may even be implemented as practical applications or may be referred to as a beginning point for another development.

The contributors of this book come from diverse backgrounds, making this book a truly international effort. This book will bring forth new frontiers with its revolutionizing research information and detailed analysis of the nascent developments around the world.

We would like to thank all the contributing authors for lending their expertise to make the book truly unique. They have played a crucial role in the development of this book. Without their invaluable contributions this book wouldn't have been possible. They have made vital efforts to compile up to date information on the varied aspects of this subject to make this book a valuable addition to the collection of many professionals and students.

This book was conceptualized with the vision of imparting up-to-date information and advanced data in this field. To ensure the same, a matchless editorial board was set up. Every individual on the board went through rigorous rounds of assessment to prove their worth. After which they invested a large part of their time researching and compiling the most relevant data for our readers.

The editorial board has been involved in producing this book since its inception. They have spent rigorous hours researching and exploring the diverse topics which have resulted in the successful publishing of this book. They have passed on their knowledge of decades through this book. To expedite this challenging task, the publisher supported the team at every step. A small team of assistant editors was also appointed to further simplify the editing procedure and attain best results for the readers.

Apart from the editorial board, the designing team has also invested a significant amount of their time in understanding the subject and creating the most relevant covers. They scrutinized every image to scout for the most suitable representation of the subject and create an appropriate cover for the book.

The publishing team has been an ardent support to the editorial, designing and production team. Their endless efforts to recruit the best for this project, has resulted in the accomplishment of this book. They are a veteran in the field of academics and their pool of knowledge is as vast as their experience in printing. Their expertise and guidance has proved useful at every step. Their uncompromising quality standards have made this book an exceptional effort. Their encouragement from time to time has been an inspiration for everyone.

The publisher and the editorial board hope that this book will prove to be a valuable piece of knowledge for researchers, students, practitioners and scholars across the globe.

List of Contributors

A. C. Kengni Jotsa and V. A. Pennati
African Institute For Mathematical Sciences (AIMS-Cameroon), PO Box 608 Limbe Crystal Gardens, South West Region
Università Degli Studi dell'Insubria, Dipartimento Di Scienza e Alta Tecnologia, Via Valleggio 11, 22100 Como, Italy

Carla Cotas, Fernando Garcia and Maria Graça Rasteiro
Chemical Engineering and Forest Products Research Centre (CIEPQPF), Chemical Engineering Department, Faculty of Sciences and Technology, University of Coimbra, Rua Sílvio Lima, Pólo II, 3030-790 Coimbra, Portugal

Dariusz Asendrych
Częstochowa University of Technology, Institute of Thermal Machinery, 42-200 Częstochowa, Poland

Pedro Faiac
Electrical and Computers Engineering Department, Faculty of Sciences and Technology, University of Coimbra, Portugal

Adetokunbo A. Adedoyin and D. Keith Walters
Department of Mechanical Engineering, Mississippi State University, Starkville, USA
Center for Advanced Vehicular Systems, Mississippi State University, Starkville, USA

Shanti Bhushan
Center for Advanced Vehicular Systems, Mississippi State University, Starkville, USA

Ke-shi He, Da-yue Chen, and Zhen-yu Huang
Laboratory of Complex Dynamics Simulation, Department of Instrument Science and Engineering, School of Electronic, Information and Electrical Engineering, Shanghai Jiao Tong University, Shanghai, China

Li-juan Sun
Institute of Crop Science, Chinese Academy of Agricultural Sciences, Beijing, China

Zheng-lu Liu
Shanghai Key Laboratory of Protected Horticultural Technology, Sunqiao Modern Agriculture Development Zone, Shanghai, China

Cheng-Hsien Lee
Department of Civil and Environmental Engineering, National University of Singapore, Singapore

Zhenhua Huang
Department of Ocean and Resources Engineering, School of Ocean and Earth Science and Technology, University of Hawaii at Manoa, Honolulu, USA

Yee-Meng Chiew
School of Civil and Environmental Engineering, Nanyang Technological University, Singapore

Joshua Katzenberg and David MacManus
Department of Power and Propulsion, School of Engineering, Cranfield University, Cranfield, MK42 0AL, Bedfordshire, UK

Z.H. Ban, K.K. Lau and A.M. Shariff
Chemical Engineering Department, Universiti Teknologi PETRONAS, Bandar Seri Iskandar, 31750 Tronoh, Perak, Malaysia

Ali Assoudi
LGM, National Engineering School of Monastir, Mechanical Engineering Department, University of Monastir, Monastir, Tunisia

Sabra Habli and Nejla Mahjoub Saïd
LGM, Preparatory Institute for Engineering Studies, Physics Department, University of Monastir, Monastir, Tunisia

Herv. Bournot and Georges Le Palec
IUSTI, UMR 7343, Faculty of Science, Mechanical Engineering Department, University of Aix-Marseille, Marseilles, France

Dong Lei and Cao Lixin
School of Mechanical Engineering, Dalian University of Technology, Dalian 116023, China

A.F. Tena and P. Casan
Área de Medicina, Departamento de Medicina, C/Julián Clavería, 6, 3a planta, Campus del Cristo, 33006 Oviedo, Asturias, España

J.F. Francos and E. Álvarez
Área de Ingeniería Hidráulica, Departamento de Energía, Universidad de Oviedo, Escuela Politécnica de Mieres. Universidad de Oviedo, C/ Gonzalo Gutiérrez Quirós s/n. 33600-Mieres, Asturias, España

Kari Bråtveit and Nils Reidar B. Olsen
Department of Hydraulic and Environmental Engineering, The Norwegian University of Science and Technology, S.P. Andersens veg 5, N-7491 Trondheim, Norway

Yun Wu and Yinghong Li
College of Aeronautical and Aerospace Engineering, Airforce Engineering University, Xi'an, China

Haideng Zhang
College of Aeronautical and Aerospace Engineering, Airforce Engineering University, Xi'an, China
College of Energy and Propulsion Engineering, Beihang University, Beijing, China

Manish K. Singh
CTFD Division, CSIR-NAL, Bangalore, India; bAerospace Engineering Department, Indian Institute of Science, Bangalore, India

V. Ramesh
CTFD Division, CSIR-NAL, Bangalore, India

N. Balakrishnan
Aerospace Engineering Department, Indian Institute of Science, Bangalore, India

Si Huang, Xianghui Su and Guangqi Qiu
School of Mechanical and Automotive Engineering, South China University of Technology, Guangzhou, China

Hamed Alemi, Seyyed Ahmad Nourbakhsh, Mehrdad Raisee and Amir Farhad Najafi
Hydraulic Machinery Research Institute, School of Mechanical Engineering, College of Engineering, University of Tehran, Tehran, 1439955961, Iran

Index